THE IMPACT OF BIOTECHNOLOGY IN AGRICULTURE

Current Plant Science and Biotechnology in Agriculture

VOLUME 8

Aims and Scope
The book series is intended for readers ranging from advanced students to senior research scientists and corporate directors interested in acquiring in-depth, state-of-the-art knowledge about research findings and techniques related to plant science and biotechnology. While the subject matter will relate more particularly to agricultural applications, timely topics in basic science and biotechnology will also be explored. Some volumes will report progress in rapidly advancing disciplines through proceedings of symposia and workshops while others will detail fundamental information of an enduring nature that will be referenced repeatedly.

The titles published in this series are listed at the end of this volume.

The Impact of Biotechnology in Agriculture

Proceedings of the International Conference:
"The Meeting Point Between Fundamental and Applied in vitro *Culture Research",*
held at Amiens (France), July 10–12, 1989

edited by

R.S. SANGWAN

and

B.S. SANGWAN-NORREEL

Université de Picardie, Faculté des Sciences,
Amiens, France

KLUWER ACADEMIC PUBLISHERS
DORDRECHT / BOSTON / LONDON

ISBN 0-7923-0741-0

Published by Kluwer Academic Publishers,
P.O. Box 17, 3300 AA Dordrecht, The Netherlands.

Kluwer Academic Publishers incorporates
the publishing programmes of
D. Reidel, Martinus Nijhoff, Dr W. Junk and MTP Press.

Sold and distributed in the U.S.A. and Canada
by Kluwer Academic Publishers,
101 Philip Drive, Norwell, MA 02061, U.S.A.

In all other countries, sold and distributed
by Kluwer Academic Publishers Group,
P.O. Box 322, 3300 AH Dordrecht, The Netherlands.

Printed on acid-free paper

Printed in the Netherlands

SCIENTIFIC COMMITTEE :
B.S. SANGWAN, R.S. SANGWAN, M. BENBADIS, M. AYNONIN, M. VAZART, M. GADAL M. DAVID, M. DESMARET, M. DESPREZ, M. BADUFLE, M. BARBE, M. VAN MONTAGU M. THOMAS, M. HARADA, M. RAGHAVAN.

LOCAL ORGANIZING COMMITTEE :
R.S. and B.S. SANGWAN, organizers
M. POIRET, C. MARQUE, A. DAVID, C. DETREZ, T. TETU, L. LAVIEVILLE, M. BELAIZI D. HEROUART, E. GONTIER, F. DUBOIS, P. MARGUERY, N'DA KONAN KOFFI N. PAWLICKI, F. FLANDRE, H. PAUL, G. VASSEUR, N. OBJOIS, E. DOYEN, E. LAINE, G. SULMONT, N. MIANNAY, J.J. LACROIX, C. LEVECQUE, G. LEVIEILLE, B. AMARO, E. RIBIER and A. DIAZ.

TECHNICAL COMMITTEE OF PICARDY :
The representatives of "Rectorat de l'Académie d'Amiens", "Conseil Régional de Picardie", "Conseil Général de la Somme", "Mairie d'Amiens", "Université de Picardie", "Chambre d'Agriculture de Picardie, Somme et Aisne", "Comité de direction du Biopôle, "Déléguation du M.R.T. de Picardie", and "Chambre de Commerce et d'Industrie d'Amiens".

FINANCIAL SUPPORT BY :
Mairie d'Amiens, Conseil Général de la Somme, Conseil Régional de Picardie, Ministère de l'Education Nationale, Crédit Agricole, Université de Picardie.

Contents

Acknowledgements ix

Introduction xi

Special essay.
Tissue culture, molecular biology and plant biotechnology – A Historical overview
S.C. Maheshwari 1

Section 1. Plant regeneration. 13
A) Micropropagation

Application of *in vitro* techniques for the production and the improvement of horticultural plants
K.C. Short 15

How important is photosynthesis in micropropagation?
M. Capellades, A. Vanderschaeghe, R. Lemeur & P. Debergh 29

In vitro radiation-induced mutagenesis in potato
B.S. Ahloowalia 39

Micropropagation status in *Beta vulgaris* L. via leaf axis tissue culture
C. Detrez, R.S. Sangwan & B.S. Sangwan-Norreel 47

Application of *in vitro* multiplication for the annual vegetable crops associated with classical breeding
S. Okhi, K. Nasuda, Y. Mori & H. Katsuta 67

B) Haploidy and somatic embryogenesis

Gene expression during anther and pollen developmental transformations in rice
V. Raghavan 85

Field experiments with DH-lines in barley with ml-o powdery mildew resistance
A. Bjørnstad 99

Anther culture of *Lolium perenne* and *Lolium multiflorum*
I. Banté, T. Sonke, R.F. Tandler, A.M.R. van den Bruel & E. M. Meyer 105

Stress-induced carrot somatic embryos and their applicability to synthetic seed
H. Harada , T. Kiyosue, H. Kamada & K. Kobayashi 129

Developmental mutants
M. Terzi & F. Lo Schiavo 159

New approach towards controlling somatic embryogenesis in certain agronomically important plants
T. Tétu, B. S. Sangwan & R.S. Sangwan-Norreel 171

In vitro culture of wheat and rice for understanding the molecular basis of somatic embryogenesis and for transformation
N. Maheshwari, K. Rajyalakshmi, C.N. Chowdry, A. Grover, A.K. Tyagi & S.C. Maheshwari 191

Calcium and calmodulin during carrot somatic embryogenesis
A.C.J. Timmers 215

Section 2. Plant genetic engineering. 235

Genomic reorganization induced by plant tissue culture
R. L. Phillips 237

Mutants in the biosynthesis of animo acids
M. Jacobs, V. Frankard & M. Ghislain 247

Genetic engineering of rice
M. R. Davey & P. T. Lynch 259

Microscopic observations of fusion process of rice and lettuce protoplasts
T. Taniguchi, T. Sato, K. Maeda & E. Marda 281

Genetic transformation and plant improvement
R.S. Sangwan & B.S. Sangwan-Norreel 299

Potential transformation systems in *Dactylis glomerata*
B.V. Conger 339

Sunflower transformation : A study of selectable markers
A. E. Escandon & G. Hahne 345

Direct gene transfer and gene rescue in sugarbeet protoplasts
K. Lindsey & P. Gallois 355

Cauliflower mosaic vins: Biology and applications
K. Gorden 381

Salinity and drought stress in rice
A. Caplan, B. Claes, R. DeKeyser & M. Van Montagu 391

Prospects for the genetic manipulation of metabolic pathways leading to secondary products
M. Holden 403

Section 3. Special topics. 419

The cytoskeleton : Importance for plant cell and protoplast research
I. Staxen & C.H. Bornman 421

Glyphosate tolerance in plant cell cultures.
M.L. Racchi 437

Section 4. Cryopreservation of plant cells. 447

Cryopreservation of plant cell cultures. The importance of pretreatments
U. Seitz, D. Banspach, E. Goldner & E. Reinhard 449

Section 5. Abstracts 459

ACKNOWLEDGEMENTS

The following local and national institutions, and private companies extended generous collaboration and assistance during the organization of the conference:

Conseil Régional de la Picardie, Mairie d'Amiens, Université de Picardie Conseil Général de la Somme, Crédit Agricole de la Somme, Ministère de l'Education Nationale, Orsan, Socotra, Sanofi. Without their cooperation and financial help this International meeting would not have materialized. We would like to express our special thanks to the Mairie d'Amiens which extended all the facilities of 'Palais des Congrès' to us. To hold the symposium within the spacious Palais des Congrès was a real pleasure, and made this conference a success. We are most grateful to Mr. De Robien, Monsieur le Maire d'Amiens and his colleagues, Mrs. Marissal; Mr. Charles Baur, Président du Conseil Régional, and his collaborate Mr. Fleury; Professor B. Nemitz, Président de l'Université de Picardie; Professor P. Personne, Vice Président, Professor P. Castellonèse, Directeur de la Faculté des Sciences; CROUS of Amiens for providing hostel facilities, and to the members of scientific and local organizing committee for their continuous support.

INTRODUCTION

In 1988, we were contacted by the "Société Botanique de France" and the French section of The International Association of Plant Tissue Culture (IAPTC) to organise a symposium on "Biotechnology and its impact in Agriculture". The committee members of these two French associations had the vision to realize that this was a time to depart from the traditional themes in plant science. Moreover, since one of the major areas of our interest for the past few years has been that of "Plant biotechnology", we welcomed the idea of organizing this meeting. Another reason for the acceptance of this challenge was the location of Amiens in Picardy, and the enthusiasm of the local governing authorities. Picardy region is one of the major agricultural zones in France, and indeed of Europe. This fact and several other aspects of Picardy Agriculture suggested that this conference would be an excellent opportunity to combine the basic aspects of plant tissue culture and genetic engineering with the applications for plant improvement.

Biotechnology is currently arousing a great deal of interest in both the developed and developing countries because of its vast and multiple ramifications in Agriculture. The research work on recombinant DNA has led scientists to consider the possibility of using the new techniques such as Ti plasmid- based gene transfer to develop improved varieties of crop plants. If successful, the "gene-revolution" will compare with the "green revolution" which was the result of improved crop varieties developed by the plant breeders. Since the theme of this conference was the new developments in plant sciences, one should attempt to define plant biotechnology. At present there is considerable confusion and lack of agreement as to its definition. The popular and semi-popular press often portrays rather negative and confused impressions of plant biotechnology. For example, plant biotechnology and recombinant DNA are considered to be the same thing. However, it can easily recognized that plant biotechnology is neither confined to recombinant DNA nor to "genetic engineering". Genetic engineering, of course, includes recombinant DNA technology and a host of other techniques for manipulating genetic material, but all of them put together still represent only a small fraction of biotechnology. Thus a simple and generally acceptable definition of plant biotechnology is rather elusive. We can say that plant biotechnology is an integrated multidisciplinary field which utilizes many different technologies and has implications in several aspects of plant improvement.

We selected several important but varied areas of plant biotechnology in this conference and then attempted to provide a feeling of inter-relationship between the fundamental and applied research. Hence, we tried to touch upon a range of disciplines across the entire spectrum, i.e. from *in vitro* plant multiplication to the use of recombinant DNA. However, we were faced with some constraints and conflicting goals - carrying out the entire spectrum of plant biotechnology in three days. Therefore, the symposium was divided into the following sections:

1. *In vitro* plant regeneration and its potential in industry. It included clonal multiplication, organogenesis, somatic embryogenesis and haploidy. We gave special importance to plant regeneration, because if recombinant DNA technology is to be of any use in plant breeding, one must have a whole plant from the transformed cells. Plant cell and tissue culture is an increasingly important aspect of plant biotechnology, and has introduced an exciting new phase in plant multiplication and breeding. Presently, clonal multiplication and haploidy are being successfully used for developing improved cultivars for agriculture and horticulture by private and public laboratories. In addition, the potential value of somaclonal variation in creating novel genetic variation is now being increasingly recognized.

2. Plant genetic engineering : This section included non-sexual gene transfer, somatic hybridation, *agrobacterium*-mediated and direct gene transfer, somaclonal variations and isolation of mutants resistant to herbicides, insects, plant pathogens etc. Plant breeding is based on the availability of suitable genetic variability. Since not all the genetic variants are available in nature and in the primitive cultivars, *in vitro* mutation induction, somaclonal variations and use of recombinant DNA technology are being promoted for obtaining additional variation.

3. Germplasm preservation and economic implications of Biotechnology in Agriculture.

The proceedings, entitled "The impact of Biotechnology in Agriculture: The meeting point between fundamental and applied *in vitro* culture research", include papers presented at the symposium held at Amiens, France, July 10-12, 1989. The preparation of the proceedings began soon after the conference and took about six months. We are sure that these proceedings will be an important contribution to the literature on new developments in plant biotechnology. The papers presented highlight increasing cross-infusion of the problem and techniques between basic and applied scientists.

The proceedings summarize the current developments in plant biotechnology in the historical perspective of the past decade and assess their potential value and present limitations. We hope that these proceedings will enable research workers to assess the value of their own field of study and to make informed judgements of the technical problems they are likely to encounter. The compilation of this volume has demanded an active participation of a number of specialists. However, the formidable task of correspondence with authors of about thirty chapters in this volume was made considerably easier by their splendid cooperation in the preparation of the manuscripts in accordance with the guidelines provided by Kluwer Academic Publishers. As editors, we would like to acknowledge our indebtedness to all the members of the organizing committee of the symposia, and specially to Drs M. BREWIS and A.C. PLAIZIER. Many persons from our laboratory participated in arranging and checking the manuscripts; however, our special thanks are due to Mrs M. Poiret for retyping certain manuscripts, and to Mr G. Vasseur, F. Flandre, N. Pawlicki, E. Doyen for arranging figures and tables in certain manuscripts, and to Dr B.S. Ahloowalia for critically reading and correcting the text.

January, 1990 R.S. SANGWAN and B.S. SANGWAN-NORREEL

TISSUE CULTURE, MOLECULAR BIOLOGY AND PLANT BIOTECHNOLOGY - A HISTORICAL OVERVIEW

S.C. MAHESHWARI
Departments of Botany and Plant Molecular Biology, University of Delhi, Delhi, India

ABSTRACT. This acticle is based on the Inaugural Lecture given in the Conference. It gives a historical overview of the major developments leading to the current studies in plant biotechnology. The researches are divided into three 'phases'. The 'First Phase', from about 1900-1960 deals with establishment of the basic tissue culture technique. The 'Second Phase' deals with the development of the techniques of anther culture and protoplast culture. The 'Third Phase' deals with the fusion of molecular biology and tissue culture techniques and the advancement of genetic engineering. A few examples are given of the current biotechnological revolution in agriculture, which is in the making.

1. INTRODUCTION

First of all I would like to take a few moments to thank all the members of Scientific and Organizing Committee, but specially Dr Rajbir Sangwan and Dr Brigitte Sangwan-Norreel, for the honour that they all have done to me by inviting me to give this inaugural talk for this very important and timely conference in Amiens. I am not quite sure that I really deserve this honour, but if I have accepted it is in all humility and because of three reasons. Firstly, it is a scientific pilgrimage for me since France is a country which has a glorious history in Plant Tissue Culture and it is here, in Paris and its environment, that pioneering studies began in plant tissue culture and in a sense also in plant biotechnology (since tissue culture is the principal base of biotechnology). The second reason is a more specific one: French scientists, namely Dr Jean P. Nitsch, Madam Collette Nitsch and their former pupils and associates, for example the Sangwans themselves, have refined and made intensive use of the anther culture technique which we first developed in Delhi. Indeed, they have done so much more that it is a good opportunity for me to get educated with the latest advances. Finally, France is one of the most beautiful countries in the world, a country also rich in history, art, and culture, which we in India love, and therefore it is a pleasure to be

R.S. Sangwan and B.S. Sangwan-Norreel (eds.), The Impact of Biotechnology in Agriculture, 1–12.

here.

Turning now to the subject proper, I have planned to give an overview of the major historical developments leading to Plant Biotechnology. I am presenting my remarks in three phases. In the First Phase, the development of basic tissue culture technique (which is the most vital component of biotechnology) in the period from 1900 to about the middle of this century will be covered. I shall then move to the Second Phase dealing with a couple of more specialized techniques. Finally, we shall come to the Third Phase dealing with plant molecular biology and molecular biotechnology. Because of the rapidity of flow of scientific discoveries, the consequent "mad rush" in science, and the pressure to accomplish as much as possible within the minimum time, there is often no time for history. Nevertheless, a historical perspective is important not only to know the accomplishments of our predecessors, but also to have a yardstick against which we can measure our own contributions to science and set our own goals in life.

2. THE FIRST PHASE

The first and foremost scientist who deserves our attention is Gottlieb Haberlandt (1902) who worked in Berlin almost a century ago. He is the one who first developed the idea that a plant cell is totipotent. Nevertheless, little was known of nutrition or of hormones essential for growth and differentiation and he could not go very far. For the next 30 years, almost nothing happened. Then, as a beacon of light came Roger Gautheret (1934) and he achieved the first real success in raising tissue cultures, here in France, by initiating research on development of callus from stem explants of cambium of trees. Another Frenchman, Paul Nobecourt (1937), also contributed significantly to the establishment of tissue culture technique. At the same time, i.e. while Gautheret and Nobecourt carried out their researches in France, Philip White (1934) worked in USA with roots and later crown-gall tissue (Braun and White 1943) and had a decisive influence on researches there. I would like to make one special remark about the relationship of work on hormones to tissue culture. Until about the thirities, i.e. the time when White and Gautheret worked, auxins represented the only known major class of growth substances and whose discovery came largely due to the efforts of Frits Went (1926) and Kenneth Thimann (1935). Later, in the fifties, however, the work of Folke Skoog and co-workers (Miller et al. 1955) led to the discovery of cytokinins as another important class of hormones. In fact, a genuine control on differentiation from calli through an adjustment of auxin and cytokinin balance was achieved only then, after the realization that a high cytokinin/auxin ratio leads to development of shoots and a low ratio to rooting. By the sixties, a fairly good understanding had been gained on the role of hormones in differentiation, although coconut milk -- whose importance had been discovered by Van Overbeek and coworkers (1941) -- also remained a popular constituent of media in many investigations.

In any case, with hormones and coconut milk in the arsenal, work on _in vitro_ culture of plant tissues and their differentiation spread to many laboratories of the world in the fifties and sixties. Representative of this effort is the work of Street in U.K. and Morel in France, Reinert in Germany and of Steward in USA. Street made important contributions regarding requirements of roots to grow in vitro (_see_ Street 1957); on the other hand, Morel made pioneering contributions to the crown gall problem as also on culture of monocots and orchids (Morel 1963, 1971). However, in my opinion, the final and decisive studies -- concluding the First Phase of Tissue Culture work -- were made by Steward and Reinert who demonstrated totipotency of single cells. Using carrot roots, Steward and coworkers (1958) showed that even root cells could form complete plants. Parallelly, in Germany, Reinert (1959) also achieved embryogenesis in vitro. But for those who still had doubts, totipotency was demonstrated in an even more rigorous sense -- _sensu stricto_ -- by my former class-fellow, Vimla Vasil, while she was working with Hildebrandt in USA (Vasil and Hildebrandt 1965). She showed that even an isolated single cell can form an entire plant -- that was demonstration of totipotency with ultimate rigour!

3. THE SECOND PHASE

The next phase of plant tissue culture can be said to have begun in sixties and the chief achievement of this period was the development of two special techniques of tissue culture -- one, relating to production of haploids by anther culture, and, another, concerning isolation and culture of protoplasts.

3.1. Anther Culture

To take up the anther culture technique of production of haploids, exactly 25 years have passed since its discovery in 1964 and we can as well celeberate the Silver Jubilee of its discovery (Guha and Maheshwari 1964, 1966)! The real credit for its discovery goes to my former postdoctoral associate, Dr Sipra Guha. Incidentally, the path to the discovery of haploids was not an easy one and has been dotted with failures. In fact, the original impetus for this work came from my father, the late Professor P. Maheshwari, who headed the Department of Botany at the University of Delhi and mounted perhaps the first serious effort anywhere in the world to develop a technique for inducing haploidy. Indeed, Mrs Nirmala Maheshwari (present here in the audience) was among the first of his students to take up tissue culture work under his guidance. The objective of her Ph.D. work was to obtain parthenogenetic haploids and at that time the goal was to culture unfertilized ovules so that the egg could be induced to divide. I also provided some assistance in the effort, although I had my own distinct research project. Unfortunately, however, the unfertilized ovules failed to grow and the problem was given up for many years (of course, ovules that had been just fertilized did grow with some effort obtaining a Ph.D. for Mrs Maheshwari!) as well as a publication which attracted some attention at that time (Maheshwari 1958). Somewhat disappointed and frustrated of failures at least in the original objective, we went to America and turned to research in Plant Molecular Biology at California Institute of Technology in the laboratory of the renowned American Plant Biochemist, James Bonner, and helped discover RNA polymerases in plants (Bonner et al. 1960, 1961; Bonner et al. 1962; Bandurski and Maheshwari 1962). When we came back to Delhi, I had different ideas -- inclined more towards solving fundamental problems of cell physiology and biochemistry. And, therefore, when in the mid-sixties, Dr Guha joined my laboratory as a postdoctoral associate and I advised her to work on Datura and initiate suspension cultures from sporogenous cells of anthers by culturing them, the idea was not so much as to raise haploids but to study the physiology and biochemistry of meiosis. But to our pleasant surprise, came the discovery of haploid embryos when we cultured anthers. To be quite honest, I had never had much faith in the tissue culture technique to obtain haploids, but it is one of those rare examples when a father's dream was achieved, practically, by a son and his associates. It is important to emphasize that, fortunately, an atmosphere of a free enquiry existed, helped by my already having an independent position of Associate Professor in the Department of Botany. In fact, had someone demanded work to produce haploids, nothing really would have come out! The lesson before us is that fundamental studies are vital for progress of science.

I would also like now to mention the pioneering contributions of Nitsch and his associates who were quick to achieve androgenesis in tobacco and even get complete plants (Nitsch and Nitsch 1969). Mrs Nitsch collaborated in much of his work. The Nitsches, together with many young people, did much work on anther culture. They refined key procedures and in particular perfected the technique of culture of isolated pollen grains (Nitsch and Norreel 1973; Sangwan and Norreel 1975). After the initial work in India, France, and Japan, work on anther culture has been taken up in many other countries of the world -- indeed the intense activity in this field has clearly exceeded our wildest dreams (Maheshwari et al. 1982). Because of lack of time, one cannot dwell on these investigations further. However, I must mention that many practical applications have come of this technique -- the Chinese have been particularly active in this area and in China haploid-derived lines of crops cover thousands of acres.

3.2. Protoplast Culture

Another major advance in tissue culture relates to the technique of isolation and culture of protoplasts. One of the pioneers in this field is Cocking who worked in England and was the first to isolate protoplasts (1960). But the credit for achievement of cell divisions and development of plants from protoplasts, first obtained in tobacco, goes to Takebe, Labib and Melchers (1971). Once success has been achieved in inducing division in protoplasts, work on somatic hybrids was to be expected and the first such hybrid was reported, again in tobacco, by Carlson, Smith and Dearing (1972), in USA. Soon, work on intergeneric hybrids was also to follow. George Melchers of Germany and his associates (1978) turned a new chapter by obtaining the so-called 'pomatoes' and 'topatoes' which are hybrids between potato and tomato.

4. THE THIRD PHASE – GENETIC ENGINEERING AND BIOTECHNOLOGY

4.1. Plant Molecular Biology and Tissue Culture Studies Come Together – A Rendezvous

The successful culture of protoplasts brings us to the Third Phase of genetic engineering and to the dawn of the new era of modern biotechnology. In a conference of this nature, no one particularly emphasizes the history of plant molecular biology, probably because the early practitioners of molecular biology were microbiologists or biochemists who worked largely on animals. However, it is important to realize that the present revolution in plant biotechnology owes a lot to people like James Bonner, who at Caltech pioneered research on plant genes and their expression, and whom I mentioned earlier. Lawrence Bogorad, of Harvard, who first cloned and sequenced chloroplast genes also contributed a great deal (McIntosh et al. 1980). Plant biotechnology is essentially an outcome of the fusion of tissue culture and molecular biology techniques.

Plant biotechnology also owes a lot to students of the crown gall problem and for the study of which the tissue culture technique has played a rather important role. Armin Braun (1950) did pioneering studies at the Rockefeller Institute and gave the idea of a tumor inducing principle which we now know is a stretch of DNA. That some kind of genetic engineering, involving DNA, is going on in plants was long anticipated even before the gene cloning revolution began. Nonetheless, it is the work of Robert Schilperoort, Joseph Schell, Marc van Montagu and Mary-dell Chilton and their colleagues in Europe and America, which really unravelled the role of the Ti plasmid in transformation of host plant cells (Chilton et al. 1974; Van Larebeke et al. 1974; Watson et al. 1975; Zaenen et al. 1974). These contributions are now already part of the classic history of plant biotechnology. Few people become legendary in their own life times, but all these scientists, who directed the work, have become so and we salute them.

Once the close relationship between Agrobacterium strains containing plasmids and tumors was shown, it was natural to look for the presence of T-DNA in host genome. I like here to mention briefly another of my former students, Narendra Yadav, who did postdoctoral studies in Mary-dell Chilton's laboratory — and along with Thomashow and others -- showed by Southern blotting that a stretch of DNA from the Ti plasmid did covalently integrate with the host genome (Yadav et al. 1980). Once this was demonstrated, it was natural to design strategies for using Ti plasmid as a vector to deliver foreign DNA. The pioneering studies in this connection were again made by Schilperoort, Schell, Montagu and Chilton and their associates (Chilton 1983; Hoekma et al. 1988; de Framond et al. 1983; Zambryski et al. 1983; Bevan 1984). Because Ti plasmids are large and difficult to manipulate, they looked for a way of employing the *E. coli* plasmids for genetic engineering. The work led to the development of special pBR322 derived plasmids which, after a step leading to incoporation of a foreign gene, can be later inserted in *Agrobacterium* so that transformation by co-cultivation of protoplast-derived cells or leaf-discs could follow (Horsch et al. 1985). There are, of course, investigators like Ingo Potrykus and their coworkers (1984, 1985) who are developing direct gene transfer methods where *Agrobacterium* is being dispensed with completely. Nevertheless, as of this date, most transformations have been made using *Agrobacterium*. Genetic engineering could be said to have come of age when Hall and co-workers (1983) in USA accomplished the remarkable feat of transferring a phaseolin gene from beans to cells of the sun-flower plant.

4.2. Some Examples of Current Plant Biotechnological Results

We have now covered the most important landmarks in the area of Plant Biotechnology until the eighties. What follows now is the contemporary scene. Genetic engineering studies have now spread to many countries. This is true at least of laboratories in countries in this part of the world where it is about to get to the fields (in India I am afraid we are somewhat isolated from the centres of active research as also of latest information and it is possible that genetically engineered crops are *already* on the field). As obvious by the Scientific Programme of this Conference, some leaders of this field will themselves lecture on their work and I do not need to go in those developments in any detail. However, I do like to show you a few vignettes or glimpses of current work that show the tremendous potential that genetic engineering has in the agriculture and forestry of tomorrow. The first of these illustrations relates to work on herbicide resistance genes. Pioneering studies of Shah, Horsch, Fraley, and others (1986) have led to introduction of herbicide resistance genes in important crop plants which promise to revolutionize agriculture. The second example relates to work on insect resistant plants e.g. by Boulter and coworkers in U.K., where transgenic plants have been made by engineering in them a gene for a bacterial protein which insects detest greatly (Hilder et al. 1987). The third illustration concerns pioneering work of Beachy and colleagues (Abel et al. 1986) in U.S.A. on plants resistant to viruses. Finally, a fourth example relates to possibilities of altering processes connected with senescense of fruits such as tomato which can lead to production of fruits that will last longer or otherwise lead to more convenient harvesting practices (Giovannoni et al. 1989). The possibilities of genetic engineering are in fact limitless. Of course, one may not be able to make a cube of an orange, but one can do many other things like introduce genes for resistance to fungi, or stress to cold and drought,or improve nutritional quality of foods.

The above account may give the impression that the future now lies all in applied research. Nevertheless, this is not correct and advances in biotechnology do require more basic research. In reality, there are several outstanding problems which demand our attention in the next one or two dacades. To give a few examples, in monocots -- where our cereal crops belong -- biotechnology is still a distant dream because tissues do not readily regenerate and it is important to understand what makes cells divide or stop dividing. How are they different from dicotyledenous cells? A second example relates to the problem of organ-specific expression of engineered gene; since heterologous promoters are used today one has little control on this matter or of the level of gene expression. A third and another outstanding example relates to the problem of targeting genes in a site-specific manner within a host genome. Ingo Potrykus and coworkers (Paszkowski et al. 1988) have begun pioneering studies in this direction, but to date most foreign genes get "lodged" in the host genome in a rather random fashion, and we must understand what causes a gene to insert at particular loci on a chromosome and how we can direct this activity.

5. CONCLUSION

Ladies and gentlemen, I now come to the end of my talk. This survey has been necessarily a sketchy and incomplete one, but even so you can see that plant biotechnology is perhaps the most challenging area of research in plant biology today and the organizers have to be congratulated for making this splendid meeting possible.

6. ACKNOWLEDGEMENTS

I have great pleasure in thanking to Dr A.K. Tyagi for his advice and assistance in preparing this manuscript. Mrs Kavita Munjal and Mrs Neelam Bedi also deserve my thanks for preparing the slides which illustrated the original talk.

REFERENCES

Abel, P.P., Nelson, R.S., De, B., Hoffmann, N., Rogers, S.G., Fraley, R.T. and Beachy, R.N. (1986) 'Delay of disease development in transgenic plants that express the tobacco mosaic virus coat protein gene'. Science 232, 738-743.

Bandurski, R.S. and Maheshwari, S.C. (1962) 'Nucleotide incorporation into nucleic acid by tobacco leaf homogenates'. Plant Physiol. 37, 556-560.

Bevan, M. (1984) 'Binary Agrobacterium vectors for plant transformation'. Nucl. Acids Res. 12, 8711-8721.

Bonner, J., Huang, R.C. and Maheshwari, N. (1960) 'Enzymatic synthesis of RNA'. Biochem. Biophys. Res. Commun. 3, 689-694.

Braun, A.C. (1950) 'Thermal inactivation studies on the tumor inducing principle in crown-gall'. Phytopathology 40, 3.

Braun, A.C. and White, P.R. (1943) 'Bacteriological sterility of tissues derived from secondary crown-gall tumors'. Phytopathology 33, 85-100.

Carlson, P.S., Smith, H.H. and Dearing, R.D. (1972) 'Parasexual interspecific plant hybridization'. Proc. Natl Acad. Sci. USA 69, 2292-2294.

Chilton, M.-D. (1983) 'A vector for introducing new genes into plants'. Sci. Amer. 248, 36-45.

Chilton, M.-D., Drummond, H.J., Merlo, D.J., Sciaky, D., Montoya, A.L., Gordon, M.P. and Nester, E.W. (1977) 'Stable incorporation of plasmid DNA into higher plant cells: The molecular basis of crown gall tumorigenesis'. Cell 11, 263-271.

Chilton, M.-D., Farrand, S.K., Eden, F.C., Currier, T.C., Bendich, A.J., Gordon, M.P. and Nester, E.W. (1974) 'Is there foreign DNA in crown gall tumor DNA?', in R. Markham, D.R. Davies, D. Hopwood and R.W. Horne (eds.), Modification of the Information Content of Plant Cells, Elsevier, New York, p. 297.

Chilton, M.-D., Saiki, R.K., Yadav, N., Gordon, M.P. and Quetier, F. (1980) 'T-DNA from Agrobacterium Ti plasmid is in the nuclear DNA fraction of crown-gall tumor cells'. Proc. Natl Acad. Sci. USA, 77, 4060-4064.

Cocking, E.C. (1960) 'A method for the isolation of plant protoplasts and vacuoles', Nature 187, 927-929.

de Framond, A.J., Barton, K.A. and Chilton, M.-D. (1983) 'Mini-Ti: a new vector strategy for plant genetic engineering'. Bio/Technology 1, 262-269.

Gautheret, R.J. (1934) 'Culture du tissus cambial'. C.R. Hebd. Seances Acad Sci. 198, 2195-2196.

Giovannoni, J.J., Dellapenna, D., Bennett, A.B. and Fischer, R.L. (1989) Expression of a chimeric polygalactouronase gene in transgenic rin (ripening inhibitor) tomato fruit results in polyuronide degradation but not fruit softening'. Plant Cell 1, 53-63.

Guha, S. and Maheshwari, S.C. (1964) '_In vitro_ production of embryos from anthers of _Datura_'. Nature 204, 497.

Guha, S. and Maheshwari, S.C. (1966) 'Cell division and differentiation of embryos in the pollen grains of _Datura_ in vitro'. Nature 212, 97-98.

Haberlandt, G. (1902) 'Kulturversuche mit isolierten Pflanzenzellen'. Sitzungsber. Akad. Wiss. Wien, Math.'. Naturwiss. Kl., Abt 1 111, 69-92.

Hilder, V.A., Gatehouse, A.M.R., Sheerman, S.E., Barker, R.F. & Boulter, D. (1987) 'A novel mechanism of insect resistance engineered into tobacco'. Nature 330, 160.

Hoekema, A., Hirsch, P.R., Hooykaas, P.J.J. and Schilperoort, R.A. (1983) 'A binary plant vector strategy based on separation of vir- and T-region of the _Agrobacterium tumefaciens_ Ti-plasmid'. Nature 303, 179-180.

Horsch, R.B., Fry, JE., Hoffmann, N.L., Eichholtz, D., Rogers, S.G. and Fraley, R.T. (1985) 'A simple and general method for transferring genes into plants'. Science 227, 1229-1231.

Maheshwari, N. (1958) 'In vitro culture of excised ovules of _Papaver somniferum_'. Science 127, 342.

Maheshwari, S.C., Rashid, A. and Tyagi, A.K. (1982) 'Haploids from pollen grains - retrospect and prospect (Special Paper)'. Am. J. Bot. 69, 865-879.

Marton, L., Wullems, G.J., Molendijk, L. and Schilperoort, R.A. (1979) 'In vitro transformation of cultured cells from _Nicotiana tabacum_ by _Agrobacterium tumefaciens_'. Nature 277, 129-130.

McIntosh, I., Paulsen, C. and Bogorad, L. (1980) 'Chloroplast gene sequence for the large subunit of ribulose bis-phosphate carboxylase of maize'. Nature 288, 556-560.

Melchers, G., Sacristan, M.D. and Holder, A. (1978) 'Somatic hybrid plants of potato and tomato regenerated from fused protoplasts'. Carlsberg Res. Commun. 43, 203-218.

Miller, C., Skoog, F., Von Saltza, M.H. and Strong, F.M. (1955) 'Kinetin, a cell division factor from deoxyribonucleic acid'. J. Am. Chem. Soc. 77, 1392.

Morel, G. (1963) 'la culture _in vitro_ du meristeme apical de certaines orchidees'. C.R. Hebd. Seances Acad. Sci. 256, 4955-4957.

Morel, G. (1971) 'Deviations du metabolisme azote des tissus de crown-gall'. Colloq. Int. C.N.R.S. 193, 463-471.

Murai, N., Sutton, D., Murray, M., Slightom, J., Merlo, D., Richert, N., Sengupta-Gopalan, C., Stock, C., Barker, R., Kemp, J. and Hall, T. (1983) 'Phaseolin gene from bean is expressed after transfer to sunflower via tumor-inducing plasmid vectors'. Science 222, 476-482.

Nitsch, C. and Norreel, B. (1973) 'Effect d'un choc thermique sur le pouvoir embroyogene du pollen de Datura innoxia cultive dans l'anthere ou isole de l'anthere'. C.R. Acad. Sci. Paris 276D, 303-306.

Nitsch, J.P. and Nitsch, C. (1969) 'Haploid plants from pollen grains'. Science 163, 85-87.

Nobecourt, P. (1937) 'Cultures en serie de tissus vegetaux sur milieu artificiel'. C.R. Hebd. Seances Acad. Sci. 200, 521-523.

Paszkowski, J., Baur, M., Bogucki, A. and Potrykus, I. (1988) 'Gene targetting in plants'. EMBO J. 7, 4021-4026.

Paszkowski, J., Schillito, R.D., Saul, M., Mendok, V., Hohn, T., Hohn, B. and Potrykus, I. (1984) 'Direct gene transfer to plants'. EMBO J. 3, 2717-2722.

Potrykus, I., Saul, M.W., Petruska, J., Paszkowski, J. and Shillito, D. (1985a) 'Direct gene transfer to cells of a graminaceous monocot'. Molec. Gen. Genet. 199, 183-188.

Reinert, J. (1959) 'Uber die Kontrolle der Morphogenesis und die Induktion von Adventiveembryonen an Gewebekulturen aus Karotten'. Planta 53, 318-333.

Reinert, J., Bachs-Husemann, D. and Zerban, H. (1971) 'Determination of embryo and root formation in tissue cultures from Daucus carota'. Colloq. Int. C.N.R.S. 193, 261-268.

Sangwan, R.S. and Norreel, B. (1975) 'Induction of plants from pollen grains of Petunia cultured in vitro'. Nature 257, 222-224.

Schell, J. and Van Montagu, M. (1977) 'On the transfer, maintenance and expression of bacterial Ti-plasmid DNA in plant cells transformed with A. tumefaciens'. Brookhaven Symp. Biol. 29, 36-49.

Schell, J., Van Montagu, M., Holsters, M., Hernalsteens, J.P., Dhaese, P., De Greve, H., Leemans, J., Joos, H., Inze, D., Willmitzer, L., Otter, L., Wortemeyer, A., Schroder, G. and Schroder, J. (1982)

'Plant cells transformed by modified Ti plasmids: a model system to study plant development'. Biochem. Differen. Morphogenesis 33, 66-73.

Shah, D.M., Horsch, R.B., Klee, H.J., Kishorer, G.M., Winter, J.A., Tumor, N.E., Hironaka, C.M., Sanders, P.R., Gasser, C.S., Aykent, S., Siegel, N.R., Rogers, S.G. and Fraley, R.T. (1986) 'Engineering herbicide tolerance in transgenic plants'. Science 233, 478.

Steward, F.C. (1958) 'Growth and organised development of cultured cells. III. Interpretations of the growth from free cell to carrot plant'. Am. J. Bot. 45, 709-713.

Steward, F.C., Kent, A.E. and Mapes, M.O. (1966) 'The culture of free plant cells and its signification for embryology and morphogenesis', in R.A. Moscona and A. Monroy (eds.), Current Topics in Developmental Biology, Academic Press, New York, pp. 113-154.

Street, H.E. (1957) 'Excised root culture'. Biol. Rev. 32, 117-155.

Takebe, I., Labib, C. and Melchers, G. (1971) 'Regeneration of whole plants from isolated mesophyll protoplasts of tobacco'. Naturwissenschaften 58, 318-320.

Thimann, K.V. (1935) 'On the plant growth hormone produced by Rhizopus suinus'. J. Biol. Chem. 109, 279-291.

Van Larebeke N, Engler, G., Holstress, M., Van den Elaracker, S., Zaenen, I., Schilperoort, R.A. and Schell, J. (1974) 'Large plasmid in Agrobacterium tumefaciens essential for crown gall inducing ability'. Nature 252, 169-170.

Van Overbeek, J., Conklin, M.E. and Blakeslee, A.F. (1941) 'Factors in coconut milk essential for growth and development of very young Datura embryos'. Science 94, 350-351.

Vasil, V. and Hildebrandt, A.C. (1965b) 'Differentiation of tobacco plants from single isolated cells in microcultures'. Science 150, 889-890.

Watson, B., Currier, T.G., Gordon, M.P., Chilton, M.-D. and Nester, E.W. (1975) 'Plasmid required for virulence of Agrobacterium tumefaciens'. J. Bacteriol. 123, 255-264.

Went, F.W. (1926) 'On growth accelerating substances in the coleoptile of Avena sativa'. Proc. K. Ned. Akad. Wet. Ser C 30, 10.

White, P.R. (1934a) 'Potentially unlimited growth of excised tomato root tips in a liquid medium'. Plant Physiol. 9, 585-600.

Yadav, N.S., Postle, K., Saiki, R.K., Thomashow, M.F. and Chilton, M.-D. (1980) 'T-DNA of a crown-gall teratoma is covalently joined to plant DNA'. Nature 287, 458-461.

Zambryski, P., Joos, H. Genetello, C., Leemans, V., Van Montagu, M. and Schell, J. (1983) 'Ti plasmid vector for the introduction of DNA into plant cells without alteration of their normal regeneration capacity'. EMBO J. 2, 2143-2150.

Zaenen, I., van Larebeke, N., Teuchy, H., Van Montagu, M. and Schell, J. (1974) 'Supercoiled circular DNA in crown-gall inducing _Agrobacterium_ strains'. J. Mol. Biol. 86, 109-127.

Section 1. Plant regeneration

APPLICATION OF *IN VITRO* TECHNIQUES FOR THE PRODUCTION AND THE IMPROVEMENT OF HORTICULTURAL PLANTS

KEITH C. SHORT
Nottingham Polytechnic
Burton Street
Nottingham NG1 4BU
United Kingdom

ABSTRACT. Plant tissue culture technology has introduced a new phase into plant multiplication and is being increasingly utilized in plant breeding. This paper reviews the current and developing biotechnological applications of *in vitro* methods for plant improvement and outlines the present state of the art of the regeneration of plants from cells and tissues. Also, the problems associated with the transfer of tissue culture derived plantlets to soil conditions are addressed. Methods are described for the *in vitro* acclimatization of cultured plantlets.

1. INTRODUCTION

Many horticulturally important plant species are difficult or impossible to manipulate by conventional propagation and breeding programmes. Consequently, propagators and breeders are seeking more effective methodologies, such as the use of tissue culture technology, with the goal of enhanced production of improved ornamental plants.

Plant tissue culture is an increasingly important aspect of plant biotechnolgy and has introduced an exciting new phase into plant multiplication and breeding. By these experimental methods isolated cells and tissues can be grown in the laboratory, under aseptic conditions, studied at leisure, subjected to various biotechnological manipulations and because of their totipotency, stimulated to undergo regeneration to whole plants.

The concept of improving plants by tissue culture methods is not new and Steward (1970) who, with his characteristic forward look for plant biology, foresaw the development of micropropagation systems and 'a sort of tissue culture genetics', all of which would be based on the totipotency of plant cells. Studies over the last 19 years have confirmed Steward's vision of the use of this methodology for the improvement and rapid multiplication of horticultural plants and shrubs (Hussey, 1986; Pennel, 1986). These procedures are now

R.S. Sangwan and B.S. Sangwan-Norreel (eds.), The Impact of Biotechnology in Agriculture, 15–27.

commonplace and routinely used by many commercial laboratories for the large-scale production of ornamentals (Harper, 1987; Constantine, 1986).

More recently, exciting developments involving the isolation, culture and regeneration of protoplasts and cells (Davey & Power, 1988; Pierik, 1987), their genetic modification by use of recombinant DNA and protoplast fusion technology (Cocking and Davey, 1987; Shillitto et al 1985) and the isolation and assessment of somaclonal variants (Brown et al 1986) has more than justified the potential of *in vitro* methods for plant improvement.

There are two distinct areas in which plant tissue culture methodology is important for horticulturalists. The first includes situations where large-scale multiplication and the maintenance of genetic stability is paramount. This comprises the micropropagation of plants, their improvement as a result of pathogen elimination and conservation in a stable form. The second concerns situations in which spontaneous and induced variation can be induced by modern cell biological methods. Both approaches, however, rely on our ability to successfully control plant cell morphogenesis and to develop reliable cell to plant regeneration systems.

2. PLANT TISSUE CULTURE TECHNOLOGY

The strategic approaches and methods used by plant tissue culture workers can be incorporated into two areas of current and developing technologies (Table 1). Advances in this area of plant biotechnology are occurring at a rapid pace and there is considerable technology transfer between fundamental and applied aspects. The developing applications listed are likely to become essential tools for horticulturalists in the future.

TABLE 1

Tissue culture in Plant Production and Breeding

Current technologies	* Clonal multiplication
	* Pathogen elimination
	* Embryo rescue
	* Haploid production
	* Genetic conservation
Developing technologies	Genetic modification of plants by:
	* Mutagenesis
	* Somaclonal techniques
	* Somatic hybridization
	* Recombinant DNA technology
	* Selection for stress and disease tolerance

3. MICROPRCPAGATION

The multiplication of plants by tissue culture - micropropagation, offers many advantages over conventional methods for the multiplication of large numbers of plants independent of climatic conditions and with conservation of space and time. Furthermore *in vitro* derived plants are frequently more vigorous and of superior quality, often justifying the term "elite" plants in comparison to those produced by *in vivo* methods. The multiplication of plants by tissue cultures centres around the formation and multiplication of shoot meristems. Two main pathways are involved, one is the regeneration of plants from existing meristems - axillary and apical meristems which results in the production of clonally stable plantlets (Figure 1). The other is by the induction of *de novo* meristems, adventitious meristems and somatic embryos in cultured organs/tissues/ callus. Whilst this *de novo* pathway is much more productive it often results in progenys which are not true to type and in the production of somaclonal variants. This phenomenon is, however, being exploited in the continual search for new improved plant varieties.

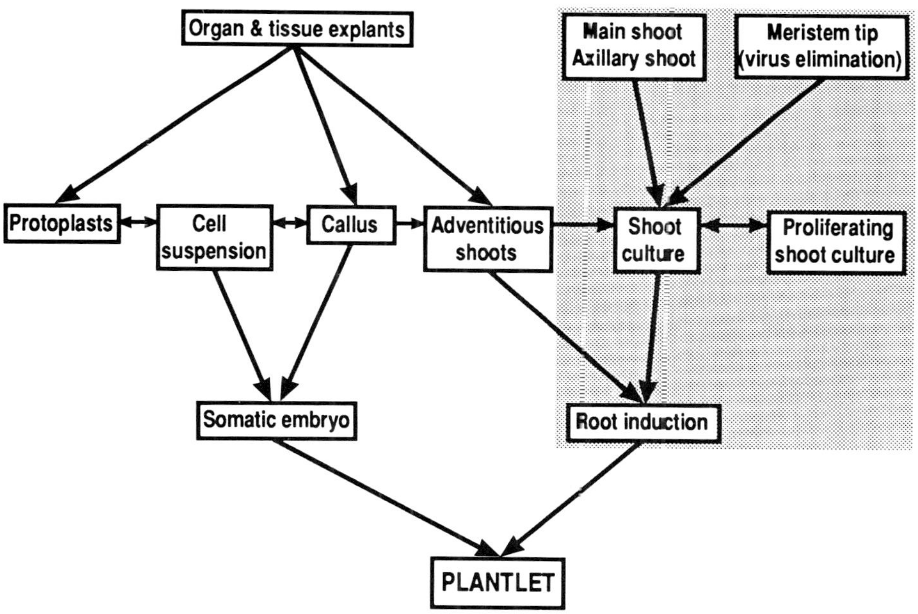

Figure 1. Schematic diagram illustrating the pathways of regeneration in plant tissue cultures.

The micropropagation of plants by tissue cultures from existing meristems or by the induction of adventitious meristems (shoots or embryos) either directly within the parent tissue or via an intermediate callus phase may be stimulated by various combinations of nutrients, hormones and environmental factors. An important determinant, however, is the physiological status of the source material and provided young, healthy, and actively growing explants are used then good responses should be obtained. In general, to maintain genetic fidelity, micropropagation should be via organised meristems (apical or axillary meristems, inflorescences, etc.) . This can be achieved by culturing shoot explants on a culture medium supplemented with cytokinin, which reduces apical dominance and allows the production of precocious numbers of axillary shoots. These can be separated and subcultured again to produce many more shoots. Following the transfer of shoots to a culture medium containing low amounts/lacking auxin, roots are initiated and subsequently the plantlets can be transferred to soil conditions.

Theoretically, explants from all plant organs and tissues cultured _in vitro_ can be induced to produce shoots by the induction of _de novo_ meristems. The formation of these organised structures is controlled by the auxin/cytokinin balance (Skoog and Miller, 1957). This is founded on experience rather than a fundamental understanding of how plant cell morphogenesis is regulated. Inevitably the empirical approach adopted means that protocols for the initiation of tissue cultures and the regeneration of plants from these have to be worked out for each species.

A number of workers have reviewed the strategic approaches which may be used in the development of reliable cell to plant regeneration systems (Short, 1986; Henshaw, 1987; Alderson, 1986; Pierik, 1987). These approaches and methods have been successfully applied for the clonal propagation of a wide range of horticultural plants and have been particularly useful for herbaceous plants (Hussey, 1983), bulbous plants (Hussey, 1982) and fruit trees (Jones, 1985).

Plant production from apial and axillary meristems may be facilitated by culturing explants on cellulose rods, Sorbarods (Baumgartner Papiers S.A., Switzerland) in liquid medium (Short et al, 1987). Studies on chrysanthemum (Roberts & Smith, 1989) and rose (Lloyd et al, 1988) revealed high rates of shoot growth with enhanced rooting in this system. Further, the system lends itself to automation. Therefore, the use of these cellulose rods has considerable potential in commercial micropropagation in that plantlets can be handled with ease and transferred directly to soil without being removed from the biodegradable Sorbarod.

Many species are unresponsive in culture and show limited rates of multiplication. In some cases, however, stressful treatments promotes shoot regeneration. The flooding of shoot cultures of trees, shrubs and herbaceous species often results in increased bud and shoot proliferation, and rooting (John and Pearson, 1986). Also, fluctuations in temperature have been found beneficial in stimulating shoot regeneration. Sugar beet explants incubated at a temperature of 5°C above that of donor cultures exhibited a 2-fold increase in shoot production (Ritchie et al, 1989a). In contrast, some organs require exposure to low temperatures before regeneration occurs e.g. anthers (Huang and Sunderland, 1982).

Organ formation in tissue cultures has been shown to be influenced by minute electric fields, of a low current applied over extended periods (Rathore & Goldsworthy, 1985). These responses were attributed to an increased polar transport of indole-3-acetic acid (IAA) into the electrically stimulated tissues which brought about a co-ordination of cellular polarities and resulted in the initiation of meristematic centres (Goldsworthy & Rathore 1985). In this way the electrical field supplied to the explant provides a framework in which morphological development can occur. High voltage electric pulses of micro- and milli-seconds duration have been shown to promote cell division and colony formation in isolated protoplasts (Rech et al, 1987) and in the cells and tissues of recalcitrant woody species (Ochatt et al, 1988).

4. PLANT IMPROVEMENT

Much interest is being shown in the emergence of new techniques and developing *in vitro* technologies with the aim of inducing variability in horticultural plants, with the objective of exploiting novel and useful genetic changes.

The regeneration of plants from protoplasts, and cells often produces plants which are not true to type (Larkin & Scowcroft, 1983). Somaclonal variants can be isolated by the routine screening of plantlets derived from adventitious meristems, somatic embryos on callus or tissue explants. Some of the variants isolated exhibit considerable physiological and morphological differences, and some may be of commercial value. For example, Cassels & Plunkett (1984) were able to isolate a range of "novel" African violets which exhibited differences in leaf shape and flower colour; Lloyd et al (1988) reported considerable variation in leaf shape in rose plantlets regenerated from callus; Khalid et al (1989) found 2 of the 6 somaclonal variants of the white genotype of Chrysanthemum cv. Early charm, were of commercial value; and Dalsou (1984) in a comprehensive study of several cultures of chrysanthemum isolated variants of yellow snowdon with 40% more flowers, variants of fandango with differences in flower structure and colour, and an early flowering variant of white snowdon.

The production and selection of horticulturally useful somaclonal variants may be increased by mutagenesis and the use of a range of selection pressures e.g. phytotoxins, herbicides, low/high temperatures etc. In some cases tolerant cell lines have been isolated but tolerance has not always been expressed in the whole plant, e.g. salt-tolerance in chrysanthemum Dalsou & Short (1986). However, using this approach *in vitro* screening has resulted in the production of salt tolerant crops (Dix, 1985), verticillium wilt resistance in hops (Connel & Heale, 1986), disease resistance in celery (Collin et al, 1987) and low temperature tolerance in African violet (Warburton et al, 1984).

Recent advances in plant protoplast technology have facilitated the *in vitro* culture and the establishment of reliable protocols for protoplast to whole plant regeneration systems in a number of important ornamental plants and fruit trees (Davey & Power, 1988; Cocking & Davey, 1987). The application of short-term electric pulses to isolated protoplasts has promoted cell division and significantly enhanced plating efficiencies (Rech et al, 1987). These effects are not temporary and have been shown to enhance long-term growth and plant regeneration of protoplast derived tissues in fruit trees (Ochatt et al, 1988). The successful regeneration of plants has been extended to protoplasts of prime ornamental value such as *Pelargonium aridum* (Yarrow et al, 1987) and to species within several genera of the Compositae e.g. *Dimorphotheca* and *Rudbeckia* (Al-Atabee and Power, 1987). An alternative approach to regeneration in recalcitrant species such as chrysanthemum can be overcome by the triggering of totipotency through protoplast fusion with an actively dividing but unrelated protoplast system. This approach has proved successful with *C. morifolium* and *Salpiglossis sinuata* (Otsuka et al, 1985). A non-conventional pathway to regeneration has been exploited in *Prunus cerasus* and *Pyrus communis* involving the isolation and culture of protoplast-derived roots (Ochatt et al, 1989) and will, no doubt, be extended to other plants of commercial value.

Somatic hybridization presents a new vista for plant improvement as it by-passes the constraints of sexual hybridization. Improved methodologies for the culture and fusion - including the use of electroporation treatments have been used to introduce RNA, DNA and intact TMV particles into plant protoplasts (Davey & Power, 1988). The use of strategies targeted at the production of intergeneric somatic hybrids is designed to facilitate gene flow between species. This has been successfully achieved in *Petunia* as a result of somatic hybridization between *Nicotiana tabacum* and *Petunia inflata* (Pental et al, 1986) and this approach may be extended with the objective of improving other ornamentals. Progress in the production of somatic hybrids in fruit tree species has been limited. However, Ochatt et al (1989) have regenerated a novel temperate fruit tree somatic hybrid of the sexually incompatible root stock genotypes - the cherry root stock, Colt (*Prunus avium* x *pseudocerasus*) and the pear root stock, wild pear (*Pyrus communis* var *pyraster* L.) As a result of improved compatibility to grafting and ability to tolerate a wide range of environmental stresses and diseases, it is possible that these somatic hybrid root stocks might provide the basis of a universal root stock in all Rosacious fruit tree species.

Until recently, gene transfer into plants was only achieved by somatic hybridization. Now, in addition, genetic manipulation may be achieved by use of a natural gene transfer system of the soil microorganism Agrobacterium tumefaciens (Gheyen et al, 1985) and A. Rhizogenes (White et al, 1982) and by the microinjection of DNA into nuclei of protoplasts (Nomoru and Komaine, 1986). Therefore, transformation studies are an important aspect of any genetic manipulation programme. The refinement of the development of plasmid vectors carrying selectable markers has facilitated numerous studies on gene expression and regulation in crop plants and resulted in the production of many transgenic plants (Davey et al, 1986). However, the use of such approaches in horticulture is limited but the potential has been highlighted for genetic improvement in fruit trees (James, 1987). Work in this area is progressing and genetic transformation of apple (Malus pumila Mill.) using a disarmed Ti-binary vector has been reported by James et al (1989a). Also, the use of this methodology has proved to be successful in the regeneration of transgenic strawberry plants (James et al, 1989b). Consequently, it is likely that these techniques will be used to transfer important agronomic genes such as those conferring pest and disease resistance into horticultural plants.

5. TRANSPLANTATION OF MICROPROPAGATED PLANTLETS

The success of any micropropagation system can only be effectively measured by the number of plants which are successfully transferred from tissue culture vessels to soil conditions. Tissue culture derived plants are similar in appearance to conventionally produced cuttings but their biochemistry, physiology and anatomy is abnormal, especially their water relations (Short et al, 1984; Ziv et al, 1983). Consequently, they need to be gradually acclimatized from the high humidity of the culture vessel to the ambient conditions found on transfer to glasshouse or field conditions. Therefore, weaning procedures involving complex acclimatization regimes and fogging devices can be a limiting factor in commercial micropropagation, particularly where individual plantlet costs are high.

The major cause of plantlet loss on transplanting is dessication through uncontrolled foliar water loss due to the absence of epicuticular wax on cultured leaves. In the absence of surface wax the control of water loss by stomata is not sufficient to maintain leaf turgor and to prevent wilting. The situation is exacerbated by the abnormal stomatal physiology of these plantlets which have large somatal apertures and guard cells which are unable to close completely (Wardle & Short, 1983).

The water-saturated atmosphere of tissue culture flasks is the prime cause of the inability of *in vitro* plantlets to control water loss after transplantation. The culture of plantlets at reduced levels of humidity has been shown to increase the deposition of epicuticular wax on the leaves of Iris (Maene & Debergh, 1986) and carnation (Ziv,

1986) and cauliflower and chrysanthemum (Short et al, 1987). A range of methods have been used to reduce the relative humidity in culture vessels including the use of dessicants (Wardle et al, 1983), increasing the concentration of agar in the culture medium (Ziv et al, 1983), the use of culture vessels with porous closures (Short et al, 1987) and cooling the bottom of the culture vessel (Maene & Debergh, 1986).

Studies on cultured chrysanthemum and sugar beet plantlets have demonstrated that the establishment of functional leaves, characterized by the typical deposition of epicuticular wax and responsive stomata, can be achieved only if the initial stages of leaf development take place under reduced levels of humidity (Ritchie et al, 1989b). Furthermore, the transfer of plantlets with leaves formed under these conditions to high humidity, for the remaining culture period, does not adversely affect their water relations on transfer to soil. The inclusion of the anti-giberellin, paclobutazol, at a concentration of 0.5 - 4.0mgl^{-1} in the rooting medium for chrysanthemum and sugar beet plantlets leads to a reduction in wilting when plantlets are transferred to soil conditions (Smith et al, 1989a; Ritchie et al, 1989b). Plantlets subjected to this treatment were characterized by smaller somatal apertures, increased epicuticular wax and the precocious development of trichomes on the surface of chrysanthemum leaves (Ritchie et al, 1989b). These procedures, therefore, may be used to facilitate the successful transfer of tissue-culture derived plants to soil conditions, obviating the need for any hardening regime. This process can be further facilitated by the use of containers which allow the humidity level of the culture vessel to be regulated by the uncovering of aseptically sealed pores in the walls of the culture vessel (Baumgartner Papiers S.A., Switzerland). A preliminary evaluation of these containers indicates that they result in the in vitro acclimatization of chrysanthemum plantlets (Smith et al, 1989b) and this is likely to extend to other horticultural plants.

6. CONCLUSION

This brief review of the application of tissue culture technology for plant production and improvement indicates that many in vitro methods have been successfully applied to the routine production and improvement of a wide range of horticultural plants. Further progress in the new developing technologies concerned with genetic manipulation and gene transfer will require a more complete understanding of the molecular biology of plant cells. Without this information it will not be possible to bring about the precisely directed changes that plant breeders wish to achieve in economically important plants. The recalcitrance of many horticultural plants has now become a problem of the past, therefore, more species can be realistically incorporated into improvement programmes based on their genetic manipulation.

7. REFERENCES

Al-Atabee, J.S. and Power, J.B. (1987). Plant regeneration from protoplasts of Dimorphotheca and Rudbeckia. Plant Cell Reports, 6, 414-16.

Alderson, P.G. (1986). Micropropagation of woody plants in Micropropagation in horticulture, practice and commercial problems, eds. Alderson, P.G. and Dullforce, W.M., University Nottingham, 37-52.

Brown, C., Lucas, J.A., Crute, I.R., Walkey, D.G.A. and Power, J.B. (1986). An assessment of genetic variability in somacloned lettuce plants (Lactuca sativa) and their offspring. Ann. Appl. Biol., 109, 391-407.

Cassels, A.C. and Plunkett, A. (1984). Production and growth analysis of plants from leaf cuttings, and from tissue cultures of disks from mature leaves and young axemic leaves of African violet (Saintpaulia ionantha Wend L). Scientia hort., 23, 361-9.

Constantine, D.R. (1986). Micropropagation in the commercial environment in Plant tissue culture and its agricultural applications, ed. Withers L.A. and Alderson P.G., Butterworths, 175-186.

Cocking E. C. and Davey M. R. (1987). Gene transfer in cereals, Science, 236, 1259-62.

Collin, H.A., Donnovan, A. and Isaac, S. (1987). Selection for disease resistance using celery tissue cultures in Cell culture applied to plant production and plant breeding, ed. Boccon-Gibod, J., Benbadis, A. and Short, K.C., ENITHP, Anger, 21-30.

Connel, S.A. and Heale, J.B. (1986). Development of an in vitro selection system for novel sources of resistance to Verticillium wilt in hops, in Plant Tissue culture its agricultural applications, ed. Withers, L.A. and Alderson, P.G., Butterworths, 451-86.

Dalsou, V. (1984). In vitro propagation and selection for sodium chloride tolerance in chrysanthemums. Ph.D. thesis, CNAA, Nottingham Polytechnic.

Dalsou, V. and Short, K.C. (1986). Selection for sodium chloride tolerance in chrysanthemum. Acta. Hort., 212, 737-40.

Davey, M.R., Gartland, K.M.A. and Mulligan, B.J. (1986). Transformation of the genomic expression of plant cells, in Plasticity in Plants Symposium, ed. Jennings, D.H. and Trewaras, A.J., Soc. Expt. Biol, 40, 85-120.

Davey, M. R. and Power, J. B. (1988). Aspects of protoplast culture and plant regeneration, Plant cell, Tissue and Organ Culture 12, 115-125.

Dix, P.J. (1985). Cell line selection. In Plant cell culture technology, ed. Yeoman, M.M., Blackwell Scientific, 141-99.

Gheejsen, G., Dhaese, P., Van Montagu, M. and Shell, J. (1985). DNA flux across genetic barriers: the grown cell phenomenon. In Genetic flux in plants, ed. John, B. and Dennis, E.S., Springer-Verlag, 220-252.

Goldsworthy, A. and Rathore, K.S. (1985). Electrical control of growth in plant tissue cultures, Biotechnology, 3, 253-4.

Harper, P. (1987). The micropropagation industry at twenty five years in Cell culture techniques applied to plant production and breeding, ed. Boccon-Gibod J., Benbadis A. and Short K.C., ENITHP, Anger, 90-97.

Henshaw, G. G. (1987). Strategies relating to <u>in vitro</u> plant production in cell culture techniques applied to plant production and plant breeding, ed. Boccon-Gibod J., Benbadis A., and Short K.C., ENITHP, Anger, 98-101.

Huang, B. and Sunderland, N. (1982). Temperature stress pretreatment in barley anther-culture. Ann. Bot., 49, 77-88.

Hussey, G. (1982). <u>In vitro</u> propagation of monocotyledonous bulbs and corms, in Plant Tissue Culture, ed. Fujiwara A., Tokyo, 677-80.

Hussey, G. (1983). <u>In vitro</u> propagation of horticultural and agricultural crops, in Plant biotechnology, eds. Mantell S.A. and Smith H., Soc. Exp. Biol., 18, Cambridge U.P., 111-138.

Hussey, G. (1986). Problems and prospects in the <u>in vitro</u> propagation of herbaceous plants, in Plant tissue culture and its agricultural applications, ed. Withers L.A. and Alderson P.G., Butterworths, 69-84.

James, D. J. (1987). Cell and tissue culture technology for the genetic manipulation of temperate fruit trees. Biotech. Gen. Eng. Rev. 5, 33-79.

James, D.J., Passey, A.J., Barbara, D.J. and Bevan, M. (1989a). Genetic transformation of apple (<u>Malus pumila</u> Mill) using a disarmed Ti-binary vector, Plant Cell Reports, 7, 658-61.

James, D.J., Passey, A.J. and Barbara, D.J. (1989b). Regeneration of transgenic strawberry plants using disarmed Ti-binary vectors. (in preparation).

John, A. and Pearson, D.J. (1986). Inducticn of vitrification in Picea sitchensis cultures, New Zealand J. Forestry Sci, 16, 328-42.

Jones, O.P. (1985). The role of growth regulators in the propagation in vitro (micropropagation) of temperate fruit trees, in Growth regulators in horticulture, eds. Menhenett R. and Jackson M.B., BCPC Monograph 13, 113-24.

Khalid, N., Davey, M. R. and Power, J. B. (1989). An assessment of somaclonal variation in chrysanthemum morifolium: the generation of plants of potential commercial value. Scientia Horticulturae 38, 278-94.

Larkin, P.J. and Scowcroft, W.R. (1983). Somaclonal variation - a novel source of variability from cell cultures for plant improvement. Theor. Appl. Genet., 60, 197-214.

Lloyd, D., Roberts, A.V. and Short, K.C. (1988). The induction in vitro of adventitious shoots in Rosa. Euphytica, 37, 31-36.

Nomoru, K. and Komanine, A. (1980). Embryogenesis from microinjected single cells in a carrot cell suspension culture, Plant Sci., 44, 53-4.

Maene, L.J. and Debergh, P.C. (1986). Optimization of plant micropropagation, Med. Fac. Londbouwv Rijksuniv Gent, 51, 1479-86.

Ochatt, S.J., Chand, P.K., Rech, E.L., Davey, M.R. and Power, J.B. (1988). Electroporation-mediated improvements of plant regeneration from colt cherry (Prunus avium x pseudocerasus) protoplasts, Plant Science, 54, 165-69.

Ochatt, S.J., Patat-Ochatt, E.M., Rech, E.L., Davey, M.R. and Power, J.B. (1989). Somatic hybridization of sexually incompatible top-fruit tree rootstocks, wild pear (Pyrus communis var. pyraster L) and colt cherry (Prunus avium x pseudocerasus). Theor. Appl. Genet., 78, 35-41.

Otsuka, H., Svematsu, N., Toda, M. (1985). The culture and plant regeneration from mesophyll protoplasts of Chrysanthemum morifolium, Bull. Shizuoka Agr. Exp. Stn., 30, 25-33.

Pennel, D. (1986). Micropropagation of glasshouse ornamentals, in Micropropagation in horticulture, practice and commercial problems, ed. by Alderson P.G. and Dullforce W.M., University Nottingham, 113-122.

Pental, D., Hamill, J.D., Pirrie, A., Cocking, E.C. (1986). Somatic hybridization of Nicotiana Tabacum and Petunia Hybridia: recovery of plants with P. Hybrida nuclear genome and N. tabacum chloroplast genome, Mol. Gen. Genet, 202, 342-7.

Pierik, R. L. M. (1987). In vitro culture of higher plants; Mertinus Nighoff publishers, 321 pages.

Rathore, K.S. and Goldsworthy, A. (1985). Electrical control of shoot regeneration in plant tissue cultures, Biotechnology, 3, 1107-9.

Rech, E.L., Ochatt, S.J., Chand, P.K., Power, J.B. and Davey, M.R. (1987). Electro-enhancement of division of plant protoplast-derived cells, Protoplasma, 141, 169-76.

Ritchie, G.A., Short, K.C. and Davey, M.R. (1989a). In vitro shoot regeneration from callus, leaf, axils and petioles of sugar beet (Beta vulgaris L). J. Exp. Bot., 40, 277-83.

Ritchie, G.A., Short, K.C. and Davey, M.R. (1989b). In vitro acclimatization of chrysanthemum and sugar beet plantlets by treatment with paclobutazol and exposure to low relative humidity, J. Exp. Bot. (in press).

Roberts, A.V., and Smith, E. (1989). The preparation in vitro of chrysanthemum for transplantation to soil, 1. Protection of roots by cellular plugs, Plant cell, Tissue and Organ culture (in press).

Shillito, R. D., Saul, M. W., Paszkowski, J., Miller, M. and Pokrykus, I. (1985). High efficiency direct gene transfer to plants, Biotechnology, 3, 1099-103.

Short, K. C. (1986). Pathways of regeneration in cultures and their control, in Micropropagation in horticulture, practice and commercial problems, ed. by Alderson, P.G. and Dullforce, W. M., University Nottingham, 15-26.

Short, K.C., Warburton, J. and Roberts, A.V. (1987). In vitro hardening of cultured cauliflower and chrysanthemum plantlets to humidity, in In vitro problems related to micropropagation of horticultural plants, Acta. Hort., 212, 329-34.

Short, K.C., Wardle, K., Gront, B.W.W., and Simpkins, I. (1984). In Vitro physiology and acclimatization of aseptically cultured plantlets, Proceedings Plant tissue and cell culture - applications to crop improvement, ed. Novak, F.J., Havel, L. and Dolezel, J., CSAV, Prague, 475-86.

Skoog, F. and Miller, C.O. (1957). Chemical regeneration of growth and organ formation in plant tissues cultured in vitro, in The biological action of growth substances, ed. by Porter, H.E., Soc. Exp. Biol, Cambridge U.P., 118-31.

Smith, E.F., Roberts, A.V. and Mottley, J. (1989a). The preparation in vitro of chrysanthemum for transplantation to soil, 2. Improved resistance to dessication conferred by paclobutrazol, Plant Cell Tissue and Organ Culture (in press).

Smith, E.F., Roberts, A.V. and Mottley, J. (1989b). The preparation in vitro of chrysanthemum for transplantation to soil, 3. Improved resistance to dessication conferred by reduced humidity. Plant Cell Tissue and Organ Culture (in press).

Steward, F.C. (1970). The silent revolution in agriculture, 4. Cloning cells and controlling the composition of crops, in Progress. (The Unilever Quarterly), 2, 44-51.

Warburton, J., Gront, B.W.W. and Short, K.C. (1984). In vitro selection for cold tolerant cell lines of Saintpaulia ionantha (Wend L). Proceedings plant tissue and cell culture - applications to crop improvement, ed. Novak, F.J., Havel, L. and Dolezel, J., CSAV, Prague, 355-6.

Wardle, K. and Short, K.C. (1983). Stomatal Responses of in vitro cultured plantlets, I. Responses in epidermal strips of Chrysanthemum to environmental factors and growth regulators, Biochemie Physiologie Pflazen, 178, 619-24.

White, F.F., Ghidossi, G., Gordon, M.P. and Nestor, E.W. (1982). Tumor induction by Agrobacterium rhizogenes involved the transfer of plasmid DNA to the plant genome, Proc. N. Acad. Sci., USA, 79, 3193-7.

Yarrow, S.A., Cocking, E.C. and Power, J.B. (1987). Plant regeneration from cultured cell-derived protoplasts of Pelargonium aridum, P x hortarium and P. peltatum, Plant Cell Reports, 6, 102-4.

Ziv, M., Meir, G., Halvey, A.H. (1983). Factors influencing the production of hardened glaucous carnation plantlets in vitro, Plant Cell Tissue Organ Culture, 2, 55-65.

HOW IMPORTANT IS PHOTOSYNTHESIS IN MICROPROPAGATION ?

Capellades M.[1], Vanderschaeghe A.[2], Lemeur R.[3] & Debergh P.[1]
[1]Lab. Horticulture, [2]I.W.O.N.L.-C.S.V.T., [3]Lab. Ecology
State University Gent
Coupure links 653
9000 GENT
Belgium

ABSTRACT. When micropropagating plants it is not always required that the final product *in vitro* is autotrophic. On the contrary, micropropagated roses with the lowest net photosynthesis yielded the best survival rate in the greenhouse, provided the culture conditions in the last stage of micropropagation were adequate. Sugar content, light intensity and controlled water retention capacity in the head space of the container are the most important factors. They can contribute to yield plants with a normal anatomy and physiology.

1. Introduction

Besides effects due to water stress, factors affecting photosynthesis may play an important role in the acclimatization and survival of tissue-cultured plantlets. Important questions are whether the low light-intensity and sugar content prevalent in most culture systems are limiting photosynthetic capacity and what is the photosynthetic capacity of plants grown under this conditions. Indeed, *in vitro* cultured plantlets are often considered to have little photosynthetic ability to provide a positive CO_2 balance and to require sugar (sucrose in most cases) as a carbon and energy source for their heterotrophic (all carbon from medium carbohydrate) or mixotrophic (carbon both from medium carbohydrate and CO_2) growth.

The growth and development of a plantlet *in vitro* may be affected by the gas environment in the tissue culture container. **Fujiwara et al.** (1987) studied different plant species in the final stages of micropropagation and they showed there was a clear photosynthetic ability (the CO_2 concentration in the culture vessel decreased sharply within 2 hours after the start of the photoperiod and increased with time during the dark period), but the plants could not achieve their full photosynthetic ability due to a sub-optimal CO_2-concentration. However, **De Proft et al.** (1985) obtained supra-optimal CO_2-concentrations (2-5%) at the end of the dark period. This comparison clearly illustrates the influence of the exchange capacity. Indeed, in the experiments of **De Proft et al.** (1985) *Magnolia* shoots were grown in hermetically closed containers, this was not the case in

R.S. Sangwan and B.S. Sangwan-Norreel (eds.), The Impact of Biotechnology in Agriculture, 29–38.

the experiments of **Fujiwara et al.** (1987). Differences in the coefficients of carbon dioxide exchange have been observed depending on the quality of the stoppers used to close the vessels: the rate of exchange of an aluminum foil cap was almost ten times smaller than from plastic cap (**Kozai et al.**, 1986).

Some authors have studied the carbon balance in micropropagated plantlets. **Grout & Donkin** (1987) observed a negative balance in cauliflower, due to a higher dark respiration than net photosynthesis rate. Moreover, a net CO_2-uptake was not measured in cauliflower until two weeks after the transfer to soil (**Grout & Aston**, 1977). **Grout & Price** (1987) reported that leaves of micropropagated strawberry plantlets did not fix enough carbon to sustain independent growth in the absence of added sucrose in the culture medium. They pointed out that the major factor contributing to poor photosynthesis was the very low activity of Rubisco, and suggested that sucrose did in some way inhibit the activity of this enzyme. It has recently been shown that the growth of *in vitro* plantlets could be promoted by CO_2 enrichment under high photon flux density (230 $\mu mol.m^{-2}.s^{-1}$) in the absence of sugar in the culture medium (**Kozai et al.**, 1987).

The main aim of this paper is to discuss the necessity of *in vitro* plants to be autotrophic or not.

2. Materials and Methods

2.1. PLANT MATERIAL AND GROWTH CONDITIONS

Elongated shootlets (Stage IIIa, **Debergh & Maene**, 1981) of *Rosa multiflora* L. cv. Montse (**Capellades et al.**, 1989) and *Pieris floribunda* were cultured *in vitro* on a basic medium with 0, 1, 3 and 5% sucrose at 23 ± 2 °C, under a 16h/day photoperiod and a photon flux density of 25 (for *Pieris* at 100% RH in the culture container) to 80 $\mu mol.m^{-2}.s^{-1}$ (for *Pieris* at 85% RH and *Rosa* under all circumstances) provided by Philips TLD-83 fluorescent lamps. Relative humidity (RH) in the container was controlled at 100% (no bottom cooling), 75% and 85% by a bottom cooling system (**Maene**, 1985; **Vanderschaeghe & Debergh**, 1987).

2.2. MEASUREMENTS OF GAS EXCHANGE

Net photosynthesis and dark respiration rates were analyzed on 4-week-old cultures placed in a plexiglass cuvette (17 x 17 x 15 cm) connected to a gas analysis circuit (CO_2 infra-red differential gas analyzer (IRGA) and air flow meter) (**Capellades**, 1989). Intensities used during the measurements ranged from 0 to ca. 800 $\mu mol.m^{-2}.s^{-1}$. All measurements were taken at 23 ± 2°C and at approximately the same RH as the plants were grown *in vitro*. The functional photosynthesis parameters were calculated by fitting rectangular hyperbola to the measured points (net photosynthesis rate, light intensity). The parameters of the equation obtained this way are listed in table 1.

For each treatment four containers were measured, and the light response curve for net photosynthesis rate was obtained from the mean value of the three best measurements - based on technical evaluations of the measurements - for each treatment. After these measurements, the plants were removed from the cuvette and their dry matter content was determined.

2.3. MEASUREMENTS OF CHLOROPHYLL FLUORESCENCE

The intensity of chlorophyll fluorescence from intact leaves was measured with a fluorometer model SF-10 (Richard Branker Research Ltd., Ottawa, Ontario, Canada). The sensor was placed over a leaf which was kept for 10 to 20 min in the dark. Then the upper surface was illuminated for 3 s with monochromatic red light with an intensity of ca. 6 $\mu mol.m^{-2}.s^{-1}$. All the measurements were done in a dark room, illuminated only with weak green light. Twenty-five leaves of each treatment were sampled and 13 readings per second were made for *Rosa*. For *Pieris*, 10 leaves of each treatment were sampled and 40 readings per second were made. A digital data acquisition unit, composed of a microprocessor with a disc drive and printer stored the fluorescence induction curve and calculated two characteristic fluorescence parameters I and P-I) of the Kautsky curve (**Kautsky and Frank**, 1943). "I", the inflection point and "P", the peak, are characteristic values of the excitation curve for fluorescence.

Since the I-level changes slightly from plant to plant the ratio (P-I)/I, which normalizes the I-level (**Aoki and Oda**, 1988), was used to express the activity of fluorescence.

Chl a fluorescence was measured at transplanting time (5 weeks on elongation medium) and again after 10 and 25 days in the greenhouse for plants cultured on 1% sucrose, and after 10, 20 and 30 days for those cultured on 5% sucrose for *Rosa*. For *Pieris* Chl a fluorescence was only measured at transplanting time (after 4 weeks on elongation medium).

3. Results

3.1. PHOTOSYNTHESIS

The results reported in table 1 and illustrated in figure 1, show that the highest net photosynthesis rate observed during our measurements corresponds to the treatment with 1% sucrose, and that it decreases as the sugar concentration in the medium increases. Moreover the difference between *Rosa* and *Pieris* is obvious.

With respect to the light compensation point, shootlets grown on a medium with 5% sucrose showed the highest values.

For light levels above 200 $\mu mol.m^{-2}.s^{-1}$ no further increase in photosynthesis rates was found. This gives an idea of the rather low light saturation point of these shootlets cultured *in vitro*. This light saturation point is obtained at lower light levels for rose shootlets (100 $\mu mol.m^{-2}.s^{-1}$) grown with 5% sucrose and *Pieris* (50 $\mu mol.m^{-2}.s^{-1}$) grown with 3 and 5% sucrose; than for rose shoots grown with 1 and 3% sucrose (200 µmol.m^{-2}.s^{-1}) and *Pieris* (150 µmol.m^{-2}.s^{-1}) grown with 0 and 1% sucrose.

For *Rosa*, the dark respiration rate was recorded highest for the shootlets grown with 1% sucrose; whereas for *Pieris* it was for those grown with 5% sucrose.

Dry matter content was significantly lower in 0 and 1%S-shootlets. Moreover, these treatments with the lowest dry matter content showed the highest respiration rates.

Shootlets developed in an atmosphere with 100% relative humidity showed a higher photosynthesis rate than those grown at 75 or 85%

TABLE 1. Values for the functional photosynthetic parameters calculated from the light response curves of net photosynthesis rate versus photon flux density. Values were obtained for Stage IIIa cultures of *Rosa* and *Pieris* grown at different relative humidities in the container and for different sucrose concentrations.

ELONGATION TREATMENT	Light comp. point µmol.m^{-2}.s^{-1}	Max.Net Phot.rate nmolCO2 (gDW.s)$^{-1}$	Dark Resp. rate nmolCO2 (gDW.s)$^{-1}$	Net CO2 exchange at different quantum flux densities (nmolCO2.(gDW)$^{-1}$.s^{-1}) µmolm-^{2}s^{-1}: 25	80	100	500	1000	Dry matter content %
Rosa									
100%RH-1%S	41.26	77	-20			17	54	64	6.8c
100%RH-3%S	12.97	71	-5			22	51	59	14.4a
100%RH-5%S	81.89	7	-16			1	5	6	12.5b
85%RH-3%S	32.13	63	-16			18	46	54	14.2a
85%RH-5%S	82.52	15	-11			1	10	12	12.5b
75%RH-3%S	31.73	46	-13			15	35	40	13.1b
75%RH-5%S	64.78	13	-14			3	10	12	14.0a
Pieris									
100%RH-0%S	20.6a*	16.5b	-5.0ab	0.8d	6.6a	7.7a	13.8ab	15.1ab	8.1cd
100%RH-1%S	39.3ab	22.5a	-11.8ab	-3.3a	5.3ab	7.1ab	17.5a	19.8a	11.4bc
100%RH-3%S	27.6ab	8.5bcd	-5.4ab	-0.2bcd	3.6bc	4.3bc	7.4bcd	7.9bc	15.4ab
100%RH-5%S	48.5b	4.4d	-21.4b	-1.3abcd	1.3cd	1.7cd	3.7de	4.0d	15.0b
85%RH-0%S	38.1ab	8.0bcd	-4.0a	-0.8bcd	2.3c	2.9c	6.4cd	7.1cd	7.8d
85%RH-1%S	41.5ab	13.5bc	-5.9ab	-1.6abc	3.2bc	4.3c	10.5bc	11.8bc	13.6ab
85%RH-3%S	42.5ab	5.3cd	-7.7ab	-1.5abcd	1.9cd	2.5c	4.6cde	5.0ce	15.7a
85%RH-5%S	-	0d	-	0 cd	0 d	0 d	0 e	0 e	-

* Means followed by the same letter are not significantly different at the 5% level.

3.2. FLUORESCENCE

At transplanting time, shootlets of *Pieris* and *Rosa* grown with 5% sucrose presented a lower (P-I)/I ratio than those grown with 1% sucrose (table 2).

For *Rosa* after the acclimatization period of 10 days, plantlets from the treatment with 5% sucrose were already rooted (ca.100%), and showed a higher (P-I)/I ratio than at transplanting time. The value of this ratio remained constant after 10, 20 and 30 days in the greenhouse. However, those which had developed on 1% sucrose were not yet rooted after 10 days, and the ratio was significantly lower (0.38). When chlorophyll fluorescence was measured on the surviving 1%-sucrose plantlets (of which only 47% survived), a very significant increase of the (P-I)/I ratio showing re-establishment of photosynthetic activity - was observed (table 2). These results show that photosynthesis was re-established after 10 days in plantlets from 5% sucrose, whereas it was inhibited in those from 1% sucrose.

TABLE 2. Influence of the sugar content in the culture medium on the fluorescence of rose and *Pieris* leaves at transplanting time and for rose leaves during the acclimatization in the greenhouse.

TREATMENT	(P-I)/I
Rosa	
A : Elongation on 1%S and 75% RH	0.67a*
A + 10 days in the greenhouse	0.38d
A + 25 days in the greenhouse	0.52c
B : Elongation on 5%S and 75% RH	0.51c
B + 10 days in the greenhouse	0.63ab
B + 20 days in the greenhouse	0.64ab
B + 30 days in the greenhouse	0.62b
Pieris	
1. 0%S and 85% RH	0.23c
2. 1%S	0.55a
3. 3%S	0.39b
4. 5%S	0.19c

* Means followed by the same letter are not significantly different at the 5% level of probability according to Duncan's Multiple Range Test.

3.3. PLANT QUALITY

Depending on the amount of sucrose added to the culture medium, differences on the quality of the plants were observed.

For instance, roses grown in a medium without sugar were chlorotic and vitrified. Those grown with 1% sucrose were light green, vitrified and exempt of anthocyanins. Plants developed on a medium with 3 or 5% sucrose were dark green, more lignified and showed anthocyanins under the leaves and at the stem level.

Pieris plants grown on a medium without sugar and with 5% sucrose showed a chlorotic state, with yellow to brown leaves; plants developed on a medium with 1% and 3% sucrose were dark green.

Also the relative humidity during the culture showed to have an influence on the quality of the *in vitro* developed roses. Plants grown under a relative humidity of ca. 100% were more susceptible to disseccation and fungi attacks (e.g. *Rhizoctonia*, *Pythium*) than those grown at 75% RH.

For *Pieris* survival rates in the greenhouse are not yet available.

4. Discussion

From the results (table 1 and 2) we can conclude that net CO_2 fixation decreases as the sucrose concentration in the culture medium increases.

Furthermore, there is a direct positive correlation between the differences in the dry matter content of the treatments and the sugar content in the culture medium (table 1). This agrees with the findings of Langford and Wainwright (1987), who observed more vitrified rose shootlets at 1% sucrose than at 2% or 4%.

We suspect that the high respiration rates recorded for plants grown with 1% sucrose, which had a higher photosynthesic rate, could be caused by the higher energy consumption, during translocation of assimilates. High respiration values and low or no photosynthesis were observed for *Pieris* with 5% sucrose, which were chlorotic due to the supra optimal sugar level.

The low light saturation point (ca. from 80 to 100 $\mu mol.m^{-2}.s^{-1}$) observed for all treatments can be explained by the low levels of irradiance under which the *in vitro* shootlets developed (ca. 25-85 $\mu mol.m^{-2}.s^{-1}$).

A hypothesis to explain the higher photosynthesis rate recorded for plants developed under 100% RH is that these plants were not submitted to any water stress in the culture container, so they did not need to spend energy to develop any strong mechanism to protect themselves against dessication. Plantlets which developed in a drier atmosphere showed higher stomatal resistance to restrict CO_2-uptake; more protruded stomata were observed in shootlets grown under 100% RH (**Capellades et al.**, 1989), which will have a reduced stomatal resistance and consequently a better CO_2 exchange.

The lower P-I/I ratio for *Rosa* and *Pieris* grown with 5% sucrose at transfer time confirms our measurements of CO_2 exchange, by revealing that shootlets developed with 1% sucrose had indeed a better photosynthetic capacity at transplanting time than those developed with 5% sucrose (**Capellades,** 1989). Also **Aoki & Oda** (1988) observed that this ratio increased proportionally to the increment of photosynthesis.

The low (P-I)/I ratio observed for 0 and 5% sucrose for *Pieris* at the moment of transfer is a clear prove for the stress situation of these plantlets.

The lower (P-I)/I ratio in 1%S-rose shootlets, 10 days after transplantation to the greenhouse, can be interpreted as a consequence of stress due to the lack of an adequate sugar supply, during the weaning and rooting period, only the fittest survived (47%).

The increase of (P-I)/I in 5%S-rose shootlets and the high survival rate (more than 90%) after a period of acclimatization in the greenhouse, can be interpreted as the re-establishment of the photosynthetic capacity due to the translocation of the stored assimilates. It was proven that shootlets of plantlets developed on 5% sucrose had a higher starch content than those developed on 1% sucrose (**Capellades**, 1989). Therefore, we could postulate that during the first days after transfer to the greenhouse, the 5%S-shootlets obtain the substrate for the synthesis of new materials from the stored assimilates. After a transition period, during which the substrate is provided by nutrient reserves and by photosynthesis, total autotrophy (when the stored sugars are exhausted and photosynthesis supplies all the energy needed by the plant) is reached. This interpretation is based on the study of **Pinto Contreras & Gaudillère** (1987) on the processes occurring during the germination of seeds.

5. Conclusions

From our results and from the discussion it is clear that high net photosynthesis (or even autotrophy) is not a pre-requisite to have the best survival rates of micropropagated plants in the greenhouse. We could conclude that more than autotrophic plants we should produce *in vitro* plants with the characteristics of a seedling. This means that enough reserves should be present to supply the plant requirements during the first days after transplanting.

Another important point that should be taken in consideration is that in clonal micropropagation of plants, a higher multiplication rate is sought as an economic priority. Therefore if we wish to produce more autotrophic plants *in vitro* by means of increasing the light intensity (230 $\mu mol.m^{-2}.s^{-1}$) and enriching the air of the culture chamber with CO_2 (**Kozai et al.**, 1987, **Desjardins et al.**, 1988); or by using a kind of micro-hydroponic propagation system (**Fujiwara et al.**, 1988) we will loose one of the principal advantages of propagating plants *in vitro* because the multiplication ratio will be affected. So that, in the case that the use of these techniques will become cheaper and they could expand to the commercial laboratories, they could be only used in the last stages of the micropropagation cycle (elongation and rooting). Moreover, till now, it remains questionable whether the economic cost incurred (cost of the installation, including light sources and air conditioning) is compensated by a greater success in the acclimatization period than when sugar is used in the culture medium.

6. References

Aoki, S., Oda, M. (1988) 'Sensing of photosynthetic capacities of seedlings of lettuce with chlorophyll fluorescence', Acta Hort. 230, 363-370.

Capellades, M. (1989) 'Histological and ecophysiological study of the changes occurring during the acclimatization of *in vitro* cultured roses', PhD dissertation, State University of Gent, Belgium, 98 pp.

Capellades, M., Carulla, C., Fontarnau, R. & Debergh, P. (1989) 'The effect of environment on anatomy of stomata and epidermal cells in tissue cultured Rosa multiflora', J. Amer. Soc. Hort. Sc., accepted for publication.

Capellades, M., Lemeur, R. & Debergh, P. (1989) 'Effects of sugar content in the culture medium on starch accumulation in the chloroplasts and photosynthesis rate of tissue cultured roses', Phys. Plant., submitted for publication.

Desjardins, Y., Laforge, F., Lussier, C. & Gosselin, A., (1988) 'Effect of CO_2 enrichment and high photosynthtic photon flux on the development of autotrophy and growth of tissue-cultured strawberry, raspberry and asparagus plants', Acta Hort., 230, 45-53.

Fujiwara, K., Kozai, T. & Watanabe, I., (1987) 'Fundamental studies on environment in plant tissue culture vessels. 3. Measurements of carbondioxide gas concentration in closed vessels containing tissue cultured plantlets and estimates of net photosynthetic rates of the plantlets', J. Agr. Met. 43, 21-30.

Grout, B.W.W. & Aston, M.J. (1977) 'Transplanting of cauliflower plants regenerated from meristem culture. I. Water loss and water transfer related to changes in leaf wax and to xylem regeneration', Hort. Res., 17, 1-7.

Grout, B.W.W. & Price, F. (1987) 'The establishment of photosynthetic independence in strawberry cultures prior to transplanting', in G. Ducaté, Jacobs, M. & Simeon, A. (Eds.), Proc. Symp. Florizel 87, Plant Micropropagation in horticultural industries, Arlon, Belgium, pp. 55-60.

Grout, B.W.W. & Donkin, M.E. (1987) 'Photosynthetic activity of cauliflower meristem cultures *in vitro* and at transplanting time', Acta Hort., 212, 323-327.

Kautsky, H. & Franck, V. (1943) 'Chlorophyll Fluoreszenz und Koh lensäureassimilation', Biocem. Z., 315, 139-232.

Kozai, T., Iwanami, Y. & Fujiwara, F. (1987) 'Environment control for masspropagation of tissue cultured plantlets. I. Effects of CO_2 enrichment on the plant growth during the multiplication stage', J. Agr. Met. 4, 22-26.

Kozai, T., Fujiwara, K. & Watanabe, I. (1986) 'Fundamental studies on environments in plant tissue culture vessels. 2. Effects of stoppers and vessels on gas exchange rates between inside and outside of vessels closed with stoppers', J. Agr. Met. 42, 119-127.

Langford, P.J. & Wainwright, H. (1987) 'Influence of sucrose concentration on the photosynthetic ability of *in vitro* grown rose shoots', Acta Horticulturae, 22, 305-308.

Maene, L.J., (1985) 'Optimalisering van de overgang van weefselteeltplantjes naar *in vivo* omstandigheden', PhD-thesis, State University Gent, Belgium, pp. 221.

Pinto Contreras, M. & Gaudillere, J.P. (1986) Plant Physiol. Biochem., 25(1), 35-42.

Vanderschaeghe, A. & Debergh, P., (1987) 'Technical aspects of the control of the relative humidity in tissue culture containers', Med. Fac. Landbouww. Rijksuniv. Gent, 52(4), 1429-1437.

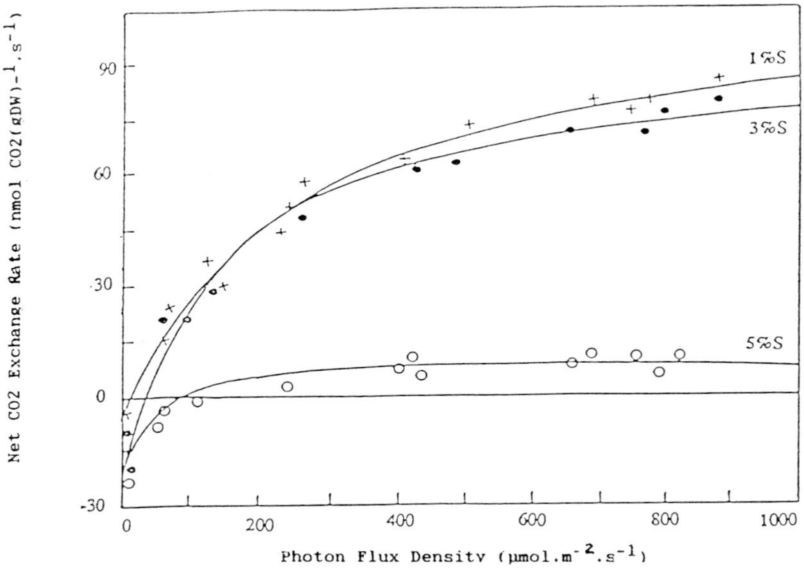

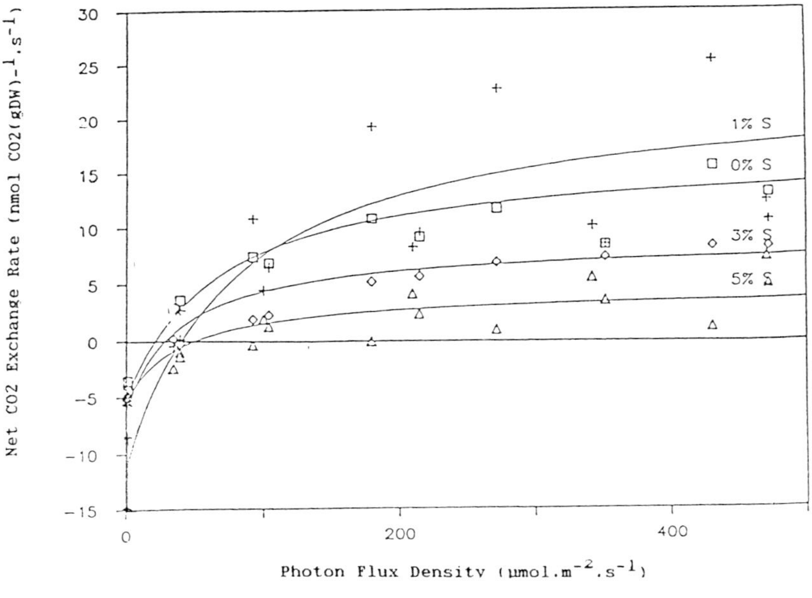

Figure 1. Photosynthetic light response curves of net CO_2 exchange rate. Each curve is fitted to the points obtained from three series of measurements. A. Roses grown with three different concentrations of sucrose (1, 3 and 5%) in the culture medium. B. *Pieris* grown at four different concentrations of sucrose (0, 1, 3 and 5%) in the culture medium.

IN VITRO RADIATION INDUCED MUTAGENESIS IN POTATO

B.S.AHLOOWALIA
Plant Biotechnology Department
Agriculture and Food Development Authority
Kinsealy Research Center, Dublin 17
Ireland

Abstract. Type and frequency of radiation induced variation for upgrading potato cultivars was investigated. Cultivars Cara, Red Cara, Kerrs Pink and Record were cultured on modified Murashige & Skoog's medium and irradiated with 2000 r gamma rays. Shoots of the irradiated and control cultures were cut into single nodes and subcultured. The proliferated stems were recultured from pieces with 2 to 4 nodes which after 6 months produced micro-tubers ca. 2 to 8 mm diameter. Of the 2450 micro-tubers harvested, 1500 were sprouted and transplanted in soil under plastic. Of these, 1085 micro-tubers produced plants, and gave 11945 macro-tubers which were scored for variation in tuber skin and eye colour, skin texture, tuber shape and eye-depth. In 'Cara', 4.8% progenies showed mutation from red-eye to white-eye. In 'Red Cara', 4.2% progenies mutated from red to white skin with red or white eyes. In 'Kerrs Pink', 5.3% plants had pale cream skin color instead of pink skin, 1.4% had very light pink skin and 5.0% had eyes shallower than the parental cultivar. In 'Record', 9.4% plants had tubers with pale light skin. Infrequently, variation in tuber shape was observed in all the cultivars. The range and frequency of observed variation suggested that in-vitro radiation is a suitable technique to upgrade potato cultivars for specific characters.

Introduction

Radiation induced mutations have been extensively used in the improvement of crop plants. A combination of in-vitro technology and radiation induced mutagenesis has been recommended to upgrade cultivars of vegetatively propagated plants (Annonymous, 1986, 1987). The International Atomic Energy Agency (IAEA) report (Annonymous, 1986) also describes terminology used for each propagation following mutagenic treatment.

R.S. Sangwan and B.S. Sangwan-Norreel (eds.), The Impact of Biotechnology in Agriculture, 39–46.

Potato (Solanum tuberosum L.) is a vegetatively propagated crop. Potato cultivars are conventionally maintained as true clones through successive planting of tubers. Potato cultivars can be easily micro-propagated from stem-cuttings and adventitious buds (Roest and Bokelmann, 1976, 1980; Ahloowalia, 1982; Sonnino et al., 1986). In-vitro cultures, usually initiated from the apical meristems, can be maintained indefinitely by successive subculture of stem-cuttings, thus avoiding infection from pathogens such as bacteria, nematodes, fungi and insect pests. Potato cultures, when proliferated for 65 to 100 days under appropriate day-length and temperature conditions, produce in-vitro micro-tubers, ca. 2 to 10 mm diameter. Micro-tubers can be cold stored, and used for further multiplication either through micro-propagation or planted in soil in the conventional manner for the production of seed potato crop. If the initial explant is obtained from an indexed tuber free from viruses and endogenous fungal and bacterial pathogens, then in-vitro culture permits multiplication of high quality, disease free true to type seed tubers of potato cultivars. Irradiation of potato cultures and their subsequent multiplication through micro-tuber production would thus permit a rapid method for producing variants of the standard cultivars. In this paper, the results of a study on in-vitro radiated potato cultivars are reported.

Materials and Methods

In-vitro cultures of four potato cultivars, Cara, Red Cara, Kerrs Pink and Record were established from shoot apical meristems of indexed Breeders' seed tubers. Cultivar Red Cara is a spontaneous red skin variant of 'Cara' which has white skin with pink splash around the eyes. Procedures for decontamination and surface sterilization of the shoots were the same as described previously (Ahloowalia, 1982). M1V1 cultures were initiated and maintained on half-strength MS (Murashige & Skoog, 1962) medium containing 1.5% sucrose. In-vitro grown, six weeks old M0V1 plantlets with single shoots having 19 to 20 buds were cultured in 25 ml capacity sterilin tubes, and irradiated in three replications with 2000 rads gamma rays at the International Atomic Energy Agency, Seibersdorf, Austria in Summer, 1987. M1V1 irradiated and control shoots were cut into 10 to 15 mm long pieces with one to two axillary buds, and cultured on the same medium in 5 mm petri-dishes by placing 4 cuttings per dish. M1V2 cuttings were scored for shoot number 5 days after subculture. The M1V2 subcultures were brought back to Ireland, and further cloned from micro-stem cuttings with two to three nodes each by placing 4 to 6 six cuttings in 150 ml capacity containers. The proliferated M1V3 cultures were grown for 5 months, and allowed to produce M1V3 micro-tubers (Fig. 1). M1V2 and M1V3 cuttings were cultured on modified RM medium based on half-strength MS salts and vitamins with 3% sucrose, and 0.2% gelrite. No growth regulators were used at any stage of in-vitro culture. During Spring, 1988, M1V3 micro-tubers were planted in

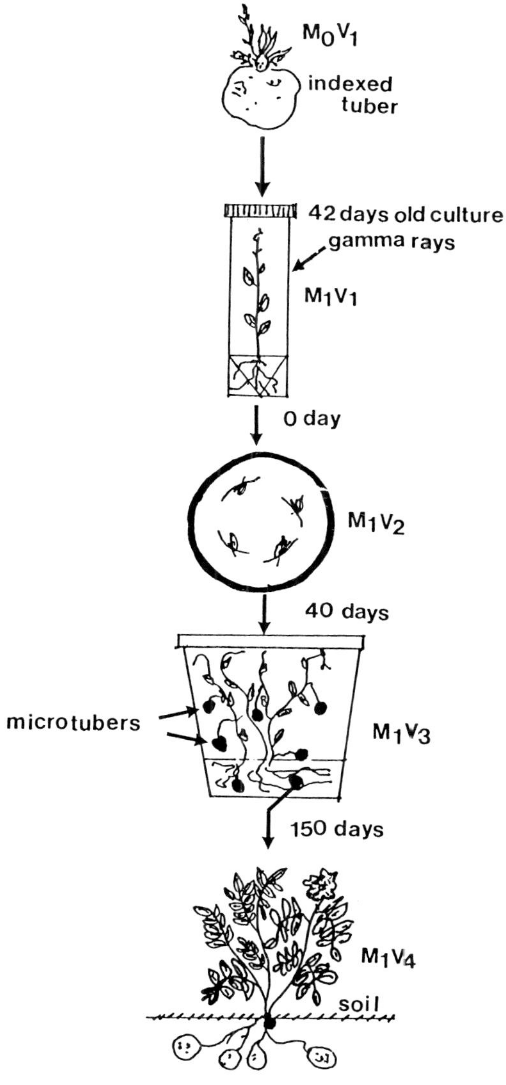

Fig. 1. Procedure for propagation of irradiated potato micro-cultures.

compost in chain type paper pots, grown for one month, and then transplanted in soil under plastic. The M1V4 plants were harvested and the resulting tubers were scored for variation in tuber number, skin color, skin texture, eye-depth and tuber shape.

Results

Irradiated (M1V1) and control (M0V1) micro-stem cultures on transfer to RM medium showed rapid growth, and produced visible M1V2 shoots from apical or axillary buds within 5 days. The M1V2 shoots after 35 days of culture proliferated to yield 4 to 6 cuttings ca. 12-15 mm length, each with 2 to 3 nodes (Table 1), which on further culture produced M1V3 micro-plants. The M1V3 plants after 5 months produced M1V3 micro-tubers ca. 2 to 10 mm in diameter. Each M1V3 micro-plant produced on average 9 micro-tubers per container (Table 1). Thus, after 255 days following irradiation, the original culture had gone through 3 successive vegetative propagations, and in most cases had amplified at least one hundred fold in-vitro. The contamination rate in the M1V3 cultures after 65 days of in-vitro growth was around 24% (Table 1).

TABLE 1. Rate of potato culture proliferation

duration (days)	42--->	0--->	5----->	35--->	65 ---> 150		
cultivar	M1V0 cultures	M1V1 cuttings	M1V1 shoots produced	M1V2 cultures produced	survived	M1V3 micro-tubers	M1V3 microtubers per culture
	---	---	---	number	---	---	---
Cara							
Irrad.	3	55	70	126	108	730	8.5
Control	1	16	15	14	22	207	9.4
Red Cara							
Irrad.	3	48	45	50	40	495	12.4
Control	1	16	9	9	7	66	9.4
Record							
Irrad.	3	48	56	42	33	318	9.6
Control	1	16	20	16	15	134	8.9
K. Pink							
Irrad.	3	62	78	68	36	401	11.1
Control	1	12	-	-	12	122	10.2

A large number of M1V3 micro-tubers failed to produce plants on soil planting. Only 56% of the micro-tubers from irradiated cultures produced plants (Table 2), the remaining either did not sprout or produced week seedlings which died before transplanting. Of the 1500 M1V3 micro-tubers sprouted, 72% gave mature plants, which produced M1V4 tubers of normal size. The plants from M1V3 micro-tubers compared with those grown from two types of controls, i.e., from micro-tubers of non- irradiated cultures and from normal sized certified seed tubers (ca. 35-55 mm diameter), did not show any difference in tuber number except in case of 'Red Cara' in which the control micro-tubers produced an excessively large number of small tubers (Table 2).

TABLE 2. Propagation of radiation induced variants and their yield

Cv.	M1V3 micro-tubers planted	plants	success	average yield of tubers per plant: micro tuber progeny M1V4	micro tuber progeny control	certified seed progeny control
	number	number	%	number	number	number
Cara	730	373	51.0	12.1	12.8	--
Red Cara	495	243	49.0	10.0	20.3	7.3
Record	318	190	59.8	11.6	10.8	13.2
K. Pink	401	279	69.6	10.0	12.3	13.5

Mutation type & frequency in M1V4 propagation

The M1V4 tubers produced by planting M1V3 micro-tubers showed variation in tuber shape, skin color, eye depth, and skin texture. The M1V4 1085 plants, gave 11945 macro-tubers. In 'Cara', 4.8% progenies showed mutation from red-eye to white-eye. In 'Red Cara', 4.2% progenies mutated from red to white skin with red or white eyes. Since 'Red Cara' is a spontaneous variant of 'Cara', these variants represented back-mutations for tuber skin colour. In 'Kerrs Pink', 5.3% plants had pale cream skin color instead of pink skin, 1.4% had very light pink skin and 5.0% had eyes shallower than the parental cultivar. In 'Record', 9.4% plants had tubers with pale light skin (Table 3). In a few cases, the progenies segregated for skin colour. For example, incase of 'Red Cara', occasionally, red and white skinned tubers were produced in the progeny of a single micro-tuber, suggesting persistence of chimeral sectors. Four progenies in 'Record' showed tubers with russett or netted skin, the controls being smooth skinned.

TABLE 3. Radiation induced variation in potato micro-tuber progeny

Cv.	microtuber progeny	M1V4 tubers scored	Variation observed frequency	type
	number	number	%	
Cara				
M1V3	373	4514	4.8	red eye ---> white eye
Control	20	255	none	
Red Cara				
M1V3	243	2452	4.2	red skin ---> white skin
Control	20	386	none	
Record				
M1V3	190	2195	9.4	dark skin --->light skin
Control	39	421	none	
Kerrs Pink				
M1V3	279	2784	5.3	pink skin --> cream skin
			5.0	deep eye --> shallow eye
Control	20	270	none	

A wide variation was observed for tuber shape among the progenies. A large number of progenies produced tubers of uneven, crooked and deformed shape. In some cases, tubers were highly elongated, In a few progenies, tubers were more uniform in shape with reduced number of shallower eyes than the parental cultivar. Both these characters are of agronomic importance in varietal improvement. Since, tuber shape to an extent is affected by the growing conditions, not much reliance could be placed on the stability of such variation.

Discussion

In-vitro multiplication of irradiated potato cultures through successive propagations has several advantages as demonstrated in the present study. The three successive propagations from M1V1 to MIV3 micro-tuber stage were obtained in 255 days which would normally take 3 years under soil propagated conditions. Moreover, all the propagated material was free from all pathogens which are numerous in potato. Savings in space, labor, time and production of healthy and infection free material clearly offer a short cut to the conventional method of mutagenesis and field propagation.

Upgrading of potato cultivars by radiation of in-vitro cultures offers several advantages over the conventional methods of plant breeding. Most potato varieties of cultivated potato (Solanum tuberosum L.) are tetraploid and highly heterozygous. Repeated back-crossing to introduce a specific character would take severel years of hybridization and selection, and may result in undesired characters being introduced along with the one wanted for upgrading. Mutation breeding offers the possibility to upgrade a cultivar in specific characters while maintaining the isogenic background. The present study clearly demonstrated that it is possible to upgrade cultivars in simple characters such as skin-color as well as complex characters such as tuber shape and eye-depth through radiation.

The observed mutation frequency (4.2 to 9.4% for various characters) suggests that small populations are sufficient to get useful variants through this technique. The observed frequency in the present study is similar to 9.32 % reported for morphological variants through in-vitro radiation of 'Desiree' (Sonnino et al., 1986, 1988). The frequency of mutation from red to white skin color reported in 'Desiree' was 4.2% which is similar to that observed in 'Red Cara' in the present study. An average mutation frequency of 74.3% was reported in plants derived from adventitious bud proliferation after x-radiation of explants such as leaf rachis, petiole, and leaf-discs of 'Desiree' (van Harten et al., 1981). However, the control plants derived from non-irradiated adventitious buds also showed mutants ranging from 12.3 to 50.3%. It seems that high mutation frequency reported by van Harten et al., (1981) was confounding radiation induced mutations with somaclonal variation.

Meristem explants (M1V1) after irradiation are periclinal chimeras of mutated and non-mutated tissues (van Harten et al., 1981; Broertjes, 1982). The separation of the two types of tissue through either adventitious bud formation or successive axillary bud proliferation should allow production of solid mutants. Rapid in-vitro multiplication where selection pressures are minimal amplifies mutated sectors several fold. Subsequent shoots and micro-tubers produced from different sectors would result in segregation and often such chimeras should produce solid mutants after a few vegetative propagations. The occurrence of either red or white tuber skin progenies in 'Red Cara' showed that such segregation of the mutated and non-mutated sectors did take place after repeated vegetative micro-propagation. However, the occasional presence of white skinned tubers among otherwise red skinned progeny of single M1V3 micro-tubers suggests that some chimeral sectors persisted for more than four propagations following irradiation. The stability of these variants in the subsequent propagations is being investigated.

Acknowledgements

I am grateful to Dr. Helmut Brunner, and Dr. F. Novak, Seibersdorf Laboratories, International Atomic Energy Agency, for irradiating micro-cultures, and for the use of laboratory facilities during my stay in Vienna. The excellent technical assistance of Mr. P. Moloney throughout the course of these investigations at Kinsealy is gratefully acknowledged. Part of this study was made possible through an award of the exchange fellowship between Royal Irish Academy and Austrian Academy of Sciences.

References

Ahloowalia, B.S. (1982) Plant regeneration from callus culture in potato. Euphytica 31, 755-759.

Annonymous. (1986) In vitro technology for mutation breeding. IAEA-TECDOC -392, International Atomic Energy Agency, Vienna, IAEA-TECDOC-392, pp. 58.

Annonymous. (1987) Improvement of root and tuber crops by induced mutations. International Atomic Energy Agency, Vienna, IAEA-TECDOC-411, pp. 48.

Broertjes, C. (1982) Significance of in vitro adventitious bud technique for mutation breeding of vegetatively propagated crops, in Induced Mutations in Vegetatively Propagated Plants, II. International Atomic Energy Agency, Vienna, pp. 1-10.

Harten, van, A.M., Bouter, H. and Broertjes, C. (1981) In vitro adventitious bud techniques for vegetative propagation and mutation breeding of potato (Solanum tuberosum L.). II. Significance for mutation breeding. Euphytica 30, 1-8.

Roest, S. and Bokelmann, G.S. (1976) Vegetative propagation of Solanum tuberosum L. in vitro. Potato Res. 19, 173-178.

Roest, S. and Bokelmann, G.S. (1980) In vitro adventitious bud techniques for vegetative propagation and mutation breeding of potato (Solanum tuberosum L.). I. Vegetative propagation in vitro through adventitious shoot formation. Potato Res. 23, 167-181.

Sonnino, A., Ancora, G. and Locardi, C. (1986) In vitro mutation breeding of potato. Use of propagation by microcuttings, in Nuclear Techniques and in vitro culture for plant improvement. Proc. Internal. Symposium on Nuclear Techniques and In Vitro Culture for Plant Improvement. International Atomic Energy Agency, Vienna, pp. 385-394.

Sonnino, A., Tavazza, R., Crino, P., Penuela, R., Martino, L. and Ancora, G. (1988) Induction and use of variation in potato: Comparison of different in vitro systems. Internal. Congress Genetic Manipulation in Plant Breeding, EUCARPIA, Helsingor, Denmark, Abst. 2.9.

MICROPROPAGATION STATUS IN *BETA VULGARIS* L. VIA LEAF AXIS TISSUE CULTURE

DETREZ, C., SANGWAN, R.S., SANGWAN-NORREEL, B.S.
UNIVERSITE DE PICARDIE - Faculté des Sciences
Laboratoire Androgenèse et Biotechnologie,
33, rue St Leu, F-80039 AMIENS cedex - FRANCE

SUMMARY

Recent progress and present status in culture of leaf axis tissues are reviewed from a genetic and breeding point of view. So adventitious caulogenesis and somatic embryogenesis process are described. Choice of leaf axis sections, hormonal and growth pattern conditions, cytological events inducing efficiency of the procedures are analysed with the aim of developping efficient method for rapid micropropagation of valuable breeding forms. Improvement of high frequency regeneration technique by successive explant culture cycles is also reported. These aspects and requirements of leaf axis explants are discussed in regard to foliar, especially lamina ontogenesis. Because of karyotypic stability observed among petiole-derived plants, adoption of leaf axis culture as a true-to-type multiplication method, but also as a candidate for developement of *Agrobacterium*-mediated gene transfert technique, is discussed.

INTRODUCTION

Sugarbeet, as a sucrose producer, is an economically important crop species. It belongs to the Chenopodiacae family and is a typical biannual plant. In respect to the cross pollinating character and to a strong inbreeding depression, improvement of this species is based in part on mass selection, and essentially on selection with progeny testing. It includes selfing and sib crossing of parental seed lines, and also, reciprocal recurrent selection procedures for improvement of two populations or lines simultaneously, with respect both to their mutual general or specific combining ability. According to the

R.S. Sangwan and B.S. Sangwan-Norreel (eds.), The Impact of Biotechnology in Agriculture, 47–65.

classical procedure, the final phase, focussed on the triploid level, restores heterozygoty and leads to an hybrid, constructed from a tetraploid pollinisator line and a diploid male sterile line.

Such conventionnal methods in breeding program rise various difficulties concerning outlay of money and time in the way of line material production. Towards such problems, fundamental insights of cell, tissue and organ culture for crop improvement as a tool for sexual work are now, however, well known. It may lead to multiplication of valuable breeding forms, isolation or creation of new characters, according to adventitious regeneration potential and clonal feasibility of the species. Historically, sugarbeet cell cultures have been recalcitrant to a wide range of plant regeneration protocols. However, to date, pre-existant shoot meristems, derived from seedlings or inflorescence pieces can be multiplicated *in vitro* by axillary branching (Margara, 1977; Hussey and Hepher, 1978; Atanassov, 1980, 1986; Coumans-Gilles *et al.*, 1981; Miedema, 1982; Saunders, 1982). This process is used for large scale clonal plant production and routine maintenance of lines. On the other hand, gynogenesis is hepful for the production of dihaploid beets (Hosemans and Bossoutrot, 1983; Bossoutrot and Hosemans, 1985; D'Halluin and Keimer, 1986; Speckmann *et al.*, 1986; Doctrinal *et al.*, 1988), used for homogeneisation of the breeding material.

In regard to recent developement of molecular biotechnology and genetic engeenering, broad array of source tissues and conditions have been tested about production of adventitious plants in sugar beet. Shoot regeneration systems based on callus step formation have been proposed and occasionally, habituated cell lines or hormone autonomous callus with subsequent shoot regeneration have been described (De Greef and Jacobs, 1979; Saunders and Doley, 1986; Saunders and Shin, 1986). Since the evidence for efficiency of such protocols is however conflicting, the most noteworthy items to date concern direct organogenesis. And especially, majors progress have been obtained since two years concerning shoot organogenesis from leaf axis explants (for review, see Table 1). Ability of such tissue has been firstly observed in 1978; our contribution precises conditions of expression leading to a strategy for high degree of shoot induction and large scale plant production, but also, analyses the karyotypic status of the regenerants.

MATERIAL AND METHODS

Large plant source population has been tested. A range of 60 individual genotypes representing a balanced sampling of available germplasm in sugarbeet, including the N, 2N, 3N, 4N ploidy level, have been so selected. This choice especially included cultivars, gynogenetic clones and inbreed lines. Moreover, sources of explants were chosen either among : i) axillary buds sprouted *in vitro* on Murashige and Skoog (MS) medium (1962) containing benzylaminopurine (BAP : 0.2 mg/l), referred to here as BAP-shoots, ii) *in vitro* rooted shoots on MS medium containing indol butyric acid (IBA : 1 mg/l), referred to here as IBA-shoots, iii)field grown plants.

For *in vitro* cultures, MS basal medium was used with 3% sucrose (W/V) and 0.9% Bacto agar. Growth regulators were added before autoclaving and pH was adjusted to 5.9. All cultures were incubated in a culture room at 24°C on 17:7 light:dark cycle.

RESULTS

1. A zonal specificity for regeneration displayed by leaf axis : choice of the explant

Our experience has shown that control of efficient shoot morphogenesis requires at first selection of suitable leaf axis sections since a zonal specificity for regeneration is displayed along petiole and midrib. For instance, *de novo* shoot primordia are highly observed in the blade petiole transition zone referred to here as BPT area (fig.1, 2), but also along petiole and primary veins. Repartition is however found as followed : 85% buds emerge from BPT and only 15% from petiolar area. Lamina fails to differentiate shoots.

It has been observed on the other hand that shoot proliferation from petiole tissue can be increased subsequently to the BPT ablation; newly formed shoots are then observed along leaf axis sides excluding axillary buds (fig.3). In all cases, closely adjacent meristems or isolated ones emerge from the upper adaxial surface of the axis.

So BPT sections, or petiolar sections from which leaf axis base is removed to eliminate any dormant axillary buds, (0.5-1cm in length), were found suitable for regeneration and were classically used for micropropagation (fig.4). Half longitudinal sections of leaf axis

Table 1 : bibliographical data concerning direct adventitious organogenesis from leaf axis tissue, in sugar beet

References	Sources plants	Numbers of genotypes analyzed	Somaclonal variation observed	Inducing factors reviewed
DETREZ et al. (1989)	in vitro BAP-shoots field grown plants	9	foliar variation, karyotypic stability	BAP, TIBA, BPT tissue
RITCHIE et al. (1989)	in vitro BAP-shoots	6 cultivars	no data	BAP, change temperatures trigger
DETREZ et al. (1988)	in vitro BAP-shoots in vitro rooted plants, axenic germinations,	2 cultivars + 8 diploid populations	no data	BAP , TIBA, BPT zone
FREYTAG et al. (1988)	in vitro BAP-shoots	6 lines	albinos plants foliar variations	amino acids and /or vitamins, BAP (0.4mg/l)
SAUNDERS (1982)	in vitro BAP-shoots	50 genotypes	no data	BAP (1.1μM)
SAUNDERS and MAHONEY 1982	greenhouse plants	2 cultivars	multifoliate leaves bumps	BAP spray (33-2000mg/l)
HUSSEY and HEPHER 1978	in vitro BAP-shoots	1 diploid population	no data	BAP (0.12 - 0.25 mg/l) horizontal position and contact with the medium
ROGOZINSKA and GOSKA 1978	in vitro BAP-shoots	anther culture-derived shoots	no variation, normal seed progeny	auxin and cytokinin contents

or thin cell layer of petiolar tissue are also competent for shoot organogenesis. In the latter case, only the subepidermal layers or combined epidermal and subepidermal tissues exhibit however buds in culture (fig.5). Organogeneis was not achieved from epidermis as monolayer or from vascular or cambial tissues. Entire petiole may be so regarded as an heterogenous explant containing a mixture of cell types with graduate competency to shoot induction conditions and with their own hormonal requirements for regeneration (table 2).

Table 2 : Data about in vitro hormonal requirements of leaf axis tissues in sugarbeet, clone G48

Source of explants	type of explants	Hormonal requirements (mg/l)
BAP-Shoots	BPT sections	BAP 1 + TIBA 0.5
	Thin cell layers	BAP 5 + NAA 0.5
In vitro rooted plants	BPT sections	BAP 1 + TIBA 0.5
	Thin cell layers	BAP 5 + NAA 0.5
Field grown plants (a)	BPT sections	Sequence of 3 media : 1° BAP 5 + TIBA 0.1 2° BAP 2.5 + TIBA 0.5 3° BAP 1 + TIBA 0.5
	Thin cell layers	no data

2. Adventitious shoot organogenesis and somatic embryogenesis on leaf axis explants

Plant regeneration from leaf axis has mostly the character of adventitious shoot organogenesis. Evidence for somatic embryogenesis has also been proved on caulogenic sections (fig. 6, 7). Embryogensis occures however at low frequency (less than 1% of responsive explants) and is often hidden by abundant shoot organogenesis. Somatic embryos with well isolated root apex have been observed, but germination and subsequent plant developements are rare because especially root apex necrosis. Nevertheless, best conditions for reproductible somatic embryogenesis in sugarbeet have been largely developped in our team and are reviewed in this book by Dr Tetu.

3. Hormonal requirements

Through our experiments, isolated petiole explants from *in vitro* rooted shoots or greenhouse plants have been proved suitable for regeneration. From various combinations of growth regulators factors, the best recovery of organogeneis from field grown plants is obtained with a sequence of three media including increasing content in triiodobenzoic acid(TIBA), an auxin transport inhibitor (Depta and Rubery, 1984; Rubery, 1987). In this case, subcultures were done every 15 days and macroscopic structures were observed 40 days after the first inoculation.
Higher regeneration frequency is however promoted by the use of plant material cultured on BAP-medium, such as axillary shoot culture. We have found that BAP elicites adventitious shoot formation when its incorporation is realised precociously into the source plant culture media (and 0.25 mg seems to be enough to induce low level of spontaneous caulogenesis on entire petiole). In the same manner, it was established that an hormone free medium is sufficient to elicite shoot formation from isolated BPT explant of BAP-shoots. Nevertheless, BAP greatly promotes a higher rate of regeneration when it is added to the petiole-culture medium, especially when BAP is combined with TIBA, according to an optimal BAP/TIBA ratio such as 1/0.5. Thus, as much as 30 shoots per explant can be produced after 2 weeks culture.

4. Cell divisions in leaf axis explants

Cytological investigations were performed on BPT explants. For light and electron microscopy, leaf axis specimens were fixed (Detrez *et al.*, 1988) then embedded in paraffin or araldite. Sections were stained with 1% aqueous toludine blue for light microscopy. Fixed specimens were also critically point dried and gold coated prior to examination under scanning electron microscop.
At any time before inoculation on shoot inducing medium, BPT explants contain quiescent buds or preformed organ primordia. Few days after culture of BPT explant on BAP/TIBA medium, mitotic activity is however observed along midrib and petiole. This ability is concentrated in subepidermal cell layers (for review, see Detrez *et al.*, 1988) and both anticlinal, periclinal or oblique divisions were observed. Few cell divisions can be also seen in the epidermis; they are then exclusively anticlinal. Callus mass as a desorganized cell proliferation was never detected in such sections. In such a way, dedifferenciation and meristematic organization in petiole explant lead up to the surrection of leaf primordia without breaking between explant epidermis and shoot

epidermis as observed in fig.8. It also induces the production of numerous epidermal hairs arround de novo buds (fig.9). These have been proposed in our teams, as markers of caulogenic structures, compared to embryogenic structures on petiole explants (Tetu *et al.*,1989).

Figure 1 : Adventitious shoot production on BPT sections with or without cut off *de novo* buds as soon as they arise from petiolar tissues of *Beta vulgaris* L. (a)

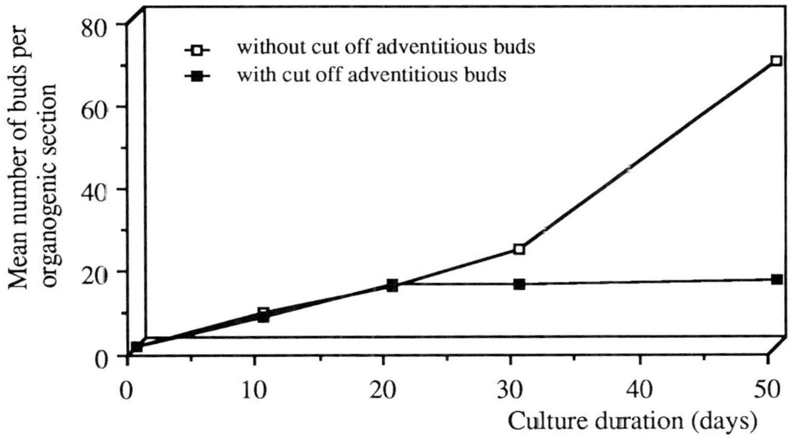

(a) : Mean data scored from 15 responding BPT sections per experiment and among 10 germplasm of sugarbeet.

5. Axillary branching and successive BPT-culture cycles

Regeneration was easily accomplished in the range of beet germplasm tested, even a batch effect on culturability and thus on the frequency of shoot production, resulting from genotype, is observed.

In the case of BAP-plant source, surrection of adventitious buds occured from BPT tissue about 10-12 days after inoculation in Petri dish and shoots can be isolated about 20 days after inoculation. From this date, referred to here as OD20 value, axillary branching arround adventitious buds is observed and increases the mass shoot. As shown in Fig.1, in a first experiment, buds were counted and cut off as soon as they arise from leaf axis explant. In a second protocol, they were counted after different culture durations (10, 20,...50 days), without isolation of *de novo* buds from the explant. It was then established that mass shoot is increased according to a factor 5 (Fig.1) by axillary buds production.

Moreover, percentage of organogenic sections and mean numbers of buds per explant have been analysed after successive BPT culture cycles, each generation of petiole-derived shoots being used for a new bud induction processus. Assessement of these two values indicates an increase with successive culture cycles (Fig.2).

Figure 3 : Frequency distribution of the relative DNA content of nuclei Hoescht 33342-treated and isolated from petiole-derived shoots in *Beta vulgaris* L.

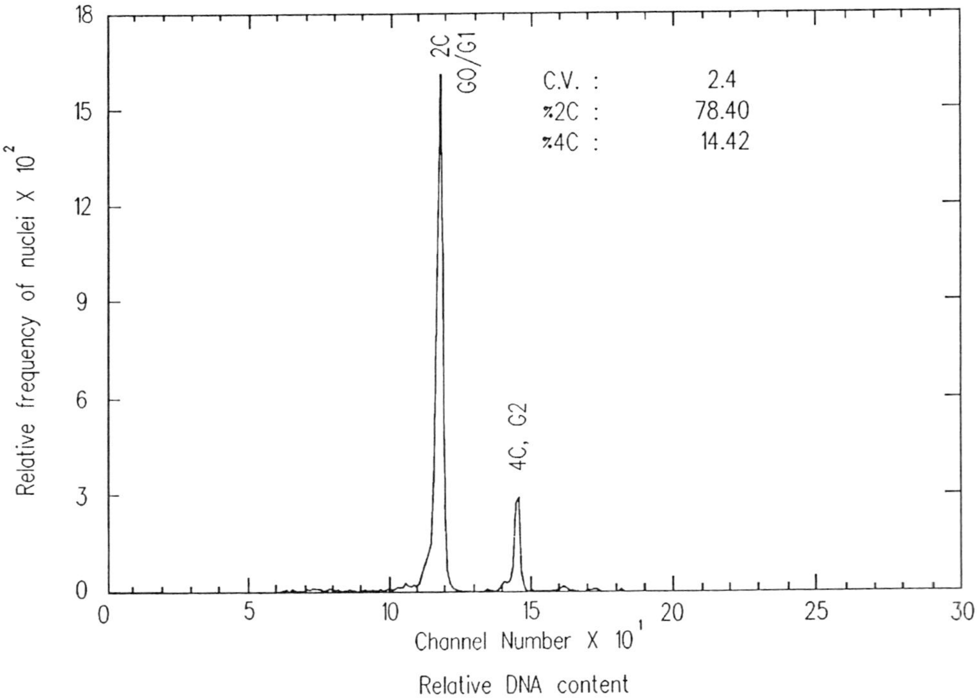

Figure 2 : Bud numbers per explant and percentage of organogenic sections (a) after 5 successive BPT culture cycles (C) in *Beta vulgaris* L. clones RI7 and G48

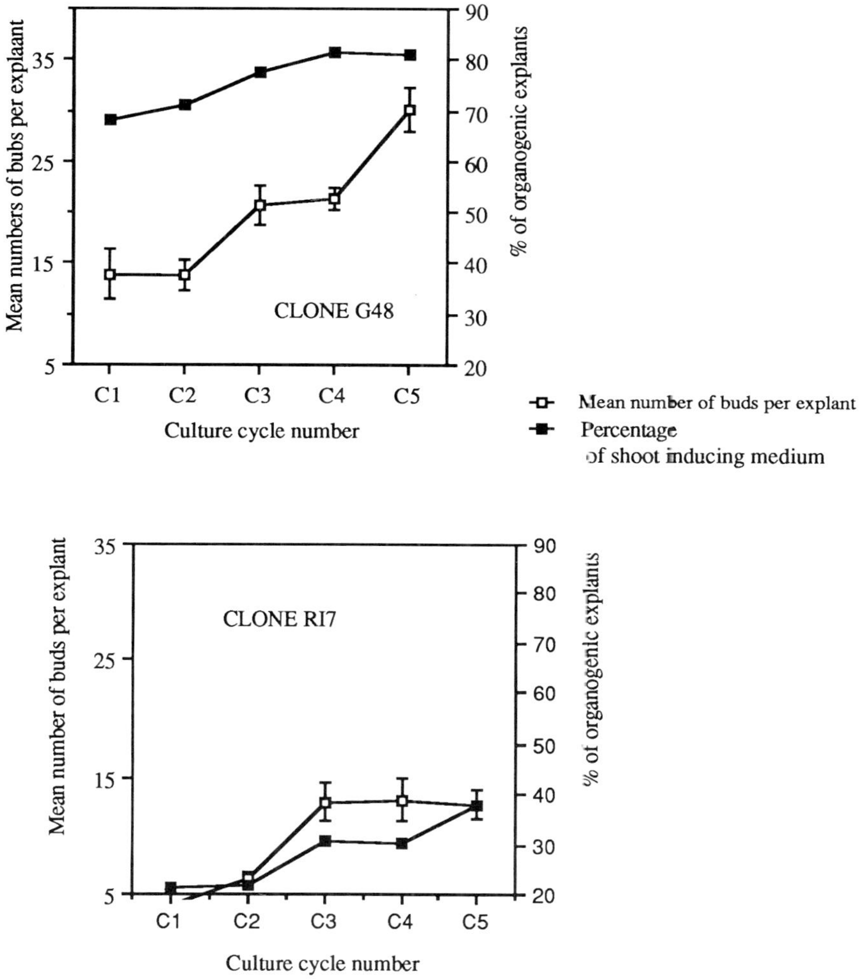

(a) : Mean data obtained from 15 organogenic sections and observed after 20 day culture on shoot inducing medium

6. Somaclonal variation

Among 7000 petiole-derived plants obtained in our teams, only disturbances in rosette growth pattern were observed (Detrez *et al.*, 1989). Variations in the morphogenetic pathways include especially leafy and tubular shoots or disturbances in phyllotaxy inducing leaf primordia three by three (fig. 10, 11). Majority of grossly abnormal shoots (fig.12) also failed to root, or at best established themselves poorly, whereas secondary buds are newly formed on BPT zone without a new subculture cycle (fig.13). However, transfer to an hormone free medium, results in the reversion to the typical morphological pattern. Thus, phenotypic alterations have been regarded as a transient response to an hormonal stress during *in vitro* culture. This hypothesis is moreover supported by karyotypic analysis.

Indeed no callus step occures during shoot formation process; in this way, disorganized cellular proliferations inducing genetic variations are reduced. However, because endoduplicated nuclei have been observed among cells forming petiole explant, and with the objective of determining genome stability, putative somaclonal variation was researched.

For testing large sets of material (3260 shoots), ploidy was judged from chloroplasts counts in the stomata guard cells. Both abnormal and normal looking plant population issued from one or three petiole culture cycles were analysed. No variation in the means of chloroplast numbers was scored.

Among phenotypic variegated shoots, chromosomes counts on young leaves and measurements of DNA content by flow cytometry analysis were also performed. Flow analysis were done on an EPICS V cytofluorometer (Coulter, Florida) with an Argon laser, using 100mV at 357nm. Nuclear isolation was realised from leaves by mechanical chopping and about 5000 nuclei, Hoescht 33342-treated, per sample were studied. As shown in Fig.3, the fluorecence histogram scored from a 2N-petiole-derived regenerant, is then resolved into Go/G1, S and G2 cell cycle compartments. It can be observed that distribution of nuclei over cycle results from a pure population of cells. The GO/G1 peak position reflect moreover a 2N ploidy state that is in good agreement with the diploid

7. Growth pattern in sugar beet and lamina hypothesis

BPT has been identified as a target site for TIBA/BAP action and as controlling efficiently steps and process for high shoot organogenesis. But in supplement to general and hormonal *in vitro* culture conditions of donor plant or explant, the explant size and growth, and also ontogenesis of organ or organelles determine subsequent behaviour in tissue culture. In a large heterogenous tissue such as leaf sections, our observations have shown that foliar primordia formation and plant regeneration is restricted to leaf axis (not to lamina) from the blade petiole transition zone. Do the zonal specificity for regeneration displayed by leaf section and the overall morphogenic pathway of the leave can be connected? Moreover, leaf shape may be profondly modified by environmental factors (i.e. normal leaf development is one of the most important aspect of photomorphogenesis) and in *Beta vulgaris,* it is also govern by ploidy level and tetraploid plants have larger organ and leaves compared to N, 2N, 3N populations). *In vitro* culture and hormonal contents also play an important part in leaf growth. In shoot apex culture, BAP-buds frequently show small leaves with narrow lamina in contrast to rooted AIB-plants or of course, fiefd grown plants which show a fully differentiated structure of the organs. Moreover among BAP-shoots, with increasing BAP levels, the comparative growth rate of lamina and the mains veins are modified compared to classical leaf ontogenesis. So, some leaves may acquire a unusual leaf shape such as multifoliate-like leaves and adventitious buds may be observed in leaflet junction.Teratological leaf structure such petiole-like leaf (fig.14) can frequently be described. Such alterations in subsequent growth of lamina have been classically ascribeb to BAP.

On the other hand, according to phyllotatic sequence, differenciation of leaf begin by petiole and veins initiation. Then few cells of the flanks of the future axis begin to growth so that constitue the marginal and sub marginal cells representing promeristematic initials which give rise to the future lamina (Wareing and Phillips, 1970; Hicks, 1980). What is the repartion of these lamina initials along the leaf axis, and specially in the blade petiole transition zone? Do BAP inhibe their divisions and subsequent differenciation. Should *in vitro* tissue culture of petiole induce re-orientation level of the original explant. Similary, chimerism, mosaicism or polyploidisation were never observed among adventitious shoots. All plants have shown the euploid chromosome numbers, both reporting the putative clonal nature of the regenerants and indicating that genetic alterations among leaf cells are not damaging to genetic organization of *de novo* buds.

of these lamina initials towards adventitious caulogenesis? Here cytological investigations and hormonal contents analysis should be required but few experimental observations are favorous to these hypothesis upside discussed.

CONCLUSIVE REMARKS, PERSPECTIVES AND CHALLENGES

Generation of suitable BPT explants from *in vitro* shoots which encourages shoot production are so well evaluated and a comprehensive culture system is now available. Perspectives and applications brought by our work include now field performance evaluation of petiole culture as a conservative clonal regeneration process.

Tissue culture is also an increasly important aspect of plant breeding since a numbers of techniques are now available which facilate the insertion of novel genetic variability into plants. However tissue culture can also be an important limitating factor since very often cells cultured *in vitro* which are capable of transformation at a high frequency (protoplasts, callus, cell suspension...) are not totipotent. In *Beta vulgaris*, transformation of protoplasts by direct uptake of vector DNA have limited applications since regenration problems from protoplast derived cells are not entirely overcamie (Bhat *et al.*, 1985). A comprehensive culture system resulting in plant regeneration at high level from callus also need improvements and futhers studies, even promising results such somatic embryo formations, have been obtained (Tetu *et al.*, 1987). In regard to look for an efficient regeneration protocole in sugar beet, leaf axis culture reported here may be the most available appoach. And so, if infection of a single undifferentiated cell is the best way for *Agrobacterium* transformation, the possibility of transforming petiole cells that subsequently give rise to adventitious multicellular shoot meristem should not be ruled out because :

i) Reasonable proportion of intact cells at the wound site of explanted tissue undergo dedifferenciation/division since rate of shoot multiplication may reach 30 buds on a single explant;

ii) adventitious shoot formation moreover seems to be a widespread phenomenon, specifically noted from *in vitro* shoot culture material but also in a few cases, with leaves of greenhouse or field grown plants, for genotypes of diverse origins. BAP-shoots have however provided the most responsive explants and have been proved useful for higher shoot production.

Tissue culture and regeneration steps are then no more limitating factors.

iii) totipotent cells in petiole explant are well localisated : no pre-existing quiescent buds or preformed organ primordia was noticed in the explant before inoculation, and *de novo* shoot primordia were observed as emerging from subepidermal cell layers of the concave adaxial side. Moreover, if such cells, competent for shoot regeneration, are deep within the leaf axis surface, thin sections of axis or thin cell layer culture render them accessible to virulent *Agrobacterium* and therefore, available for transformation. Large wound sites, required to provide bacterial access and to stimulate the production of wound associated compounds such as acetoseryngone, are on the other hand, present.

Acknowledgements

This work was supported by GIE Betterave Industrielle-France which is gratefully acknowledged. We also appreciate the gift of M. POIRET for correcting the manuscript and T. TETU for his discussions.

References

ATANASSOV, A.I., 1980 - Method for continuous bud formation in tissue cultures of sugarbeet (*Beta vulgaris* L.). Z. Pflanzenzüchtg., **84** : 23-29.

ATANASSOV, A.I., 1986 - Sugarbeet (*Beta vulgaris* L.). In "Biotechnology in Agriculture and Forestry", Vol.2, Crops I, pp 462-470, Y.P.S. Bajaj Ed., Springer-Verlag Berlin Heidelberg.

BHAT, S.R., FORD LLOYD, B.V., CALLOW, J.A., 1986 - Isolation and culture of mesophyll protoplasts of garden, fodder and sugarbeet using a nurse culture system : callus formation and organogenesis. J. Plant Physiol., **124** : 419-423.

BOSSOUTROT, D., HOSEMANS, D., 1985 - Gynogenesis in *Beta vulgaris* L. : from *in vitro* culture of unpollinated ovules to the production of doubled haploids plants in soil. Plant Cell Reports, **4** : 300-303.

COUMANS, M., COUMANS-GILLES, M.F., MENARD, D., KEVERS, C., CEULEMANS, E., 1982 - Micropropagation of sugarbeet : possible ways. In "Plant tissue culture". Proc. 5th Intl. Cong. Plant Tissue & Cell Culture", pp 689-690.

COUMANS-GILLES, M.F., KEVERS, C., COUMANS, M., CEULEMANS, E., GASPAR, T., 1981 - Vegetative multiplication of sugarbeet through a *vitro* culture of inflorescence pieces. Plant Cell Tissue Organ Culture, **1** : 93-101.

DE GREEF, W., JACOBS, M., 1979 - *In vitro* culture of the sugarbeet : description of a cell line with high regeneration capacity. Plant Science Letters, **17** : 55-61.

DEPTA, H., RUBERY, P.H., 1984 - A Comparative study of carrier participation in the transport of 2,3,5-triiodobenzoic acid, indole-3-acetic acid, and 2,4-dichlorophenoxyacetic acid by *Cucurbita pepo* L. hypocotyl segments. J. Plant Physiol., **115** : 371- 387.

DETREZ, C., TETU, T., SANGWAN, R.S., SANGWAN-NORREEL, B.S., 1988 - Direct organogenesis from petiole and thin cell layer explants in Sugarbeet cultured *in vitro* . J. Exp. Bot., **39** : 917-926.

DETREZ, C., SANGWAN, R.S., SANGWAN-NORREEL, B.S., 1989 - Phenotypic and karyotypic status of *Beta vulgaris* plants regenerated from direct organogenesis in petiole culture. Theor. Appl. Genet., **77** : 462-468.

D'HALLUIN, K., KEIMER, B., 1986 - Production of haploid sugarbeets (*Beta vulgaris* L.) by ovule culture. In "Genetic manipulation in plant breeding", pp 307-309, Walter de Gruyter *et al*, Berlin, New york.

DOCTRINAL, M., SANGWAN, R.S., SANGWAN-NORREEL, BS., 1988 - *In vitro* gynogenesis in *Beta vulgaris L.* : effects of plant growth regulators, temperatures genotypes and seasons. Plant cell, Tissue and organ culture, **17** : 1-12.

FREYTAG, A.H., ANAND, S.C., RAO-ARELLI A.P., OWENS L.D., 1988 - An improved medium for adventitious shoot formation and callus induction in *Beta vulgaris* L. *in vitro* . Plant Cell Report, **7** : 30-34.

HICKS, G.S., 1980 - Patterns of organ development in plant tissue culture and the problem of organ determination. The Botanical Review, **46** : 1-23.

HOSEMANS, D., BOUSSOUTROT, D., 1983 - Induction of haploids plants from *in vitro* culture of unpollinated beet ovules (*Beta vulgaris* L.). Z. Pflanzenzüchtg., **91** : 74-77.

HUSSEY, G., HEPHER, A., 1978 - Clonal propagation of sugarbeet plants and the formation of polyploids by tissue culture. Ann. Bot., **42** : 477-479.

MARGARA, J., 1977 - La multiplication végétative de la Betterave (*Beta vulgaris* L.) en culture *in vitro* . C. R. Acad. Sc. Paris, **285** : 1041-1044.

MIEDEMA, P., 1982 - A tissue culture technique for vegetative propagation and low temperature preservation of *Beta vulgaris*. Euphytica, **31** : 635-643.

MURASHIGE , TY., SKOOG, F., 1962 - A revised medium for rapid growth and bio-assays with tobacco tissue cultures. Physiol. Plant., **15** : 473-495.

RITCHIE, G.A., SHORT, K.C., DAVEY, M.R., 1989 - *In vitro* shoot regeneration from callus, leaf axils and petioles of Sugarbeet (*Beta vulgaris* L.). J. Exp. Bot., **40** : 277-284.

ROGOZINSKA, J.H., GOSKA, M., 1978 - Induction of differentiation and plant formation in isolated sugarbeet leaves. Bull. Acad. Pol Sci., **26** : 343-345.

RUBERY, P.H., 1987 - Auxin transport. In : "PLant hormones and their role in plant growth and development", pp 341-36, Davies P.J. (ed), Martinus Nijhoff Publishers.

SAUNDERS, J.W., 1982 - A flexible *in vitro* shoot culture propagation system for sugarbeet that includes rapid floral induction of ramets. Crop Science, **22** : 1102-1105.

SAUNDERS, J.W., DAUB, M.E., 1984 - Shoot regeneration from hormone-autonomous callus from shoot cultures of several sugarbeet (*Beta vulgaris* L.) genotypes. Plant Science Letters, **34** : 219-223.

SAUNDERS, J.W., DOLEY, W.P., 1986 - One step shoot regeneration from callus of whole plant leaf explants of sugarbeet lines and a somaclonal variant for *in vitro* behaviour. J. Plant Physiol., **124** : 473-479.

SAUNDERS, J.W., MAHONEY, M.D., 1982 - Benzyladenine induces foliar adventitious shoot formation on young plants of two sugarbeet (*Beta vulgaris* L.) cultivars. Euphytica, **31** : 801-804.

SAUNDERS, J. W., SHIN, K., 1986 - Germplasm and physiologic effects on induction of high-frequency hormone autonomous callus and subsequent shoot regeneration in sugarbeet. Crop Science, **26**: 1240-1245.

SPECKMANN, G.J., VAN GEYT, J.P.C., JACOBS, M., 1986 - The induction of haploids of sugarbeet (*Beta vulgaris* L.) using anther and free pollen culture or ovule and ovary culture. In "Genetic Manipulation in Plant Breeding", pp 351-353, Walter de Gruyter *et al.*, Berlin, New york.

TETU, T., SANGWAN, R.S., SANGWAN-NORREEL, B.S., 1987 - Hormonal control of organogenesis and somatic embryogenesis in *Beta vulgaris* callus. J. exp. Bot., **38** : 506-517.

TETU, T., SANGWAN R.S., SANGWAN-NORREEL B.S., 1989 - Epidermal hair cells and protein bodies as specific markers of meristematic and embryogenic areas in sugarbeet tissue cultures. (Submitted).

VAN GEYT, J.P.C., JACOBS, M., 1985 - Suspension culture of sugarbeet (*Beta vulgaris* L.). Induction and habituation of dedifferentiated and self-regenerating cell lines. Plant cell reports, **4** : 66-69.

WAREING, P.F., PHILLIPS, I.D.J., 1970 - Patterns of growth and differenciation. In: "The control of growth and differenciation in plants", Chap. 2, pp 22-47, Pergamon Press, Oxford.

Illustrations

1 - *De novo* shoot primordia (arrows) arising along leaf axis from *in vitro* BAP-shoot, **2** - Adventitious shoot observed in the blade petiole transition (BPT) zone, **3** - Shoot proliferation (arrows) from basal leaf axis side subsequently to the BPT ablation, **4** - Shoot mass proliferation from isolated BPT tissue after 17 days culture on bud inducing medium, **5** - Bud formation from combined epidermal and subepidermal cell layers (TCL) of BPT tissue in sugar beet, **6** - Somatic embryo-like structure isolated from BPT section, **7** -Scanning electron micrography of a somatic embryo at torpido stage observed on a BPT caulogenic section, **8** - Tissular relation between lower or superior explant epidermis (respectively epi or eps) and arising bud-epidermis, **9** - Production of numerous epidermal hairs arround adventitious buds, **10** - Normal rosette-like growth pattern among adventitious buds, **11** - Disturbance in phyllotaxic pathways among adventitious buds : initiation of leaf primordia (p) 3 by 3, **12** - Phenotypic foliar variation among BPT-derived shoots : petiole-like structure, **13** - Spontaneous adventitious shoot formation (arrows) on variegated BPT-regenerant, **14** - Teratological structure of leaf among BAP-shoot, with evident enlargement of leaf axis (arrows).
L: lamina, P : petiole, BPT : blade petiole transition zone

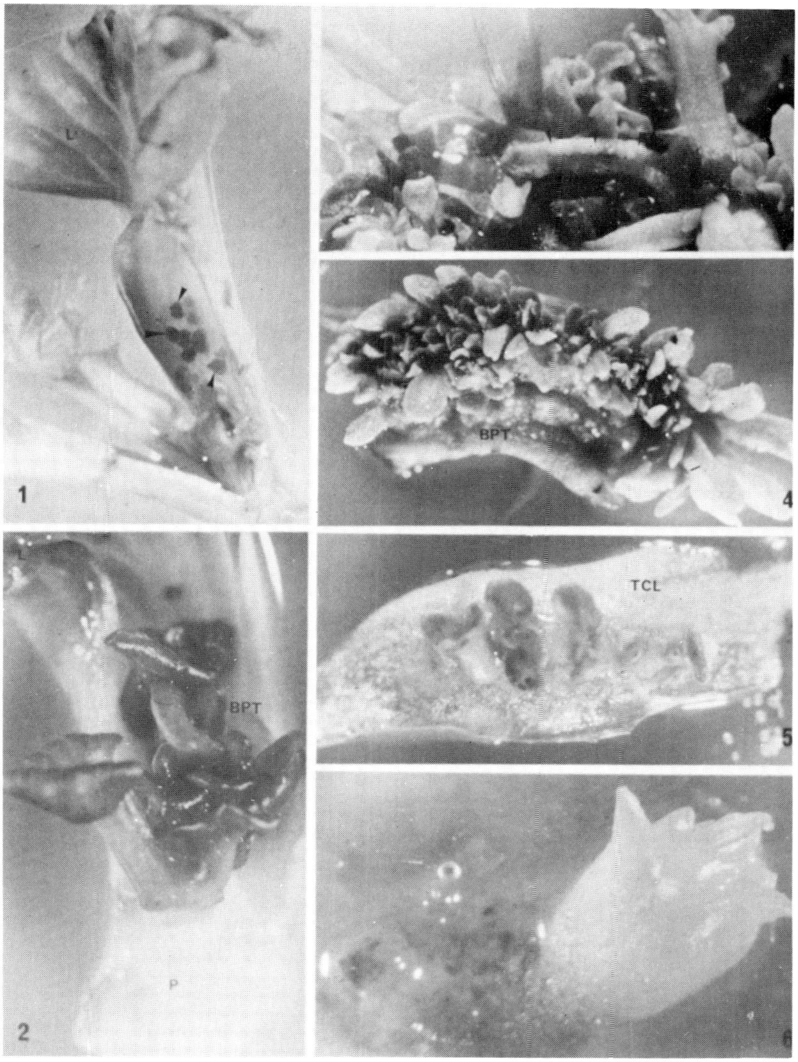

L
1
L
BPT
P
2
BPT
4
TCL
5

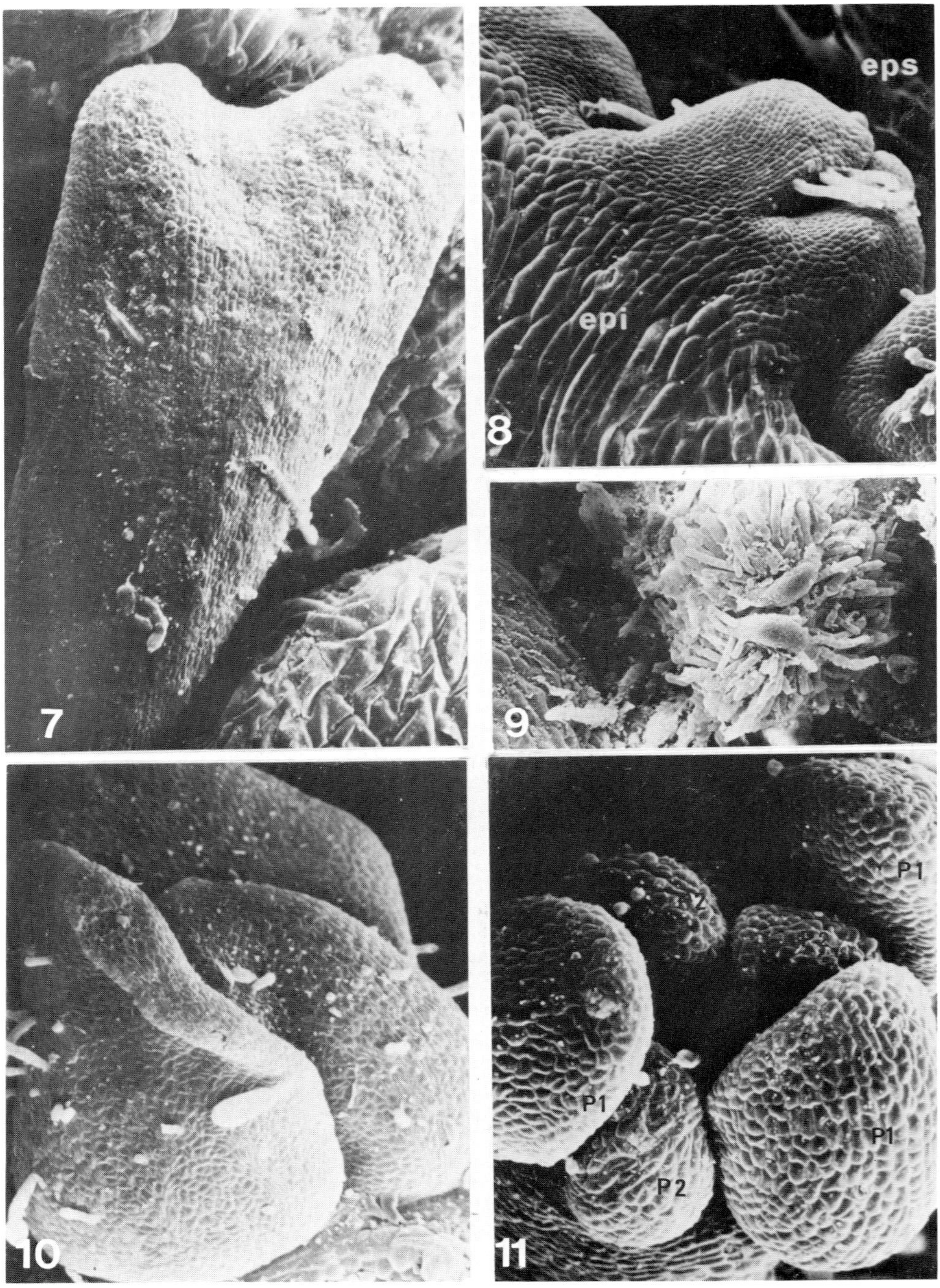
7
eps
epi
8
9
10
P1
P1
P2
P1
11

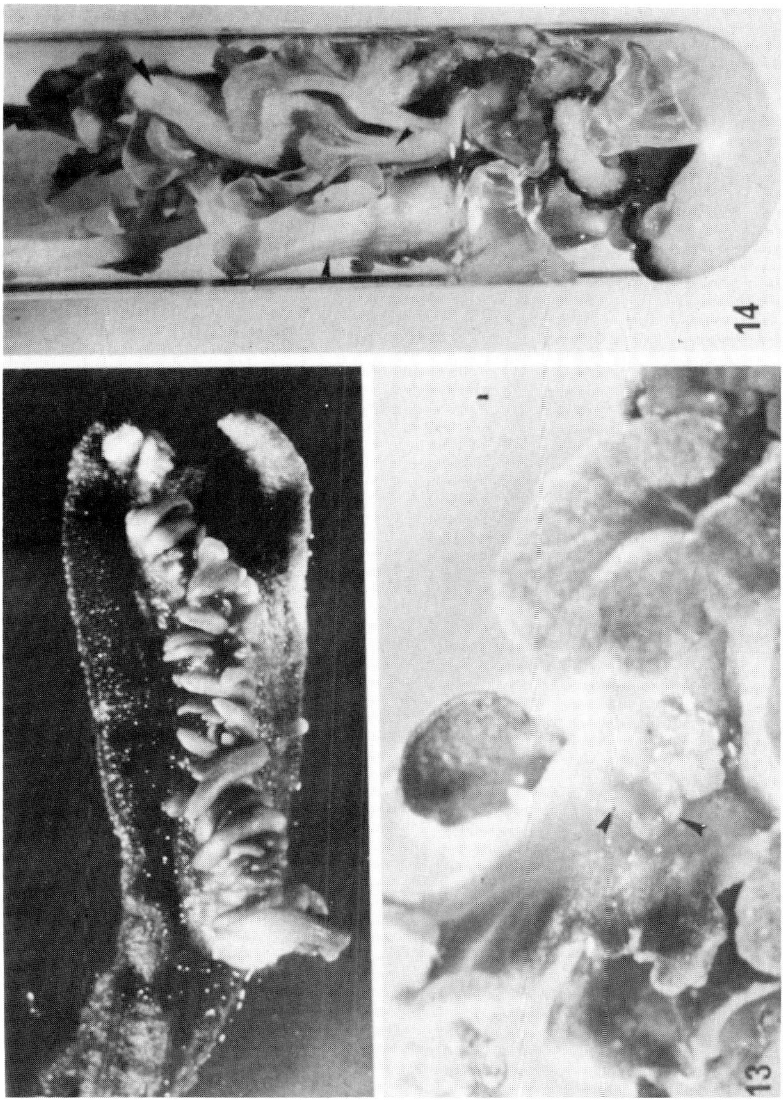
14
13

Application of *in vitro* Multiplication for the Annual Vegetable Crops Associated with Classical Breeding

S.OHKI*, K.NASUDA**, Y.MORI** and H.KATSUTA**
Laboratory of Horticulture* and Experimental Farm**
Fukui Prefectural College
97-21-3, Obatake-cho
Fukui 910
Japan.

ABSTRACT. Application of the tissue culture propagation technique was attempted in annual vegetable crops. In the course of the hybridization, we selected several high quality melon plants which possessed disease resistance and produced good quality fruits. From the second filial generation of hybrid between netted melon and Asiatic cultivar, we could not select any good quality fruit bearing plants. After backcrossing with netted melon, four high quality melon plants fitting to our breeding objectives were obtained from 104 plants examined. These selected plants were propagated by culturing nodal sections. For the multiplication of axiallary buds, low concentration of 6-benzylaminopurine was necessary. In order to avoid vitrification, plastic boxes with holes sealed with gas permeable filters were used. *In vitro* propagated plants were successfully cultured in the greenhouse and showed disease resistance and uniform growth. This *in vitro* nursery system combined with classical breeding systems will be useful for the annual vegetable crops unless planting density is very high and price of products is too cheap.

1. INTRODUCTION

1.1 How we can use *in vitro* propagation for annual crops?

It was in 1939 that White reported the development of leafy buds from his tissue culture of *Nicotiana glauca* × *N.langsdorffi*. After that, numerous scientists have been working on the differentiation *in vitro* and now we can obtain entire plants in so many species of higher plants. This technique has been used in agriculture and horticulture for the clonal multiplication in the past twenty years, but the application is restricted to perennial plants. Perennial plants are normally propagated by vegetative methods and often infected by pathogens such as viruses transmitted during the propagation process. The elimination of these pathogens and increasing multiplication efficiency are the main advantage of *in vitro* propagation.

Annual vegetable crops are propagated by seed, because most of them

R.S. Sangwan and B.S. Sangwan-Norreel (eds.), The Impact of Biotechnology in Agriculture, 67–84.

are cultured on a large scale and uniform growth are required. Many efforts are made by breeders to obtain high quality cultivars. To fix several useful characters for seed propagation, F_1 breeding system is now mainly used. However if we want to fix several recessive characters, it takes many years and often it is impossible. On the other hand, the breeders can find high quality plants expressing plural recessive characters during their hybridization program, especially after the second filial generation according to the Mendel's law of inheritance. We considered that if we can apply the tissue culture technique for these obtained plants to nursery plants production, we need not fix their useful characters.

1.2 Melon

The origin of melon is not clear but European cultivars, which are classified as *Cucumis melo* var. *reticulatus* Naudin or var. *cantalupensis* Naudin, are difficult to culture in rainy and humid zones. On the other hand, Asiatic cultivars belonging to *C. melo* var. *makuwa* Makino or var. *conomon* Makino have resistance to diseases such as gummy stem blight or downy mildew. However, the quality of the fruits of these Asiatic cultivars is poor. Many Japanese breeders are trying to introduce the disease resistance of Asiatic cultivars to European cultivars but the resistance to disease, although some of them are dominant, is difficult to fix complexly. Tamai *et al.*(1962) reported that it took 9 generations to create one cultivar with powdery mildew resistance which is considered as incomplete dominant. Moreover, as shown in Table 1, many desirable characters of fruit are recessive.

We considered that it is possible to obtain several plants expressing disease resistance and bearing fruit with high marketable values, so we tried to create the *in vitro* nursery system to propagate these plants. Recently in Japan, many of melon growers are using F_1 cultivars and these seeds are quite expensive. Thus the cost for tissue culture propagation is not considerably high compared with that for seed propagation.

TABLE 1. Inheritance in the muskmelon. (Summarized from Hayes and Garber, 1921, Sagared, 1926, Shoda and Sadaoka, 1936,Hagiwara and Uemura, 1937)

Characters	Dominant or incomplete dominant	Recessive
Color of skin	Yellow	Green
Form of fruit	Round	Elliptical
Ribbing	Deep	None
Netting[1]	Netted	Smooth
Size of seed	Large	Small
Size of fruit	Large	Small
Taste	Sour	Sweet

[1] Netting was recessive in the cross var.*reticulatus* × *makuwa* (Shoda and Sadaoka, 1936).

2. Materials and Methods

2.1 Breeding and Selection Methods

The breeding program is shown in Figure 1. In our previous report, we reported that the selection of good quality plants for tissue culture propagation should be carried out in F_2 or F_3 generations (Ohki *et al.*, 1988). So the selection was realized in these generations. The cultivars and lines used in this report were 'AF', 'No.80' and 'Ochiuri' (Figure 2). 'AF' and 'No.80' are *reticulatus* type fixed progeny lines of F_1 cultivar 'Amusu' and ('Earl's Favourite'×'Supaisii')×'Kosakku' respectively. 'Ochiuri' is a *makuwa* type Japanese old cultivar.

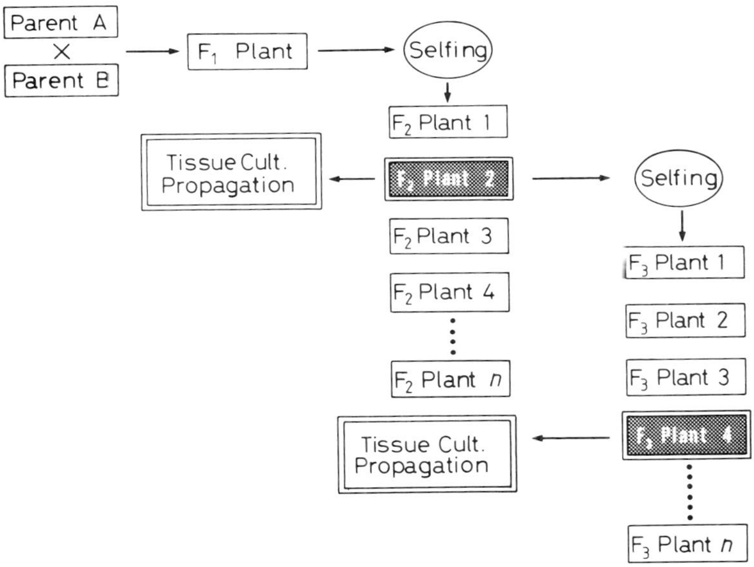

Figure 1. Schema of breeding and selection system for *in vitro* nursery system.

All plants were cultured in the College Farm's plastic greenhouse highly infected by *Fusarium oxysporum*. The selection for the F_2 of 'AF'×'Ochiuri' and backcrossing ('AF'×'Ochiuri')×'No.80' were carried out in the spring croppings of 1987 and 1988 respectively.

2.2 Tissue Culture

2.2.1 Plant Materials

The following cultivars of melon were used:

reticulatus types

'Melody No.2', 'Honey Dew', 'Melody No.2'×'Earl's Favourite'
'15-6'(selected line)
hybrids with *makuwa* type
'AF'×'Ochiuri', 'Honey Dew'×'Ginsen'
cantalupensis type
'Cantor'

In our preliminary experiments, among the organs tested, nodal section was suitable for *in vitro* multiplication of melon. The upper 3 nodes of the lateral branches were sterilized for 10 minutes in 1% NaClO containing 0.1% of polyoxyethylene sorbitan monolaurate, and rinsed three times in sterile tap water. As shown in Figure 3, nodal sections were cut into approximately 5 mm length under aseptic conditions and placed horizontally on the culture medium. Nodal parts of shoots obtained in the initial culture were cut into 3 mm length and were subcultured.

Figure 2. Cultivar and lines of melon used. (A) line 'AF' (B) line 'No.80' (C) cultivar 'Ochiuri'.

Figure 3. Preparation of explants. (A) Lateral branch of melon. Nodes are arrowed. (B) Nodal sections used as explants.

2.2.2 Culture Media and Vessels

The basal culture medium consisted of Murashige and Skoog's macro-elements and Fe-EDTA (1962), micro- and organic-elements of Ringe and Nitsch (1968), sucrose 30 g/ℓ and jelling agent (agar: 6.5g/ℓ or Gellun Gum: 3g/ℓ). As growth regulators, indole-3-acetic acid (IAA), indole-3-butylic acid (IBA), α-naphthaleneacetic acid (NAA), 6-benzylaminopurine (BAP), kinetin (KN) and 6-(γ,γ-dimethylallylamino)-purine (2iP) were used. The pH of the medium was adjusted to 5.8 with 0.1N NaOH prior to the addition of the jelling agent. The following culturing vessels were used:

- Test tube of 25 mm diameter and 150 mm length with metal cap
  20 mℓ of medium per tube unless indicated
- Plastic box of 70×70×95(H) mm (Plant Box : Figure 4A)
  50 mℓ of medium per box
  Aeration : 2 holes of 10 mm diameter were sealed with gas permeable membrane filter (Milliseal©)
- Cylindrical plastic pot of 100 mm diameter and 105 mm height (Bio Pot : Figure 4B)
  100 mℓ of medium per pot
  Aeration : One or two holes of 15 mm diameter were sealed with gas permeable filter paper.
- Plastic box of 110×100×100(H) mm (Culture Bottle : Figure 4C)
  100 mℓ of medium per box
  Aeration : From one end of the box filter sterilized air is flowed 2 min/hr at a rate of 3ℓ/min.

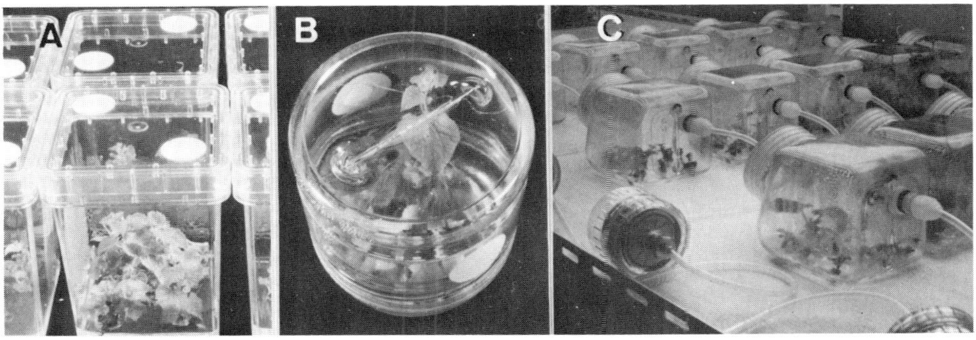

Figure 4. Culture vessels used. (A) Plant Box (B) Bio Pot (C) Culture Bottle.

The media were autoclaved for 15 minutes at 120 °C. The cultures were placed in the culture room (12 hr day-length, 75 or 150 μmol $m^{-2}s^{-1}$ using white fluorescent lamps, 23±2°C) or in the growth chamber (12 hr day-length, 60 μmol $m^{-2}s^{-1}$ using white fluorescent lamps, 23±0.5°C, 65±5% RH).

2.3 Culture of *in vitro* Propagated Plants

2.3.1 Grafting of *in vitro* Propagated Shoots

Proliferating shoots were subcultured on the medium without or with 0.5 μM of NAA. After 20 to 25 days in culture, well developed shoots were directly grafted on the 10-day-old seedlings of rootstock cultivars of melon. These grafted plants were cultured in the humidified growth chamber for about 10 days (Figure 5).

Figure 5. Acclimatization stage of *in vitro* propagated shoots grafted plants in the humidified growth chamber.

2.3.2 Experimental Culture of *in vitro* Propagated Melon

Non-grafted and grafted nursery plants were cultured in a plastic greenhouse or in a hydroponic system in the greenhouse. In the hydroponic system, 15 plants were cultured using 800ℓ of culturing medium. Elongation of the shoots was measured every week. Fruits were yielded 50 to 55 days after the pollination. Brix was measured by reflactometer at the innermost part of the mesocarp tissue.

3. Results

3.1 Breeding and Selection

The results on F_2 generation of 'AF'×'Ochiuri' are shown in Table 2. Vigorousness, big fruits size heavier than 1.3 kg and green mesocarp fruits appeared at a high percentage. However they showed a strong tendency to express *makuwa* phenotypes, e.g. elliptical less netted and less aromatic fruits. The number of plants having resistance to downy mildew attributable to *Pseudoperonospora cubensis* was also small. Finally we could not select any plants showing sufficient desirable characteristics.

TABLE 2. Appearace of desirable characters in F2 generation of crossing 'AF'×'Ochiuri'. 40 plants were investigated.

Selection objectives	% of fitting plants	Min - Max
Vigorousness	85.0	-
Resist. to downy mildew	15.0	-
Resist. to powedery mildew	27.5	-
Round fruits	5.0	-
Big fruits[1]	70.0	1.10 - 1.43 kg
Well netted	7.5	-
Green mesocarp	77.5	-
Thick mesocarp[2]	17.5	2.2 - 3.6 cm
Aroma	7.5	-
High Brix[3]	32.5	9.8 - 15.3 %

[1] hevier than 1.3 kg
[2] thicker than 3.5 cm
[3] higher than 14.0 %

So we backcrossed *reticulatus* type fixed line 'No.80'. Table 3 shows the result of the selection of this backcrossing from 104 plants. The resistance to downy mildew and powdery mildew was not found abundantly, but we could obtain more *reticulatus* type fruit bearing plants than in the former crossing (Figure 6). Four plants fitting to all breeding objectives were selected, therefore selection efficiency was 3.8 %.

TABLE 3. Appearace of desirable characters in the backcrossing ('AF'×'Ochiuri')×'No.80'. 104 plants were investigated.

Selection Objectives	% of fitting plants	Min - Max
Vigorousness	65.4	-
Resist. to downy mildew	12.5	-
Resist. to powedery mildew	18.3	-
Round fruits	34.6	-
Big fruits[1]	46.2	0.96 - 1.68 kg
Well netted	50.0	-
Green mesocarp	29.8	-
Thick mesocarp[2]	31.9	2.1 - 4.3 cm
Aroma	10.6	-
High Brix[3]	40.4	11.1 - 16.4 %

[1,2,3] see Table 2

Figure 6. Fruit of selected line 'K-16' from backcrossing ('AF'×'Ochiuri')×'No.80'.

3.2 Tissue Culture Propagation

3.2.1 Effect of Cytokinins on Shoot Formation from Nodal Sections

Table 4 shows the effect of three different cytokinins on shoot and callus formation from nodal sections of melon cv. 'Melody No.2'. With KN and 2iP, no shoot was developed. 2iP promoted callus development. On the medium with BAP at 5.0 μM, 63.3% of explants formed shoots.

TABLE 4. Effect of three cytokinins on percentage of explants formed shoots (SP), number of shoots per explant (SN) and callus dry weight (CW) after 30 days in culture.

		Cytokinin Concentration (μM)			
		0	0.5	1.0	5.0
BAP	SP(%)	18.2	30.0	25.0	63.3
	SN	0.2±0.12	0.4±0.22	0.3±0.19	1.0±0.30
	CW(mg)	26.5±1.8	38.0±14.0	57.0±11.2	119.6±13.0
KN	SP(%)	0.0	0.0	0.0	0.0
	CW(mg)	37.4±6.4	30.3± 1.7	43.3± 8.8	74.7±13.6
2iP	SP(%)	0.0	0.0	0.0	0.0
	CW(mg)	40.3±4.2	64.5±10.0	96.2±15.9	174.2±10.5

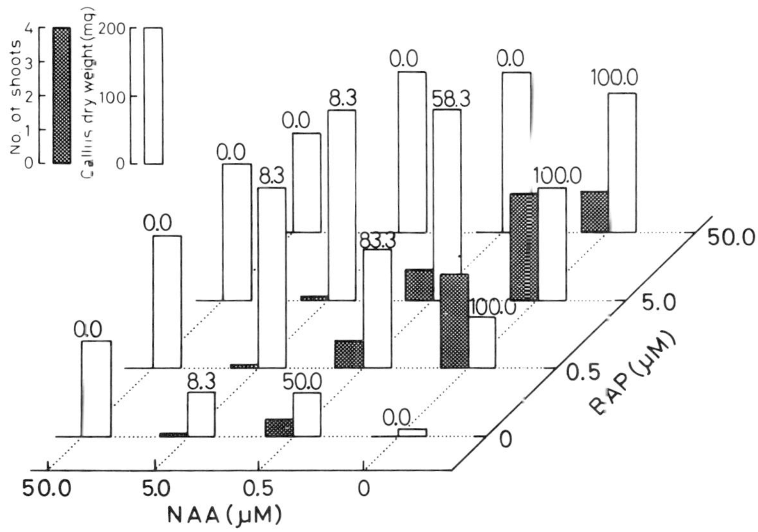

Figure 7. Effects of BAP and NAA on shoot regeneration and callus formation from nodal sections of melon hybrid 'AF'×'Ochiuri' 35 days in culture. Figures above each bar indicate percentage of explants with regenerated shoots.

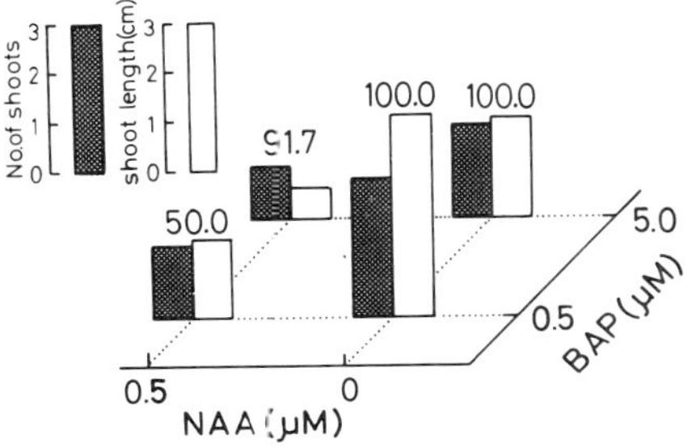

Figure 8. Effects of BAP and NAA on shoot regeneration from nodal sections of melon hybrid 'Melody No.2'×'Earl's Favourite' 35 days in culture. Figures above each bar indicate percentage of explants with regenerated shoots.

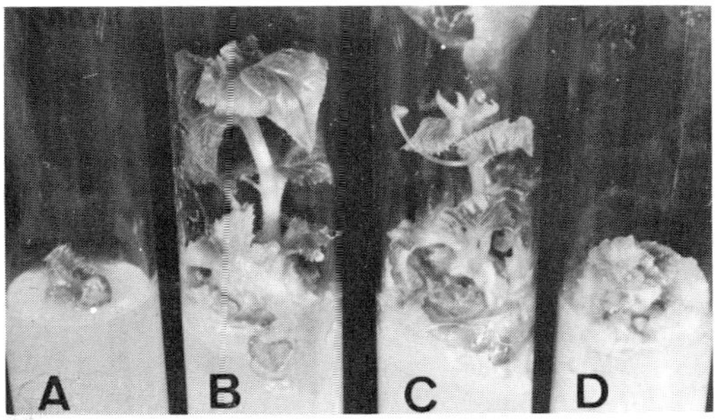

Figure 9. Shoot regeneration from nodal sections of melon cultured on the media supplemented with BAP (left to right : 0, 0.5, 5.0, 50.0 μM) for 35 days.

3.2.2 Effect of Cytokinins and Auxins on Shoot Formation

A screening test was carried out with 18 combinations of BAP and NAA concentrations using hybrids 'AF'×'Ochiuri' (Figure 7) and 'Melody No.2'×'Earl's Favourite' (Figure 8). For these hybrids, BAP was essential for shoot formation (Figure 9). With only NAA, very few shoots were formed. BAP concentration higher than 5 µM caused malformation of shoots. Abundant callus was produced when both BAP and NAA were added, that inhibited shoot differentiation. 2 to 3 shoots per explant were obtained on the medium supplemented with 0.5 to 1.0 µM of BAP alone.

3.2.3 Difference in Shoot Formation Among Cultivars

Nodal sections of the cultivars 'Ginsen', 'Honey Dew','Cantor' and hybrid 'Honey Dew'×'Ginsen' were cultured on the medium with BAP 0.5 µM. The results are shown in Table 5 and Figure 10. The percentage of explants that formed shoots was high in 'Ginsen' and 'Honey Dew', and lowest in 'Cantor'. However the number of shoots per explant and development of the shoot formed were superior in 'Honey Dew'×'Ginsen' and 'Cantor'. 'Honey Dew' was inferior to its hybrid.

TABLE 5. Percentage of shoot formed explants (SP), mean shoots number per explant (SN), and mean maximum shoot length (SL) in 3 cultivars and their hybrid of melon 30 days in culture on the medium with 0.5 µM of BAP.

Cultivars	SP(%)	SN	SL(cm)
Ginsen (GIN)	100.0	1.7±0.2[1]ab[2]	14.8±2.4 ab
Honey Dew (HD)	85.7	1.3±0.1 a	8.4±2.1 a
HD × GIN	70.3	2.1±0.3 b	22.4±3.9 b
Cantor	64.0	1.9±0.2 b	19.7±3.9 b

[1] Each value represents the mean ± s.e. of 12 replications.
[2] Mean separation within columns by Duncan's multiple range test, 5% level.

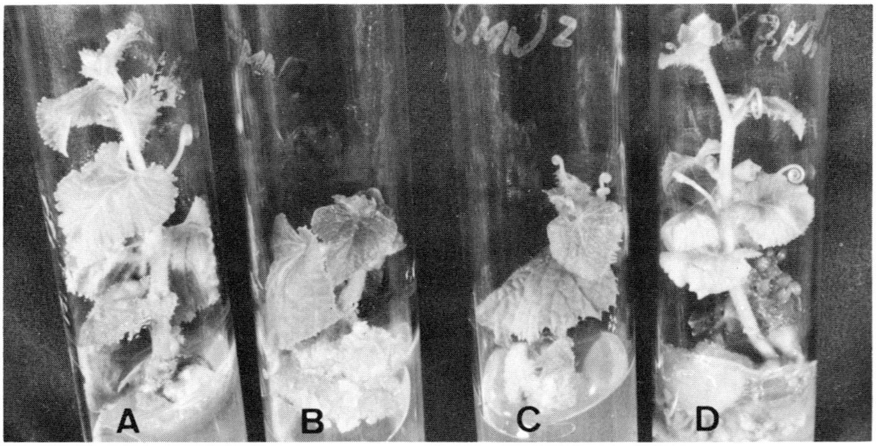

Figure 10. Cultures of 3 cultivars and their hybrid of melon. 30 days in culture on the medium with 0.5 µM BAP. (A) 'Ginsen', (B) 'Honey Dew', (C) 'Honey Dew'×'Ginsen', (D) 'Cantor'

3.2.4 Subculture of the differentiated shoot

We subcultured the nodal sections of differentiated shoots in the initial culture. Like in the initial culture, BAP at a concentration of 0.5 µM was suitable for shoot proliferation (2.8±0.4 shoots/explant). In the subculture, vitrification was apt to occur. In order to know the effect of size of the culture vessels and of aeration of the vessels, we compared four different vessels.

The nodal sections of easy to propagate selected line '15-6', which were excised from the shoots regenerated in the initial culture, were cultured on the medium with 1.0 µM of BAP. We subcultured one explant for test tube, five for Plant Box, ten for Bio Pot and Culture Bottle. Quantity of the medium is adjusted to 10 mℓ per explant. Results are shown in Table 6. Percentage of vitrificated shoots was highest in the test tubes and lowest in Bio Pot. The number of shoots per explant were similar in all vessels tested. Callus formation was promoted in the test tubes. The dry weight ratio of shoots was slightly higher in Culture Bottles and Bio Pots.

TABLE 6. Effect of different culture vessels on vitrification, shoot proliferation and callus formation in melon line '15-6' after 25 days in subculture.

Culture vessels	Vitrification (%)	Shoot number per explant	Shoot fresh weight(mg)	Callus fresh weight(mg)	Shoot dry weight ratio(%)
Test tube	80.0	5.7 a[1]	631.3 a	1635.3 a	4.4 a
Plant Box	66.7	5.4 a	382.0 b	1284.0 b	5.0 a
Bio Pot	53.0	5.9 a	795.3 a	1100.0 b	4.9 bc
Culture Bottle	33.3	5.1 a	366.0 b	1310.7 b	4.4 ac

[1] Mean separation within columns by Duncan's multiple range test, 5% level.

Rooting from the shoots obtained was achieved on the medium with NAA or IBA at a concentration of 0.1 to 0.5 µM. NAA at the same concentration also promoted root formation but higher concentration of NAA resulted in malformation of roots. There was considerable difference in root formation among cultivars and in general rooting was not abundant. To improve root formation and development, we tested the solid half strength basal media containing 1.0 µM of IBA solidified with agar (0.65%) or Gellun gum (0.2%), and the same liquid media with support means (rock wool or vermiculite). As shown in Table 7, Gellun gum was the best in the incidence of root formation and their development. Gellun gum is less sticky and it was rather easy to pull out the root system. On vermiculite, although the number of rooted shoots was small, root hair development was promoted. This root hair development makes it easy for us to acclimatize.

TABLE 7. Effect of media type on root formation from subcultured shoots of melon (F_3 of 'Melody No.2' × 'Fukamidori') 40 days in subculture. Half strength basal media supplemented with 5 μM of IBA were used.

Media type	Jelling agents or supporting means	Percentage of explants rooted	Number of roots	Maximum shoot length
Solid	Agar	75.0 (%)	2.2 a[1]	2.8 a (cm)
	Gellun gum	91.7	2.9 a	3.4 a
Liquid	Rock Wool	58.3	1.5 ab	2.3 a
	Vermiculite	41.7	0.6 b	1.3 a

[1] Mean separation within columns by Duncan's multiple range test, 5% level.

3.3 Culture of *in vitro* Propagated Plants

3.3.1 Homogeneity of the Growth

To know the variation of growth rate among the *in vitro* propagated plants, 14 *in vitro* propagated grafted plants of line '15-6' were cultured in the hydroponics. As a control 16 plants of F_1 cultivar 'Arôme' were also cultured in the hydroponics. The results on stem elongation measurement and its coefficient of variation are shown in Figure 11. The stem length of F_1 plants was small just after the planting but it reached the same height as tissue cultured plants within one month. The coefficient of variation of *in vitro* plants was lower than that of F_1 plants. In tissue cultured plants, the number of days required to set first fruits after planting was more uniform and Brix of fruits was higher (Table 8).

TABLE 8. The fruits quality and its coefficient of variation of *in vitro* propagated line '15-6' (grafted) and F_1 cultivar 'Arôme' of melon cultured in hydroponics. 14 plants for '15-6' and 16 plants for 'Arôme' were used.

Line and cultivar	Days to pollination[1]		Fruit weight(g)		Brix (%)	
	Mean	C.V.[2] (%)	Mean	C.V.(%)	Mean	C.V.(%)
15-6	30.5	4.2	1207.8	15.0	15.2	3.0
Arôme	34.9	17.2	1744.4	18.3	13.9	5.9

[1] Days from planting to pollination
[2] Coefficient of variation

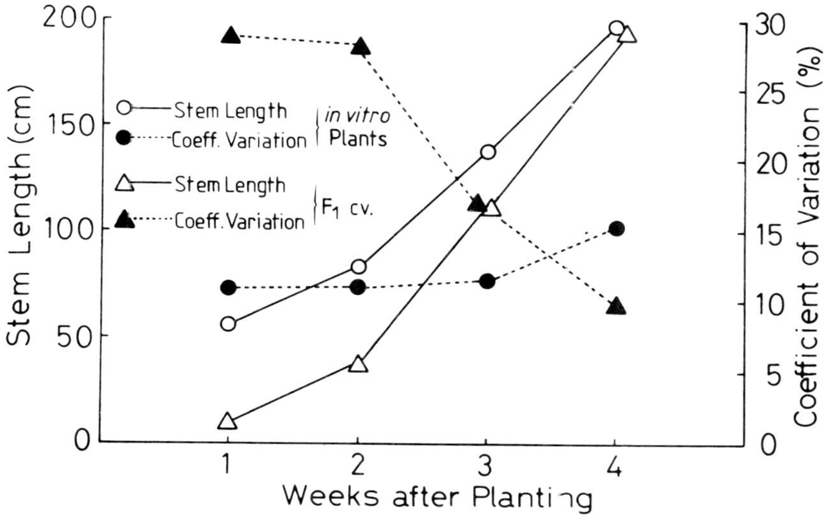

Figure 11. Stem elongation and its coefficient of variation of *in vitro* propagated line '15-6' and F_1 cultivar 'Arôme' of melon cultured in hydroponics. 14 plants for '15-6' and 16 plants for 'Arôme' were used.

3.3.2 Experimental Culture of *in vitro* Propagated Plants

Selected and tissue culture propagated line '965-8' was cultured with F_1 cultivar in the same plastic greenhouse. All plants of F_1 cultivar were diseased by gummy stem blight and completely withered before ripening fruits (Figure 12).

Figure 12. *In vitro* propagated line '965-8' (Left) and F_1 cultivar (Right) cultured in the plastic greenhouse.

The tissue culture propagated four selected line from the backcrossing ('AF' ×'Ochiuri') ×'No.80 were cultured in the plastic greenhouse in the spring cropping of 1989. Eight plants of each line were used. In Table 9, comparative data with the original selected mother plants are shown. All plants grew vigorously. The fruits weight was slightly lighter than that of mother plants, but the thickness of mesocarp tissue and the fruits appearance were almost identical with the mother plants. Brix was higher than the mother plants especially in line 'K-1'.

TABLE 9. Comparison of fruits quality of four *in vitro* propagated lines with original mother plants. Selection was realized from the backcrossing ('AF' ×'Ochiuri') ×'No.80.

Lines	Fruits weight (g)		Mesocarp thickness (cm)		Brix (%)	
	Original[1]	T.C.P[2]	Original	T.C.P	Original	T.C.P
K- 1	1780	1420±22.9	3.5	3.6±0.0	12.7	15.4±0.2
K- 11	1530	1380±22.1	4.2	3.9±0.1	16.8	16.2±0.2
K- 16	1430	1360±22.6	3.4	3.9±0.1	13.9	14.6±0.2

[1] value of one original mother plant
[2] average value of 8 tissue culture propagated plants

4. Discussion

4.1 Breeding and Selection

The main purpose of the hybridization with *makuwa* type melon is to introduce disease resistance in order to culture melon plants in the open field even in the humid zone. In our previous experiment (Ohki *et al.*, 1988), we tried to select excellent plants from the crossing among several *reticulatus* type muskmelon cultivars. In the case of hybrid between cv. 'Melody No.2' and 'Earl's Favorite', the percentage of plants expressing resistance to gummy stem blight, setting fruits of heavy and high soluble solid content declined as the generation advanced. Therefore we concluded that the selection of good quality plants for tissue culture propagation should be carried out in F_2 and F_3 generations.

We expected the same tendency in the crossing reported here, but expression of characters of *makuwa* was too strong and we could not select any excellent plant in F_2 generation. In particular the form of the fruit was elliptical, even though it is reported that round shaped fruit is dominant as shown in Table 1. The mode of inheritance may be different in the crossing with *makuwa* type melon. After the backcrossing with netted muskmelon we could successfully select *reticulatus* type melon. However, a slight characteristic odor of *makuwa* melon consistently appear in all selected fruits. Compared with F_1 cultivars with transduction of *makuwa* type, our selected plants were more resistant to downy mildew and with much less unfavorable characters of *makuwa*.

Therefore, our selection system could be the only solution to create muskmelon type cultivars after hybridization with *makuwa* type melon.

4.2 Tissue Culture Propagation

Several authors have reported on the shoot regeneration in melon tissue culture. Moreno *et al.*(1985) obtained shoots from 90% of calli derived from cotyledonary explants on the medium containing 1.5 mg/ℓ IAA and 6.0 mg/ℓ KN. They also observed embryogenesis from hypocotyl derived callus on the medium with 3 mg/ℓ IAA. Trulson and Shahin (1986) observed embryogenesis from cotyledonary tissue using the medium with 2,4-D (1.0 mg/ℓ), NAA (1.0 mg/ℓ) and BAP (0.5 mg/ℓ). Kathal *et al.*(1986) reported shoot regeneration from hypocotyl callus. Furthermore Niedz *et al.*(1989) analyzed several factors influencing shoot regeneration. However all these reports presented the results of cotyledon and hypocotyl culture. We are not interested in such juvenile organs, because our culture must be started after the fruits quality is estimated. In the preliminary experiment not reported here, we cultured young leaves, but only callus was formed. This callus did not show any shoot regeneration. Therefore we applied node culture (single-node method) for the multiplication.

Among the cytokinins and auxins tested, BAP alone gave the best result for the shoot formation from nodal tissue. Niedz *et al.*(1989) compared BAP, kinetin, 2iP and zeatin in their cotyledon culture, and concluded that BAP at a concentration of 5 μM is the best for shoot regeneration from cotyledonary explants. However, they used 5 μM IAA combined with BAP. Moreno *et al.*(1985) also used IAA but combined with KN for hypocotyl culture. In our node culture, when auxin was added abundant callus was formed and shoot formation was suppressed. This different tendency of shoot differentiation in these organs may be due to the age of them.

The difference in capacity of shoot formation among cultivars and lines used was clear . Bouabdallah and Branchard (1986) reported this difference between var. *cantalupensis* cv. 'Cantaloup Charantais T' and var. *makuwa* cv. 'Ogon No.9' in the cotyledonary culture. In this case, the former showed a higher capacity than the latter. Niedz *et al.*(1989) also noted a different response to cytokinins in shoot regeneration from cotyledonary tissue among four cultivars. From a long series of the breeding of melon, which has lasted more than five years, we selected not a smallnumber of excellent plants. But not all selected plants were propagated easily by tissue culture. Line '15-6' selected from the progeny of the crossing between *reticulatus* type cultivars was only one of which propagation was quite easy. Concerning the heredity of shoot forming capacity in tomato, the maternal effect and heterosis is reported (Ohki *et al.*,1978). According to Frankenberger et al. (1981), recessive genes were associated with high shoot-forming capacity from leaf disc explants in high shoot-forming genotypes of tomato. So when we raise new lines for tissue culture propagation, these findings should be considered.

During the *in vitro* propagation process, vitrification troubled us. Several factors influencing this physiological disease are reported (Pierik, 1987). In our case, the high cytokinin level in the medium, excessive water uptake and high humidity in the culturing vessel might be the main factors. The maximum BAP concentration should be 1.0 μM. Debergh *et al.* (1983) concluded that they only succeeded in overcoming vitrification in glove artichoke mass propagation by raising the agar concentration to 1.1% in stead of 0.6%. We also increased the concentration of Gellun Gum to 0.3% and it resulted in slight improvement. Therefore lowering the humidity in the culture vessel was the best way to decrease vitrification. Forced aeration by filter sterilized air flow gave the best result, but the installation was complicated. So for commercial propagation, a vessel sealed with the gas permeable filter will be recommended. Kozai *et al.* (1988) emphasized the importance of photoautotrophic growth before acclimatization. They also reported that in the closed culture vessel CO_2 concentration was 350 vpm lower than exterior air 1 or 2 hours after the beginning of light period (Kozai *et al.*, 1987). Slight natural gas exchange via filter and high light intensity will be efficacious for sound shoot proliferation without vitrification.

The rate of *in vitro* multiplication using single-node culture is not very high. Breaking apical dominance with cytokinin is the best way to increase this rate (Pierik, 1987), but for melon the vitrification limited increasing cytokinin concentration. For the easy to propagate line '15-6', applying gas exchangeable vessel, at least 5 shoots per inoculum were obtained 25 days after subculturing. Therefore it will be possible to get 5^{14} shoots by continuous subculture. This number will be enough for commercial nursery plants production.

4.3 Culture of *in vitro* Propagated Plants

Rooting from *in vitro* propagated shoots was not very easy and development of the root system after planting in the greenhouse was limited. As a result, when these plants were cultured in the field with unfavorable soil conditions, fruits were small and less sweet although they preserved disease resistance and appearance of fruits was excellent (Ohki *et al.*, 1988). In order to remove these defect, we grafted these shoots directly on pumpkin or melon cultivar for rootstock. As a normal grafting, about one week was required for successful union in a humidified incubator. Therefore the rooting and acclimatization process of propagated shoots was not necessary.

We did not find any morphological disorder that is considered to occur during *in vitro* culture. And disease resistance was well retained through the tissue culture process. *Lycopersicon* plants regenerated from cotyledon explants retained their resistance to root-knot nematode (Ammati *et al*, 1984). However, Shoemaker and Swartz (1985) reported that tissue culture propagated strawberry plants were less resistant to *Phytophthora fragariae* than runner-propagated plants. These shoots were proliferated on the medium with relatively high BAP concentration. To avoid such a variation during the propagation stage, a single-node culture using medium supplemented with very low concentration of cytokinin may be suitable though the multiplication rate is not very high.

5. Conclusion

The advantage of our in vitro nursery system for melon was confirmed. For this system, the classical breeding process is indispensable, but it is possible to raise new better cultivars more easily within shorter periods than in the case for seed production. We also tried to apply this system to the bacterial wilt resistant tomato (Ohki *et al.*, 1988) and high soluble solid content fruits bearing tomato. To distribute these tissue culture propagated plants commercially, an effort to reduce the cost per plant should be made.

Acknowledgements. This work was partly supported by a Grant-in-Aid for Developmental Scientific Research (No.60860005) of the Ministry of Education, Science and Culture of Japan. The authors thank Mr.K.Yamazaki, Ms.T.Miyakoshi and Ms.A.Nakano for their technical assistance.

6. REFERENCES

Ammati, M., Murashige, T. and Thomason, I.J. (1984) 'Retention of resistance to the root-knot nematode, *Meloidogyne incognita*, by *Lycopersicon* plants reproduced through tissue culture', Plant Sci. Lett. 35, 247-250.

Bouabdallah, L., Branchard, M. (1986) 'Regeneration of plants from callus cultures of *Cucumis melo* L.', Z.Pflanzenzuchtg. 96, 82-85.

Debergh, P., Harbaoui, Y., Lemeur, R. (1981) 'Mass propagation of globe artichoke (*Cynara scolymus*): Evaluation of different hypotheses to overcome vitrification with special reference to water potential', Physiol. Plant. 53, 181-187.

Frankenberger, E.A., Hasegawa, P.M. and Tigchelaar, E.C. (1981) 'Diallel analysis of shoot-forming capacity among selected tomato genotypes', Z.Pflanzenphysiol. 102, 233-242.

Hagiwara, T., Kamimura, K. (1937) 'Genetic studies in *Cucumis*', Japan J. Breed. 13, 71-79.

Hayes, H. K., Garber, R. J. (1921) 'Muskmelon', in Breeding Crop Plants, McGraw-Hill Book Comp. Inc., New York, pp.257-258.

Kathal, R., Bhatnagar, S.P. and Bhojwani, S.S. (1986) 'Regeneration of shoots from hypocotyl callus of *Cucumis melo* cv. Pusa sharbati', J. Plant Physiol. 126, 59-62.

Kozai, T., Hayashi, M., Hirosawa, Y., Kodama, T., Watanabe, I. (1987) 'Environmental control for acclimatization of *in vitro* cultured plantlets. (1) Development of the acclimatization unit for accelerating the plantlet growth and the test cultivation, J. Agr. Met. 43, 349-358.

Kozai, T., Iwanami, Y. (1988) 'Effects of CO_2 enrichment and sucrose concentration under high photon fluxes on plantlets growth of carnation (*Dianthus caryophyllus* L.) in tissue culture during the preparation stage', J. Japan. Soc. Hort. Sci. 57, 279-288.

Moreno, V., Garcia-Sogo, M., Granell, I., Garcia-Sogo, B. and Roig, L.A. (1985) 'Plant regeneration from calli of melon (*Cucumis melo* L., cv.'Amarillo Oro')', Plant Cell Tiss. Org. Cult. 5, 139-146.
Murashige, T., Skoog, F. (1962) 'A revised medium for rapid growth and bioassays with tobacco tissue cultures', Physiol. Plant. 15, 473-497.
Niedz, R.P., Schiller Smith, S., Dunbar, K.B., Stephens, C.T. and Murakishi, H.H. (1989) 'Factors influencing shoot regeneration from cotyledonary explants of *Cucumis melo*', Plant Cell Tiss. Org. Cult. 18, 313-319.
Ohki, S., Bigot, C., Mousseau, J. (1978) 'Analysis of shoot-forming capacity in vitro in two lines of tomato (*Lycopersicon esculentum* Mill.) and their hybrids', Plant & Cell Physiol. 19, 27-42.
Ohki, S., Nasuda, K., Mori, Y., Katsuta, H. (1988) 'Establishment of breeding=*in vitro* nursery system for vegetable crops', Acta Hort. 230, 89-96.
Oridate, T., Oosawa, K. (1986) 'Somatic embryogenesis and plant regeneration from suspension callus culture in melon (*Cucumis melo* L.)', Japan J. Breed. 36, 424-428.
Pierik, R.L.M. (1987) In vitro Culture of Higher Plants, Martinus Nijhoff Publishers, Dordrecht.
Ringe, F., Nitsch, J.P. (1968) 'Conditions leading to flower formation on excised *Begonia* fragments cultured *in vitro*', Plant & Cell Physiol. 9, 639-652.
Shoemaker, N.P. and Swartz, H.J. (1985) 'Cultivar dependent variation in pathogen resistance due to tissue culture-propagation of strawberries', HortSci. 20, 253-254.
Shoda, S., Sadaoka, S. (1936) 'Genetic studies of Occidental and Oriental varieties of melon (2)', Agr. & Hort. 11, 2846-2852.
Tamai, T., Kamido, S., Tomari, I., Shinohara, K. (1962) 'Raising disease resistant melon cultivar', *ibid.* 37, 557-558.
Trulson, A.J., Shahin, E.A. (1986) 'In vitro plant regeneration in the genus *Cucumis*', Plant Sci. 47, 35-43.

GENE EXPRESSION DURING ANTHER AND POLLEN DEVELOPMENTAL TRANSFORMATIONS IN RICE

V. RAGHAVAN
Department of Botany
The Ohio State University
Columbus, Ohio
U.S.A.

ABSTRACT. *In situ* hybridization of anther sections of rice (*Oryza sativa*) with a radioactively labeled RNA probe for rice histone 3 gene showed that transcripts complementary to histone mRNA were expressed only in the endothecium of the anther wall and in the bicellular pollen grains. Intense binding of the probe however occurred in the embryogenically determined uninucleate pollen grains and in the vegetative cell of potentially embryogenic bicellular pollen grains of cultured anthers. Transcripts of histone 3 gene were also found to accumulate in the cells of early division phase pollen embryoids. These results suggest a role for histone gene expression in the induction of embryogenic divisions in pollen grains of cultured anthers of rice.

1. Introduction

In the reproductive biology of modern angiosperms, the anther is reckoned as a highly reduced structure of the flower consisting of four elongated microsporangia or pollen sacs held together by the connective. Initially, the microsporangium is filled with a number of diploid microsporocytes, each of which undergoes meiosis to produce a tetrad of haploid microspores. The microspores separate from one another, synthesize characteristically patterned walls and become pollen grains. Concurrently, an asymmetric division ensues in the pollen grain to give rise to a large vegetative cell and a small generative cell. During the terminal phase of differentiation of the pollen grain, the generative cell is partitioned into two sperm which participate in 'double fertilization' in the embryo sac. In the majority of plants the generative cell divides after the pollen grain lands on the stigmatic surface of a compatible flower; however, in some plants, this division occurs within the confines of the microsporangium so that the pollen is shed with three nuclei.

Although pollen grains of angiosperms are thus programmed for terminal differentiation, culture of anthers of certain plants at an appropriate stage of

R.S. Sangwan and B.S. Sangwan-Norreel (eds.), The Impact of Biotechnology in Agriculture, 85–98.

development in a medium with or without hormonal supplements has been shown to induce multicellularity in a small number of the enclosed pollen grains leading to the formation of embryo-like structures (embryoids) or calluses with the haploid set of chromosomes. This discovery is generally credited to Guha and Maheshwari (1964) arising out of their work with cultured anthers of *Datura innoxia*. According to recent surveys (Vasil, 1980; Maheshwari et al., 1982; Bajaj, 1983), embryogenic or callus type of growth has been induced by *in vitro* culture techniques in about 170 species of plants including some hybrids, distributed within 68 genera and 28 families and the number continues to grow.

The normal development of the anther and pollen grain terminating in male gametogenesis and the induction of embryogenic type of divisions in the pollen grain leading to immortality are obviously complex phenomena with gene control acting together with unknown metabolites or hormones. Although this assumption may well be true, we have only minimum evidence based on modern molecular approaches to support it. Previous studies (Raghavan, 1987 for review) have shown that the early phase of pollen development in various angiosperms is characterized by the synthesis of ribosomes and mRNA associated with gametophytic growth, with a general decline in nucleic acid synthesis during pollen maturation. Recently, Stinson et al. (1987) have followed the expression of two groups of pollen-specific mRNAs during the entire spectrum of pollen grain development in *Zea mays* and *Tradescantia paludosa*. These studies have shown that genes of one group are activated after the first haploid mitosis and increase in concentration during pollen maturation. In contrast, an actin mRNA is first detected prior to mitosis, reaching a maximum at pollen interphase and decreases in the mature pollen grain. By *in situ* hybridization, the vegetative cell of *Z. mays* pollen grain has been identified as the site of accumulation of a pollen-specific gene of the first group (Hanson et al., 1989).

The fundamental premise of gene regulation during pollen embryogenesis is the deflection of the normal developmental program of a single-celled pollen grain into a pathway of continued divisions and increasingly complex morphology of an embryoid. From this perspective, some information is available on the pattern of accumulation and synthesis of total RNA (Bhojwani et al., 1973; Sangwan-Norreel, 1978; Raghavan, 1979a, b) and mRNA (Raghavan, 1981a) during embryogenic development of pollen grains of certain plants. These results have led to the general conclusion that as a result of the trauma of excision and culture of anthers, a small proportion of the enclosed pollen grains change their program by the synthesis of mRNAs which probably code for the proteins of the first haploid mitosis in culture. Subsequent divisions of the pollen grain in the embryogenic pathway are mediated by the synthesis of additional or perhaps new mRNAs by the constituent pollen cells (Raghavan, 1981b). However, the specific genes that are activated during embryogenic episode of pollen grains of cultured anthers have not been characterized.

Using *in situ* hybridization methods, I have recently monitored the temporal and spatial localization of a rice histone 3 gene during developmental transformations in the anther and pollen grains of rice (*Oryza sativa*). The results of some of these studies, described in detail elsewhere (Raghavan, 1989), are summarized here.

2. Developmental Cytology of the Anther

For orientation purposes, it is worthwhile to decribe briefly the development of the anther and pollen grains of rice and the changes that occur in cultured anthers. For details of anther and pollen ontogeny, see Raghavan (1988).

2.1. RICE ANTHER DEVELOPMENT

The anther is a highly reduced structure of the rice floret. A fully mature anther of IR-30 rice used in these studies attains a maximum length of about 2.6 mm and a maximum width of 0.4 mm. As seen in a transverse section, at the early stage of its development, the anther consists of a homogenous mass of cells surrounded by a well-defined layer of isodiametric cells constituting the epidermis. Later, as the anther becomes four-lobed, a single row of hypodermal cells differentiates through the entire length of each lobe (Fig. 1). These cells designated as archesporial initials, divide periclinally to give rise to an inner primary sporogenous cell and an outer primary parietal cell (Fig. 2). The former divides twice mitotically to generate four microsporocytes which undergo reduction division to yield pollen grains. The anther wall is formed by a series of anticlinal and periclinal divisions of the primary parietal cell. The mature anther consists of a wall made up of three concentric layers of cells constituting (from outside to inside) the epidermis, endothecium and middle layer and a layer of cells of the tapetum, within which the pollen grains are housed (Fig. 3, 4).

Microspores liberated from the callose wall of the tetrad have a centrally placed nucleus surrounded by a lightly staining granular cytoplasm (Fig. 5). The microspore soon expands in volume without accompanying cytoplasmic increase resulting in vacuolate cells (Fig. 6). Deposition of wall material leading to the delimitation of an exine is also evident in the microspores at this stage of development. The first haploid mitosis occurs in the vacuolate pollen grain to form a large vegetative cell and a small generative cell (Fig. 7). Pollen maturation is associated with the accumulation of starch grains, a process known as engorgement (Fig. 8). The final phase of pollen maturation involves the division of the generative cell to form two sperm. Since this division takes place before anthesis, pollen grains of rice are shed at the three-nucleate stage (Fig. 9).

2.2. CYTOLOGY OF EMBRYOGENIC DEVELOPMENT OF POLLEN GRAINS

Anthers of rice are most vulnerable for embryogenic divisions when they are cultured at the vacuolate uninucleate pollen grain stage. In the protocol followed in my laboratory, culture of anthers in a medium containing 1.0 mg/l kinetin and 1.0 mg/l ß-naphthaleneacetic acid, followed by 4 days at 4 C triggers embryogenic development of pollen grains during a subsequent period at 25 C. In about 8 days after culture of the anther, most of the pollen grains display signs of degeneration such as collapse of vacuoles and cytoplasm, while the small number of 'embryogenically determined' pollen grains remain healthy and divide to form vegetative and generative cells (Fig. 10). These bicellular pollen grains are potentially embryogenic and during further periods in culture, they continue the gametophytic program by engorgement or become embryogenic by continued mitotic activity. Multicellularity is attained by repeated divisions of the vegetative cell, while the generative cell remains inactive or undergoes one or two divisions before disintegrating. Figs. 11-13 illustrate some stages of formation of multicellular units by pollen grains of cultured anthers. The induction of embryogenic growth of pollen grains of rice by divisions of the vegetative cell has previously been described (Chen, 1977; Chen and Wu, 1983; Kim and Raghavan, 1988). In general, these observations reveal the existence of two developmental blocks in the large-scale transformation of pollen grains of cultured anthers of rice in the embryogenic pathway: (1) failure of a large majority of the uninucleate pollen grains to divide asymmetrically and (2) mitotic failure of the vegetative cell of potentially embryogenic pollen grains.

3. Temporal and Spatial Patterns of Histone 3 Gene Expression

Information relating to the intracellular distribution of specific genes is important for an understanding of the organization and function of cells of the developing anther and pollen grains of rice. In particular, localization of a specific mRNA provides insight into the site of synthesis of the corresponding protein. For example, is the protein synthesized close to the site of its function, as determined by localization of the corresponding mRNA, or is it synthesized at another site and transported across membranes to the functional site? In this study, one could follow the changes in the expression of a host of genes, for example, genes for cytoskeletal proteins, actin and tubulin, genes for metabolic enzymes etc. The use of histone 3 gene for the present study was dictated by the fact that some investigators have implicated histones in nuclear differentiation during pollen development (Sauter, 1969; Reznikova et al., 1978; Sangwan-Norreel, 1978; Bednarska, 1981). Moreover, since embryogenic development of pollen grains involves increased mitotic activity, it is reasonable to expect a need for histones to complex with the large amount of DNA synthesized.

3.1. PROBE PREPARATION

Histone 3 gene used is a 1.3 kb insert isolated from a genomic library of 10-day old seedlings of rice by Peng and Wu (1986) and kindly supplied by Dr Ray Wu (Cornell University) for these investigations. The insert which is ligated into the *Bam*H1 site of pBR322 includes a 405 bp coding sequence that starts with ATG and terminates with a stop codon TGA (Fig. 14A). The general strategy to monitor the changes in histone mRNA expression is by *in situ* hybridization of anther sections with a ^{3}H-labeled RNA probe made from the gene.

Asymmetric RNA probes were prepared by ligating the insert into the transcription vector pBS M13$^+$ (originally known as Bluescribe). For this purpose, rice histone gene sequence was excised via the adjacent *Bam*H1 sites in the original vector and inserted in the *Bam*H1 site of pBS M13$^+$ by ligation in the presence of T4 DNA ligase (Fig. 14B). Following transformation of *Escherichia coli* JM101, recombinants with histone gene inserts were identified by restriction enzyme analysis. Plasmid DNA was isolated from an amplified recombinant by banding in cesium chloride density gradient and linear templates of antisense (non-coding) and sense (coding) strands were obtained by truncation of plasmid DNA with restriction enzymes. RNA was transcribed from the truncated templates in 50 μl reaction mixtures containing ^{3}H-UTP, ^{3}H-CTP, ^{3}H-ATP and GTP and T3 or T7 RNA polymerase, as appropriate. After purification by phenol/chloroform extraction and ethanol precipitation, RNA probes were dissolved in Denhardt's solution and used directly for *in situ* hybridization of sections of anthers of different stages of development and after different periods of culture. Annealing sites were identified by autoradiography.

3.2. DEVELOPMENTAL EXPRESSION OF HISTONE 3 GENE

Autoradiographs of sections of anthers annealed with the antisense strand histone RNA probe showed no binding above background in the cells of the anther primordium, or in the archesporial cell, the primary parietal cell, the primary sporogenous cell, microsporocytes and tapetum; the autoradiograph shown in Fig. 15 is representative of the annealing pattern of the cloned gene with early stage anther sections. Duplex RNA molecules formed between the antisense transcript and complementary mRNA sequences were first detected in the endothecium of the premeiotic anther (Fig. 16). However, appearance of histone transcripts in the endothecium was transient, as, following meiosis there was a decrease in the number of annealing sites in this tissue (Fig. 17).

As the endothecium became radially thickened, ^{3}H-histone labeling completely disappeared from this tissue. During pollen development, expression of the gene was noted rather abruptly in the starch-filled bicellular pollen grains (Fig. 18) and it continued at a reduced level in the mature pollen grains. In the former, autoradiographic silver grains were uniformly distributed in the cytoplasm of the generative and vegetative cells.

The most striking changes in gene expression were seen in the pollen grains of cultured anthers. Embryogenically determined pollen grains which were identified in anthers cultured for 8 to 10 days gave strong hybridization signals over both nucleus and cytoplasm (Fig. 19). After the first haploid mitosis, potentially embryogenic pollen grains revealed label in the vegetative cell, but not in the generative cell (Fig. 20A, B). This is consistent with the cytological observations showing division of the vegetative cell in the embryogenic pathway. Although I have not followed the expression of histone 3 gene in the bicellular pollen grain during its transformation into an embryoid, *in situ* hybridization of multicellular pollen grains with the antisense strand has revealed the presence of complemenatry transcripts in all newly formed cells (Fig. 21). As regards its intracellular distribution, the label was present uniformly or in small clusters over tne cytoplasm and nucleus (Fig. 22).

Two approaches were used to verify that the pattern of silver grain distribution observed in rice anther sections after *in situ* hybridization with histone 3 probe is not an artifact either of hybridization or of autoradiographic detection of ^{3}H-labeled probes. First, RNase-treated sections of anthers hybridized with the probe showed no annealing signals; second, when sections were hybridized with sense RNA probe, no annealing occurred.

4. Conclusions

It is clear from these results that rice anther provides an interesting system for examining the control of gene expression during development and differentiation. Included in the progression of events are mitotic proliferation of undifferentiated cells; differentiation of specific cell types such as endothecium, middle layer and tapetum; meiosis in the sporogenous cells; formation of cells endowed with specific functions such as the vegetative and generative cells and morphological transformation of the unicellular pollen grain into multicellular entity. This study also shows that *in situ* hybridization is a powerful tool for mapping and detection of specific mRNA transcripts in individual cells. The usefulness of this method becomes all the more significant when the target cells are inaccessible or cannot be obtained in enriched fractions. The anther tissues and pollen grains of rice belong to this group. In the long run, results of this study may give new ideas to increase the yield of embryoids in cultured anthers of rice. The objective of this work was as much to know about gene expression in those pollen which become embryogenically determined, as it was to monitor gene repression in the vast majority of nonembryogenic pollen grains of cultured anthers. It is a truism that unless we identify the cytological, biochemical and molecular changes that occur in the anther and pollen grains during their normal ontogeny and during embryogenic transformation, it will be impossible to control developmental events to our advantage and increase the yield of embryoids and calluses in cultured anthers.

5. Acknowledgement

Part of the work described in this article was supported by grants from the Rockefeller Foundation (GA AS 8510 and 8604) and the National Science Foundation (DCB 8709092).

6. References

Bajaj, Y.P.S. (1983) '*In vitro* production of haploids, in D. A. Evans, W. R. Sharp, P. V. Ammirato and Y. Yamada (eds.), Handbook of Plant Cell Culture, Vol. 1, Macmillan Publishing Co., New York, pp. 228-287

Bednarska, E. (1981) 'Autoradiographic studies of DNA and histone synthesis in successive differentiation stages of pollen grain in *Hyacinthus orientalis* L.; Acta Soc. Bot. Pol. 50, 367-380.

Bhojwani, S.S., Dunwell, J.M. and Sunderland, N. (1973) 'Nucleic-acid and protein contents of embryogenic tobacco pollen'. J. Expt. Bot. 24, 863-871.

Chen C. (1977) '*In vitro* development of plants from microspores of rice'. In Vitro 13, 484-489.

Chen, C. and Wu, Y. (1983) 'Segmentations in microspores of rice during anther culture'. Proc. Natl. Sci. Counc. B, ROC 7, 151-157.

Guha, S. and Maheshwari, S.C. (1964) '*In vitro* production of embryos from anthers of *Datura*'. Nature 204, 497.

Hanson, D.D., Hamilton, D.A., Travis, J.L., Bashe, D.M. and Mascarenhas, J.P. (1989) 'Characterization of a pollen-specific cDNA clone from *Zea mays* and its expression'. Plant Cell 1, 173-179.

Kim, M.Z. and Raghavan, V. (1988) 'Induction of pollen plantlets in rice by spikelet culture'. Plant Cell Rep. 7, 560-563.

Maheshwari, S.C., Rashid, A. and Tyagi, A.K. (1982) 'Haploids from pollen grains - Retrospect and prospect'. Am. J. Bot. 69, 865-879.

Peng, Z. and Wu, R. (1986) 'A simple and rapid nucleotide sequencing strategy and its application in analyzing a rice histone 3 gene'. Gene 45, 247-252.

Raghavan, V. (1979a) 'Embryogenic determination and ribonucleic acid synthesis in pollen grains of *Hyoscyamus niger* (henbane)'. Am. J. Bot. 66, 36-39.

Raghavan, V. (1979b) 'An autoradiographic study of RNA synthesis during pollen embryogenesis in *Hyoscyamus niger* (henbane)'. Am. J. Bot. 66, 784-795.

Raghavan, V. (1981a) 'Distribution of poly(A)-containing RNA during normal pollen development and during induced pollen embryogenesis in *Hyoscyamus niger*'. J. Cell Biol. 89, 593-606.

Raghavan, V. (1981b) 'Pollen embryogenesis in *Hyoscyamus niger*: A review', in A. N. Rao (ed.), Tissue Culture of Economically Important Plants, COSTED, Singapore, pp. 262-268.

Raghavan, V. (1987) 'Developmental strategies of the angiosperm pollen: A biochemical perspective'. Cell Diffn. 21, 213-226.

Raghavan, V. (1988) 'Anther and pollen development in rice (*Oryza sativa L)*'. Am. J. Bot. 75, 183-196.

Raghavan, V. (1989) 'mRNAs and a cloned histone gene are differentially expressed during anther and pollen developmet in rice (*Oryza sativa L.)*'. J. Cell Sci. 92, 217-229.

Reznikova, S., Bugara, A. and Erokhina, A. (1978) 'A cytochemical study of DNA and histone in the developing *Lilium* anther'. Bull. Bot. Soc. Fr. Actual. Bot. 125, 51-56.

Sangwan-Norreel, B.S. (1978) 'Cytochemical and ultrastructural peculiarities of embryogenic pollen grains and of young androgenic embryos in *Datura innoxia*'. Can. J. Bot. 56, 805-817.

Sauter, J.J. (1969) 'Cytochemische Untersuchung der Histone in Zellen mit unterschiedlicher RNS- und Protein-Synthese. Zeits'. Pflanzenphysiol. 60, 434-449.

Stinson, J.R., Eisenberg, A.J., Willing, R.P., Pe, M.E., Hanson, D.D. and Mascarenhas, J.P. (1987) 'Genes expressed in the male gametophyte of flowering plants and their isolation'. Plant Physiol. 83, 442-447.

Vasil, I.K. (1980) 'Androgenetic haploids'. Int. Rev. Cytol. Suppl. 11A, 195-223.

Figs. 1-4. Anther development in rice. Fig. 1. Transverse section of a floret of rice showing the anther primordia. Arrows point to the archesporial initial. Fig. 2. Transverse section of a four-lobed anther primordium showing the division of the archesporial initial into the primary sporogenous cell (sc) and the primary parietal (pc) cell. Fig. 3, 4. Longitudinal and transverse sections, respectively, of premeiotic anthers showing the wall layers (e, epidermis; en, endothecium; m, middle layer), tapetum (t) and microsporocytes (ms). Scale bars=10μm.

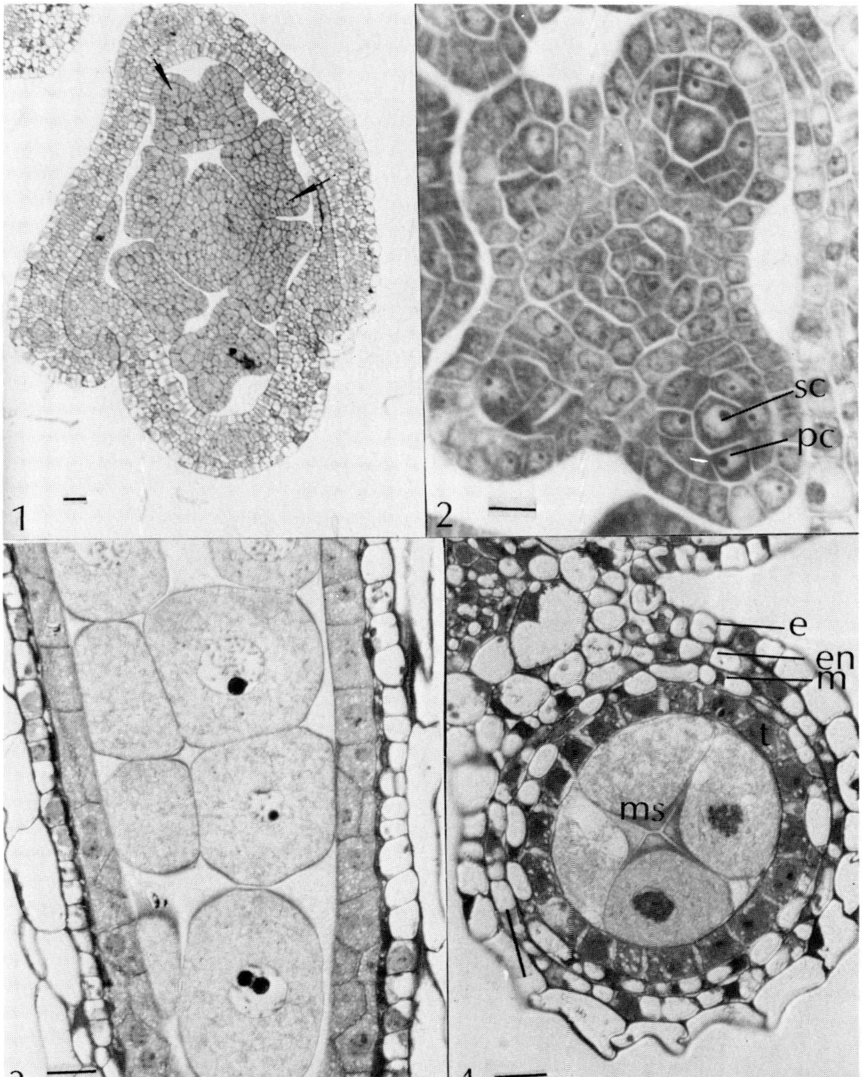

Figs. 5-9. Pollen development in rice. Fig. 5. Transverse section of an anther lobe showing microspores after release from the tetrad. Arrows point to the nucleus. Fig. 6. Longitudinal section of an anther showing the vacuolate pollen grains. Fig. 7. Bicellular pollen grains. (g, generative cell; v, vegetative cell). Fig. 8. A starch-filled pollen grain (gn, nucleus of the generative cell; vn, nucleus of the vegetative cell). Fig. 9. A three-nucleate pollen grain (sn, sperm nuclei; vn, nucleus of the vegetative cell). Scale bars=10μm.

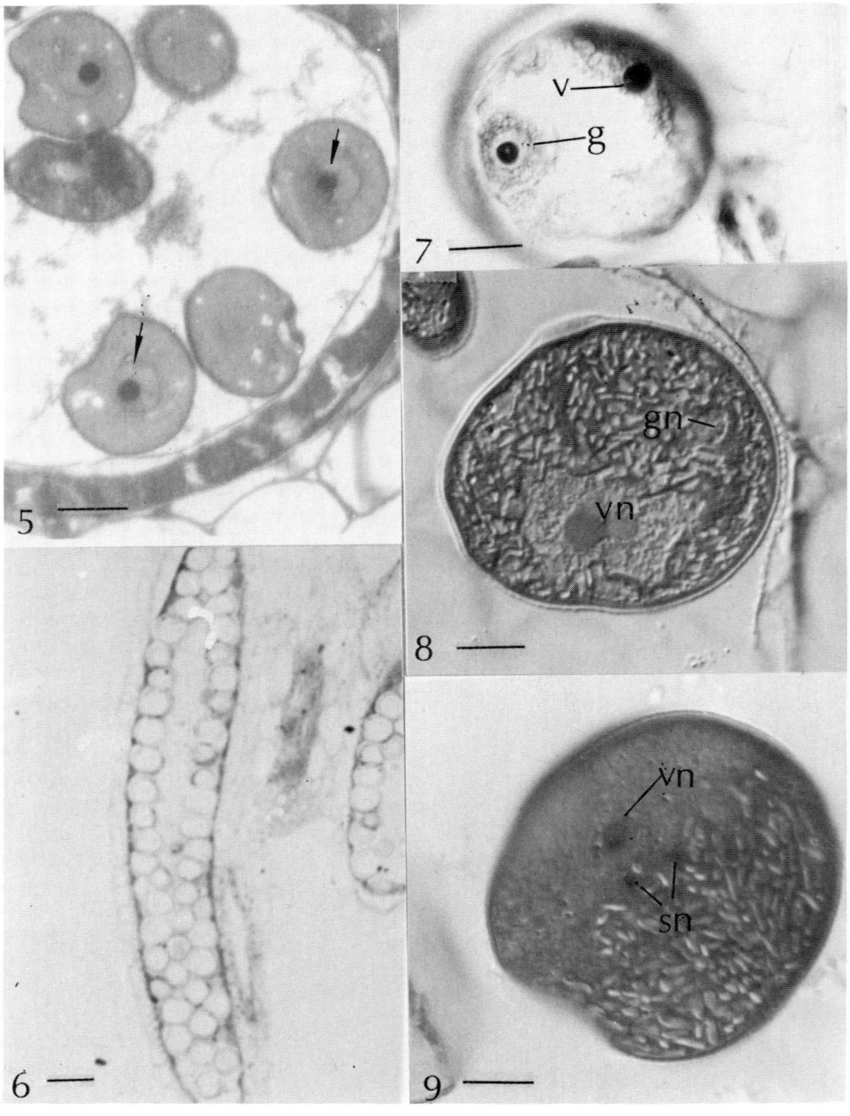

Figs. 10-13. Embryogenic development of pollen grains in cultured anthers of rice. Fig. 10. An embryogenic (em) and a nonembryogenic (ne) pollen grain from a section of an anther, 10 days after culture. Fig. 11. An early stage of formation of a multicellular pollen; 15 days after culture. Fig. 12. A later stage of a multicellular pollen; 16 days after culture. Fig. 13. A pollen embryoid; 20 days after culture. Scale bars=10μm.

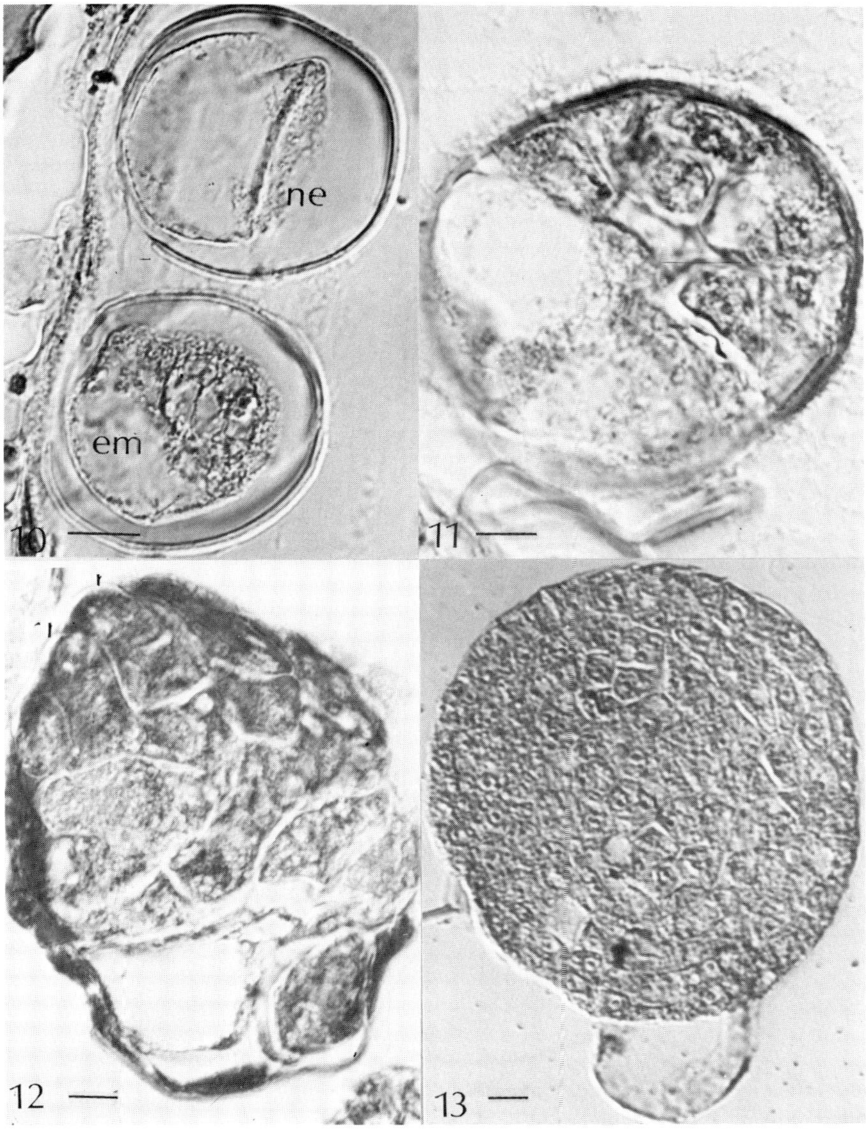

Fig. 14A. Restriction map of rice histone 3 gene ligated into pBR322. B. Ligation of histone 3 gene into the *Bam*H1 site of pBS M13^{+} for RNA transcription.

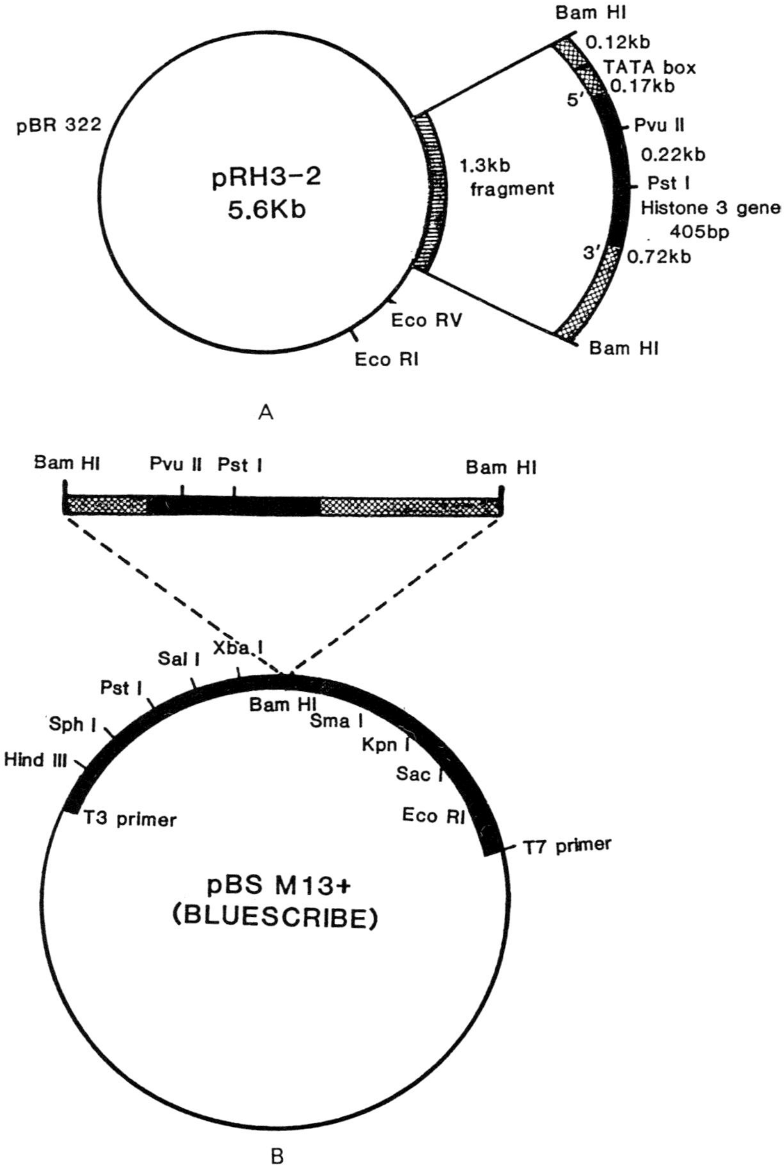

Figs. 15-18. *In situ* hybridization of sections of rice anthers of different ages with antisense histone 3 RNA probe. Fig. 15. Transverse section of an anther primordium. Fig. 16. Longitudinal section of a premeiotic anther (en, endothecium). Fig. 17. Longitudinal section of a post-meiotic anther (en, endothecium). Fig. 18. Section of a bicellular pollen grain. Scale bars=10μm.

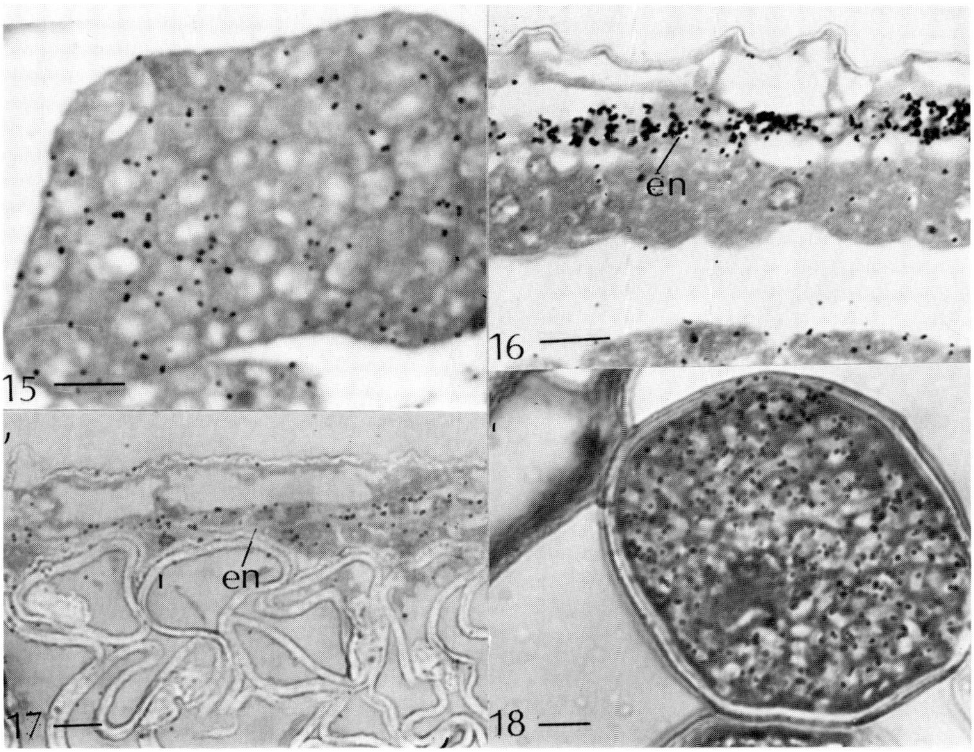

Figs. 19-22. *In situ* hybridization of sections of cultured rice anthers with antisense histone 3 RNA probe. Fig. 19. Embryogenically determined pollen grain, 8 days after culture. Fig. 20A. Annealing of the probe to a potentially embryogenic bicellular pollen; 12 days after culture. B. Same pollen with focus on nuclei (gn, nucleus of the generative cell; vn, nucleus of the vegetative cell). Fig. 21. A pollen embryoid, 24 days after culture. Fig. 22. Magnified view of cells of the embryoid showing intracellular localization of the label. Scale bars=10μm.

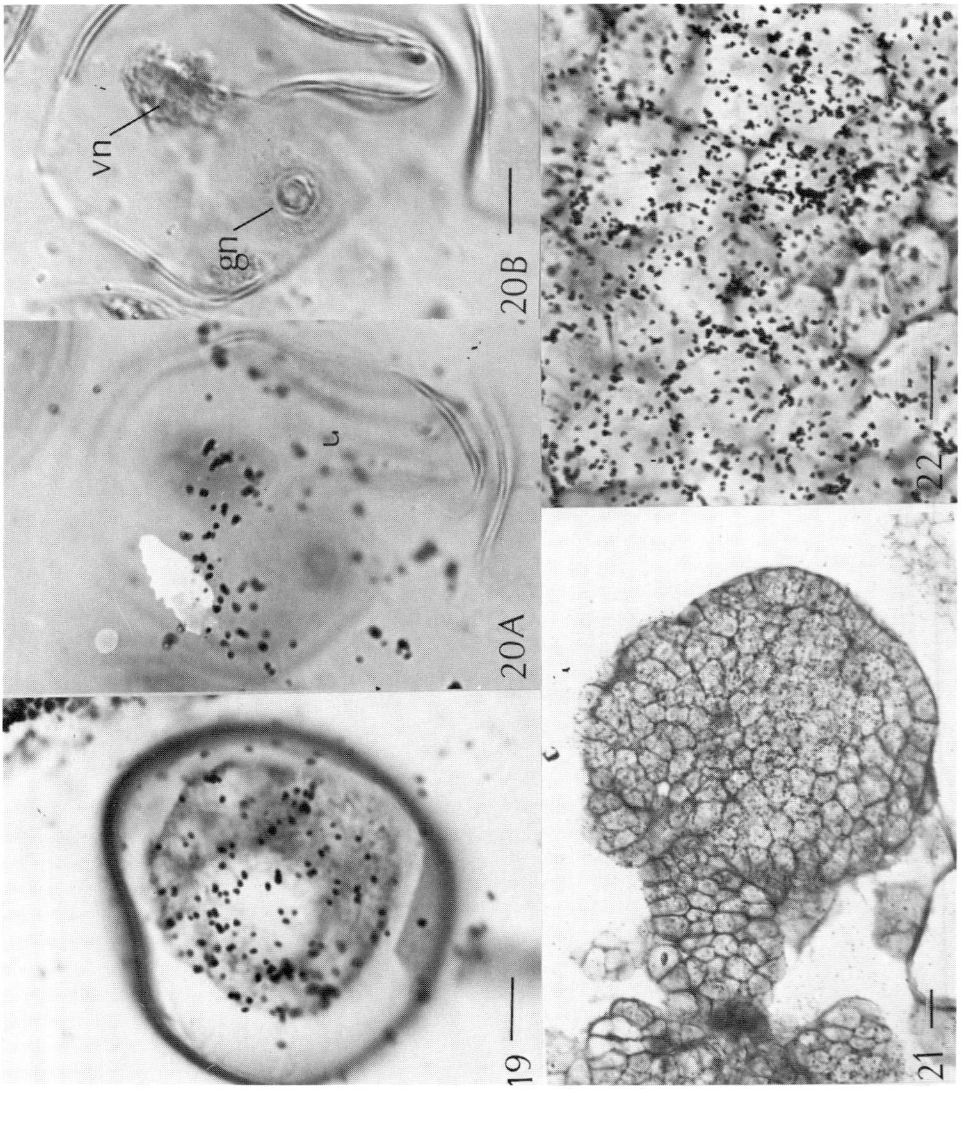

FIELD EXPERIMENTS WITH DH-LINES IN BARLEY WITH ML-O POWDERY MILDEW RESISTANCE

ÅSMUND BJØRNSTAD
Department of Genetics and Plant Breeding
Agricultural University of Norway
N-1432 Aas-NLH
Norway

ABSTRACT. Doubled-haploid (DH) lines of barley (*Hordeum vulgare* L.) produced by the *H. bulbosum*-method have been used to evaluate the performance of the *ml-o* resistance gene against barley powdery mildew (*Erysiphe graminis* DC f.sp. *hordei* Em. Marchal). The effect of background genes interacting with the major gene is shown to be of importance in alleviating the negative pleiotropic effects on grain yield and/or 1000-grain weight.

INTRODUCTION

The title of this short presentation outlines its theme. Since this meeting primarily is on plant biotechnology and not barley pathology and breeding, I will try to emphasize points pertaining to the interface between these fields. My own priority of work has been to use *in vitro*-methods in genetics and breeding research of wheat and barley, not to develop or study the tissue culture process *per se* (although this is badly needed). A full account of the mildew work will appear in Euphytica (Bjørnstad and Aastveit, in press).

In genetical studies we are concerned with:
(1) The genetic architecture of traits like resistance to disease (reported here) or lodging (see below);
(2) Comparisons of DH-lines of wheat and barley with conventional homozygous lines. (This year we test almost 1500 DH-lines of spring wheat in replicated trials, in comparisons with SSD lines).
(3) Testing methods for cross prediction (the possibility of early screening of crosses by means of DH-lines).

In general, haploidization may be regarded as a conservative breeding method, particularly suited to advanced, well-adapted genotypes. Its unique feature is to reduce recombination and possibly to facilitate combination of characters from different parents. One example may be lodging resistance. We have in Norway some very straw-stiff barley cultivars, but this trait seems difficult to transfer in crosses. A possible reason may be that the resistance is due to a build-up of a coupling/associationphase of alleles enhancing

R.S. Sangwan and B.S. Sangwan-Norreel (eds.), The Impact of Biotechnology in Agriculture, 99–104.

the character. Genetic theory would then predict a breakdown of such a system during recombination, at a rate depending on the tightness of linkage. Although this is a difficult character to study, we are trying to test this hypothesis in barley. - It may be mentioned that the control of recombination in DH-lines could also be utilized in RFLP-studies.

MLO-RESISTANCE

The present account deals with our work on powdery mildew resistance in barley. The locus ml-o confers a general resistance against all known races of the pathogen, and since it is a single locus, it does not conform to the virulence-gene/resistance-gene type of dialectics. Jørgensen (1987) has called it a "third kind" of resistance, beside the classical "specific" or "non-specific" types. The locus is known to have 11 functionally identical recessive alleles, of which one is spontaneous (see below), the others are induced mutants. They act by preventing entry of the pathogen through excessive deposition of callose in affected cells. This hypersensitive response may also be elicited by mechanical injury, e. g. by aphids. As a consequence, cells where callose is deposited tend to turn necrotic, a fact often visible on infected leaves. This, in turn, is probably responsible for the pleiotropic effects of resistance: a decrease in grain yield and specific (e.g. 1000-) grain weight. Although ml-o mutants have been known for more than 40 years, this has siginificantly slowed the incorporation of the gene into commercial cultivars. However, Schwarzbach (1976) reported that the genetic "background" may significantly modify the degree of necrotic spots on leaves. An Ethiopian barley carrying the spontaneous allele ml-o11 was reported almost to be free of necroses, and this landrace has since been used in the production of a few currently grown cultivars.

It may be mentioned, in passing, that the phenomenon of pleiotropy is familiar to classical genetics and breeders. It seems, however, a bit forgotten by molecular biologists and people striving with genetical transformation of crop plants, which may be regarded as a kind of "directed mutagenesis". It may be anticipated that any introduction of major mono- or oligogenic factors into a genome will have pleiotropic effects, which it may take a long time to overcome. This is particularly so because of our inability to target genes into the "good" areas of the chromosomes. Also, the phenomenon of "background" effects, or in biometrical terms, gene interaction, will need to be dealt with when the molecular transformation part has been done.

RESULTS

A first part of our work was to identify well-adapted genotypes that were able to reduce necrotic spots. An example of two contrasting types is shown in Figure 1. The two-rowed cultivar 'Pernilla' has a poor ability, whereas the six-rowed cultivars 'Agneta' and 'Bamse' (sister lines) are well suited for the production of high yielding barley lines. From F_3-plants with both these lines in their pedigree, a small sample of 74 resistant lines were produced in 1985 and tested in the field during the three successive years. The results from a 1988 3-location, 4 replicate trial of the 4 best lines compared with 4 check genotypes or cultivars, showed that the DH-lines were quite competitive, although somewhat later maturing than the 'Bamse' parent. Also, the possibility of running a multi-location, large plot trial only 3 years after the lines were made, shows the potential of the method.

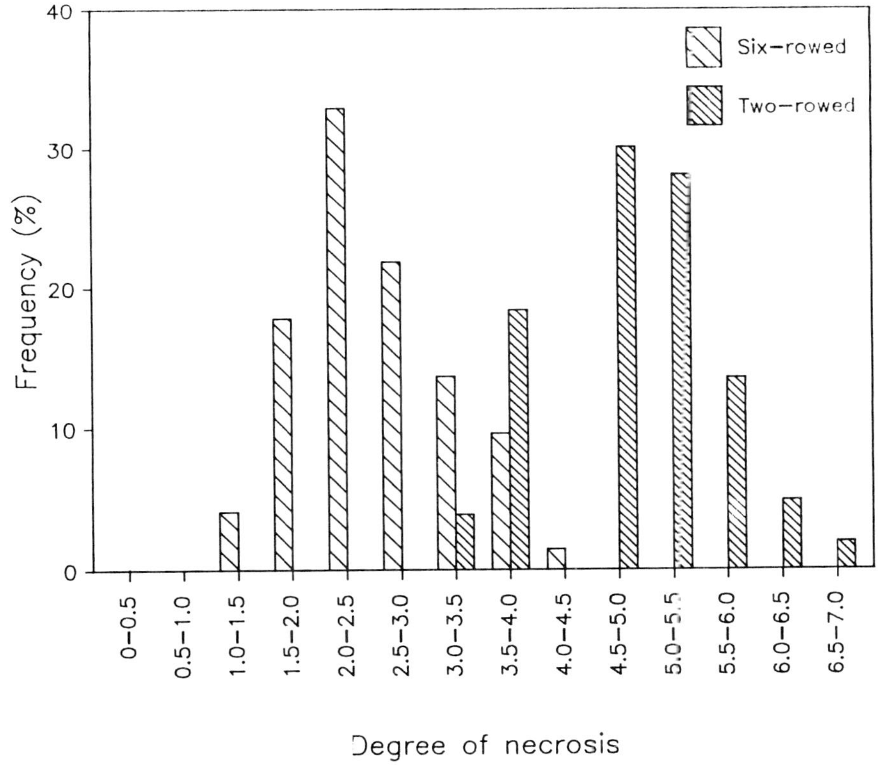

Figure 1. Degree of leaf necroses in ml-o resistant DH-lines from two crosses, with six-rowed (recurrent parent 'Bamse') or two-rowed (recurrent parent 'Pernilla') backgrounds. Scale 0-10. (The results are not <u>general</u> for six- or two-rowed barleys).

We wanted to assess the quantitative effect of the ml-o locus in more detail. To do this we have developed several DH-populations from F_1 crosses segregating at the locus. The comparison of a large number of mutant (ml-o) and wild type (Ml-o) DH-lines from a given cross may allow such estimates. All other genes will be randomized, except for those that are closely linked. DH-lines are ideal for such studies, since large plot sizes with homozygous lines may be used.

The results reported here comes from a cross in two-rowed barley, from which more than 400 DH-lines were tested in the field in 1987 and 1988. The segregation ratio of resistant to susceptible lines was close to 2:1, instead of the expected 1:1. (This is unusual in barley lines produced by the Hordeum bulbosum-method, but quite common in anther cultures (cf. Powell et al. 1986). Such distorted segregation may seriously influence genetic studies). The 1987 experiment was one with small (0.2 m^2) plots with 4 replicates and a strong mildew infection. The 1988 experiment was sown with large plots (3 m^2) and 2 replicates and only had a weak mildew infection.

The mean results are shown in Table 1. In 1987 there was no significant difference in grain or total yield between the resistant and susceptible classes, although the resistant yielded slightly less (-1.2 %). The effects in 1988 were, however, pronounced, with a mean grain yield reduction of 5.9 %. Evidently, the mildew infection in 1987 depressed the yield potential of susceptible lines almost as much as the presence of the ml-o allele. The total yield was, however, not different in the two classes, but the average harvest index was reduced with 3.5 %, pointing to a decreased translocation to the grains. Contrary to expectation, the 1000 grain weight was increased, not reduced, in the resistant lines. This may mean that the "genetic background" in a cross may independently modify grain yield and 1000 grain weight. Kjær et al. (1989) have, in a paralell study, found a reduction in both these parameters in a powdery mildew free environment. Evidently more crosses need to be studied to assess the frequency of positive "background" effects. Our hypothesis is that "normal" resistance genes, either specific or non-specific, may be involved.

Many high yielding resistant lines could be found in this cross. The mean yield of the cross was, however, significantly inferior to the parental mean and also the check cultivars. The theory of cross prediction states that a "good" cross should have both a high mean and a large additive genetic variance. In this case only the latter requirement was fullfilled. The practical conclusion from this study would be to adopt a two-step procedure: (1) To test crosses for low levels of necroses in F_2 and hence (2) to develop DH-lines from the selected crosses.

TABLE 1. Means of grain yield and 1000 grain weight in ml-o (resistant) and Ml-o (susceptible) classes in DH-lines from a barley cross tested in the field in 1987 and 1988

		1987	1988
Grain yield	ml-o	99^{ns}	94^{***}
	Ml-o	100	100
1000-grain weight	ml-o	100	100
	Ml-o	93^{**}	97^{ns}

DISCUSSION

At last, I would put up a few challenges to tissue culturists. Our knowledge of the anther culture process is largely empirical. Genotype x method interactions are common. If you compare the anther culture response of the winter barley genotype 'Igri' with many spring barleys, the regeneration of green plants is strikingly higher. Why is this so ? Has 'Igri' been used so much by tissue culturists because of its responsiveness, that the protocol is actually adapted to this cultivar ? Another hypothesis may be that the winter habit may help to cope with the enormous stresses imposed by a cold treatment at 4 degrees and subsequently an incubation at 25 degrees on a medium. However, winter barleys also seem to differ widely in response, although at a higher average efficiency. To my mind, the stresses involved should be minimized. The recent improvements of anther cultures of barley (Hunter *et al*. 1988) and apple (Lars Johansson, pers.comm.) by replacing sucrose with their natural carbon sources (maltose or sorbitol, respectively), is a case to be learnt from. The beneficial effects of maltose in widely differing species like *Allocasuarina verticillata* (Cao *et al*., this volume), *Vitis* (Monnier *et al*., this volume) and rye-grass (Bante *et al*., this volume), may point to a common effect that hopefully will be clarified in the near future. - At a more primitive stage, "staggering" hormones like 2,4-D were necessary to get growth in *Brassica* anthers, today the protocol is hormone-free. Still 2,4-D is necessary to get embryoid induction in wheat. Who can find the key from tissue nature to tissue culture ?

REFERENCES

Hunter, C. P., Loose, R. W., Clerk, S. P., Lyne, R. L. (1988) Maltose - the preferred carbon source for barley anther culture, Abstract, Eucarpia symposium "Genetic manipulation in plant breeding", Elsinore, 11.-16.9. 1988.

Jørgensen, J. H. (1987) Three kinds of powdery mildew resistance in barley, Barley Genetics V, 583-592.

Kjær, B., Jensen, H. P., Jensen, J., and Jørgensen, J. H. (1989) Associations between three ml-o powdery mildew resistance genes and agronomic traits in barley, Euphytica (in press).

Powell, W., Borrino, E. M., Allison, M. S., Griffiths, D. W., Asher, M. J. C., and Dunwell, J. M. (1986) Genetical analysis of microspore derived plants of barley (*Hordeum vulgare*), Theor. Appl. Genet. 72, 619-626.

Schwarzbach, E. (1976) The pleiotropic effects of the *ml-o* gene and their implications in breeding, Barley Genetics *III*, 440-445.

ANTHER CULTURE OF *LOLIUM PERENNE* AND *LOLIUM MULTIFLORUM*

I. Bante, T. Sonke, R.F. Tandler, A.M.R. van den Bruel and E.M. Meyer

DSM Research, Bio-organic chemistry section, P.O.Box 18, 6160 MD Geleen, the Netherlands

SUMMARY

Albino as well as green plantlets were obtained by anther culture of *Lolium perenne* L. and *L.multiflorum* L. Isozyme and flow cytometric analyses showed the microspore origin of these plantlets. There is a clear genotype effect on anther culture response and on the ratio of green to albino regenerants. The majority of regenerated plantlets were albinos, although green plants were obtained from different *L.perenne* and *L.multiflorum* varieties. After transfer to soil, the microspore derived green plants grew vigorously under field conditions.
Pretreatment of the spikelets at 4°C before plating the anthers positively influences the anther response for both *L.multiflorum* and *L.perenne*; the optimum duration of pretreatment is 3 and 2 weeks respectively. Reduction of the ammonium nitrate concentration in the induction medium resulted in a significant increase in *Lolium* anther response, callus and embryo growth and plantlet regeneration. For *L.perenne*, addition of potassium nitrate to an induction medium containing ammonium sulfate (1mM) and glutamine (5mM) decreased the culture efficiency. For *L.multiflorum*, anther response decreased when the potassium nitrate concentration exceeded 10mM; the influence upon the culture efficiency was not clear. Replacement of sucrose in the induction medium by other sugars such as maltose, trehalose and maltotriose dramatically increased the culture efficiency as compared with sucrose for both *L.multiflorum* and *L.perenne*.

INTRODUCTION

Methods for the production of haploid and homozygous doubled haploid plants by means of anther culture have been considerably improved during recent years. Nowadays anther culture appears to be a particularly powerful tool when integrated with conventional breeding techniques; the practical value of haploids in plant breeding has been il-

R.S. Sangwan and B.S. Sangwan-Norreel (eds.), The Impact of Biotechnology in Agriculture, 105–127.

lustrated by several investigators [11,12,38]. There is much interest in the development of anther culture techniques, particularly for the production of doubled haploid plants for use as homozygous breeding lines. The true breeding lines can be used for selection of specific characteristics and for hybrid breeding. The homozygous plants can be produced in a shortened time frame compared to the numerous cycles of inbreeding or backcrossing needed to obtain pure lines in conventional breeding programmes.

In contrast with the major research efforts directed to anther culture of cereal species [11], much less information is available for the forage species of the *Gramineae* [1,64]. Although ryegrasses are the most important grass species for West-European agriculture, results with anther culture in these species are still poor. Albino haploid plantlets were first recovered from *Lolium multiflorum* L. by Clapham [5]. Other researchers report a very low frequency of green plant regeneration in this species [39,41,46]. The first results with anther culture of perennial ryegrass, *Lolium perenne* L., were reported by Zenkteler [66] and Niizeki [39] who obtained multicellular microspores and calluses, respectively. Microspore-derived albino plantlets of *L.perenne* were obtained by Stanis et al.[56]. Stanis and Butenko [54, 55] and Olesen et al.[43] published the successful regeneration of both green and albino plantlets in this species. Recently, Rose et al. [51] obtained albino plantlets from the annual species *L.temulentum* L. As to intergeneric *Lolium-Festuca* hybrids, albino and green plantlets have been obtained from anther culture of *Lolium multiflorum* x *Festuca arundinacea* [40], *L.multiflorum* x *F.pratensis* [41,51], *L.perenne* x *F.pratensis* and *L.temulentum* x *F.pratensis* [51].

A prerequisite for the application of anther culture to *Lolium* breeding programmes is the establishment of culture protocols which result in efficient haploid plant production in a wide range of commercial varieties. However, a general, simple, reproducible method applicable to a wide range of varieties of subspecies in *Lolium* is still lacking. Current limitations to the use of anther culture as a method for the production of homozygous lines in *Lolium* are, mainly, the low frequency of embryoid or callus formation from cultured anthers and the difficulties associated with regeneration of green plants. The present study was conducted with *L.multiflorum* and *L.perenne* and was designed to evaluate the applicability of anther culture to these species.

MATERIALS AND METHODS

PRECULTURE CONDITIONS

Diploid and tetraploid *Lolium perenne* and *L.multiflorum* plants were obtained from Barenbrug Holland BV. The plants were grown in a greenhouse with supplemental lighting (24h photoperiod and a minimum temperature of 16°C) or in a growth chamber (24h photoperiod, 15°C). Plants were induced to flower by artificial vernalization of seedlings as described by Buhring and Neubert [3]. Spikes were harvested and test-anthers were removed from spikelets in the middle region of each spike. The developmental stage of the microspores was determined by means of acetocarmine staining [57,60]. Spikelets with anthers having unicellular microspores were surface sterilized in 1% (w/v) sodium hypochlorite for 5min and rinsed three times with sterile water. After sterilization, the spikelets were incubated at 4°C for three weeks (*L.multiflorum*) or two weeks (*L.perenne*) in darkness in two compartment Petri dishes containing a drop of sterile water,which were sealed with Parafilm to maintain humidity. The optimum duration of cold pretreatment was determined in a specific experiment. Each plant was numbered and a record was maintained of the source of each spike harvested.

CULTURE CONDITIONS

A comparison of treatment effects can only be made between samples consisting of anthers from the same spike. As the three anthers in each floret are usually synchronized, six anthers out of the first and second floret of one spikelet were randomly divided between the different treatments in each experiment.

Anthers were cultured at a density of 25 anthers per 3ml aliquot of medium in plastic Petri dishes (Falcon 1006, 50x9mm) sealed with Parafilm. To prevent dessication, a group of 6 Petri dishes with anthers, plus one containing water, were placed inside a larger Petri dish (Falcon 1058, 150x15mm) which also was sealed with Parafilm. The anthers were incubated at 26°C in darkness in 2% CO_2 on modified Linsmaier and Skoog [33] medium (LS-1 medium, Table 1). The effect of elevated CO_2 concentration during the incubation was studied in a separate experiment. After 5 and 7 weeks of incubation, the percentage of anthers giving calluses and/or embryoids was recorded (anther response). To induce shoot regeneration, responding anthers were transferred to differentiation medium 190-2 [63; Table 1] and incubated at 26°C in light with a 16h photoperiod. After one month, the regeneration frequency (percentage anthers with calluses/embryoids forming shoots) and the culture efficiency (percentage anthers forming shoots) were recorded. Shoots were subsequently transferred to rooting medium (LS-6, Table 1) and incubated at 26°C in light (16h photoperiod). Rooted plantlets were formed on this medium after a few weeks. Well established green plants were transferred to soil.

The media were autoclaved (110°C for 30min) after adjusting the pH to 5.8, except for the vitamins, casein hydrolysate, glycine and hormones which were added as filter sterilized stocks (pH 5.4). Medium modifications include different types of nitrogen sources (ammonium nitrate, potassium nitrate and glutamine), carbon sources (sucrose, maltose, glucose, raffinose, trehalose, cellobiose, melizitose, melibiose and maltotriose), and growth regulators (2,4-D [2,4-dichlorophenoxy acetic acid], IAA [indoleacetic acid], NAA [naphthaleneacetic acid], kinetine and BAP [benzylaminopurine]).

TABLE 1: Composition of induction media (LS-1,2,3,4), regeneration media (LS-5 and 190-2), and rooting medium (LS-6) for *Lolium* anther culture (concentrations in $mg.l^{-1}$).

component	induction				regeneration		rooting
	LS-1	LS-2	LS-3	LS-4	LS-5	190-2	LS-6
KNO_3	1900	1900	1900	var.	1900	1267	950
NH_4NO_3	1650	165	165	-	1650	1100	825
$(NH_4)_2SO_4$	-	-	-	132	-	-	-
glutamine	-	1000	-	1000	-	-	-
$MgSO_4.7H_2O$	370	370	370	370	370	247	185
$CaCl_2\cdot 2H_2O$	440	440	440	440	440	293	220
KH_2PO_4	170	170	170	170	170	113	85
$MnSO_4\cdot 4H_2O$	22.3	22.3	22.3	22.3	22.3	14.9	11.2
KI	0.8	0.8	0.8	0.8	0.8	0.6	0.4
H_3BO_3	6.2	6.2	6.2	6.2	6.2	4.1	3.1
$ZnSO_4\cdot 7H_2O$	8.6	8.6	8.6	8.6	8.6	5.7	4.3
$CuSO_4\cdot 5H_2O$	0.03	0.03	0.03	0.03	0.03	0.02	0.01
$Na_2MoO_42H_2O$	0.25	0.25	0.25	0.25	0.25	0.17	0.12
$CoCl_2\cdot 6H_2O$	0.03	0.03	0.03	0.03	0.03	0.02	0.01
$FeSO_4\cdot 7H_2O$	27.9	27.9	27.9	27.9	27.9	18.6	13.9
$Na_2\cdot$EDTA	37.3	37.3	37.3	37.3	37.3	24.8	18.6
glycine	20	20	20	20	20	1.2	20
nicotinic acid	5	5	5	5	5	0.3	5
pyridoxine·HCL	5	5	5	5	5	0.3	5
thiamine·HCL	1	1	1	1	1	0.06	1
myo-inositol	100	100	100	100	100	60	100
agarose	6000	6000	6000	6000	6000	6000	6000
cas.hydrol.	-	-	-	-	500	1000	-
sucrose	90000	90000	90000	90000	20000	14000	20000
2,4-D	1.5	1.5	1.5	1.5	-	-	-
IAA	-	-	-	-	0.4	1.2	-
kinetine	-	-	-	-	-	0.2	-
BAP	-	-	-	-	0.4	-	-

ANALYSIS

The ploidy level of the plantlets was determined by flow cytometric analysis, performed by Mr. G. Geenen, Plant Cytometry Services, Schijndel, the Netherlands.
Isozyme analyses were performed with polyacrylamide gel electrophoresis (gel density 12%) for phosphoglucoisomerase (PGI).

RESULTS

EFFECT OF GROWTH CONDITIONS OF ANTHER-DONOR PLANTS

Several factors associated with the donor plants are known to affect the production of calluses/embryoids and plants from cultured anthers: plant genotype, physiological state of the anther-donor plants, plant age. Studies on barley [15], oilseed rape [61], turnip [29] and wheat [32] have shown that the temperature at which the donor plants are grown can markedly affect the culture response. The growing conditions for the *Lolium perenne* anther-donor plants have also some effect on the results obtained in anther culture, judging from the fact that field-grown material responded about twice as well as greenhouse-grown material [43]. However, it is not possible to make general recommendations about optimum growth-conditions since they seem to vary between species.
To investigate the influence of growth temperature of anther-donor plants, four *L.multiflorum* plants (four genotypes of variety Lm-1) were vegetatively multiplied *in vitro*. During spring of 1988 clones of the same genotype were grown both in a greenhouse with supplemental lighting (24h photoperiod, temperature between 16 and 35°C depending on weather conditions) or in a growth chamber (24h photoperiod, 15°C). Although spikes and anthers from growth-chamber grown donor plants looked better and were bigger, the anther response, the culture efficiency and the ratio of green to albino plants were somewhat higher for the anthers coming from the greenhouse-grown donor plants (Table 2). Plant age was also an important factor because anther-response was greater at the beginning of the flowering period than towards the end.

EFFECT OF COLD PRETREATMENT

It has been known for some time that the number of embryoids produced by anther cultures can be increased if the anthers are exposed to a shock before cultivation. The method most frequently used is a cold pretreatment of the spikes before plating the anthers, which has been found to increase the yields substantially in many species. Cold pretreatment resulted in a higher frequency of embryoid production in cereals such as barley [20,47], rice [16], maize [62], wheat [17] and grasses [27,46,51,54]. The optimum temperature and duration of pretreatment vary with species. In the case of *Lolium* a single most effective procedure has not been published yet. Both high and low temperatures have been used [54].

TABLE 2: Influence of greenhouse-grown and growth-chamber grown *L.multiflorum* donor plants on anther culture.

Growth condition	number of anthers tested	anther response (%)	regeneration frequency (%)	culture efficiency (%)	anthers with green shoots
greenhouse					
genotype-5	475	32	19	6	8
genotype-7	550	28	10	3	0
genotype-10	75	3	0	0	0
genotype-11	650	12	8	1	0
total	1750	22	13	3	8
growth chamber					
genotype-5	750	18	15	3	2
genotype-7	1100	21	8	2	1
genotype-10	400	12	11	1	0
genotype-11	1075	11	14	1	0
total	3325	16	12	2	3

To investigate this phenomenon for *L.perenne* and *L.multiflorum* anther culture, spikes were for 0, 1, 2, 3, or 4 weeks kept at 4°C in darkness before plating on LS-1 induction medium supplemented with $2mg.l^{-1}$ NAA and 12% sucrose (instead of 2,4-D and 9% sucrose). 1200 *L.perenne* and 4000 *L.multiflorum* anthers from different varieties were used in this experiment. Comparisons of pretreatment effects were made between samples consisting of anthers harvested simultaneously from the same anther-donor plant (same genotype). Results from all genotypes of one variety have been pooled.

A cold pretreatment period significantly increased the anther response in all varieties tested. For *L.perenne* variety Lp-1, anther response increased with increased duration of spikelet pretreatment at 4°C up to two weeks and then declined. For *L.multiflorum* variety Lm-1, there was an optimum anther response at a cold pretreatment of 3 weeks (Fig 1). *L.multiflorum* variety Lm-2 showed a trend similar to that of Lm-1 (Results not shown). Clearly, pretreatment of spikelets at 4°C has a beneficial effect on the number of anthers responding.

EFFECT OF ELEVATED CO_2 CONCENTRATION

The gaseous environment is a factor which has seldom been investigated in anther culture. Horner et al.[18] reported on ethylene production in anther cultures of *Nicotiana tabacum* and Dunwell [10] showed that the composition of the gas mixture that surrounds the anthers had a

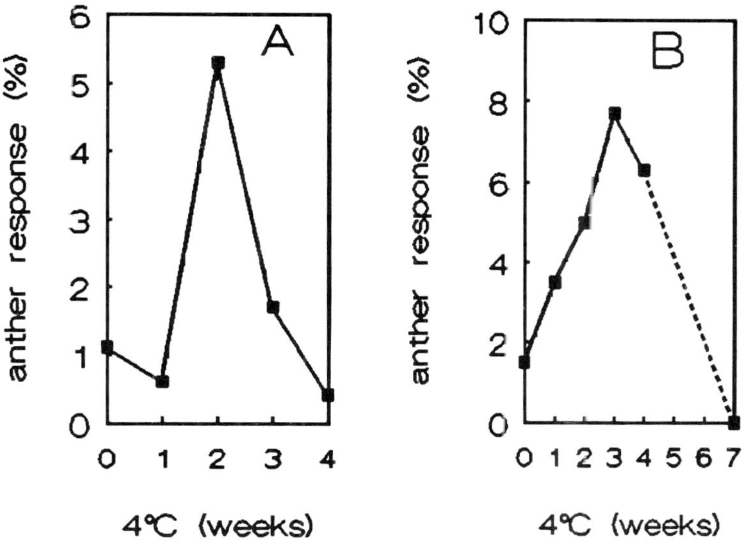

FIGURE 1: Influence of duration of pretreatment of spikelets at 4°C on anther response for the *L.perenne Lp-1* (A) and *L.multiflorum Lm-1* (B) varieties.

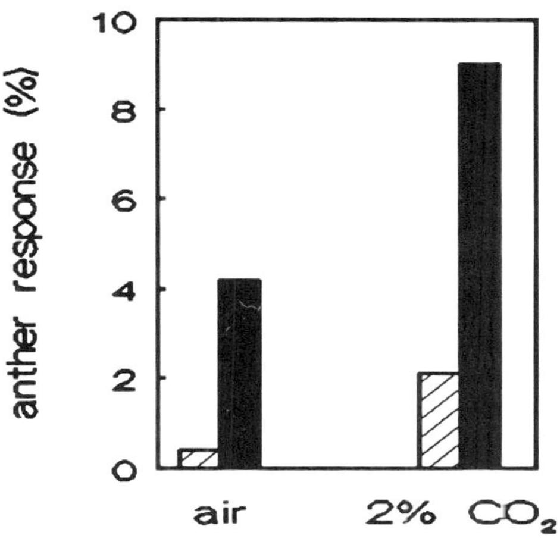

FIGURE 2: Influence of the gaseous environment during callus/embryoid induction on the average anther response of *L.perenne* (▨) and *L.multiflorum* (■).

A : atmospheric CO_2 concentration

B : 2% CO_2 concentration

great influence on the number of embryoids produced in anther cultures of this species. The removal of CO_2 from the culture atmosphere through KOH absorption leads to a decrease in anther response, but anther treatment in reduced atmospheric pressure or in a nitrogen atmosphere with or without oxygen [22] allowed better yields of embryoids in anther culture of the same species. Incubation of anthers at elevated CO_2 concentrations has a positive effect on embryogenesis in some species of the genera *Papaver*, *Anemone*, *Clematis* and *Nicotiana* [23,24, 25]. However, a negative effect due to an elevated CO_2 concentration has been found for *Zea mays* [8].

In order to investigate possible effects for *Lolium*, 3000 *L.perenne* and 3000 *L.multiflorum* anthers were plated on LS-1 induction medium and incubated at atmospheric or 2% CO_2 concentrations. Results from 5 *L.perenne* varieties (Lp-1, Lp-2, Lp-7, Lp-8, Lp-9) and 2 *L.multiflorum* varieties (Lm-1, Lm-3) have been pooled because they all showed a similar trend. Incubation of anthers at elevated CO_2 concentrations increases the anther response for both *L.perenne* and *L.multiflorum* (Fig 2). Incubation at 2% CO_2 instead of atmospheric CO_2 concentration had no significant effect on pH in the culture medium. After 7 weeks of incubation of *Lolium* anthers, the mean pH varied between 5.5 (atmospheric CO_2) and 5.7 (2% CO_2).

EFFECT OF GROWTH REGULATORS

The role of growth regulators in androgenesis and regeneration is not clear and contradictory results are often reported. In cereal anther culture, both auxins and cytokinins have been used routinely. It has been claimed that addition of auxins to the culture medium can have both positive [50] and negative [51] effects on the anther response of *Sorghum bicolor* and *L.temulentum*, respectively.
The effects of different hormones on *Lolium* anther culture were investigated. 2300 *L.perenne* and 3800 *L.multiflorum* anthers were plated on LS-1 induction medium supplemented with 1.5 $mg.l^{-1}$ 2,4-D, or 1.0 $mg.l^{-1}$ BAP, or 1.5 $mg.l^{-1}$ 2,4-D plus 0.5 $mg.l^{-1}$ kinetine. Embryoid formation was not influenced by the type of growth regulators used. Calluses and embryoids were formed after plating the anthers on induction medium supplemented with all three hormone combinations tested. However, the medium with 1.5 $mg.l^{-1}$ 2,4-D was the most suitable for anther response (Results not shown).

To determine the optimum auxin concentration, anthers were plated on LS-1 induction medium supplemented with 2,4-D in different concentrations. 600 *L.perenne* anthers (varieties Lp-1, Lp-2 and Lp-3) and 300 *L.multiflorum* anthers (variety Lm-3) were tested for each concentration. The curves in Fig 3 show that a hormone supplement in the LS induction medium has little influence on anther response. Callus and embryoid formation on LS-1 induction medium is possible for both *L.perenne* and *L.multiflorum* without a hormone in the medium, although a small amount seems to have a positive effect on the anther response of *L.perenne*.

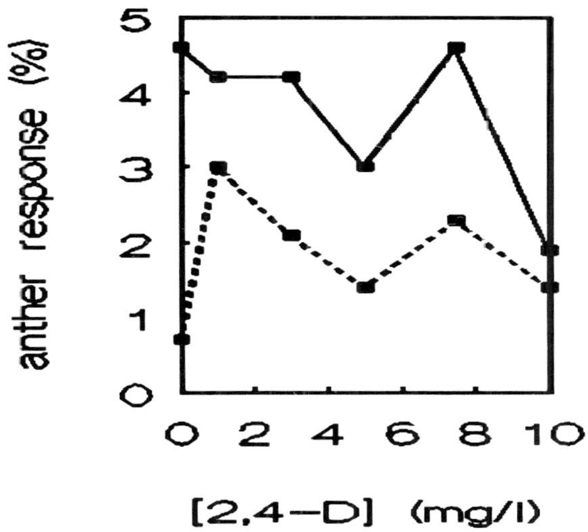

FIGURE 3: Influence of the 2,4-D concentration on the mean anther response of *L.perenne* (----) and *L.multiflorum* (——).

INFLUENCE OF THE NITROGEN SOURCE

Nitrogen in the medium is an important factor in anther culture of *Gramineae* species. Different types and amounts of nitrogen clearly influence the growth and differentiation of the cells *in vitro*. Clapham [6] and Chu et al.[4] found that high concentrations of ammonium ions could inhibit formation of microspore-derived calluses and embryoids in barley and rice anther culture. In maize, the ammonium concentration affects the embryoid induction from microspores *in vitro* too [14]. Henry and DeBuyser [17] have shown that glutamine can replace potato-extract in wheat anther culture. Olsen [44] reported improvement of the yield of green barley plants as a result of a decrease of the concentration of ammonium nitrate in the induction medium and the use of glutamine as a non-toxic nitrogen source.

Based on these results, a modified LS induction medium with a low ammonium nitrate concentration (2mM instead of the 20mM originally used in the LS medium), but supplemented with 5 mM glutamine (LS-2, Table 1) was tested. Results from 3 *L.perenne* (Lp-1, Lp-2 and Lp-4) and 2 *L.multiflorum* (Lm-1 and Lm-3) varieties have been pooled because they all showed a similar trend; 800 *L.perenne* and 1500 *L.multiflorum* anthers were tested for both LS-1 and LS-2 medium. Use of LS-2 induction medium led to a significant increase in *Lolium* anther response, regeneration frequency and culture efficiency (Fig 4). Callus growth on LS-2 induction medium was much better than on LS-1 induction medium. The mean size of microspore-derived calluses and embryos after

5 weeks of growth on induction medium was 1.5mm for LS-1 medium and 2.4mm for LS-2 medium.

Use of LS-2 induction medium containing 2mM ammonium nitrate, 18mM potassium nitrate and 5mM glutamine resulted in the same anther response and a higher culture efficiency for *L.multiflorum* compared with medium without glutamine (LS-3). Glutamine in the LS-induction medium appeared to have a negative effect on the anther response and the subsequent regeneration of *L.perenne* (Fig 5). It can be concluded that the yield improvement shown in Fig 4 was due to the reduction of the ammonium nitrate concentration and not to the addition of glutamine in the induction medium. During this experiment, 800 *L.perenne* anthers (varieties Lp-1 and Lp-5) and 1400 *L.multiflorum* anthers (varieties Lm-1, Lm-2, Lm-3 and Lm-4) were tested for both media.

Both nitrate and ammonium are currently used as nitrogen sources for *Gramineae* anther culture. It has been reported that the potassium nitrate concentration in the induction medium significantly affects the anther culture of wheat [14,36] and rice [4]. To investigate the influence of potassium nitrate for *Lolium* anther culture, additions of different KNO_3 concentrations to LS-4 induction medium (Table 1) containing ammonium sulphate (1mM) and glutamine (5mM) were tested. 750 *L.perenne* anthers (varieties Lp-1, Lp-5 and Lp-6) and 1200 *L.multiflorum* anthers (varieties Lm-1, Lm-3, Lm-4 and Lm-5) were tested for each KNO_3 concentration.

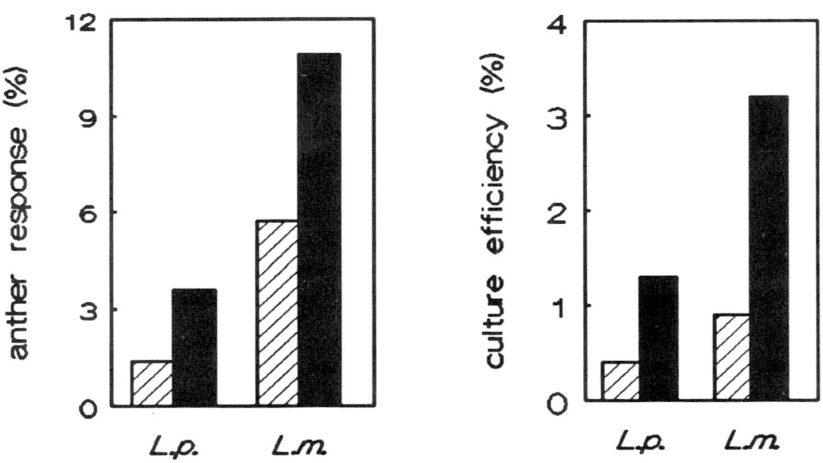

FIGURE 4: Effect of reduced ammonium nitrate concentration and glutamine addition in LS induction medium on *Lolium* anther response and culture efficiency.

▨ LS-1 : 20mM NH_4NO_3, no glutamine , 18mM KNO_3

■ LS-2 : 2mM NH_4NO_3, 5mM glutamine, 18mM KNO_3

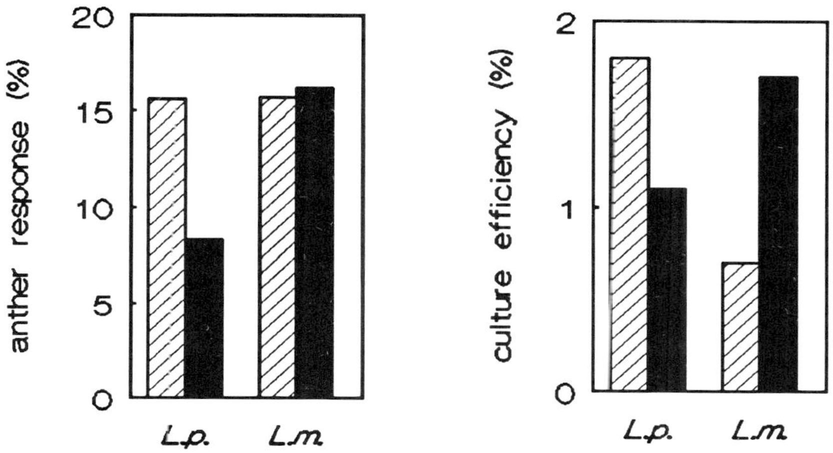

FIGURE 5 : Effect of glutamine addition to LS induction medium on *Lolium* anther response and culture efficiency.
LS-3 : 2mM NH_4NO_3 , no glutamine, 18mM KNO_3
LS-2 : 2mM NH_4NO_3 , 5mM glutamine, 18mM KNO_3

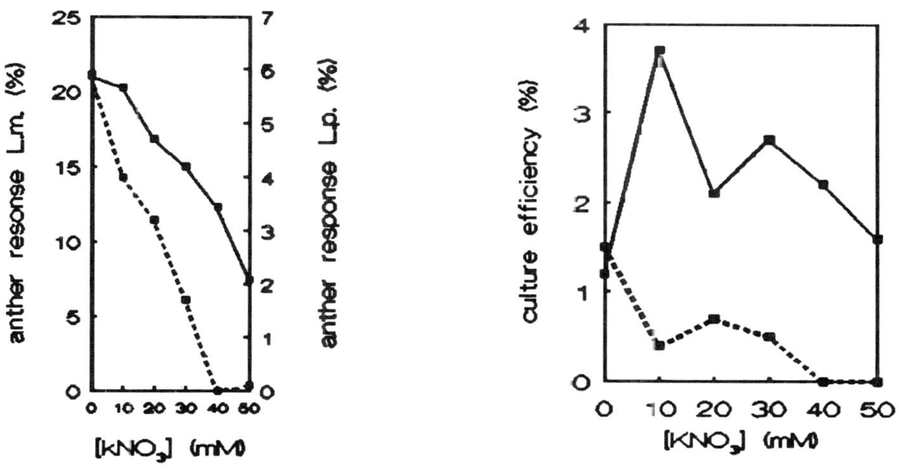

FIGURE 6: Effect of KNO_3 concentration in LS induction medium (with 1mM $(NH_4)_2SO_4$ and 5mM glutamine) on *L.perenne* (*L.p.*- - -) and *L.multiflorum* (*L.m.* ——) anther response and culture efficiency.

For *L.perenne*, any addition of KNO_3 decreased the anther response and culture efficiency. For *L.multiflorum*, addition of KNO_3 decreased the anther response, but the influence of the potassium nitrate concentration in the induction medium upon the culture efficiency was not clear. A low KNO_3 addition seems to be useful (Fig 6).

INFLUENCE OF CARBON SOURCE

The media on which plant tissue cultures are grown contain a carbohydrate component as a major nutrient. The preferred carbohydrate for anther culture of cereals has been sucrose, which acts as a carbon and energy source and as an osmoticum. For wheat and barley, 9% (w/v) sucrose is appropriate for anther culture [57], and for maize 6 to 9% [62]. According to Clapham [5,6] and Niizeki [39], *Lolium* anther culture required 12% sucrose. Olesen [42] reported an optimum sucrose content of 9% for *L.perenne*. Our results showed that 6, 9 and 12% sucrose can be used for *L.multiflorum* and *L.perenne* anther culture, but use of 9% sucrose resulted in the best anther response (Results not shown).

A number of studies have been carried out to quantify sucrose requirement in anther culture of various species, but only a few attempts have been made to replace it by other suitable carbohydrates. Oono [45] showed that microspore-derived rice calluses could be obtained by using glucose, maltose, lactose, fructose and galactose as carbon source. Misoo and Matsubayashi [37] obtained microspore-derived tobacco plantlets on fructose. Maltose, glucose, lactose and fructose have been used successfully in anther culture of *Petunia* [48]. Also Keller et al.[28], Babbar and Gupta [2] and Hunter [21] reported about replacement of sucrose by other carbohydrates.

To investigate the effects of various carbon sources on *Lolium* anther culture, different carbohydrates were incorporated at a concentration of 0.26M into LS-2 induction medium instead of sucrose. A comparison was made for *L.perenne* and *L.multiflorum* anther culture using sucrose, maltose, glucose, melibiose, trehalose, raffinose, cellobiose, melizitose and maltotriose as carbon source. The results demonstrated the superiority of maltose, trehalose and maltotriose over sucrose for anther response. Glucose, melibiose, raffinose, cellobiose and melizitose were very poor carbon sources for *Lolium* anther culture at the concentrations tested (Results not shown). Green plants were obtained in comparable numbers with the carbon sources sucrose, maltose, trehalose, melizitose and maltotriose in the induction medium.

A high frequency of anther response does not directly correlate with a high regeneration frequency. Anther incubation conditions can have a strong effect on the subsequent differentiation process. Therefore, in another experiment the effect of different carbon sources in the induction medium upon the subsequent regeneration (regeneration medium 190-2 with sucrose as carbon source) was investigated. Maltose, maltotriose and trehalose appeared not only to increase the anther response

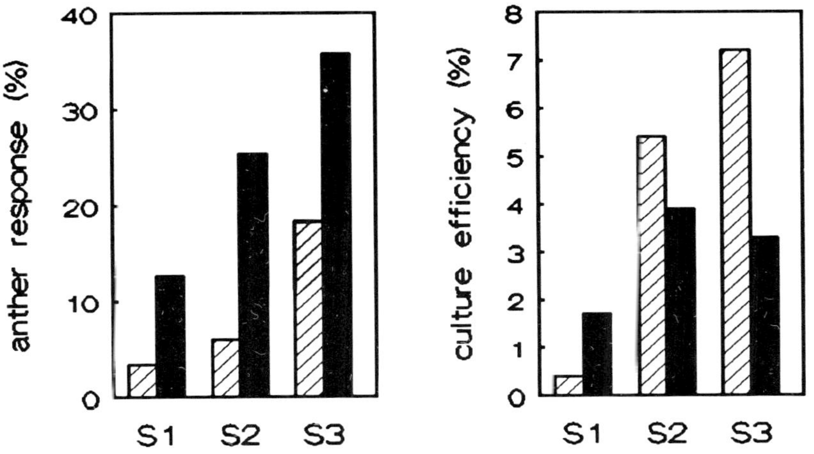

FIGURE 7: Influence of sucrose (S1), maltose (S2), and maltotriose (S3) in the induction medium on *L.perenne* (▨) and *L.multiflorum* (■) anther response and culture efficiency.

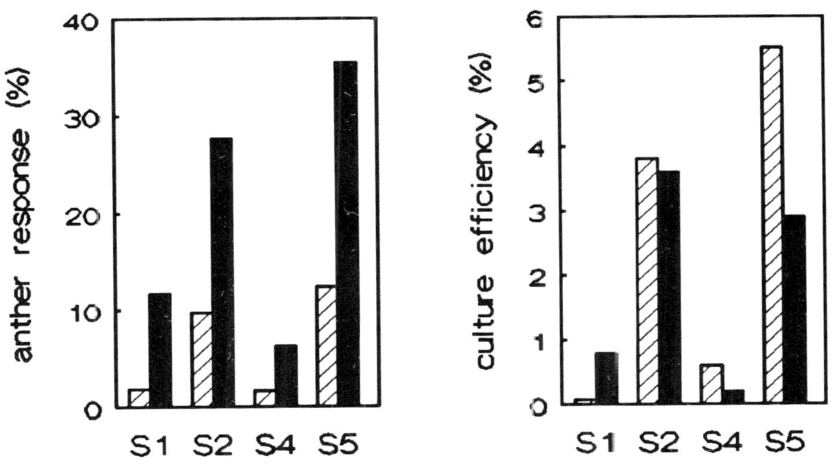

FIGURE 8: Influence of sucrose (S1), maltose (S2), glucose (S4) and trehalose (S5) in the induction medium on *L.perenne* (▨) and *L.multiflorum* (■) anther response and culture efficiency.

but also the overall culture efficiency for both *L.multiflorum* and *L.perenne* (Fig 7 and 8). For each type of sugar, 675 (Fig 7) / 525 (Fig 8) *L.perenne* anthers and 1250 (Fig 7) / 600 (Fig 8) *L.multiflorum* anthers were tested. The number of regenerated green plants was low. With regard to the number of responding anthers as well as the yield of plantlets, the carbohydrates maltose, trehalose and maltotriose in the induction medium appeared to be particularly effective in *Lolium* anther culture.

REGENERATION AND PLOIDY ANALYSIS

Shoot regeneration was achieved by transferring the responding anthers to the regeneration media LS-5 and 190-2 (Table 1). After transfer to the regeneration media, calluses/embryoids either failed to grow, proliferated as callus or formed shoots. For *L.multiflorum*, shoots were produced on both regeneration media tested, but is was not possible to regenerate *L.perenne* shoots on LS-5 medium. Well growing *L.perenne* and *L.multiflorum* shoots were formed on the 190-2 regeneration medium only. Albinos as well as green plants were produced out of *L.perenne* and *L.multiflorum* microspores. Albino plantlets were obtained from almost all *Lolium* varieties tested, and green plants were obtained from 4 different *L.perenne* (Lp-1 [2n], Lp-3 [2n], Lp-6 [4n], Lp-10 [4n]) and 5 different *L.multiflorum* (Lm-1 [4n], Lm-2 [2n], Lm-4 [4n], Lm-5 [4n], Lm-6 [4n]) varieties. Some genotypes yielded many green plants, while others only produced albino plantlets. There is a clear genotype effect on anther culture response and on the ratio of green to albino regenerants.

TABLE 3: Ploidy level of microspore-derived *Lolium* plantlets, determined by flow cytometric analysis.
A : one green plant
B : six green plants
nd : not detected

donor plant	plantlets tested	ploidy level n	2n	4n	5n	6n	8n
L.perenne							
2n	44	12	24	7	nd	1	nd
4n	40	nd	14A	24A	1	nd	1
L.multiflorum							
2n	67	6	49A	12	nd	nd	nd
4n	89	nd	42B	44A	nd	nd	4A

The ploidy level of 234 anther derived *Lolium* plantlets was determined by flow cytometric analysis. 30% of the tested plantlets had the gametes' ploidy level. The others mainly consisted of diploid and tetraploid plantlets, but one pentaploid, one hexaploid, five octaploid, and a few aneuploid plantlets were found, too (Table 3). Two mixaploid plantlets were obtained, which showed the chromosome counts corresponding to the haploid and diploid levels. About 70% of the plantlets derived from diploid anther-donor plants were diploid. However, their microspore origin was confirmed by PGI-2 isozyme analysis. This indicated that spontaneous diploidization took place during callus growth or subsequent plantlet development. Intact green plants of both *L.perenne* and *L.multiflorum* were acclimated to greenhouse conditions with a 100 % success rate. After transfer to the soil, the microspore-derived green plants grew vigorously under field conditions.

DISCUSSION

The use of homozygous doubled haploid plants obtained from anther culture for crop breeding is a powerful approach when integrated with conventional breeding. However, grass species have generally been difficult to manipulate *in vitro*. The present work indicates the possibility of obtaining microspore-derived plants of ryegrass in anther culture. Two pathways of appearance of these plantlets were observed: formation of embryoids from individual microspores, which develop into plants, and callus formation from the pollen with subsequent regeneration of sporophytes from this callus. The embryoid route is a direct one and the probability of occurence of chromosomal aberrations during the callus phase is lower. Further efforts are needed to eliminate callus phase and encourage the formation of embryoids directly from microspores.

Several physical and chemical factors affect *Lolium* anther response and culture efficiency. Physiological states of the donor plants affected by photoperiod, light intensity and temperature have been reported to influence subsequent anther culture efficiency [9,58]. The temperature under which the donor plants are grown can markedly affect the culture response [15,28,32,58]. Our results show that anthers derived from donor plants grown in a greenhouse with temperature fluctuations between 16 and 35°C result in a higher anther culture efficiency than anthers derived from growth-chamber grown donor plants (constant temperature of 15°C). Olesen et al.[43] reported that donor plants grown in the field responded better than greenhouse-grown material. Both these and our observations are consistent with the view that nutritional [58] and environmental stress could lead to relaxation of the gametophytic control system, which renders the pollen more susceptible to culture. If the crucial environmental factors can be identified, better growing conditions for the donor plants might be designed for material grown in greenhouse or growth chamber.

A temperature-stress pretreatment is a method to enhance the production of microspore-derived plants from cultured anthers. In many grass species it has been found that anther culture efficiency can be increased by pretreatment of the spikes prior to anther incubation on induction medium. Pagniez and Demarly [46] used pretreatments between 4 and 10 days at 3°C for *L.multiflorum*, while Kasperbauer et al. [27] used pretreatment times up to 25 days at 5°C with *Festuca arundinacea*. Stanis and Butenko [54] reported that pretreatments of both 18 days at 4°C and 4 days at 38°C pretreatment stimulated the yield of *L.perenne* callus compared with an unpretreated control. Rose et al. [51] found maximum increase of *L.temulentum* anther response after pretreatments of 28 days at 7°C or 35 days at 4°C. So, the optimum temperature and duration of pretreatment vary with species. In the case of *L.perenne* and *L.multiflorum*, two to three weeks pretreatment of the spikelets at 4°C had a beneficial effect on the anther response.

Huang and Sunderland [20] reported that the regeneration potential of barley microspore-derived calluses was markedly enhanced by a pretreatment. Maximal regeneration occurred after several weeks of pretreatment at 4°C. Most of the calluses obtained without pretreatment failed to grow when transferred to the regeneration medium, while calluses obtained after pretreatment of several weeks at 4°C gave rise to plantlets. The ratio of green to albino plantlets was also better after a pretreatment. An explanation for the effects of this "shock" pretreatment can be that a number of pollen grains in this way are forced to deviate from the normal path of development. Sunderland [58, 59,60] suggested that the switch from gametophytic to sporophytic development takes place during this stress pretreatment. This switch can be a result of the reduction in the level of endogenous hormones during the cold pretreatment [23,25]. The mechanisms attributed to the effects of cold pretreatment in barley include slow degradation of the tapetum and locular matrix. One effect of cold pretreatment in species like *Nicotiana tabacum* is faster anther senescence. At low temperature, more microspores are able to complete the sequence of events that lead to the first embryogenetic division. Embryos derived from chilled material are more advanced and divisions start sooner [59].

Very few experiments have been carried out to study the influence of the gas mixture surrounding the cultures. The culture atmosphere has proved to be very important in the induction of androgenesis [10,34]. Ethylene is released and accumulated in the culture atmosphere [18]. Removal of the ethylene by flushing [18] or by absorbants [10,49] influenced the anther response. Dunwell suggested that the presence of ethylene and the absence of CO_2 may inhibit embryogenesis. CO_2 competitively inhibits the action of ethylene [26]. Application of elevated CO_2 concentrations during anther incubation increased anther response for several species [23,24,25], including *L.perenne* and *L.multiflorum* in our experiments (Fig 2). No explanation is available as to the effects of the CO_2 treatment. It is possible that CO_2 inhibits ethylene induced senescence in anthers. Another possible explanation is that elevated CO_2 concentrations promote dark-fixation of CO_2, enhancing development of embryogenic microspores [23]. Johansson found a synergistic effect of CO_2 with cold pretreatment. A combination of cold pretreatment of the spikelets and subsequent incubation of the anthers in elevated CO_2 concentrations was beneficial for the production of embryos in anther culture of some species; for other species cold pretreatment sometimes partly inhibits the effects of CO_2 [25].

The type and the amount of nitrogen may be critical for growth and morphogenesis of plant cells. The pattern of cellular development has been shown to change by the ratio and concentration of ammonium and nitrate supplements [35]. Using *N.tabacum*, Kyo and Harada [31] showed that pollen grains can be directed to either the gametophytic or the sporophytic pathway by regulating their culture conditions, particularly the supply of glutamine. Addition of 5mM glutamine to the induction medium increased the number of embryoids and the overall yield of green barley plants [44]. Henry and DeBuyser [17] have shown that the optimum glutamine concentration in liquid medium for barley anther culture was 3.4mM, whereas concentrations of 6.8 and 13.6mM glutamine resulted in a reduction of the anther response. However, Xu and Sunderland [65] reported that glutamine at 1.1 or 5.5mM was either insufficient or inhibitory for barley pollen callus production. Our results show that 5mM glutamine in *L.perenne* anther culture was inhibitory for anther culture efficiency but increased the anther culture efficiency for *L.multiflorum*.

Both nitrate and ammonium are currently used as nitrogen sources for *Gramineae* anther culture. The development of rice microspore-derived calluses requires both ammonium and nitrate, but a high ammonium concentration would inhibit the formation and growth of microspore-derived calluses [4]. Clapham [6], Chu et al.[4], and Olsen [44] found that a high concentration of ammonium ions could inhibit anther response for barley and rice. For *Lolium* anther culture, reduction of the ammonium nitrate concentration resulted in an increase in culture efficiency, too. The ammonium concentration in the induction medium influenced not only the anther response, but also the subsequent differentiation. The regeneration ability can be influenced during the induction process! The nitrate concentration was found to affect the anther culture of wheat [36]. Calluses from the induction medium with higher potassium nitrate concentrations produced more green plantlets than those from medium with lower potassium nitrate concentrations. The albino plantlet regeneration frequency was only slightly affected by potassium nitrate concentration. However, for *Lolium* there was no effect of the potassium nitrate concentration upon the yield of green plants, but there was a clear effect upon the production of albino plantlets (Fig 6).

Sucrose is the most commonly used carbohydrate in plant tissue culture media, acting as an energy source and as an osmoticum. Most investigators agree that sucrose provides the best carbohydrate source for a wide range of isolated plant tissues. Sharp et al.[52] and Clapham [6] showed that there was an apparent correlation between the concentration of sucrose in the culture medium and the ratio of green to albino tobacco and barley plantlets obtained by anther culture. When the anthers were incubated on a medium with a high concentration of sucrose, a high frequency of albino plantlets was observed. Chlorophyll synthesis in carrot-callus is suppressed specifically by sucrose but not by glucose or fructose. The "sucrose effect" on greening is due to a reduction of both chloroplast complexity and chloroplast number per cell [13]. A sucrose concentration of at least 9% in the induction medium is required for efficient *Lolium* anther culture [42; our results] but probably increases the ratio of albino to green regenerants. So, replacement of sucrose by another carbohydrate is a possible way of overcoming this "sucrose effect".

In plant tissue and cell cultures, sucrose can often be replaced by other carbohydrates such as glucose, fructose, maltose, lactose, mannose, cellobiose, trehalose, raffinose and galactose [53]. Kinnersly [30] reported improved cell differentiation achieved by culturing wild carrot and tobacco tissues in a medium wherein the carbohydrates comprise a mixture of maltose and glucose. Only a few attempts have been made to replace sucrose in anther culture by other suitable carbohydrates [2,21,28,37,45,48]. Raquin [48] demonstrated the higher effectiveness of maltose in comparison with sucrose and glucose. Hunter [21] found that maltose, trehalose (an isomer of maltose) and cellobiose were particularly effective in barley anther culture. Our results demonstrated the superiority of maltose, trehalose and maltotriose over sucrose, glucose, melibiose, raffinose, cellobiose and melizitose in *Lolium* anther culture. However, there was no effect of the carbohydrate used in the induction medium on the ratio of green to albino plantlets formed. To make a valid comparison of the effectiveness of different sugars, it is necessary to specify the optimum dose for each of these sugars and to test the effects of replacement of sucrose by these sugars in the regeneration medium, too.

One of the major problems to be overcome before the anther culture technique can be widely applied in *Lolium* breeding programmes is the occurrence of large numbers of albino regenerants. Many factors have been found to affect the degree of albinism, such as development stage of microspores, culture temperature [19], cold pretreatment [20], sucrose concentration [6,52], and nitrogen source [14]. The molecular basis of albinism is not known. The chloroplast genomes in albino wheat and barley plantlets show large deletions [7]. Genetic factors influence the occurrence of albinism in grasses [51]. Also the anther response is determined to a large extent by the genotype of the donor plants [32,43]. Two of the basic components of culture efficiency, the anther response and the regeneration frequency, appear to be genetically determined independently. A genetic approach to the problem of how to increase the anther culture efficiency was applied with several plant species where responsive genotypes were crossed with recalcitrant types to produce lines with higher androgenetic capabilities [64]. *Lolium* genotypes vary in anther culturability. This problem of genotypic variability is a major factor limiting a wide application of anther culture technology to crop improvement. The successful utilisation of anther culture in crop improvement is dependent upon the development of culture protocols which are applicable to a wide range of commercial available genotypes. It is demonstrated that, using the present technique, green plants can be obtained from microspores of *L.multiflorum* and *L.perenne*. It is now important to refine these techniques to improve the output. Anther culture can improve the breeding efficiency, as it would greatly reduce the time required for the production of inbreds. Homozygous plants, which express both dominant and recessive traits, can be produced quickly in only one single generation. The haploid plants offer a unique possibility for gamete selection. Chromosome doubled haploids might be used as inbred lines for the production of hybrid varieties utilizing genetic self incompatibility already inherent in the species. Selection for self fertility, as would occur with conventional methods of inbreeding, is not likely to occur during the anther culture technique.

ACKNOWLEDGEMENTS

This project is part of a research co-operation with Barenbrug Holland B.V., Duitse Kampweg 60, 6874 BX Wolfheze, the Netherlands.

We are grateful to Dr.H.J.Huizing (Foundation for Agricultural Plant Breeding, Wageningen, the Netherlands) for stimulating discussions.

REFERENCES

1. Ahloowalia BS (1984) "Forage grasses", in Ammirato PV, Evans DA, Sharp WR, Yamada Y (eds.), Handbook of plant cell culture, Macmillan, New York **3**, pp. 91-125
2. Babbar SB and Gupta SC (1986) "Effect of carbon source on *Datura metel* microspore embryogenesis and the growth of callus raised from microspore-derived embryos", Biochem Physiol Pflanzen **181**, 331-338
3. Buhring J and Neubert K (1977) "Results regarding the accelerated production of complete generations of *Lolium perenne* L. and *Dactylis glomerata* L. under artificial climate", in Wo-Jah E and Thons E (eds.), Proceedings XIII Int Grassland Congress, Akademie Verlag, Berlin **1**, 407-409
4. Chu CC, Wang CC, Sun CS, Hsu C, Yin KC, Chu CY, Bi FY (1975) "Establishment of an efficient medium for anther culture of rice through comparative experiments on the nitrogen sources", Sci Sin **18**, 559-668
5. Clapham D (1971) "*In vitro* development of callus from the pollen of *Lolium* and *Hordeum*", Z Pflanzenzuchtg **65**, 285-292
6. Clapham D (1973) "Haploid *Hordeum* plants from anthers *in vitro*", Z Pflanzenzuchtg **69**, 142-155
7. Day A and Ellis THN (1985) "Deleted forms of plastid DNA in albino plants from cereal anther culture", Current Genet **9**, 671-678.
8. Dieu P and Beckert M (1986) "The effect of an elevated CO_2 concentration in combination with cold treatments in maize (*Zea mays* L.)

anther culture", in Horn et al (ed.), Genetic manipulation in plant breeding, Walter de Gruyter & Co., Berlin, pp. 291-293

9. Dunwell JM (1976) "A comparative study of environmental and developmental factors which influence embryo induction and growth in cultured anthers of *Nicotiana tabacum*", Env Exp Bot **16**, 109-118

10. Dunwell JM (1979) "Anther culture in *Nicotiana tabacum*: the role of the culture vessel atmosphere in pollen embryo induction and growth", J Exp Bot **30**, 419-428.

11. Dunwell JM (1985) "Anther and ovary culture", in Bright SWJ and Jones MGK (eds.), Advances in agricultural biotechnology: Cereal tissue and cell culture, Martinus Nijhoff, Dordrecht, pp.1-44

12. Dunwell JM (1986) "Pollen, ovule and embryo culture as tools in plant breeding", in Withers LA and Anderson PG (eds.), Plant tissue culture and its agricultural applications, Butterworth, London, pp. 375-404.

13. Edelman J and Hanson AD (1972) "Sucrose suppression of chlorophyll synthesis in carrot-tissue cultures", J Exp Bot **23**, 469-478

14. Feng and Ouyang (1988) "The effects of KNO_3 concentration in callus induction medium for wheat anther culture", Plant cell, tissue and organ culture **12**, 3-12

15. Foroughi-Wher B, Mix G, Gaul H, Wilson HM (1976) "Plant production from cultured anthers of *Hordeum vulgare* L.", Z Pflanzenzuchtg **77**, 198-204

16. Genovesi AD and Magill CW (1979) " Improved rate of callus and green plant production from rice anther culture following cold shock", Crop Sci **19**, 662-664

17. Henry Y and DeBuyser J (1981) "Float culture of wheat anthers", Theor Appl Genet **60**, 77-79

18. Horner M, McComb JA, McComb AJ, Street HE (1977) "Ethylene production and plantlet formation by *Nicotiana tabacum* anthers cultured in the presence and absence of charcoal", J Exp Bot **28**, 1365-1372.

19. Huang H (1984) "The relative efficiency of microspore culture and chromosome elimination as methods of haploid production in *Hordeum vulgare* L.", Z Pflanzenzuchtg **92**, 22-29

20. Huang B and Sunderland N (1982) "Temperature-stress pretreatment in barley anther culture", Ann Bot **49**, 77-88

21. Hunter CP (1987) "Plant generation method", Shell Int Research Maatschappij B.V., Patent EP 0.245.898

22. Imamura J and Harada H (1981) "Stimulation of tobacco pollen embryogenesis by anaerobic treatments", Z Pflanzenphysiol **103**, 259-263

23. Johansson LB (1986) "Effects of activated charcoal, cold treatment and elevated CO_2-concentrations on embryogenesis in anther cultures", in Horn et al (eds.), Genetic manipulation in plant breeding, Walter de Gruyter & Co, Berlin, pp. 257-264

24. Johansson L and Eriksson T (1984) "Effects of carbon dioxide in anther cultures", Physiol Plant **60**, 26-30

25. Johansson L, Andersson B, Eriksson T (1982) "Improvement of anther culture technique: activated charcoal bound in agar medium in combination with liquid medium and elevated CO_2 concentration", Physiol Plant **54**, 24-30

26.Kang BG, Yocum CS, Burg SP, Ray PM (1967) "Ethylene and carbon dioxide: Mediation of hypocotyl hook-opening response", Science **156**, 958-959
27.Kasperbauer MJ, Buckner RC, Springer WD (1980) "Haploid plants by anther-panicle culture of tall fescue", Crop Sci **20**, 103-107
28.Keller WA, Rajhathy T, Lacapra J (1975) "*In vitro* production of plants from pollen in *Brassica campestris*", Can J Genet Cytol **17**, 655-666.
29.Keller WA, Armstrong KC, DeLaRoche AI (1983) "The production and utilization of microspore-derived haploids in *Brassica* crops", in Sen SK and Giles KL (eds.), Plant cell culture in crop improvement, Plenum Press, New York, pp. 169-183
30.Kinnersly AM (1987) "Method and composition for plant tissue and cell culture", CPC Int Inc Patent EP-0.249.772
31.Kyo M and Harada H (1986) "Control of the developmental pathway of tobacco pollen *in vitro*", Planta **168**, 427-432
32.Lazar MD, Schaeffer GW, Baenziger PS (1984) "Cultivar and cultivar x environment effects on the development of callus and polyhaploid plants from anther cultures of wheat", Theor Appl Genet **67**, 273-277
33.Linsmaier EM and Skoog F (1965) "Organic growth factor requirements of tobacco tissue cultures", Physiol Plant **18**, 100-127
34.Lyne RL, Bennett RI, Hunter CP (1984) "Embryoid and plant production from cultured barley anthers", in Withers LA and Alderson PG, Plant tissue culture and its agricultural application, Butterworth pub, Guildford, pp. 405-411
35.Masuda K, Kikuta Y, Okazawa Y (1981) "Regulation of development by glutamine and nitrate in Parsley endosperm culture", Japan J Crop Sci **50**, 125-130
36.Miao SH (1980) "Effects on NH_4^+ on embryoid generation from pollens of maize", Acta Bot Sinica **22**, 356-359
37.Misoo S and Matsubayashi M (1978) "Studies on the mechanism of pollen embryogenesis. 2. Effects of varied saccharide concentrations on the plantlet formation in tobacco anther culture", Sci Rept Fac Agr Kobe Univ **13**, 19-28
38.Morrison RA and Evans DA (1988) "Haploid plants from tissue culture: new plant varieties in a shortened time frame", Bio/Technology **6**, 684-690
39.Niizeki M (1977) "Haploid, polyploid and aneuploid plants from cultured anthers and calluses in species of *Nicotiana* and forage crops", J Faculty Agric **58**, 343-466
40.Nitzsche W (1970) "Herstellung haploider Pflanzen aus *Festuca-Lolium* Bastarden", Naturwissenschaften **57**, 199-200
41.Nitzsche W and Wenzel G (1977) "Haploids in plant breeding", Fortschr Pflanzenzuchtg **8**, 1-80
42.Olesen A (1987) "Anther culture of perennial ryegrass (*Lolium perenne* L.): The production of haploid plants and their potential in relation to traditional breeding strategies", Ph.D. Thesis in plant breeding, Dep Crop Sci of Royal Vet Agr Univ, Copenhagen
43.Olesen A, Andersen SB, Due IK (1988) "Anther culture response in perennial ryegrass (Lolium perenne L.)", Plant Breeding **101**, 60-65

44. Olsen FL (1987) "Induction of microspore embryogenesis in cultured anthers of *Hordeum vulgare*. The effects of ammonium nitrate, glutamine and asparagine as nitrogen sources", Carlsberg Res Commun **52**, 393-404
45. Oono K (1975) "Production of haploid plants of rice (*Oryza sativa*) by anther culture and their use for breeding", Bull Natl Inst Agric Sci, Tokyo, D**26**, 139-222
46. Pagniez M and Demarly Y (1979) "Obtention d'individus androgenetiques par culture *in vitro* d'antheres de Ray-grass d'italie (*Lolium multiflorum* Lam.)", Ann Amelior Plantes **29**, 631-637
47. Powell W (1988) "The influence of genotype and temperature pretreatment on anther culture response in barley (*Hordeum vulgare* L.)", Plant cell, tissue and organ culture **12**, 291-297
48. Raquin C (1983) "Utilization of different sugars as carbon source for *in vitro* anther culture of *Petunia*", Z Pflanzenphysiol Bd **111**, S.453-457
49. Reynolds TL (1987) "A possible role for ethylene during IAA-induced pollen embryogenesis in anther culture of *Solanum carolinense* L.", American J Bot **74**, 967-969
50. Rose JB, Dunwell JM, Sunderland N (1986) "Anther culture of *Sorghum bicolor* (L.) Moench. 1. Effect of panicle pretreatment, anther incubation temperature and 2,4-D concentration", Plant cell tissue organ culture **6**, 15-22
51. Rose JB, Dunwell JM, Sunderland N (1987) "Anther culture of *Lolium temulentum, Festuca pratensis* and *Lolium* x *Festuca* hybrids. 1. Influence of pretreatment, culture medium and culture incubation conditions on callus production and differentiation", Annals Bot **60**, 191-201
52. Sharp WR, Dougall DK, Paddock EF (1971) "Haploid plantlets and callus from immature pollen grains of *Nicotiana* and *Lycopersicon*", Bull Torrey Bot Club **98**, 219-222
53. Smith MM and Stone BA (1973) "Studies on *Lolium multiflorum* endosperm in tissue culture", Aust J Biol Sci **26**, 123-133
54. Stanis VA and Butenko RG (1984a) "Physiological and genetical conditions for haploid regeneration in common ryegrass anther culture",Sel'skokhozyaistvennaya Biologiya (**4**), 73-75
55. Stanis VA and Butenko RG (1984b) "Developing viable haploid plants in anther culture of ryegrass", Doklady Biol Sci **275**, 249-251
56. Stanis VA, Slesaravichus AK, Kaleda VA (1983) "Regenerarion of *Lolium perenne* and *Trifolium pratensis* plants by culturing tissue of different origin", Sel'skokhozyaistvennaya Biologiya (**5**), 29-31
57. Sunderland N (1974) " Anther culture as a means of haploid induction", in Kasha KJ (ed.), Haploids in higher plants: Advances and potential, University of Guelph Press, Guelph, pp. 91-122.
58. Sunderland N (1978) "Strategies in the improvement of yields in anther culture", in Proceedings of symposium on plant tissue culture, Science Press, Peking, pp. 65-86.
59. Sunderland N (1979) "Comparative studies on anther and pollen culture", in Sharp WR, Larsen PO, Paddock EF, Raghavan V, Plant cell and tissue culture: Principles and applications, Ohio State Univ Press, Columbus, pp. 203-219

60. Sunderland, N (1980) "Anther and pollen culture 1974-1979:, in Davies DR and Hopwoods DA (eds.), John Innes Charity, Norwich, pp. 171-183
61. Thurling N and Chay PM (1984) "The influence of donor plant genotype and environment on production of multicellular microspores in cultured anthers of *Brassica napus* ssp *oleifera*". Ann Bot **54**, 681-693
62. Tsay HS, Miao SH, Widholm JM (1986) "Factors affecting haploid plant regeneration from maize anther culture", J Plant Physiol **126**, 33-40
63. Wang XZ and Hu H (1984) "The effect of potato 11 medium for triticale anther culture", Plant Sci Lett **36**, 237-239
64. Wenzel G and Foroughi-Wehr B (1984) "Anther culture of cereals and grasses", in Vasil IK (ed.), Cell culture and somatic cell genetics of plants, Academic Press, New York, **1**, pp. 311-327.
65. Xu ZH and Sunderland N (1981) "Glutamine, inositol and the conditioning factor in the production of barkley pollen callus *in vitro*", Plant Sci Lett **23**, 161-168
66. Zenkteler M (1977) "Induction of haploid plants from anthers cultured *in vitro*", in Novak FJ (ed.), Use of tissue cultures in plant breeding, Czechoslovak acad sci inst of exp botany, Prague, pp. 337-354

Stress-Induced Carrot Somatic Embryos and their Applicability to Synthetic Seed.

Hiroshi HARADA, Tomohiro KIYOSUE, Hiroshi KAMADA and Katsunori KOBAYASHI

Institute of Biochemical Sciences, University of Tsukuba
Tsukuba-shi, Ibaraki-ken 305 JAPAN

Summary

When apical meristems or cotyledons of carrot (*Daucus carota* L. cv. US-Harumakigosum) seedlings were cultured on hormone-free Murashige and Skoog's (MS) medium with 0.3 - 0.7 M sucrose or 0.1 - 0.4 M NaCl or one of the heavy metal ions, such as cadmium, cobalt, nickel and zinc at a concentration ranging from 0.25 to 1.0 mM, somatic embryos formed on the surface of the explants without visible callus formation. Somatic embryo formation was also induced on malformed seedlings, when carrot seeds were treated with hypochlorite solution at a high concentration and sown on MS msdium. These somatic embryos can be fractionated into several groups of different sizes by passing through stainless steel sieves with different pore sizes and encapsulated in calcium alginate gel to make synthetic seeds. These synthetic seeds germinated 1 to 2 weeks after sowing. Frequency of the seeds which developed both a radicle and a green bud was about 30 - 50 % in the case of large embryos (0.5 - 2.0 mm) induced by the treatment with sucrose, cadmium or sodium hypochlorite, but less than 15% in the case of 2,4-D induced embryos. When 2,4-D induced embryos were sown, numerous secondary and tertiary embryos were formed disturbing further development to form normal seedlings. Our results indicated that the stress-induced somatic embryos possessed higher quality than auxin-induced ones in the use for synthtic seed production.

Introduction

Somatic embryogenesis is known as the process of expressing the totipotency of the plant cell and as the useful tool for plant micropropagation.

R.S. Sangwan and B.S. Sangwan-Norreel (eds.), The Impact of Biotechnology in Agriculture, 129–157.

There are a number of reports describing successful induction of somatic embryogenesis in various plant species (1, 2, 3, 4, 5). Until recent time, only auxin has been known to play an essential role in the induction of somatic embryogenesis, for somatic embryogenesis can be induced by transferring explants from high auxin to low auxin or to auxin-free medium in almost all cases. Using this procedure, physiological, biochemical and molecular analysis on it have been carried out with a limited number of plant species, including carrot (6). However, it is difficult to analyse biochemical aspects directly concerned with the induction of embryogenic potential in somatic cells only with this procedure, because auxin regulates various physiological events taking place in plant tissues, such as stimulation of ethlene biosynthesis (7), cell expansion, cell division (8), adventitious root formation (9), etc... Therefore, in order to analyse the mechanism concerning the process during which somatic cells become embryogenic and form somatic embryos, it is likely to be necessary to develop new methods capable of inducing somatic embryogenesis without any application of plant hormones and which allow to compare biochemical changes between non-embryogenic and embryogenic cells.

On the other hand, in the field of plant micropropagation, somatic embryos occupy an important position and attract the attention as the plant materials for artificial seed production (10). It is well known that somatic embryos are easily and effectively induced by auxin treatment in a number of plant species, however, they tend to have genetic variations in different degrees, so called "somaclonal variations". This is thought due to that high amont of exogenously applied auxin not only induce embryogenic cells but also induce somaclonal variations. Thus, to work out a new somatic embryo induction method which do not induce somaclonal variation is needed.

In this paper, we report stimulative effects of different stesses on carrot somatic embryogenesis, and discuss about advantages of stress-induced embryos for synthetic seed production.

Materials and Methods

- Sucrose treatment

One-week-old seedlings of *Daucus carota* L. cv. US-Harumakigosun grown on vermiculite were surface sterilized with 10% sodium hypochlorite solution (available chlorite concentration of ca. 1%) for 5 min, then rinsed three times with sterilised distilled water. Cotyledons, hypocotyls and apical tip segments (5 mm) were exised from the seedlings. These explants were cultured in Petri-dishes (6 cm in diameter) containing 10 ml of Murashige and Skoog's agar (1%) medium to which no plant growth regulator was added and sucrose concentration was varied from 0.1 to 0.7 M. After 3 weeks in culture, one half of the explants were transfered onto MS medium in which the concentration of sucrose was decreased to 0.09 M. The remaining half of the explants were continuously cultured on the same medium with a high concentration of sucrose. Cultures were placed under 16 h light / 8 h dark photoperiod (approx. 3,000 lux) at 25°C. Experiments were conducted with at least ten explants for each treatment and repeated three times. Otherwise mentioned, same experimental procedure has been applied in the following treatments.

- NaCl Treatment

The explants were cultured in Petri-dishes (5 explants / dish) containing 8 ml of MS agar (0.8%) medium with 0.1 M sucrose to which NaCl was supplemented at a concentration of 0.1M to 0.4 M. After 1 to 3 weeks in culture, the explants were transferred onto hormone-free MS medium without NaCl.

- Heavy metal treatment

Apical tips (ca. 5 mm) were excised from the sedlings and cultured on MS Gelrite (0.2%) medium containing 0.25, 0.5, 0.75 and 1 mM of heavy metal chlorides, such as $MnCl_2$, $FeCl_2$, $CoCl_2$, $NiCl_2$, $CuCl_2$, $ZnCl_2$ and $CdCl_2$. After 1 to 2 weeks of the culture, the explants were transfered to MS medium without heavy metals. Morpholgical responses were examined 6 weeks after the transfer. In each experiment, 100 explants were used and the experiment was repeated twice.

-Hypochlorite treatment

Carrot seeds were treated with one of the following three hypochlorite solutions: sodium hypochlorite (available chlorite concentration 10%), potassium hypochlorite (available chlorite concentration 5%) and calcium hypochlorite solutions (filtrate of 6% (w/v) calcium hypochlorite solution). Seeds were completely soaked for 15 to 120 min in each of the hypochlorite solutions mentioned above. In another experiment, seeds were soaked for 45 min in variously diluted sodium hypochlorite solutions. The treated seeds were rinsed 3 times with sterilised distilled water, and placed in Petri-dishes (five seeds per dish) containing 8 ml of hormone-free MS agar (0.8%) medim with 0.1 M sucrose. After culturing for 4 days under 16h light (1,600 lux) / 8h dark photoperiod at 25°C, randomly selected non-contaminated seeds (100 seeds per treatment) were transferred to fresh MS medium and cultured for 7 weeks under the same conditions. The germinating rate of non-treated seeds was 90% in average.

-Induction of somatic embryogenesis for synthetic seed production

Apical tips were exiced from the seedlings and cultured in Petri-dishes containing 10ml each of MS Gelrite (0.2%) medium. In a treatment with osmotic stress, sucrose concentration was increased to 0.7 M and the explants cultured on this medium for 2 weeks were then transferred to MS medium with 0.1M sucrose. In the case of heavy metal treatment, the explants were cultured on 0.5 mM $CdCl_2$ containing MS medium for 2 weeks and then transferred to MS medium without $CdCl_2$.

As for the treatment with hypochlorite, carrot seeds were immersed in a sodium hypochlorite solution (available chlorite concentration was 6%) for 1 hour, washed 3 times with sterilised distilled water, then sown on MS agar medium.

One month after the start of culture, explants with somatic embryos were suspended in liquid MS medium and subcultured every 2 to 4 weeks on a gyratory shaker (70 rpm). Somatic embryos thus obtained were used for artificial seed production.

In order to compare the adaptability for synthetic seed of embryos produced by the 3 methods mentioned above with those produced by 2,4-D treatment, the following lot has also been prepared.

Apical tips obtained from 1-week-old seedlings were cultured on MS medium containing 2,4-D (1 mg/l) and subcultured at 2-week intervals. Cell clumps of 63 -3 7μm in diameter were obtained by passing the suspension through stainless steel sieves (37 and 63 μm in pore size) and washed twice with fresh MS medium. The cell clumps (0.1 ml of packed cell volume at 100 x g) were resuspended in 100 ml of MS medium and cultured for 2 to 4 weeks to obtain somatic embryos.

- Synthetic seed production

Somatic embryos obtained by the 4 different treatments mentioned above were

divided into 3 categories of sizes (0.25 - 0.5, 0.5 - 1 and 1 - 2 mm in diameter) by passing them through stainless steel sieves with appropriate pore sizes. Somatic embryos were washed twice with fresh MS medium and collected by centrifugation at 100 x g. They were mixed with 3% (w/v) sodium alginate solution and the mixture was added drop by drop into 100 mM $CaCl_2$ solution with a pipette. After 30 to 60 min, the drops were gelled completely and the capsules were washed twice with fresh MS medium. After 30 min immersion of the capsules in the MS medium, only those with a single embryo were collected and put into Petri-dishes (9 cm in diameter) under aseptic conditions.
All cultures were placed under 16 h light (6,000 lux) / 8h dak conditions at 25°C, except suspension cultures which were carried out in total darkness at 25°C. All experiments were conducted with 20 to 100 sededs per treatment and repeated at least twice.

- Histological observation

For histolgical observation, cultured explants were embedded in O.C.T. compound (Miles Scientific Lab., USA), frozen, and then cut into 20 µm sections by using COLDTOME (SAKURA Seiki Co., Japan, MODEL CM-41). The specimens thus prepared were observed under a microscope and microphotographs were taken.

- SEM observation

For scanning electron microscopic observation, apical tip segments producing somatic embryos were prefixed in 3% glutaraldehyde in 0.1 M phosphate buffer (pH 7.0) for 2 h at 4°C. The specimens were post fixed for 12 h in similarly buffered 1% osmium tetraoxide, dehydrated with a graded ethanol series, subjected to critical point drying, and coated with gold. Observation was made with a JEOL Scanning Electron Microscope type JSM-T20.

Results

- Induction of somatic embryogenesis by osmotic stress (salt stress)

Cotyledon explants cultured on MS medium with 0.5 M or 0.7 M sucrose produced somatic embryos 3 to 6 weeks after transfer to MS medium with 0.09 M sucrose (Tab.1). Somatic embryos appeared directly at a cut-end of explants without visible callus formation (Fig. 1A). When apical tip segments were used as explants, the first and second leaves protruded and elongated during the culture on MS medium containing sucrose at low concentrations (0.1 and 0.3 M), but did not at high concentrations (0.5 and 0.7 M). When explants cultured on 0.3M sucrose medium were transferred to 0.09 M sucrose medium, somatic embryos were formed on tip-end surfaces of elongated leaves without visible callus formation 3 to 6 weeks after the transfer (Fig. 1B). When apical tip segments cultured on 0.5 M or 0.7M sucrose medium were transferred to 0. 09 M sucrose medium, the first leaves were slightly elongated after the transfer and somatic embryos were formed on the surface of the elongated leaves without visible callus formation (Fig. C and D). On the other hand, when hypocotyl segments were used as explants, they elongated slightly, but did not produce somatic embryos in any of the media used (Tab.1).

Table 1. Effects of sucrose concentration on somatic embryo formation in carrot. Explants of cotyledons, hypocotyls and apical tips were cultured for 3 weeks on hormone-free MS medium with sucrose at the concentrations indicated, then half of the explants were transferred to hormone-free MS medium with 0.09 M sucrose (3-week treatment). The remainder of the explants were continuously cultured on the same medium (continuous treatment). Each value indicates a percentage of explants producing somatic embryos 9 weeks after the start of culture. Experiments were conducted with at least ten replicates for each treatment and all experiments were repeated at least three times.

Treatment with sucrose		Explant source		
concentration	duration	cotyledon	hypocotyl	apical tip
0.1M	3 weeks	0.0 (%)	0.0 (%)	0.0 (%)
0.3M	3 weeks	0.0	0.0	39.6
0.5M	3 weeks	20.5	0.0	35.3
0.7M	3 weeks	24.7	0.0	62.8
0.1M	continuous	0.0	0.0	0.0
0.3M	continuous	0.0	0.0	22.2
0.5M	continuous	15.8	0.0	23.6
0.7M	continuous	20.8	0.0	39.6

Frequency of somatic embryo formation obtained in apical meristems was higher than that in cotyledons (Tab. 1). The treatment with high concentrations of sucrose was also effective for somatic embryo formation (Tab. 1). When coyledons or apical meristems were cultured continuously on high sucrose medium (0.3 to 0.7 M), somatic embryos were also formed during 9-week culture, but the frequency was generally lower than that in the transfer experiment (Tab. 1). These embryos did not develop further on the high sucrose medium but developed to plantlets by transferring them to 0.09 M sucrose medium.

In the case of NaCl treatment, apical tips being used as explants, the first and second leaves protruded and elongated during 1 to 2 week culture on MS medium containing NaCl at a low concentration (0.1M), but they did not form somatic embryos before and even after the transfer to MS medium without NaCl. On the other hand, apical tips cultured with a high concentration (0.2 to 0.4) of NaCl or those cultured for 3 weeks with a low NaCl concentration (0.1 M), turned to red and formed somatic embryos on the surface of the explants and/or elongated leaves without visible callus formation after the transfer to MS medium without NaCl (Fig. 2A and 2C).

When cotyledon and hypocotyl explants were cultured on MS medium with 0.1M to 0.4 M NaCl, their color turned to brown or white during the culture ad no somatic embryo was observed before as well as after the transfer to MS medium without NaCl.

Frequency of somatic embryo formation obtained in apical tip segments was summarized in Fig. 3. The higher concentration of NaCl, the shorter treatment was required for the somatic embryo formation and the higher frequency was observed. However, in the case of 0.4 M NaCl treatment, many of the explants died (data not shown) rendering the frequency of somatic embryo formation quite low (ca. 10%).

- Induction of somatic embryogenesis by heavy metals

Apical tip segments of carrot are the most responsive parts for the induction of somatic embryos by stress as indicated above. Apical tips cultured on MS medium with $CoCl_2$, $NiCl_2$, $ZnCl_2$ or $CdCl_2$ for 1 or 2 weeks and then transferred to MS medium showed emergence of true leaves and/or formation of somatic embryos on the surface of expanded leaves and/or apical segments without visible callus formation (Fig. 4A,B).

Frequencies of somatic embryo formation were varied depending on the kind and the concentrations of heavy metals and on the duration of the treatment (Tab. 2).

Table 2. Effects of combined treatment with 0.61 M mannitol and 0.09 M sucrose on somatic embryo formation in carrot. Explants of cotyledons and apical tips were cultured during 3 weeks on hormone-free MS medium with 0.61 M mannitol and 0.09 M sucrose, then half of the explants were transferred to hormone-free MS medium with 0.09 M sucrose (3-week treatment). The remainder of the explants were continuously cultured on the same medium (continuous treatment). Each value indicates a percentage of explants producing somatic embryos 9 weeks after the start of culture. Experiments were conducted with at least ten replicates for each treatment and all experiments were repearted at least three times.

Duration of treatment	Explant source	
	cotyledon	apical tip
3 weeks	12.5 (%)	38.4 (%)
continuous	0.0	0.0

Among several heavy metals tested, $CdCl_2$ showed the maximum effect on somatic embryo induction. The highest frequency of embryogenesis was obtained by the treatment with 500 μM CdC12 for 2 weeks and the survival rate of explants was 62% with this treatment. The number of survived explants decreased as the concentration of the heavy metals became higher and/or the duration of the treatment was prolongated.

In the case of $CoCl_2$, $NiCl_2$ and $ZnCl_2$, somatic embryos could also be formed but with slightly lower frequencies (max. 9%, 14% and 5%, respectively). When the explants were treated with $ZnCl_2$, the percentage of survival of explants remained constantly high, but the frequency of somatic embryo formation was quite low.

Treatment with $CuCl_2$ for 1 or 2 weeks, at a concentration of 250 μM $CuCl_2$ for 3 weeks induced embryo formation at a low frequency (ca. 4%) after the transfer to MS medium. Incorporation of $CuCl_2$ at a high concentration (e.g. over 500 μM) inhibited gel solidification and caused precipitation of the ion (probably as the form of $Cu(OH)_2$). Somatic embryogenesis was not induced by the treatment with 250 to 500 μM $FeCl_2$ for 1 week or 2 weeks, but when treated continuously for 9 weeks with 1 mM $FeCl_2$, 20% of the explants produced somatic embryos. The treatment with $MnCl_2$ for 1 to 3 weeks at a concentration ranging from 250 μM to 1 mM induced no embryo formation.

- Induction of somatic embryogenesis by hypochlorite treatment of seeds

When seeds were treated with concentrated NaClO solution (available chlorite concentration 10%) for 60 min and planted on hormone-free MS medium, somatic embryos were formed on the surface of slightly germinating seeds (Fig. 5A - D) as well as on the surface of cotyledons and hypocotyls of the seedlings without visible callus formation (Fig. 5E). These somatic embryos developed into young plantlets by subsequent transfer onto MS medium (Fig. 5F).

Effect of NaClO (available chlorite 10%) treatments with various durations showed lower germination rate with longer treatment duration (Fig. 6). On the other hand, the frequency of somatic embryo formation increased with the prolonged duration of trearment. Average values of the frequency were 7.5% and 15.3% by the treatment for 15 min and 120 min, respectively. In the case of 60 min treatment, somatic embryos were formed on cotyledons, hypocotyls and true leaves of yound seedlings with different frequencies of 64%, 15% and 2%, respectively. These somatic embryos could develop into young plantlets (Fig. 5F). Whereas, multiple globular embryo-like stuctures were formed at the emerging sites of zygotic embryos. However, most of these structures and did not develop to whole lantlets during subsequent 6-week culture and even after a transfer onto new medium (Fig. 5G). Formation of these structures occured scarcely by a shorter treatment but observed frequently by a longer treatment. Similar phenomena were noted in the cases of NaClO treatment for 45 min with high concentrations and of the treatments with other hypochlorites.

In another experiment, we examined the effects of NaClO concentrations keeping the duration of treatment for 45 min (Fig. 7). With increasing concentrations of NaClO, the frequency of somatic embryo formation increased and germination frequency decreased gradually. In order to clarify whether hypochlorite concentrations or metal elements were essential to induce somatic embryogenesis, we also examined effects of KClO and $Ca(ClO)_2$ solutions. Somatic embryos were formed by the treatment with $Ca(ClO)_2$ solution and the frequency increased with longer treatment (Fig. 8). In the case of KClO treatment (Fig. 9), germination frequency and the frequency of somatic embryo formation were slightly higher in the short treatments of 15 and 45 min, respectively, and remained at constant values in longer treatments (60 to 120 min).

- Production and germination of synthetic seeds

Carrot synthetic seeds made by us were round and about 5 mm in diameter (Fig. 10A). The somatic embryos produced by 4 different methods (2,4-D, sucrose, cadmium and hypochlorite) showed different developmental aspects and germination rates depending on the initial treatment. Radicles and green buds emerged 1 week after the sowing and the number of synthetic seeds with a radicle and/or a green bud increased rapidly with time (data not shown). After 4 weeks of the cultivation, the number of synthetic seeds having both an elongated radicle and a green bud was higher with the somatic embryos produced by sucrose, cadmium or sodium hypochlorite treatment (Tab. 3 and Fig. 10B) as compared to that

Table 3. Conversion rates of synthetic seed. Somatic embryos induced by 4 different methods indicated were devided into 3 categories of size and encapsulated in alginate gells. The synthetic seeds were put into Petri dishes and cultured under sterile conditions. Each value indicates a percentage of synthetic seeds producing radicles and/or greeen buds 4 weeks after the start of cultivation. Experiments were conducted with at least 100 replicates for each treatment and all experiments were repeated at least twice.

Induction method of somatic embryos	Size of somatic embryos (mm)	% of seeds with radicle	green bud	radicle and green bud
2,4-D	0.25 - 0.5	10	20	8
	0.5 - 1.0	19	35	14
	1.0 - 2.0	14	44	11
Sucrose	0.25 - 0.5	13	12	9
	0.5 - 1.0	34	40	27
	1.0 - 2.0	43	56	35
Sodium Hypochlorite	0.25 - 0.5	37	35	27
	0.5 - 1.0	54	62	47
	1.0 - 2.0	59	68	51
Cadomium Ion	0.25 - 0.5	8	9	5
	0.5 - 1.0	50	49	40
	1.0 - 2.0	66	64	51

produced by the treatment with 2,4-D. When larger somatic embryos (1.0 to 2.0 mm) were encapsulated to make synthetic seeds, the frequency of that with a green bud and/or a radicle was high (Tab. 3). This tendency was observed in all the cases regardless of the four different somatic embryo induction methods. When somatic embryos induced by 2,4-D were used, several small embryos developed together as a mass from individual synthetic seeds (Fig. 10C). Even when these embryos developed to small seedlings with a green bud, numerous secondary and tertiary embryos differentiated from the seedlings without further development. On the other hand, when somatic embryos induced by the treatment with sucrose, cadmium or sodium hypochlorite were used, more than 50% of the embryos developed to seedlings with a green bud, but a part of the seedlings were transplanted and cultivated with vermiculite under non-sterile conditions and they were able to develop further (Fig. 10D)

Discussion

In this report, we showed that carrot somatic embryogenesis could be induced by the treatment of apical tip segments with 0.3 to 0.7 M sucrose or 0.1 to 0.4 M NaCl or 0.25 to 1.0 mM one of the heavy metals, $CdCl_2$, $CoCl_2$, $NiCl_2$, $ZnCl_2$, and of seeds with concentrated hypochlorite solutions. These treatments are known to induce stress responses to plant cells. Besides these facts, it is also known that 2-4 D is one of the most effective compound to induce somatic embryogenesis in many plant species and an inducer of stress protiens (11). Thus, it seems likely that different physiological stresses trigger the induction of somatic embryogenesis. On the other hand, somatic embryogenesis in *Zea mays* , *Carica papaya, Helianthus annuus* can be induced by treatment with auxin under high osmotic conditions (12, 13), and shoot formation in tobacco callus requires both osmotic stress and hormonal treatment (14). These facts indicate that the induction of somatic embryogenesis by physiological stress must be mediated through a mechanism other than that directly related to hormonal treatment.

Clonal propagation through shoot culture can be used in a number of plants but is often labour consuming because several steps must be cleared before obtaining plants which can grow in soil. These difficulties may be overcome by using synthetic seeds with somatic embryos. Several researchers are working on the production of synthetic seeds, and successful encapsulation of somatic embryos with sodium alginate or other gelling reagents was made (15, 16, 17). However, frequency of synthetic seeds growing to normal seedlings (conversion rate) was rather low when somatic embryos induced by 2-4 D treatment were used without further improvement (18). One of the reason for such a low conversion rate was tha quality of somatic embryos. Somatic embryos unduced by 2-4 D at a high concentration (50 μM) rarely contained seed storage protein of 11 S and only a small part of them grew into normal seedlings. On the other hand, embryos induced by 2-4 D at a low concentration (10 μM) contained seed storage protein of 11 S just as that of zygotic embryos and grew readily into normal seedlings (18). In the course of studies on the stress effects on somatic embryo induction in carrot, we found that the somatic embryos induced by various stresses grew more readily into normal seedlings than those produced by 2-4 D treatment. Therefore, we examined the conversion rate of synthetic seeds produced by 4 different methods. The results indicated that the conversion rate was high in the synthetis seeds containing somatic embryos induced by osmotic stress, cadmium or sodium hypochlorite in comparison with those induced by 2-4 D. Thus, stress-induced somatic embryos would provide valuable material for synthetic seed production.

Acknowledgement: Tha authors thanks Dr. V.S. Jaiswal for his linguistic suggestions.

References

1. Desai, H.V., Bhatt, P.N., Mehta, A.R. (1986). Plant Cell Reports. **3**: 190-191.
2. Fujimura, T., Komanine, A. (1979). Plant Physiol. **64**: 162-164.
3. He, D.G., Tanner, G., Scott, K.J. (1986). Plant Science. **45**: 119-124.
4. Li, B.J., Langridge, W.H.R., Szalay, A.A. (1985). Plant Cell Reports. **4**: 344-347.
5. Yeh, M., Chang, W. (1986). Theor. Appl. Genet. **73**: 161-163.
6. Nomura, K., Komanine, A. (1986). Oxford Surv. PLant Mol. cell Biol. **3**: 456-466.
7. Imaseki, H., Yoshii, H., Todaka, L.(1982). PLant Growth Substances. Wareing, P.F.(ed) Academic Press, London, New-York, pp 259-277.
8. Gamburg, K.Z. (1982). Plant Growth Substances. Wareign, P.F. (ed) Academic Press, London, New-York, pp 59-70.

9. Kamada, H., Harada, H.(1979). Z. Pflanzenphysiol. **91**: 255-266.

10. Thorpe, T.A. (1988). ISI Atlas of Science: Animal and Plant sciences, pp 81-88.

11. Czarnecka, E., Edelman, L., Scoffl, F., Key, J.L. (1984). Plant Mol. Biol. **3**: 45-58.

12. Finer, J.J. (1987). Plant Cell Reports. **6**: 372-374.

13. Litz, R.E. (1986). J. Am. Soc. Hort. Sci. **111**: 969-972.

14. Brown, D.C.W., Leung, D.W.M., Thorpe, T.A. (1979). Physiol. plant. **46**: 36-41.

15. Fufii, J.A., Slade, D.T., Redenbaugh, K., Walker, K.A. (1987). Tibtech. 5: 335-339.

16. Gray, D.J. (1987). Hort. Science. **22**: 795-814.

17. Redenbaugh, K., Paasch, B.D., Nichol, J.W., Kossler, M., Viss, P.R., Walker, K.A. (1986). Bio/technol. **4**: 781-797.

18. Stuart, D.A., Nelsen, J., Strickland, S.G., Nichol, J.W. (1985). Tissue culture in forestry and agriculture. New-York: Plenum Press. pp 59-73.

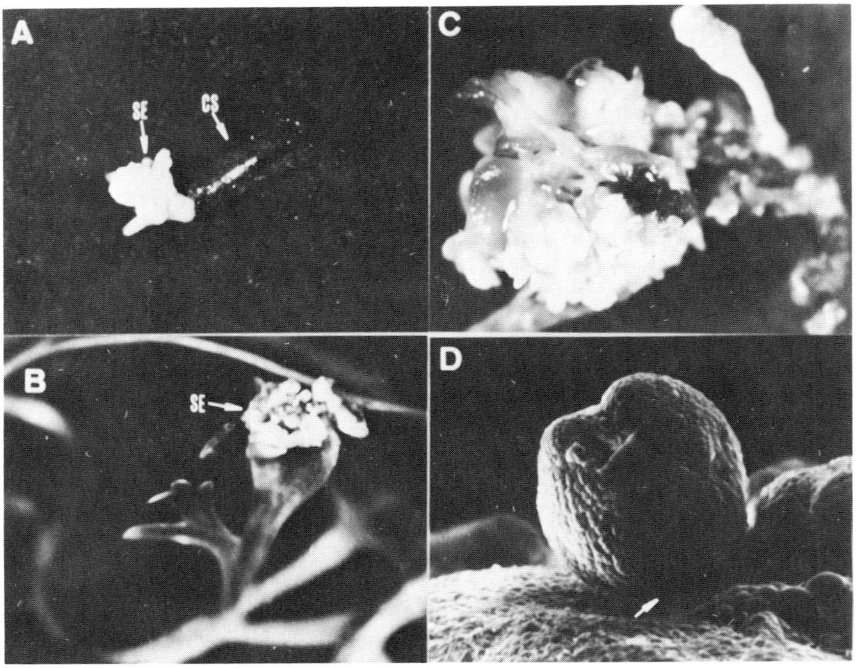
A
SE
CS
B
SE
C
D

Fig.1. Somatic embryogenesis induced by osmotic stress in carrot.

(A). Somatic embryos (SE) formed at a cut end of a cotyledon segment (CS) without visible callus formation. The cotyledon segment was cultured for 3 weeks on MS medium with 0.7 M sucrose, then transferred to MS medium with 0.09 M sucrose. Photograph taken 4 weeks after the transfer (x 7).

(B). Somatic embryos (SE) formed directly on a tip-end surface of the first leaf grown from an apical meristem. The apical tip was cultured on MS medium with 0.3 M sucrose for 3 weeks, then transferred to MS medium with 0.09 M sucrose. Photograph taken 4 weeks after the transfer (x 3.5).

(C). Numerous somatic embryos formed directly on the surface of the first leaf slightly grown from an apical meristem. The apical tip was cultured for 3 weeks on MS medium with 0.7 M sucrose, then transferred to MS medium with 0.09 M sucrose. Photograph taken 6 weeks after the transfer. Some embryos already started to form plantlets (x 20).

(D). Scanning electron micrograph of a somatic embryo arising directly from the first leaf. The embryo is attached directly to the leaf by a suspensor-like structure (arrow) without callus formation. The apical tip was treated in a same manner as in (C). Photograph taken 4 weeks after the transfer (x 100).

Fig. 2A

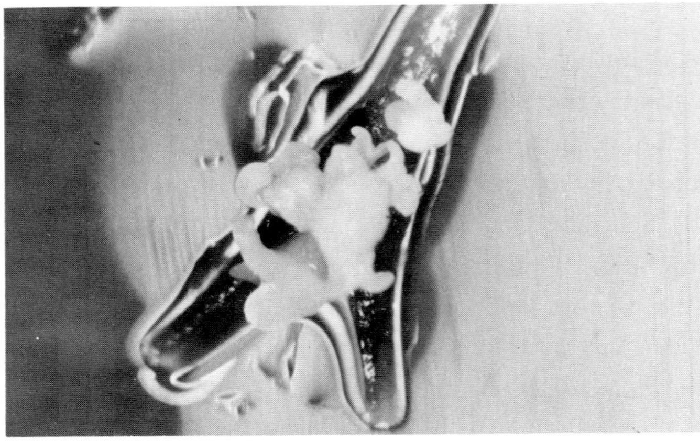

Fig. 2B

Fig. 2C

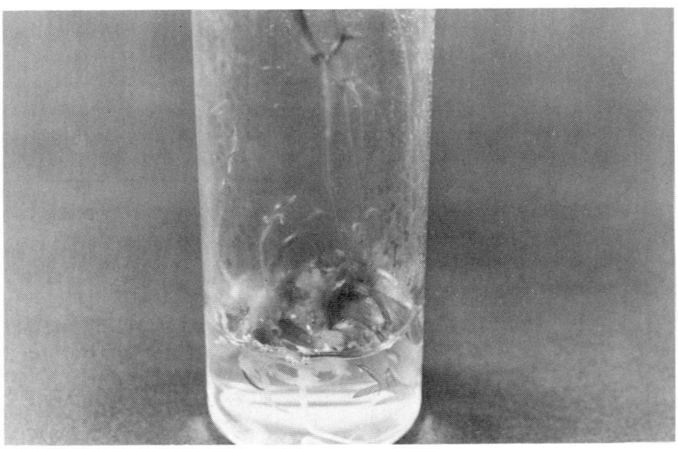

Fig. 2A. Somatic embryos induced by salt stress in carrot. Apical tip segments were cultured for 3 weeks on MS medium with 0.3 M NaCl, then transferred onto MS medium without NaCl. Somatic embryos were formed on the surface of the explants without visible callus formation. Photograph taken 6 weeks after the transfer (x 20).

Fig. 2B. Somatic embryos formed directly on the surface of the first leaf grown from an apical tip segment . The apical tip segment was cultured under the same conditions as described in Fig. 2A. (x 60).

Fig. 2C. Young plantlets derived from somatic embryos which were induced by a 3-weeks treatment with 0.3 M NaCl. Photograph taken 10 weeks after the transfer to MS medium without NaCl (x 1.5).

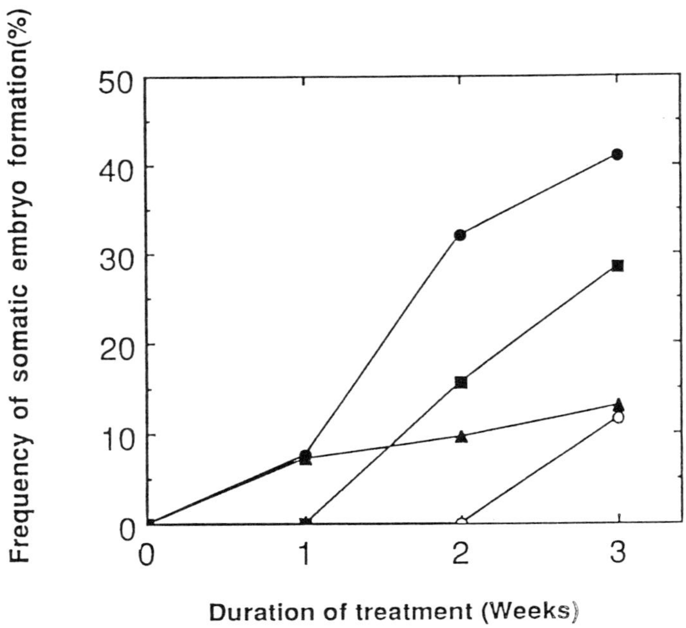

Fig. 3. Effects of NaCl on somatic embryo formation in apical tip segments of carrot. Apical tip segments were cultured on MS medium with NaCl for 1 to 3 weeks and then transferred to MS medium without NaCl. The concentration of NaCl used were 0.1 M (○), 0.2 M (■), 0.3 M (●), and 0.4 M (▲). Frequency of somatic embryo formation (%) was calculated by applying the following formula :

$$\frac{\text{No. of segments producing somatic embryos}}{\text{No. of treated segments}} \times 10$$

Fig. 4A

Fig. 4B

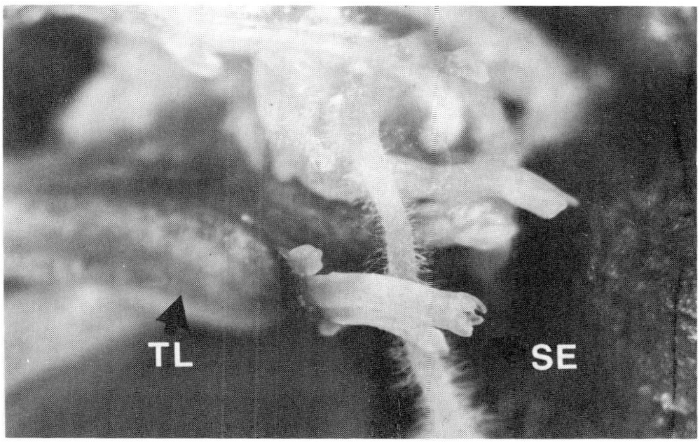

Fig. 4C

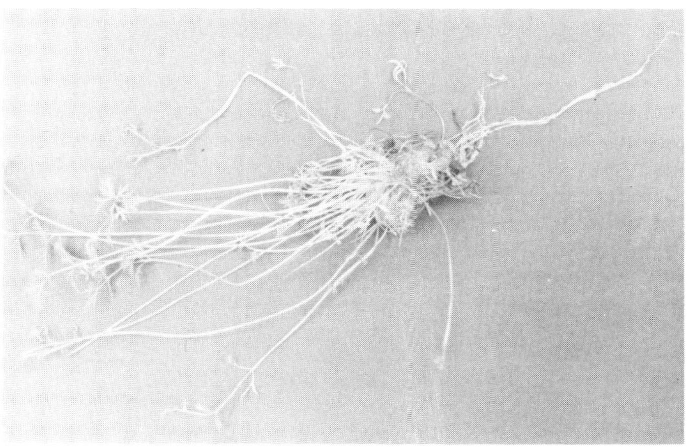

Fig. 4A. Somatic embryos induced by heavy metal stress in carrot. Apical tip segments (AT) were cultured for 2 weeks on MS medium with 500 µM $NiCl_2$, and then transferred onto MS medium without $NiCl_2$. Somatic embryos (SE) were formed on the surface of the explants without visible callus formation. Photograph taken 6 weeks after the transfer (x 20).

Fig. 4B. A magnified photograph of somatic embryos (SE) formed directly on the surface of the true leaf (TL) growing from the apical tip segment. The segment was precultured on MS medium with 250 µM of $CoCl_2$ for 2 weeks and then cultured on MS medium for 6 weeks (x 60).

Fig. 4C. Young plantlets derived from somatic embryos which were induced by a 2-weeks pre-culture with 250 µM $CdCl_2$. Photograph taken 6 weeks after the transfer onto MS medium without $CdCl_2$ (x 0.5).

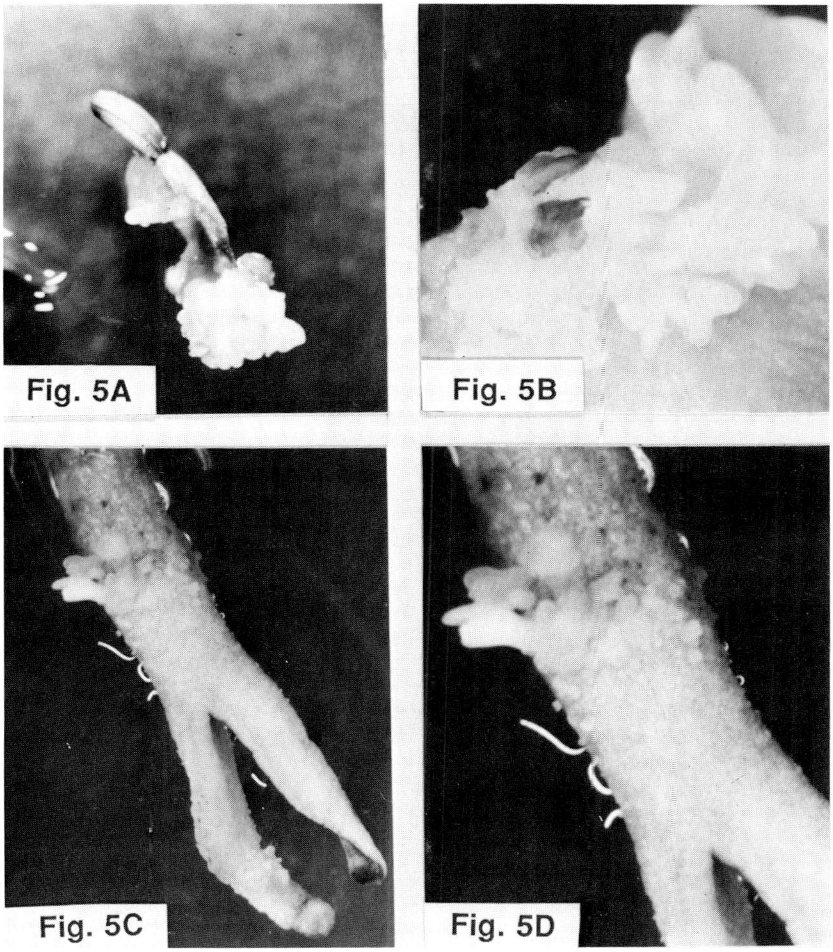

Fig. 5A Fig. 5B Fig. 5C Fig. 5D

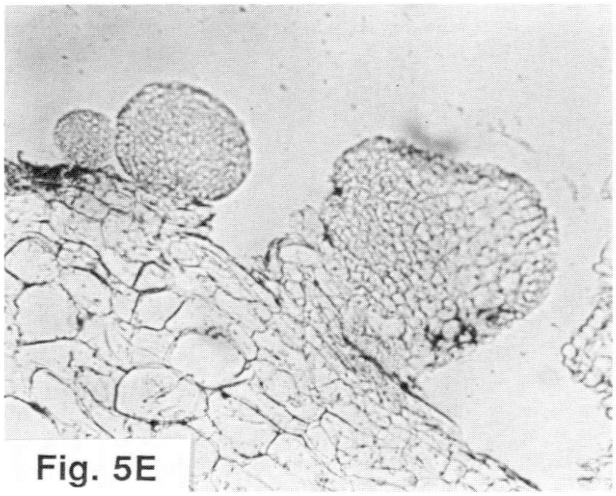

Fig. 5E

Fig. 5F

Fig. 5G

Fig. 5A. Somatic embryos induced by the treatment with NaClO on the surface of cotyledons of carrot. Carrot seed was treated with NaClO solution (available chlorite concentration of ca. 10%) for 60 min. and then cultured on MS medium for 7 weeks (x 20).

Fig. 5B. A magnified photograph of somatic embryos formed directly on the surface of the cotyledons which were emerged from a seed The seed was treated with NaClO solution for 60 min. and then cultured on MS medium for 7 weeks (x 60).

Fig. 5C. Somatic embryos formed on the surface of a hypocotyl which were induced by the same treatment described in Fig. 5A. (x 20).

Fig. 5D. A magnified photograph of somatic embryos in Fig. 5C. (x 40).

Fig. 5E. A histological section of somatic embryos induced by the NaClO treatment (x 150).

Fig. 5F. Young plantlets derived from somatic embryos which were induced by the NaClO treatment as in Fig. 5B. Photograph taken 10 weeks after the transfer onto MS medium (x 1.2).

Fig. 5G. Somatic embryo-like structures from a seed treated with NaClO solution (available chlorite concentration of ca. 10%) for 120 min. and then cultured on MS medium for 7 weeks.

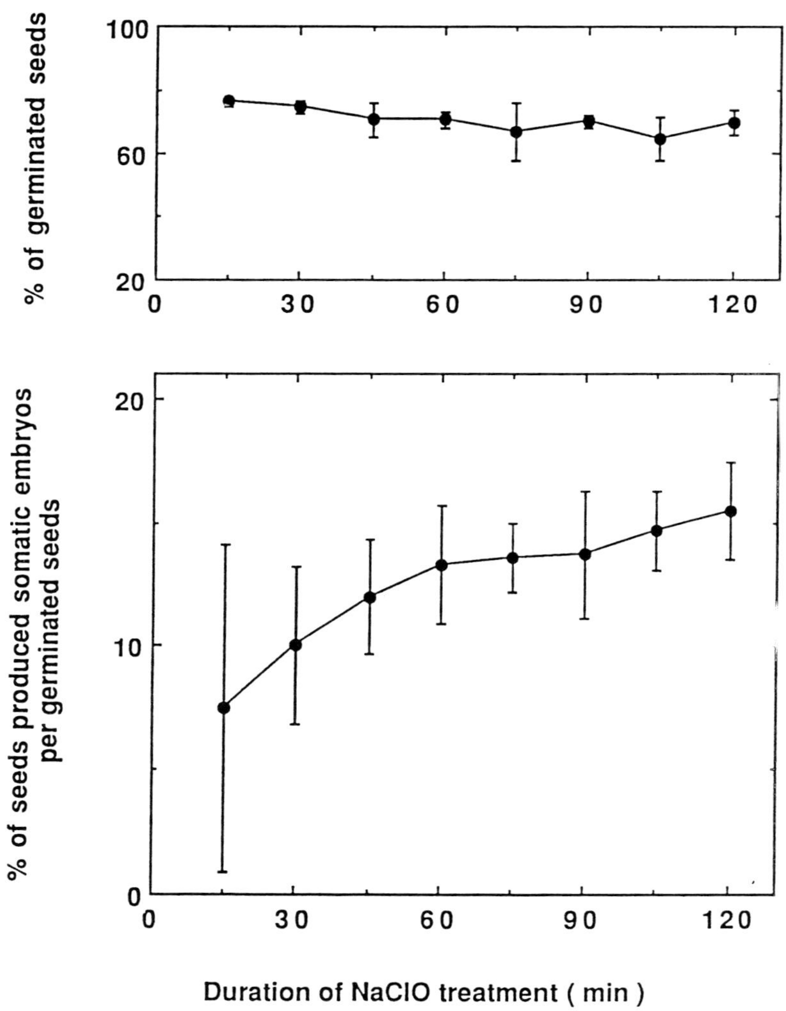

Fig. 6. Effects of the duration of NaClO treatment on seed germination and somatic embryo formation in carrot. Seeds were treated with NaClO solution (available chlorite concentration of ca. 10%) for 15 to 120 min. and then planted on MS medium. Both frequencies were determined after 7 weeks of the culture. Error bars represent standard error of the mean

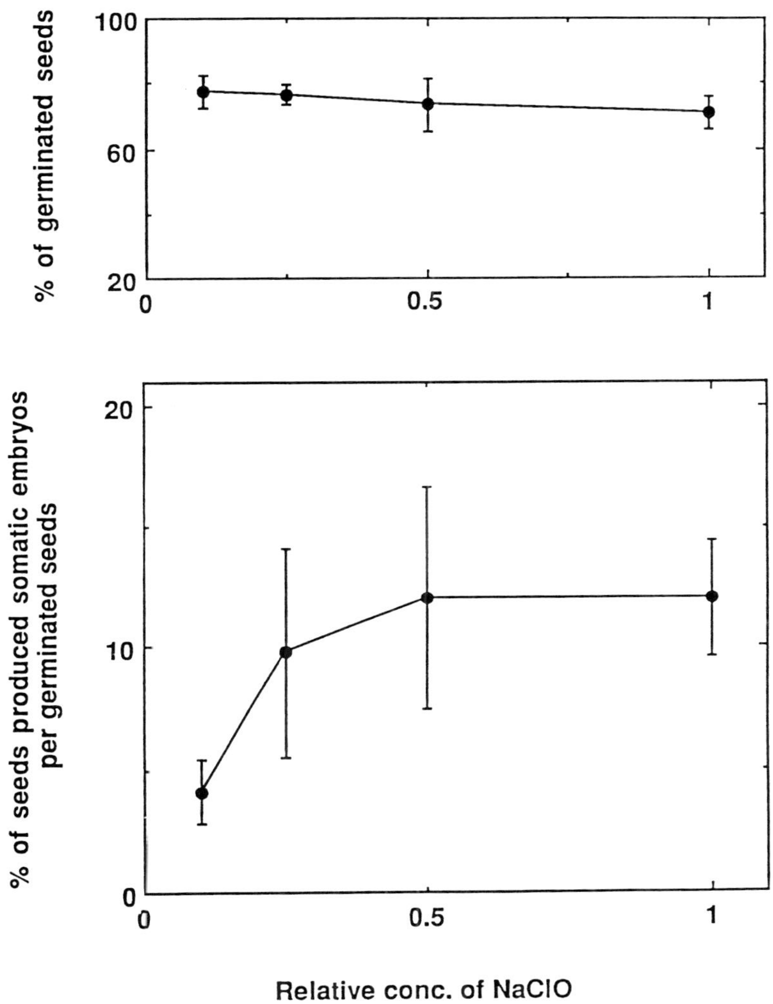

Fig. 7. Effects of NaClO concentrations on seed germination and somatic embryo formation. Seeds were treated with commercially available NaClO solution (available chlorite ca. 1 to 10%) for 45 min. and then planted on MS medium. Each value was determined as described in Fig. 6. Error bars represent standard error of the mean.

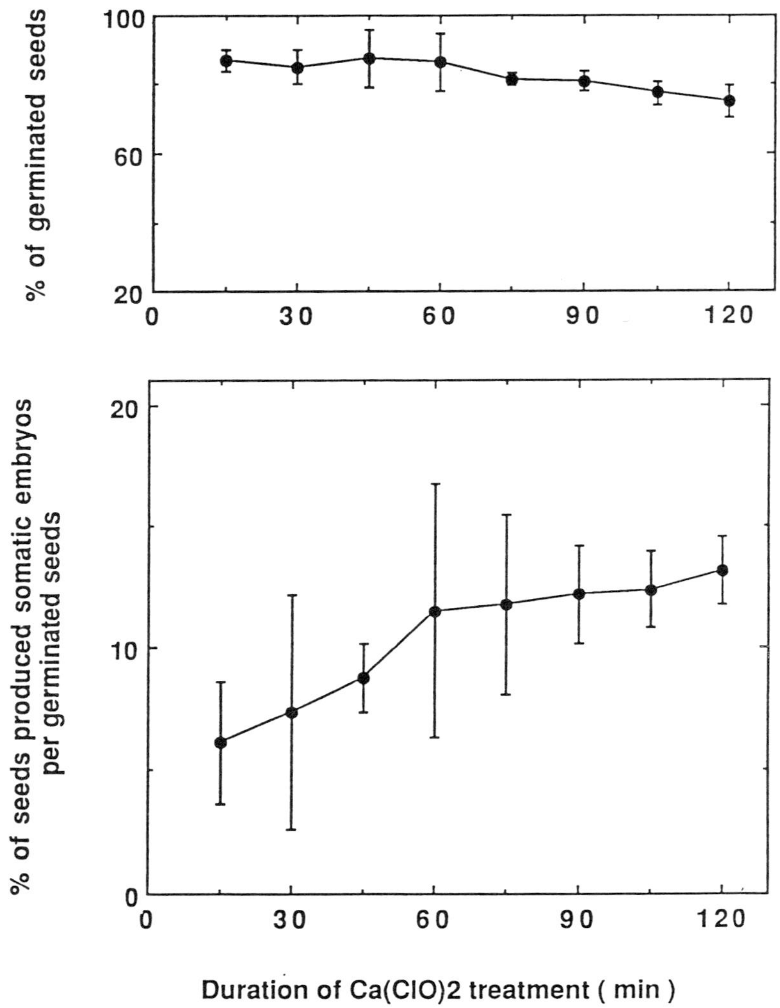

Fig. 8. Effects of the duration of Ca (ClO) $_2$ treatment on seed germination and somatic embryo formation. Seeds were treated with the filtrate of 6% (w/v) calcium hypochlorite solution for 15 to 120 min. and then planted on MS medium. Each value was determined as determined as described in Fig. 6. Error bars represent standard error of the mean.

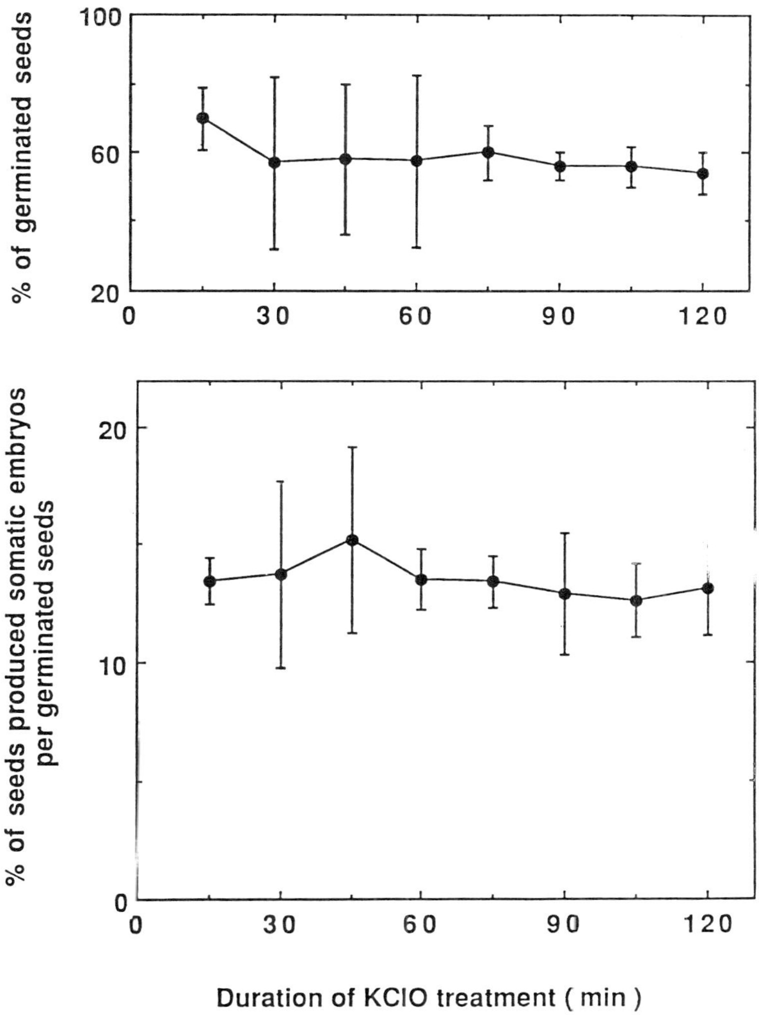

Fig.9. Effects of the duration of KClO treatment on seed germination and somatic embryo formation. Seeds were treated with KClO solution (available chlorite ca. 5%) for 15 to 120 min. and then planted on MS medium. Each value was determined as described in Fig. 6. Error bars represent standard error of the mean.

Fig. 10A Fig. 10B Fig. 10C Fig. 10D

Fig. 10A. A synthetic seed containing a single carrot somatic embryo (x 12).

Fig. 10B. Carrot synthetic seeds having an elongated radicle and/or a green bud (x 1).

Fig. 10C. Masses of small embryos developed from carrot synthetic seeds in which 2-4 D induced somatic embryos were encapsulated (x 2.6).

Fig. 10D. Carrot seedlings developing from synthetic seeds, each encapsulating a single somatic embryo, on vermiculite under non-sterile conditions (x 0.3).

DEVELOPMENTAL MUTANTS

Mario TERZI and Fiorella LO SCHIAVO
Department of Genetics, General and Molecular Biology,
University of Naples,
Via Mezzocannone, 8
80134 Napoli, Italy.

Studies on plant development are much behind the animal counterpart. This is no doubt due to the lack of accessibility of the embryo embedded in the seed protective coat. Therefore we are tempted to use the intellectual frame used in the animal studies and to make analogies for the experiments that could not be performed.Unfortunately albeit we know little on plant development, we know enough to say that many important features are different in the two kingdoms, so that proceeding by analogy is rather risky (Walbot 1985).

Development of the flowering plants can be divided in two parts: the first corresponds to morphogenesis and formation of the apical meristems; from this stage we pass -after dessiccation and the preparation to dormancy- to germination and growth, that corresponds to development of the apical meristems, i.e. the continuous production of new plant organs, shoots, leaves, roots. The way meristematic development and embryogenesis proper are tackled are rather different, so that they will be dealt with in different sections.

MUTANTS IN VIVO

In order to study early development, the microsurgical experiments, used with such a success in animal embryology, could not be performed, due to the difficulty posed by the seed protective coats.It was thought that, perhaps, a genetic approach might be more fruitful. In a series of works published in the 20's, Mangelsdorf (1923, 1926) described 14 maize strains with a recessive lethal causing seed abortion. All but two of these factors were non-allelic. Other mutants affecting seed development in maize were isolated by, among others, Jones (1920) andDemerec (1924). More references to the early work can be found in a review by Marx (1983)

More recently the problem was taken up by Sheridan and his co-workers (Sheridan and Neuffer,1980, 1982, Neuffer and Sheridan,1980, Sheridan andClark 1984, Sheridan, 1984) who isolated 855 mutant lines segregating 25% defective kernels upon self-pollination. These lines were divided in 5 classes:

1. abnormal endosperm and non-viable embryo (432 mutants)
2. abnormal endosperm and viable but abnormal embryo (59 mutants)

R.S. Sangwan and B.S. Sangwan-Norreel (eds.), The Impact of Biotechnology in Agriculture, 159–169.

3. abnormal endosperm and normal embryo (147 mutants)
4. normal endosperm and non-viable embryo (3 mutants)
5. non-classifiable or inadequately tested (214 mutants)

89 mutants of class 1 were assigned to specific chromosome arms by using the B-A translocation technique. With the same technique, as indicated by Robertson (1952), embryo-endosperm interactions can be studied. This was done on 19 mutants and the results indicated that a normal endosperm tissue was unable to rescue a mutant embryo, hence lethality is determined by the genotype of the embryo and not of the endosperm.

The various mutants showed developmental arrest at different stages (mostly dealing with embryo maturation, those affecting early stages of morphogenesis being rare). Histological studies were also carried out (Clark and Sheridan 1986, Faccio-Dolfini and Sparvoli 1988). In some cases embryo development could be completed in vitro giving rise to homozygous mutant plants (nutritional mutants) .

Arabidopsis, the Drosophila of the plant kingdom, was also used to select embryo-lethal mutants by Muller (1963) and his group. They characterized 6O mutants in terms of time of developmental arrest (they were mostly arrested at globular or heart stage, see below) or colour of seedlings.

More recently, Meinke (1985) isolated about 40 new mutants that were characterized in terms of stage of developmental arrest, percentage and distribution of aborted seeds in heterozygous siliques, regeneration in vitro. All arrested stages are now represented. The non-random distribution of aborted seeds indicates gametophytic expression of mutant genes. Several mutants with abnormal suspensor have been identified. The formation of protein bodies and the regulation of storage protein synthesis was also investigated. More information about these mutants may be found in a recent review by Meinke (1986).

SOMATIC EMBRYOGENESIS

An alternative approach to the study of development can be made by using somatic embryogenesis. If an embryogenic culture is diluted and the medium is deprived of hormones, plant cell cultures (starting from single cells or, more frequently from clumps, which however, have been shown to derive from single cells) give rise to embryos that, morphologically, follow the same pathway of the zygotic embryo, i.e. they pass through the stages of globular, heart-shaped, torpedo-shaped and plantula (for a general reference see Sung et al. 1984). Differences from the zygotic process concern the suspensor (which is reduced or absent) and associated tissues. But the fact that both processes proceed through the same sequence of developmental stages indicates that similar patterns of gene expression and similar regulatory mechanisms operate in both cases. Moreover, the fact that embryo development can occur in culturewithout addition of hormones is an indication that the embryo is autonomous and self-sufficient in terms of developmental programme. The adiacent tissues within the seed are only needed for nutritional reasons. All this is supposed to be true as a general rule in flowering plants. It should be stressed however, that most studies were carried out in carrot, which was the first and still is the model system. Therefore we are not sure that what we are going to say in the next few pages (which comes entirely from carrot studies) has general validity.

The autonomy of somatic embryo is surprising because it means that callus cells, at variance with adult, differentiated cells, have acquired the ground state (Sung, 1985) i.e. the developmental state typical of the unfertilized egg, that permits faithful execution of morphogenesis during embryogeny. This embryonic state, or ground state, which in animals is maintained in germ line cells, in plants has to be regained. In fact, floral meristems develop from vegetative meristems which consist of somatic cells no longer embryogenic. Instead, during floral development, meiotic products that are embryogenic arise.On the other hand, it cannot be rigorously excluded that a few germ line cells are sequestered in the shoot meristem and serve as the mother cells for gametophytic development.

Although the developmental origin of the unfertilized egg requires further clarification, it is clear that the embryogenic state can be acquired by cells which have departed from the embryonic state. In fact, in carrot, the embryonic state can be regained soon after explantation of several plant body parts (e.g. phloem cells, hypocotyl trissue etc.) in a proper medium. What happens is largely unknown, but we know that treatment with auxin is necessary and several days are needed for the embryogenic potential to be acquired. In an embryogenic hypocotyl-derived suspension culture, the appearance of newly-formed pro-embryogenic masses (PEM, Halperin 1966) from which somatic embryos develop under appropriate conditions, has been followed. After approx. 20 days, a cell population emerges that contains, among others, clusters consisting of between 10 and 20 small, densely cytoplasmic cells, tightly adhering to each other (PEM). The fraction of the cell population containing PEM slowly increases over a period of six weeks (de Vries et al.1988a) reaching a frequency which is, in different lines, between 1 and 10%.

These PEM can be concentrated in a Percoll gradient after sieving and/or picked manually and it was possible to obtain a subpopulation of single cells whose embryogenic efficiency was 90%. These single cells underwent unequal division, forming a cytoplasm-rich cell and a highly-vacuolated one. Further cell divisions occurred preferentially in the cytoplasm-rich cells resulting in the formation of a polarized cell cluster showing, e.g. polarized DNA synthesis (Nomura and Komamine 1986a).

Therefore from the initial explant to an established embryogenic culture we may have three distinct stages:

First stage: erasure of the differentiative tissue-specific programme of those cells which respond to auxin and begin to proliferate. The molecular events behind this stage are not defined. Recently it has been shown that treatment with exogenous auxin promotes DNA hypermethylation and the hypothesis was made that auxin is necessary whenever re-programming is to be made, because hypermethylation erases the previous differentiative programme (LoSchiavo et al.1989)

Cells autotrophic for auxin (habituated cells) whether of genetic or epigenetic origin, do not hypermethylate under auxin and cannot embryogenize. An intermediate case has been reported of a mutant which can regenerate at low efficiency; it shows the presence of tumours associated with the regenerated plantlets. Curiously enough it does not show hypermethylation in the presence of auxin, nor hypomethylation in the presence of azacytidine (Lo Schiavo et al.1989).

To this first stage, a second one follows that corresponds to acquisition and maintenance of totipotency. This stage is characterized by the appearance of PEM in the cell population. The embryogenic capacity of some cells might depend on a specific molecular event -that we ignore- that may occur at the end of the first stage, or it may be the automatic consequence of the erasure of the tissue-specific programme, whereby a fraction of the cells reach the ground state, i. e. totipotency.

The third stage correspond to the expression of the embryonic programme: When the auxin is removed and perhaps the cell density lowered, embryos are formed with the typical progression of globular, heart and torpedo. Whereas early forms require auxin, embryogenesis cannot proceed beyond globular stage if auxin is present.

Numerous workers have approached somatic embryogenesis from a biochemical point of view analyzing the third stage and recording differences between embryo and callus cells or using drugs inhibiting certain steps in order to create phenocopies. Although the earliest stages remain very interesting, owing to the experimental difficulties, the data which have accumulated concern practically only the third stage.

The rate of protein synthesis was shown to increase as did the rate of poly(A) + RNA synthesis (see Nomura and Komamine 1986b). DNA synthesis also increased during embryogenesis. Polyamine metabolism changed as polyamine levels increased several times over the non-embryogenic controls (Fienberg et al.1984). By analyzing the quality of the macromolecules synthesized during embryogenesis, differences were found in the 2-dimensional pattern of polyacrylamide gel electrophoresis by Sung and Okimoto (1981) and Choi and Sung (1984). Thomas and Wilde (1985) analyzed a cDNA library and found a few clones specific for the embryos. Monoclonal antibodies were obtained against antigens unique to somatic embryos by Smith et al.(1985).

All these techniques were shown to work in the system but a paradox remains, that the amount of embryo-specific functions is very small (1-2% in the case of proteins, O.5-1% in the case of cDNA) so that the problem remains on how such a complex developmental programme can be carried out by such a limited genomic information. One possible answer is that the callus cells from embryogenic lines are already early embryos; in other words, the cells that regained totipotency have expressed at least partially their developmental programme so that when we compare, say, globular embryos with cells we are not comparing differentiated vs. undifferentiated cells, but rather two successive stages of differentiation. In fact, exponentially growing cells look more similar (according to the parameters investigated above) to early embryos than to old senescent non-embryogenic cells.

More results on this third stage come from another type of approach, i.e. to look at special classes of products (or functions) that were deemed more likely to be differentially expressed during development.

In fact, in the class of heat-shock proteins, several were found to be specifically associated with different stages of development (Pitto et al.1983). In other cases, sensitivity to drugs (5-bromo-deoxyuridine, Dudits et al.1979, ethanol, Perata et al.1986, cycloheximide, Sung et al.1987, tunicamycin, LoSchiavo et al.1986) was also found to be associated with developmental stage and, in two more cases (azacytidine-ECP and a-amanitin) resistant mutants were isolated and characterized (Lo Schiavo et al.1989, Pitto et al.1985). Interesting enough, some of the resistances and the embryo-specific proteins were coordinately expressed, but different cell lines were expressing more or fewer functions at the callus stage, as if callus were a blocked embryonal stage and the block could occur earlier or later in the different lines. Instrumental for this type of analysis has been the purification of different embryo stages which can be achieved with different techniques (discussed by Lo Schiavo, 1984).

Another approach that may contribute to the understanding of the developmental programme comes from experimental embryology, of the type used on eggs from sea-urchins and Amphibia, which gave rise to concepts such as mosaic and regulative development, pattern formation etc. These experiments which have now the status of classics are very difficult on zygotic embryos. Not so much so on somatic ones. In fact, carrot somatic heart and torpedo-stage embryos were severed (Schiavone, 1988) at their midlengths to produce two halves. These pieces were grafted or kept separate. Grafted embryos developped normally and matured earlier than uncut control embryos. The halves kept separate did also grow and showed release from apical control.

TEMPERATURE-SENSITIVE MUTANTS

Another,more productive approach that permits to overcome some of the limits of biochemistry consists in disrupting development from within, by making conditional mutants that, in non-permissive conditions, were arrested at some more or less specific stage. A search for such mutants, of the temperature-sensitive type, was done on haploid (Breton and Sung 1982) or diploid (Terzi et al.1982, Schnall et al.1988) cell lines, with (Terzi et al.1982) or without (Breton and Sung 1982, Schnall et al.1988) mutagenic treatment. The permissive temperature was always 24-25°C and the non-permissive one 31-33°C. Scores of mutants were isolated in each case and even after disposing of the unstable or of the leaky mutants, many more were isolated than characterized in detail.The procedures, slightly different in the details, have in common a negative selection, based on disposing of the embryos grown at the non-permissive temperature.(The successive positive selection may be done on colonies or embryos and as such may select for different types of mutants) The negative selection based on the incorporation at high temperature of drugs such as bromo-deoxyuridine, used with success in animal cells, had only very limited application (Malmberg, 1979) in the isolation of plant developmental mutants, as that drug stops somatic embryogenesis (Dudits et al.1979). In all cases variants were found blocked at different stages as judged morphologically (Breton and Sung 1982) or by studying the temperature-sensitive period (Terzi et al.1982, Schnall et al.1988).

The mutants isolated by Breton and Sung (1982) belonged to six different classes: *absolute* - or *ts-emb*$^-$, incapable of forming embryos, at any temperature, or at the non-permissive one, respectively; *absolute-* or *ts-growth*$^-$, both callus and embryo development poor at any temperature, or only at the non-permissive one, respectively; *habituated* lines, capable of growth in embryogenic medium; *2,4-D resistant* lines, capable of forming initial embryonal stages in the presence of hormones.

In general the most frequent temperature-sensitive periods in all three laboratorioes was found to be between globular and torpedo stage, with a few variants being affected at very early stages (pro-embryogenic masses).It should be mentioned however, that pre-treatment with high temperature during growth had, for some mutants, an effect on subsequent embryogenesis allowed to occur at permissive temperature. This was thought to indicate that functions necessary for embryogenesis have in some cases to be performed during callus proliferation (Giuliano et al.1984). Embryonal functions were also found to be expressed at the callus level in Northern blots analyzed by Wilde et al.(1988).

A dominant mutation (ts59) was also isolated (LoSchiavo et al.1988) which points to another advantage with the use of somatic vs. zygotic embryogenesis.

There are no indications that the number of mutants isolated in the different laboratories havein any way saturated the map of developmental mutants. It is possible to get some indications on the total number of loci involved in functions necessary for somatic embryogenesis (that can give rise to mutations of the ts type). This is based on the total frequency of mutation for ts-emb$^-$ mutants, compared to the frequency in known loci. This was done for different loci, with or without mutagenesis and the number of loci indicated by this rough approach turned out to be in the range 100-600. Another indication came out however, on the excess frequency of mutation in diploid, whose frequency of recessive mutants is far from being the square of the frequency found in haploid lines. This is common knowledge for both plant and animal somatic cell geneticists and the most commonly accepted interpretation lies in a sort of functional haploidization of a large number of loci (citare Terzi, 1974 , Terzi and Sung 1986).

Two mutants have been characterized in molecular terms. One, ts59 is dominant in somatic cell hybrids, shows two temperature-sensitive periods, at pre-globular and heart stage and reveals a defect in the phosphorylation of a heat-shock protein. This is interesting because it assigns a role in plant cells to highly-specific kinases (that, besides self-phosphorylation, may act on one or two more peptides) in activating a function. In addition, as the peptides to be phosphorylated are of the heat-shock class, it confirms the earlier suggestion (Bensaude, 1983) that heat-shock proteins are involved in development. This mutant therefore may provide a handle in tackling some of these problems.

The other mutant, ts11c, (LoSchiavo et al.1985 and *in preparation*) is recessive in somatic hybrids and at high temperature cannot acquire polarity, thus progressing from globular to heart stage. It is peculiar in that it can be complemented with medium conditioned by wild type embryos.The factor present in wild type conditioned medium turns out to be a glycoprotein secreted in the medium. The medium contains a relatively high number (perhaps greater than 5O) of proteins ranging in M_r from 13 to 7O kDa or more and they have to be properly glycosylated to perform their function, which is important and varied (de Vries et al.1988b) . Important because if they are not properly glycosylated (as in the case of tunicamycin-treated embryos, or, if the glycans cannot be modified from the high-mannose to complex type,as in the case of ts11c) embryogenesis cannot proceed; varied because the glycoprotein necessary for the onset of embryogenesis is different from the one(s) necessary for the transition from globular to heart, and there may be others as yet unspecified. It is well possible that also in plant embryo development, as in other animal systems, secreted proteins are a way of communicating between the cells sharing a common endeavour, and of controlling and coordinating their expression. The fact that the ts mutants characterized so far are deficient in a mechanism rather than in the stability of one particular peptide and are therefore pleiotropic, may be more than a coincidence; more observations are needed to build on a firmer ground.

SEED FORMATION

Seed formation and embryogenesis are not entirely overlapping processes. In parallel with the formation of embryo and controlled by it we have production of seed storage proteins. These proteins accumulate during embryogenesis, are stored in the dry seed (and as such they are important for human and animal consumption) have a largely unknown function although one of the functions must be to provide carbon and nitrogen to the germinating seedling (Higgins, 1984)

The functions present here are numerous. More than 9O% of the 15,000 diverse mRNA present in mid-maturation stage embryos are represented in fully-differentiated mature embryos as well as in post-germination cotyledons and mature plant leaf, which means that most mRNA sequences persist throughout maturation, are stored in the mature seed and are represented in both the seedling and the mature plant (Goldberg et al.1989).It should be recalled here however, that the in vitro translation products of somatic torpedo-shaped embryos differ substantially from those of hypocotyl and leaves (Wilde et al.1988).

Mutations affecting seed development are known since pre-Mendelian times. They affect production and proper distribution of pigments in the seed and/or production of storage material in the cotyledons and endosperm tissue (see Marx, 1984). Non-lethal mutations affecting seed development may also affect accumulation of carbohydrates. Mutants defective in the production of specific seed proteins were isolated whose mutation lied in the seed protein gene itself (Goldberg et al.1983) or in an unlinked regulatory locus (Larkins et al.1984).

So far those mutants have not contributed much to the understanding of development, possibly because they are non- lethal; some of them however, have been important for studying regulation of gene expression and the mutants accumulated so far are likely to be analyzed in the future, thus contributing in the same way as perhaps some of the Drosophila mutants have been "rescued" several years after their isolation.

Of immediate physiological interest are mutants that are defective in the preparation of dormancy and therefore appear as viviparous. Those in maize (Robertson, 1955) are particularly important for the relationship they uncovered between embryonic maturation, germination, abscisic acid content and carotenoid pigments.

PLANT DEVELOPMENT AND THE APICAL MERISTEMS

Shoot and root meristems are specified during embryogenesis. However mature plant leaf, stem and root differentiate continuosly from the apical meristems. Major morphogenetic events occur post-embryonically, after seed germination. Even floral meristems, with the reproductive organ systems, derive from a reprogramming of vegetative meristems (for general reference see Sussex, 1989)

The shoot meristem, contrary to the seed, is amenable to microsurgery and in fact the experimental embryology of the shoot apical meristem is fairly advanced. The shoot apical meristem of dicotyledonous species (and some monocots) consists of three layers called, from surface to interior, L_1,L_2,L_3. Studies with chimeraes have shown that the epidermis is generated exclusively from L_1. This means that all cell divisions in L_1 are anticlinal. Occasionally, a periclinal division may occur and the inner cell gets displaced into the L_2 layer Occasionally one or few cells from L_2 may intrude into L_1 but then they differentiate as epidermal cells. This means that the fate of the cell is determined by its position rather than lineage; it also means that cell committment (to its final fate) occurs late in development.

With chimeraes it is also possible to define the number of initial cells for each layer. The most likely number turns out to be three cells for each layer.

The meristem may be subjected to various microsurgical procedures, but the part left intact (even if it is only 1/2O of the total) is still capable of reconstituting the whole meristem.

From the shoot meristem vegetative shoots are generated. They are organized as a succession of units called *phytomeres* (Gray, 1879). Each phytomer consists of a leaf, node, internode,axillary bud (which may produce a lateral shoot). This unit seems to constitute a *compartment* in the sense that mutations generated within the meristem segregate as sectors that generally have their boundaries at the boundaries of phytomers. In analogy with animal compartments, that does not mean that the developmental fate is determined in each cell unequivocally.

The number of phytomers that a meristem can generate is, in some species at least, a fixed number. After so many phytomers, the remaining part of the meristem becomes a floral meristem and generates reproductive organs.

The meristem is autonomous in the sense that it can be explanted in vitro and still grow and develop, the only growth regulator necessary being IAA. The fact that it is autonomous does not mean that it is insensitive to signals originating outside the meristem. For instance, it may be converted into a flower(in response to a signal sent by newly-expanded leaves that "measure" day-length, or responding to other signals it may convert into a thorn, or become dormant.

A few mutants have been isolated that affect the meristem or its function. In tomato, the *reduced* form of *lanceolate* lacks a shoot apical meristem (Caruso, 1964) *Vegetative* in pea fails to initiate flowering (Reid and Murfet, 1984). In *anantha* (tomato) meristems become determined for floral function but fail to initiate floral organs (Paddock and Alexander, 1952).

The plant height is frequently modified by mutation, and "dwarf" mutants are known and have been analyzed in several species. In general the character depends on a defect in the metabolism of gibberellins. But other genes may produce the same effect and they interact in complex ways. Perhaps they are involved in the perception of the gibberellin signal (Potts et al.1982). Of course, final height depends not only on internode length, but also on internode number. The number of vegetative and reproductive internodes that a plant produces is controlled by another set of genes, those governing flowering and photoperiodism.

Virtually every plant part can be subject to modifications caused by genetic variation. In tomato, lateral roots are not initiated in *diageotropic* (Zobel 1974) nor axillary buds in *lateral suppressor* (Malayer and Guard, 1964). The modification may concern the architecture (pattern formation) of the organ by changing its shape or dimensions, or it may be of the *homeotic* type, i.e. causes convertion of one body part into another.

Let us consider the organ "leaf" in Pisum and the mutants that modify this system (see Marx, 1987). This organ is rather simple as it consists of three cell types (dermal, vascular and ground tissue) and few cell layers. The leaves are compound and odd pinnate, consisting of two large stipules, one or more pairs of leaflets and terminal tendrils. The size and shape of the leaves are modified by mutants like *st* (stipules reduced), *sil* (stipules and leaflet margins undulate or sinuate) *cri* (stipules and leaflets distorted and revolute). Mutants that can be considered homeotic are *af* (tendrils in lieu of leaflets) *tl* (laminar leaflets instead of tendrils), *tac* (laminar leraflet, subterminal tendrils).

Homeotic mutants affecting various parts of the flower (sepals replaced by leaves, petals replaced by stamens, or petals replaced by sepals and stamens replaced by carpels) or petals lacking and sepals converted to carpels were isolated in Arabidopsis by Mayerowitz and Pruitt (1988) and Komati et al.(1988). Homeotic mutants as well as mutants affecting the morphology and pattern formation of the flower were isolated in Petunia (Gerats et al.,1984) and tobacco (Malmberg et al.1985).

These mutants, if experimentally pursued, are likely to lead to some fresh insights and, with time, to a full comprehension of the general principles of plant development.

Acknowledgement

Work supported by the programme "Sviluppo di Tecnologie avanzate" of the Italian Ministry of Agriculture (M.A.F.) and MURST.

References

Bensaude O, Babinet C, Morange M, Jacob F.(1983) Heat shock proteins, first major product of zygotic activity in mouse embryo. Nature 305:331-33

Breton AM, Sung ZR.(1982) Temperature-sensitive carrot variants impaired in somatic embryogenesis. Dev.Biol.90:58-66

Caruso JL.(1968) Morphogenetic aspects of a leafless mutant in tomato. I. General patterns of development. Am.J.Bot.55:1169-1176

Choi JH, Sung ZR.(1984) Two-dimensional gel analysis of carrot somatic embryonic proteins. Plant Mol.Biol.Rep.2:19-25

Clark JK, Sheridan WF.(1986) Developmental properties of the maize embryo-lethal mutants dek22 and dek23. J.Hered.77:83-92

Demerec M.(1923) Heritable characters of maize. XV. Germless seeds. J.Hered.14:297-300

de Vries SC, Booj H, Meyerink P, Huisman G, Wilde HD, Thomas TL, van Kammen A. (1988a) Acquisition of embryogenic potential in carrot cell-suspension cultures. Planta 176:196-204

de Vries SC, Booj H, Janssens R, Vogels R, Saris L, Lo Schiavo F, Terzi M, van Kammen A. (1988b) Carrot somatic embryogenesis depends on the phytohormone-controlled presence of correctly glycosylated extracellular proteins. Genes & Dev.2:462-476

Dudits D, Lazar G, Bajszar G.(1979) Reversible inhibition of somatic embryo differentiation by bromodeoxyuridine in cultured cells of Daucus carota L. Cell Diff.8:135-144

Faccio-Dolfini SF, Sparvoli F.(1988) Cytological characterization of the embryo-lethal mutant dek-1 of maize. Protoplasma 144:142-148

Fienberg AA, Choi JH, Lubich WP, Sung ZR.(1984) Developmental regulation of polyamine metabolism in growth and differentiation of carrot culture. Planta 162:532-539

Gerats AGM, Kaye C, Collins C, Malmberg RL.(1988) Polyamine levels in Petunia genotypes with normal and abnormal floral morphologies. Plant Physiol.86:390-393

Gray A.(1879) Structural botany or organography on the basis of morphology. Ivison, Blakemann & Co., New York

Giuliano G, Lo Schiavo F, Terzi M.(1984) Isolation and developmental characterization of temperature-sensitive carrot cell variants.Theor.Appl.Genet.67:179-183

Goldberg RB, Hoschek G, Vodkin LO.(1983) An insertion sequence blocks the expression of aa soybean lectin gene. Cell 33:465-475

Goldberg RB, Barker SJ, Perez-Grau L.(1989) Regulation of gene expression during plant embryogenesis. Cell 56:149-160

Halperin W.(1966) Alternative merphogenetic events in cell suspensions. Am.J.Bot.53:443-453

Higgins TJV.(1984) Synthesis and regulation of major proteins in seeds. Ann.Rev.Plant Physiol.35:191-221

Komaki MK, Okada K, Nishino E, Shimura Y.(1988) Isolation and characterization of novel mutants of Arabidopsis thaliana defective in flower development. Development 104:195-203

Jones DF.(1920) Heritable characters in maize. IV. A lethal factor: defective seeds. J.Hered.11:161-167

Larkins BA, Pedersen K, Marks MD, Wilson DR.(1984) The zein proteins of maize endosperm. Trends Biochem.Sci.9:306-308

LoSchiavo F.(1984) A critical review of the procedures for embryo purification. Plant Mol.Biol.Rep.2:15-18

LoSchiavo F, Giuliano G, Terzi M.(1985) Pattern of polypeptides excreted in the conditioned medium and its alteration in a mutant ts for embryogenesis. In "Somatic embryogenesis" (Terzi

M, Pitto L, Sung ZR.eds).IPRA, Rome pp.32-34
LoSchiavo F, Quesada-Alluè LA, Sung ZR. (1986) Tunicamycin affects somatic embryogenesis but not cell proliferation in carrot. Plant Sci.44:65-71
LoSchiavo F, Giuliano G, Sung ZR.(1988) Characterization of a temperature-sensitive carrot cell mutant impaired in somatic embryogenesis. Plant Sci.54:157-164
LoSchiavo F, Pitto L, Giuliano G, Torti G, Nuti-Ronchi V, Marazziti D, Vergara R, Orselli S, Terzi M.(1989) DNA methylation of embryogenic carrot cell cultures and its variations as caused by mutation, differentiation, hormones and hypomethylating drugs. Theor.Appl.Genet.77:325-331
Malayer JC,Guard AT.(1964) A comparative developmental study of the mutant sideshootless and normal tomato plants. Am.J.Bot.51:140-143
Malmberg RL.(1979) Temperature-sensitive variants of Nicotiana tabacum isolated from somatic cell culture. Genetics 92:215-220
Malmberg RL, McIndoo J, Hiatt AC, Lowe BA.(1985) Genetics of polyamine synthesis in tobacco: developmental switches in the flower. Cold Spring Harbor lab.Symp.40:475-482
Mangelsdorf PC.(1923) The inheritance of defective seeds in maize. J.Hered.14:119-125
Mangelsdorf PC.(1926) The expression of Mendelian factors in the gametophyte of maize. Genetics 11:423-455
Marx GA.(1983) Developmental mutants in some annual seed plants. Ann.Rev.Plant Physiol.34:389-417
Marx GA.(1987) A suite of mutants that modify pattern formation in Pea leaves. Plant Mol.Biol.Rep.5:311-335
Mayerowitz E, Pruitt RE.(1985) Arabidopsis thaliana and plant molecular genetics. Science 229:1214-1218
Meinke DW.(1985) Embryo-lethal mutants of Arabidopsis thaliana: Analysis of mutants with a wide range of lethal phases. Theor.Appl.Genet.69:543-552
Meinke DW.(1986) Embryo-lethal mutants and the study of plant embryo development. Oxford Surv.Plant Mol.Cell.Biol.3:122-165
Muller AJ.(1963) Embryonentest zum Nachweis rezessiver Letalfaktoren bei Arabidopsis thaliana. Biol.Zentralbl.82:133-163
Neuffer MG, Sheridan WF.(1980) Defective kernel mutants of maize. I.. Genetic and lethality studies. Genetics 95:929-944
Nomura K, Komamine A.(1986a) Polarized DNA synthesis and cell division in cell clusters during somatic embryogenesis from single carrot cells. New Phytol.104:25-32
Nomura K, Komamine A.(1986b) Molecular mechanisms of somatic embryogenesis. Oxford Surv.Mol.Cell.Biol.3:456-466
Paddock EF, Alexander LJ.(1952) Cauliflower, a new recessive mutation in tomato. Ohio J.Sci.52:327-334
Perata P, Alpi A, Lo Schiavo F.(1986) Influence of ethanol on plant cells and tissues. J.plant Physiol.126:181-186
Pitto L, Lo Schiavo F, Giuliano G, Terzi M.(1983) Analysis of heat shock protein pattern during embryogenesis of carrot. Plant Mol.Biol.2:231-237
Pitto L, Lo Schiavo F, Terzi M.(1985) Alfa-amanitin resistance is developmentally regulated in carrot. Proc.Nat.Acad.Sci.US 82:2799-2803
Potts WC, Ingram TJ, Reid JB, Murfet IC. (1982) Internode length in Pisum genotype and the involvement of gibbrellins. Plant Physiol.69S:Abstr.133
Reid JB, Murfet IC. (1984) Flowering in Pisum: a fifth locus veg. Ann.Bot.53:369-382

Robertson DS. (1952) The genotype of the endosperm and embryo as it influences vivipary in maize. Proc.Nat.Acad.Sci.US 38:580-583
Robertson DS. (1955) The genetics of vivipary in maize. Genetics 40:745-760
Schiavone MF.(1988) Microamputation of somatic embryos of the domestic carrot reveals apical control of axis elengation and root regeneration. Development 103:657-664
Schnall JA, Cooke TJ, Cress DE. (1988) Genetic analysis of somatic embryogenesis in carrot cell cultures: initial characterization pf six classes of temperature-sensitive variants. Dev.Genet.9:49-67
Sheridan WF.(1984) Isolation of Ds induced dek mutants. Maize Genet.Coop.Newslett.58:95-96
Sheridan WF, Clark JK.(1984) Testing lethals for allelism using a double pollination technique. Maize Genet.Coop.Newslett.58:92-95
Sheridan WF, Neuffer MG. (1980) Defective kernel mutants of maize. II. Morphological and embryo culture studies. Genetics 95:945--960
Sheridan WF, Neuffer MG. (1982) Maize developmental mutants. J.Hered.73:318-329
Smith JA, Choi JH, Krauss M, Karu AE, Sung ZR. (1985) Monoclonal antibodies that recognize proteins unique to somatic embryos in Daucus carota. In "Somatic embryogenesis" (Terzi M., Pitto L, Sung ZR. eds.) IPRA, Rome pp.86-94
Sung ZR.(1985) Developmental states of embryogenic culture. In "Somatic embryogenesis" (Terzi M, Pitto L, Sung ZR, eds.) IPRA, Rome, pp.117-120
Sung ZR, Lazar G, Dudits D. (1981) Cycloheximide resistance in carrot culture; a differentiated function. Plant Physiol.68:261-264
Sung ZR, Fienberg A, Chorneau R, Borkird C, Furner I, Smith J, Terzi M, Lo Schiavo F, Giuliano G, Pitto L, Nuti-Ronchi V. (1984) Developmental biology of embryogenesis from carrot culture. Plant Mol.Biol.Rep.2:3-14
Sung ZR, Okimoto R. (1981) Embryonic proteins in somatic embryos of carrot. Proc.Nat.Acad.Sci.US 78:3683-3687
Sussex IM. (1989) Developmental programming of the shoot meristem. Cell 56:225-229
Terzi M. (1974) Genetics and the animal cell. J.Wiley-Interscience, Chichester and New York
Terzi M, Giuliano G, Lo Schiavo F, Nuti-Ronchi V.(1982) Studies on plant cell lines showing temperature-sensitive embryogenesis. In: Embryonic development.Part B: Cellular aspects (Burger M.M. ed.) Alan R.Liss, New York pp.521-534
Terzi M, Sung ZR. (1986) Plant somatic cell genetics. CRC crit.Rev.Biotechnol.3:303-330
Thomas TL, Wilde HD. (1985) Analysis of gene expression in carrot somatic embryos. In "Somatic embryogenesis" (Terzi M, Pitto L,Sung ZR.eds.) IPRA, Rome pp.77-85
Walbot V. (1985) On the life strategies of plants and animals. Trends Genet.1:165-169
Wilde HD, Nelson WS, Booij H, de Vries SC, Thomas TL. (1989) Gene expression programs in embryogenic and non-embryogenic carrot cultures. Planta 176:205-211§
Zobel R. (1974) Control of morphogenesis in the ethylene-requiring tomato mutant, giageotropic. Can.J.Bot.52:735-741.

NEW APPROACH TOWARDS CONTROLLING SOMATIC EMBRYOGENESIS IN CERTAIN AGRONOMICALLY IMPORTANT PLANTS.

T., TETU; B.S., SANGWAN and R.S., SANGWAN.
Androgenèse et Biotechnologie; U.F.R. de Sciences.
Université de Picardie. 33, Rue St Leu.
AMIENS 80039, FRANCE.

ABSTRACT :

Different systems of plantlet regeneration *via* somatic embryogenesis or organogenesis in sugarbeet (i), soybean (ii) and pea (iii) tissue cultures have been described. Organogenic sugarbeet calli were obtained preferentially from inflorescence apices with an association of zeatin and triiodobenzoïc acid. Somatic embryogenesis from calli or from primary explants needed multiple hormonal sequences and depended from the genotypes used. The liquid consistency of the endosperm was used as a marker for screening the convenient embryogenic zygotic embryos. Interactions between the two morphogenic pathways; i.e, organogenesis and somatic embryogenesis have been scored. Incompletion of the somatic embryos lead to morphogenic developmental phases identical to those observed during adventitious buds formation. However, factors controlling the development of somatic embryos could not be completly elucidated. Scanning electron microscope studies showed the presence of numerous hair cells in the original adventitious buds but not in the dormant somatic embryos. On the contrary, ultrastructural observations revealed a large amount of protein bodies in the embryogenic calli, but not in the organogenic ones. In soybean and pea, direct adventitious somatic embryos were initiated from cultured immature zygotic embryos on Picloram and/or NAA suplemented media. Embryogenic capacity was significantly influenced by genotypes and sizes of the immature embryos at the time of culture. The effect of subculture frequencies were also investigated. Somatic embryos were transferred sequentially to different media until germination. Normal development and elongation was enhanced by reducing the auxin level and by increasing the nitrate level in the germination medium. Finally, these different regeneration systems are discussed as a prerogative for the production of transgenic plants in these three crop plants.

INTRODUCTION :

Recent progress in cell culture and molecular biology of higher plants, which are key components of plant biotechnology, have stimulated a great deal of interest. However, most of the success achieved so far, has been with model plant species; i.e, Nicotiana, Arabidopsis, Carrot etc.; and the transfer of these new technologies to major crop plants, that are the main targets of biotechnology has been slow and difficult, or non-existent. In order to have any meaningful impact on agriculture, the developping biotechnology must be equally and readily applicable to important crop species. For this purpose, we have tried to develop various plant regeneration systems, according to their supposed tendency for genetic transformation *via Agrobacterium* infections, in three

R.S. Sangwan and B.S. Sangwan-Norreel (eds.), The Impact of Biotechnology in Agriculture, 171–189.

important crops; i.e, regeneration process in Sugarbeet, Pea and Soybean tissue cultures.

(i) In sugarbeet, different technics of plant regenerations have been well established, and especially directly from petiole explants (Detrez et *al,* 1988; Freytag *et al*, 1988) or *via* a callus phase (Tétu *et al*, 1987; Saunders and Doley, 1986). The first one have been used for large scale multiplication (Detrez, 1988; Tétu, 1989) systems involved in breeding strategies; as stability of regenerants have been demonstrated (Detrez et *al*, 1988). In contrast, regeneration of viable plantlets using a callus stage was observed at a lower frequency (Tétu *et al,* 1987; Saunders and Daub, 1984; Van Geyt and Jacobs, 1985). Recently, in sugarbeet, successive ways of regeneration *via* somatic embryogenesis (Tétu, 1989) have been observed. However, in few cases, incomplete development of somatic embryos led to a bud-like development, i.e, development of a rosette like structure. Some cytological differences have also been observed between the two morphogenic pathways, i.e, somatic embryogenesis and organogenesis.

(ii) In soybean, several reports have described the formation of somatic embryos and/or plants from *in vitro* cultures (Ranch *et al*, 1985; Wright *et al*, 1986; Barwale *et al*, 1986; Tétu *et al*, 1987; Komatsuda and Ohyama, 1988) using epicotyls, cotyledonary nodes and immature embryos. Our topic on soybean was to find a regeneration process using European cultivars. Indeed, most of the success achieved so far, have been established from US cultivars. These data have permitted the production of transgenic soybean plants using cotyledon explants (Hinchee *et al*, 1988) *via* direct organogenesis.

(iii) Recent progress in pea concerned mainly the regeneration of plantlets *via* organogenesis (Mrozinski and Kartha, 1981; Rubluo *et al*, 1984; Tétu, 1989). However, the induction of somatic embryos (Jacobsen aned Kisely, 1984), and their subsequent development have been recently obtained from pre-existing meristems (Kisely *et al*, 1987). Picloram appeared to be the most effective hormone for the initiation of somatic embryos from calli, using axillary nodes (Kisely *et al*, 1987), but also from undifferentiated tissues as well as immature cotyledons (Tétu *et al*, 1989). Elsewhere, a specific balance between NAA and organic addenda has permitted the production of somatic embryos (Tétu *et al*; 1989) from immature zygotic embryos, without a callus phase.

MATERIAL AND METHODS :

Plant material :

(i) In Sugarbeet, two triploïd monogerm cultivars (Monosvalof and Veronique), eight diploïd breeding-lines (PL 120, T21, T130, SW107, SW108, T36, M2, FD1), and inflorescence apices from diploïd and tetraploïd field grown plants have been tested. Additionnal and personal breedings between fertile diploïd populations were realized in order to check regularly immature zygotic embryos. Plant material was supplied by G.I.E. Betterave Industrielle, France. (ii) In soybean, immature embryos were taken from pods of three cultivars (Osso, Sito and Verdon); sterilization procedures have been reported earlier (Tétu *et al*, 1987). Cultivars were provided by TOURNEUR frères, France. (iii) In pea, nine cultivars (Bémol, Mini, Menuet, Bonnaire, Stampède, Goya, Atlas, Davina, Sonar) and two regenerated bud-populations (P 7 and P 10) issued from three cycles of adventitious bud formation (Tétu *et al*, 1989; Tétu, 1989) have been used. These genotypes were chosen in order to allow an extensive study, envelopping many of the existing variations in pea; e.g, normal or asila foliage, light green green or dark green seeds, position of the flowering nodes, etc... These cultivars were kindly provided by ASGROW; N.E.R.S, France.

Media and culture conditions :

(i) Mature and immature zygotic embryos of sugarbeet, at different stages of development, were excised from monogerm and polygerm glomerules. The consistency of the endosperm, liquid or solid, have been selected for the assessment of the zygotic embryos development. Sterilization procedures have been done according to previous data (Tétu *et al*, 1987). Bacterial contaminations of sugarbeet field grown plants have been reported earlier (Jacobs *et al*, 1985; Tétu *et al*, 1985). The time of sterilization was increased to 60 mn and the concentration of the calcium hypochlorite solution to 14%. Inflorescences apices and immature zygotic embryos were cultured respectively in tubes and plates, at 24°C. The different media used in sugarbeet experiments are indicated in Table 1.

Table 1 : Composition of media used for somatic embryogenesis in Sugarbeet tissue cultures.

Nutritive media							
Composition	References						
	E1 (x)	**E2** (x)	**E3** (x)	**ED4** (y)	**ED5** (y)	**ED6** (y)	**ED7*** (y)
Basal medium	PGo	PGo	PGo	MS	MS	MS	MS
+ Vitamines	PGo	PGo	PGo	MS	RV*	RV*	RV*
+ Hormones (µM))							
BAP		0.4					1.7
NAA	5.4	5.4		43			
IBA							
GA3				0,1			0.5
Picloram					0.9	0.2	
+ Organic addenda (µM)							
Inositol				2700			
Adenine sulfate				22			
L-Arginine					23	23	23
L-Asparagine							30
L-Glutamine							41
L-Phenylalanine							12
L-Tryptophane							19
Thiamine					0.05	0.05	
Nicotinic acid					0.2	0.2	
Glycine							0.3
Casein hydrolysat (mg/l)				400			
+ sucrose (g/l)				30			
+ Agar (g/l)				7			

(ii) Soybean and (iii) pea immature zygotic embryos were excised from seed pods which have been surface-sterilized with calcium hypochlorite 7% (w/v) for 30 mn and rinsed three times in distilled water. Between 6 and 8 of these were plated at 27°C in Petri dishes (9. cm diameter) containing the various media presented in Table 2.

Table 2 : Composition of media used for somatic embryogenesis in pea and soybean tissue cultures.

Composition	Nutritive media						
	References						
MS basal medium	**L1** (p)(s)	**L2** (p)	**L3** (p)	**L4** (p)	**L5** (s)	**L6** (s)	**L7** (s)
+ Vitamines :	MS	MS	RV *	RV *	MS	MS	MS
+ Hormones (µM):							
BAP		2.2					
NAA	43				43	5.4	
IBA		14					2
GA3							0.25
Picloram			5	0,1			
2 IP							1
+ Organic addenda (µM)							
L-Glutamine					3500		
L-Arginine	60		240	240	240		
Nicotinic acid	40		160	160	1000		
Thiamine	5		20	20	5		
+ KNO3 (g/l)		10					10
+ Sucrose (g/l)	30-60	30	30	30	30	30	30
+ Agar (g/l)	7						

(p) and (s) : Media used respectively for pea and soybean tissue cultures.

All cultures were maintained under a photon flux density of 10-20 µE m^{-2}. s^{-1} provided by Sylvania cool-white, life time, fluorescent light on a 17 : 7 light/dark cycle. Various concentrations of separates auxins such as IAA, IBA, NAA, picloram or together with a cytokinin such as BAP ; and for certain media, an inhibitor of the endogenous auxin transport such as TIBA were used in the experiments. All media were adjusted to pH 5.9 with 1 N NaOH before adding the agar (7 g/l Difco Bacto Agar) and autoclaving (120°C, 20 mn). Aqueous stock solutions of vitamins and/or amino acids were filter-sterilized (o.22µM Millipore filters) and added individually to autoclaved medium at the required concentrations.

Histological studies :

Sugarbeet calli, pea and soybean cotyledons were used for infrastructural and ultrastructural observations. Specific markers of direct organogenesis in Sugarbeet tissue cultures was done under a Philips scanning electronic microscope. Complete technical details of histological procedures have been reported earlier (Sangwan and Sangwan-Norreel, 1987; Tétu, 1989).

RESULTS :

Two main morphogenic pathways viz : organogenesis and somatic embryogenesis, were observed in Sugarbeet, Pea and Soybean tissue cultures. The

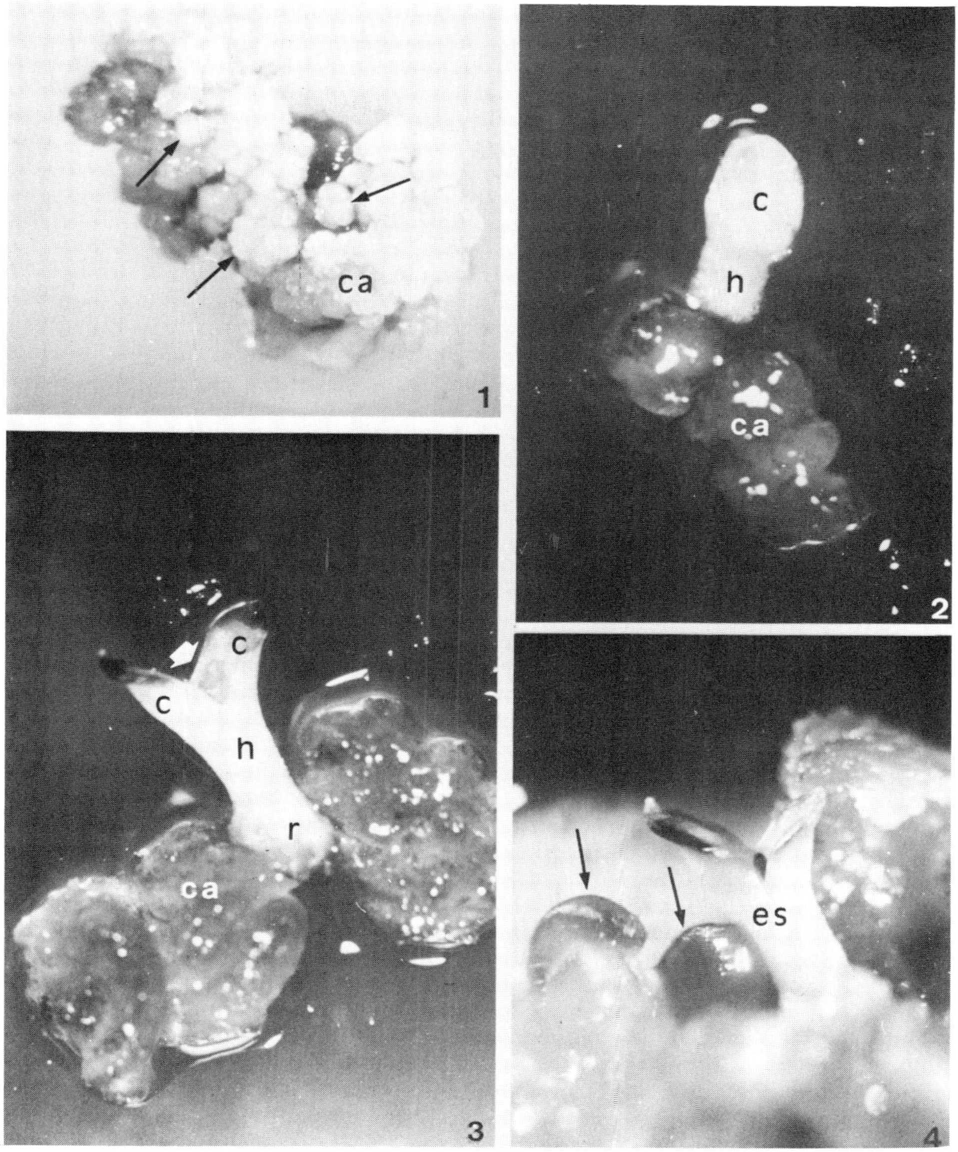

PLATE I : Somatic embryogenesis from sugarbeet calli derivated from inflorescences apices.

Fig. 1 : Embryogenic callus (ca) initiated in darkness on E2 medium. The formation of embryogenic nodules (arrows) occurred after 4 weeks in culture; (x6). Fig. 2 : Somatic embryo showing two welded cotyledons (c), note the elongation of the hypocotyl region (h). Fig. 3 : complete somatic embryo development (ca) with 2 cotyledons (c), hypocotyle (h), and radicle (r); (x 15). Fig. 4 : Attendant emergence of somatic embryos and buds (arrows); (x 8).

present paper describes the different factors involved *in vitro* cultures and especially, in somatic embryogenesis; and give some specific recognizing markers between the two morphogenetic pathways, e.g, organogenesis and embryogenesis.

SUGARBEET TISSUE CULTURES :

Simultaneously to the experiments of organogenesis from petiole cultures (Tétu, 1989; Detrez et *al*, 1988) or from calli (Tétu *et al*, 1987), somatic embryos could be observed spontaneously, at a very low frequency, indicated that somatic embryogenesis was possible in Sugarbeet. Subsequent experiments have shown two main morphogenetic pathways, viz : somatic embryogenesis from calli derivated from flower explants or directly from primary explants (hypocotyls or petiole explants).

1/ Somatic embryogenesis from calli :

No somatic embryogenesis occurred from primary calli derivated from explants such as petioles, roots, cotyledons, leaves, (Tétu et *al*, 1987) with any of the auxin tested i.e, 2.4-D, NAA and IAA. Similar results have been scored with experiments using cytokinins (kinetin, zeatin and BAP). First positive embryogenic responses were obtained with inflorescence apices as explant sources. These were cultivated on MS medium supplemented with 9 µM BAP and 0.5 µM GA_3. After 4 weeks culture, axillary buds were excised for establisment of vitroclones. Those clones, obtained from inflorescence apices, developped vegetative apex, like a rosette structure. This phenomenon is already known on sugarbeet (Margara, 1982) and reflect the reversion of flower meristems to the vegetative ones. After six weeks culture on a 1 µM BAP supplemented medium, petiole explants were excised from shoots having 6-8 leaves, and plated on E1 medium (Table 1) for induction of somatic embryogenesis. Then, as soon as calli were initiated, they were subcultured on E2 medium for the production of nodular and yellowish calli (Plate I, Fig.1). These calli formed somatic embryoïds after transfer in darkness on E3 medium (Plate I, Fig.2). Embryoids which had been separated from nodular calli produced globular and heart stages embryos after 4 weeks culture on E2 medium. Complete development of somatic embryos (Plate I, Fig.3) was achieved in a 17/7 light-dark cycle on E3 medium, without growth regulator. Rarely, both somatic embryos and buds were obtained simultaneously (Plate I, Fig. 4).

Cytological observations have indicated that the first step in cell differentiation, i.e, proembryos organization, occurred from superficial areas of the calli (Plate II, Fig. 1) and was identical to the meristematic differentiation in organogenic calli (Tétu, 1989). From superficial cells of the embryogenic calli emerged numerous globular somatic embryos (Plate II, Figs. 2 and 3). Ultrastructural studies from organogenic and embryogenic calli revealed that embryogenic calli were mainly composed of cells with a large amount of protein bodies whereas cells of organogenic calli were highly vacuolated with dense and juvenile cytoplasmic organisation without protein bodies (Tétu, 1989). Biochemical studies in Carrot and Pea embryogenic calli cultures revealed two different specific proteins (Stirn and Jacobsen, 1987; Sung and Okimoto, 1981) of similar molecular weights.

2/ Somatic embryogenesis from the zygotic embryos :

Zygotic immature embryos were excised from polygerm glomerules coming from three diploïd fertile populations (T21, T130, M2). The nature (coenocytic or cellular) or

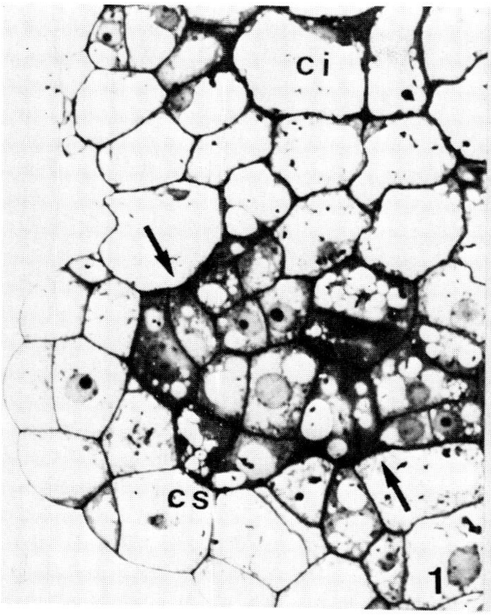

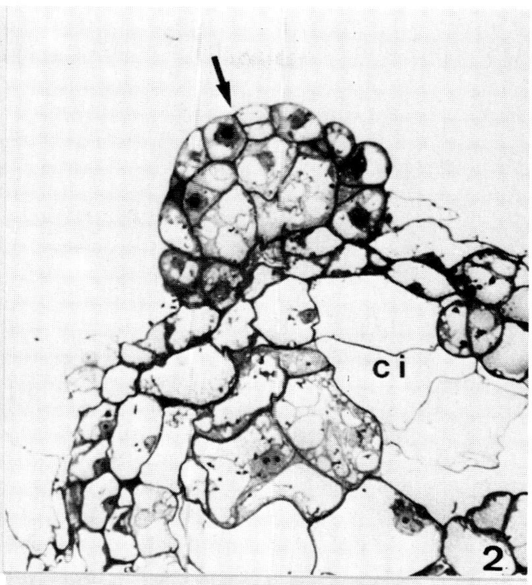

PLATE II : Cytological observations from embryogenic sugarbeet calli.

Fig. 1 : Internal (ci) and superficial (cs) cell layers; differentiation of meristematic cells occurred in the superficial cell layers. Fig. 2 : Proembryo emergency (arrow) from external cells; internal cells (ci) are undifferentiated. fig. 3 : Actives areas showing globular somatic embryos (eg); angular cells (arrows) links the embryos with undifferentiated cells.

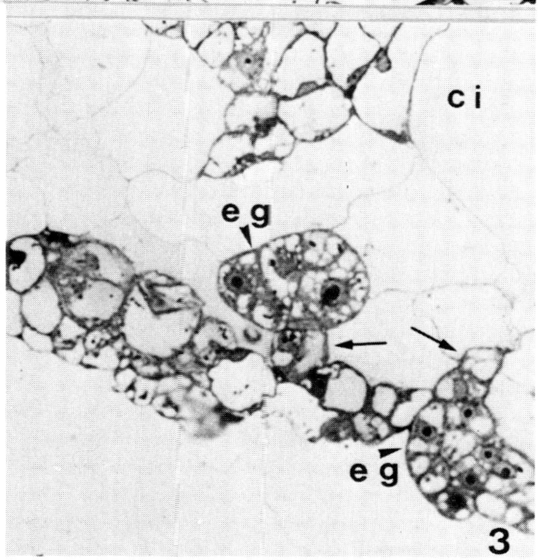

the consistency of the endosperm (liquid or solid) was selected as a marker for the evaluation of the zygotic embryos development. Entire immature zygotic embryos were plated on ED4 medium (Table 1) in darkness at 31°C. Results are given in Table 3.

Table 3 : Role of the endosperm for screening embryogenic zygotic embryos in three sugarbeet diploïd populations.

Consistency of the endosperm (a)	T21	T130	M2
Liquid	(144) 5	(118) 4	(158)7
Solid	(82) 0*	(64) 0*	(102) 0*

(a) : The stages of development of the zygotic embryos were evaluated according to the consistency of the endosperm; i.e, liquid or solid endosperm.

The number of zygotic embryos plated in culture are indicated in parenthesis; those underlined represents the number of embryogenic responses.

* : Adventitious bud formation (rosette like structure).

Positive responses have been obtained only from zygotic embryos excised from seeds with a liquid endosperm, but not from those with a solid one. At the cellular level, the liquid endosperm indicates a coenocytic organisation while the solid endosperm reflect the cellularisation of the endosperm (Raghavan, 1976). So, embryogenic responses appeared correlated to the morphogenic competences according to specific stages of development of the zygotic embryos, i.e, the physiological age of the explant. Embryogenic frequencies were very low and were about 3.8% (16/520) with the convenient zygotic embryos with liquid endosperm. Embryogenic formations occurred directly from hypocotyls of the entire zygotic embryos or via a callus phase, after a minimum of 4 - 6 weeks cultures. These embryoïds did not develop further than the cotyledonary stages, and finally necrosed without growth of the root axis.

3/ Adventive somatic embryogenesis from petiole explants :

Two triploïd seed lots, cv, Monosvalof and Veronique, were cultured on ED7 medium (Table 4), according to Freytag *et al*, (1988). Petiole explants (50 per separate experiment) were excised and plated on ED5, ED6 and ED7 media. Results are scored in Table 4. No somatic embryogenesis occurred on the Freytag medium (ED7) but just adventitious bud formation. On the other hand, from the two picloram supplemented media, i.e, ED5 and ED6, adventitious somatic embryos have been observed at a very low frequency, < 2%, either from the excised and non excised zones of the petiole explants.

Table 4 : Influence of nutritive sequences on somatic embryogenesis from sugarbeet hypocotyl explants

Nutrient media (a)	Sequences n°1	n°2	n°3
Shoot cultures	ED7	ED7	ED7
Petiole cultures	ED5	ED6	ED7
Somatic embryogenesis (%)	1,4	2,2	0 *

(a) : Vitroclones were established from 20 - 30 seeds of each Monosvalof and Veronique cv, on ED7 medium (according to Freytag *et al*, 1988; Table 1). 500 petiole explants were excised and plated on ED7 medium, or subcultured on picloram media (ED5, ED6; Table 1).

(*) : Adventitious bud formation.

Most of the somatic embryos did not germinate normally due to the necrosis of the root axis but just developed their stem apices, with a subsequent bud-like development. In order to screen original adventitious buds from uncomplete developped somatic embryos, scanning electron microscopic studies have been done, and revealed that original adventitious buds showed numerous hair cells (Tétu, 1989).

PEA AND SOYBEAN TISSUE CULTURES :

For these two species, the developmental stage of the zygotic embryos was very critical for somatic embryos initiation (Table 5) and most of the embryogenic responses were observed with 3 to 4 mm zygotic embryos. Those longer than 6 mm germinated precociously, and those smaller than 2 mm, turned brown in cultures.

Table 5 : Influence of embryo-size on embryogenesis from immature pea and soybean zygotic embryos.

Size (mm)	Embryogenesis (%)			
	PEA (a)		SOYBEAN (b)	
	Bonnaire	Davina	Osso	Sito
0 - 1	0	0	0	0
1 - 2	2	5	26	12
3 - 4	35	48	60	32
5 - 6	32	22	18	4
7 - 8	4	6	0	0

(a) and (b) : Initiation of somatic embryogenesis was achieved on L1 medium.

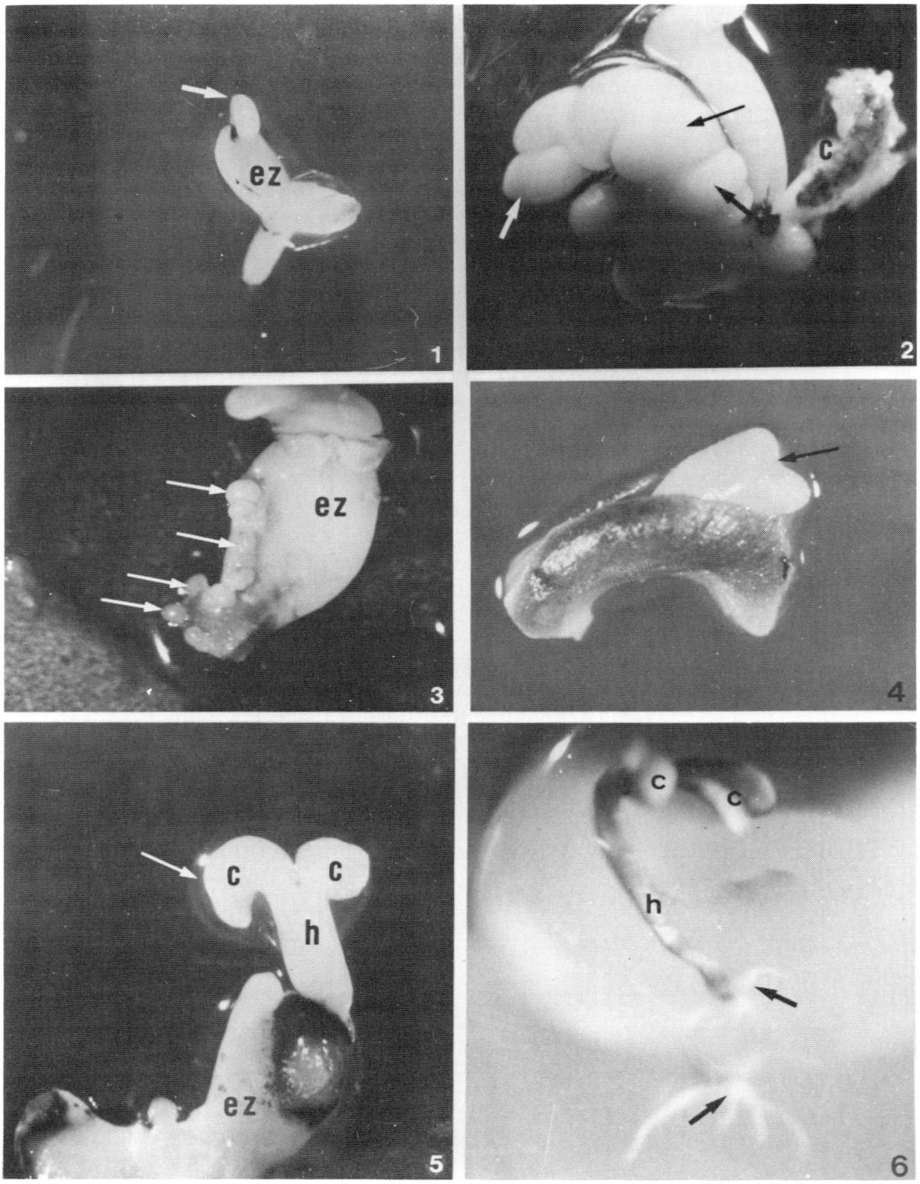

PLATE III : Adventive somatic embryogenesis from zygotic soybean embryos.

Fig. 1 : Embryogenic nodule initiation from cotyledon (c) of an entire immature zygotic embryos (ez) after 4 weeks culture; (x8). Numerous globular (fig. 3) and heart (Fig. 4) stages somatic embryos from immature zygotic embryos of cv Verdon. Fig. 5 : complete somatic embryos showing cotyledons (c), and elongation of the hypocotyl region (h). Fig. 6 : somatic embryo germination with development of the root axis (arrows); (h) hypocotyl; (c) cotyledons.

Observations were made 4-5 weeks after the cultures were initiated, using 50/60 embryos per treatment.

Results are scored in percentage of immature embryos forming embryogenic cultures.

1/ Somatic embryogenesis in soybean :

The soybean somatic embryos did not develop spontaneously on the initiation media L1, and for complete plant regeneration, different successive nutritive media were needed (Table 6).

Table 6 : Influence of organic addenda and hormones on somatic embryos initiation from the cultured immature zygotic soybean embryos cultures, cv Sito.

Induction medium of somatic embryos + amino acids or vitamins	embryogenic zygotic embryos (%)	Intensity (a)	Remarks
Control *	60	+ +	Adventitious somatic embryogenesis from immature cotyledons
+ **L-Arginine** (2.9 mM)	60	+ +	Adventitious somatic embryogenesis from immature cotyledons.
+ **L-Glutamine** (3.4 mM)	90	+ + +	Direct somatic embryogenesis. High number of somatic embryos per immature embryos (15/20)
+ **Nicotinic acid** (1 mM)	80	+ + +	Calli formation. Direct and indirect somatic embryogenesis.
+ **L-Glycine** (6.7 mM)	20	+	Undifferentiated calli
+ **L-Valine** (4.3 mM)	10	+	Abnormal somatic embryos development
+ **L-Proline** (4.35 mM)	60	+ +	Adventitious somatic embryogenesis from immature cotyledons

(a) : 40 zygotic embryos of 4 mm in length plated in darkness at 31°C were used per experiment. Results were scored after 4-5 weeks culture. Each cotyledon showing new expanded tissues was considered as positive response.

+ : low (< 4) ++ : moderate (5-10) +++ : high (> 10)

* : L1 medium (Table 2).

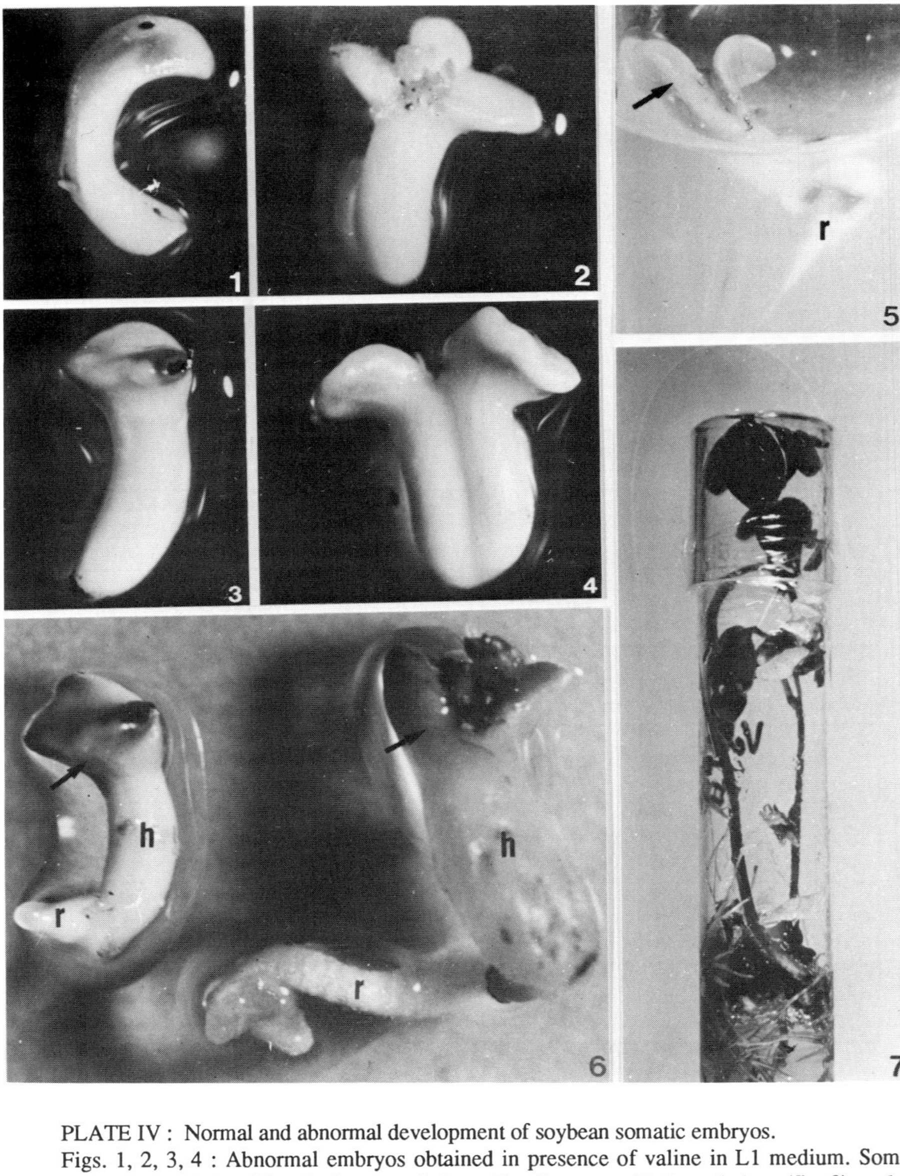

PLATE IV : Normal and abnormal development of soybean somatic embryos.

Figs. 1, 2, 3, 4 : Abnormal embryos obtained in presence of valine in L1 medium. Somatic embryo with one cotyledon (Fig. 1), three cotyledons (Fig. 2), two welded cotyledons (fig. 3), and two cotyledons welded by the hypocotyl region (fig. 4). Fig.5 : precocious germination of a somatic embryo. Fig. 6 : abnormal germination with hypertrophy of the hypocotyl (h). Fig. 7 : normal germination leading to complete plant regeneration, cv, Verdon.

First step in the embryogenic pathway was the formation of expanded zones at the upper surface of the cotyledons (Plate III, Figs.1, 2, 3). After a minimum of 4 weeks, somatic embryoïds developed from globular (Plate III, Fig.3), heart (Plate III, Fig. 4), and cotyledonary stages (Plate III, Fig. 5). Germination of the somatic embryos (Plate III, Fig.6) was obtained after transfer in a 17/7 light/dark cycle.

Histological observations showed that the superficial cells of the cotyledons giving somatic embryos were composed of meristematic and vacuolated cells (Tétu, 1989), with big nuclei and juvenile cytoplasmic organization. When vitamin and/or amino acids were added to the medium, the frequency of embryogenic responses was increased (Table 7), e. g, with glutamine and nicotinic acid, the number of somatic embryos per immature zygotic embryo was considerably increased.

Table 7 : Effect of nutritive media on the development of embryogenic tissues.

Sequences of nutritive media *	Embryogenic zygotic embryos (%)	% of somatic embryos developped to the cotyledonary stage.	% of somatic embryos developped to complete plant regeneration.
A	60	4	0
B	90	6	0
C	90	45	20
D	90	50	60
E	90	26	10

* : The composition of media are given in table 1.

A : L1 medium, **B** : L5 medium, **C** : L5 + L6 media; **D** : L5 + L6 + L7 media, **E** : L5 + L7 media. Zygotic embryos (5 mm in length) are plated in darkness at 31°C; then, young somatic embryos (at the cotyledonary stage) are transferred directly (**A, B**) or after subculture (**C, D, E**) to a 16/8 light /dark cycle.

Proline and arginine at the concentration used, had no effect on embryoids formation, while valine or glycine apeared to be inhibitory for somatic embryogenesis. Moreover, with 4.3 mM of valine, numerous abnormal somatic embryos were obtained (Plate IV, Figs. 1, 2, 3,4). The effect of amino acids and/or vitamins on somatic embryogenesis is a well known fact, and has been demonstrated on several different species (Sangwan, 1983; Trigiano and Conger, 1987; Ronchi *et al*, 1984; Duncan *et al,* 1985; Armstrong et Green, 1985). Nevertheless, The role of auxin used, was the second factor affecting somatic embryogenesis (Table 6). Indeed, we have observed that three successive media were necessary for complete plant regeneration. This observation was correlated to the auxin level in the medium.Somatic embryos did not developped further

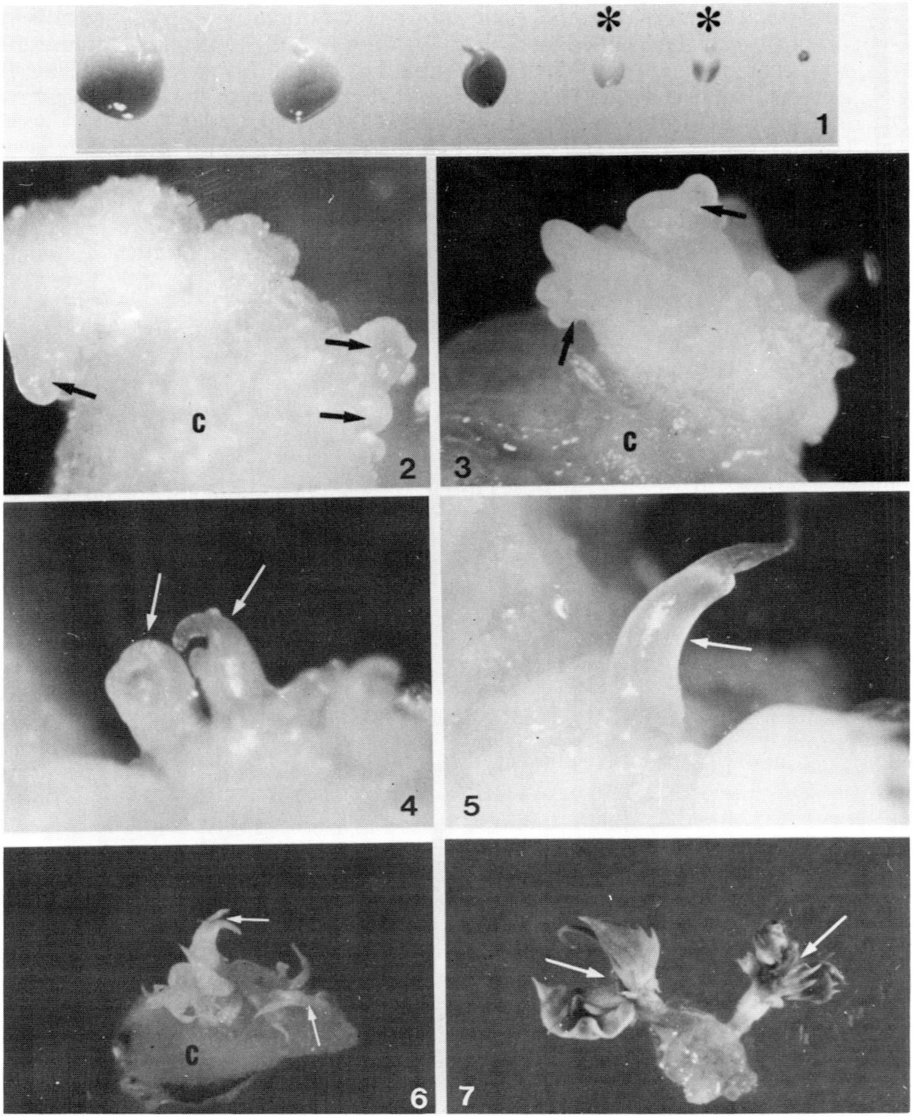

PLATE V : Adventive somatic embryogenesis in Pisum sativum.

Fig. 1 : Immature zygotic embryos used in somatic embryogenesis; embryogenic responses were obtained for 3 - 4 mm zygotic embryos (x 2.5). Fig. 2 : Initiation of embryogenic tissues (arrows) in darkness, from immature cotyledons (c) (x 20). Fig. 3 : Starting of somatic embryos individualization on L1 medium, (c) cotyledon (x 20). Fig.4 : individualization of two somatic embryos (x10). Fig.5 : somatic embryos showing the elongation of the hypocotyl region (x10). development of multiple somatic embryos from an entire cotyledon (x 10). Fig. 7 : somatic embryos germination on L6 medium after subculture in light/dark condititons. Stem apices development (arrows) (x 10).

than the heart stages on the initiation media (A, B sequences). When somatic embryos were transferred at the young stages to a less concentrated auxin medium (C and D sequences), 45/50 % of the somatic embryos developped to the cotyledonary stages. However, for obtaining the development of stem apices, a third subculture was needed; this, has permitted to increase the percentages of complete somatic embryos development from 20 % (C sequence) to 60 % (D sequences). Independantly from the media, precocious germination occurred (Plate IV, Figs. 5, 6). A nitrogen supplement incorporated in the germination medium have given vigourous dark greenish plants with normal axillary buddings at the cotyledonary node, while without nitrate supplement, stunted and fragile regenerated plants were obtained. The use of nitrogen supplement as well as, potassium or ammonium nitrate, have also been used for development of somatic embryos in other species (Duncan *et al*, 1985; Von Arnold, 1986).

2/ Embryogenesis in Pea :

Analogous experimental conditions to those used for Soybean have permitted somatic embryogenesis from immature zygotic embryos. The size of the explants (Plate V, Fig. 1) used were found to influence significantly embryogenic responses, as discussed above (Table 5). The data presented here, showed the embryogenic responses obtained, using NAA for the initiation of somatic embryos. However, we have obtained somatic embryogenesis using the picloram (L3 and L4 media, Table 2). Amino acids and/or vitamins increased significantly embryogenesis (Table 6).

Table 6 : Effect of amino acids and/or vitamins on percentage of immature embryos forming somatic embryos.

Media **MS basal medium**	**Embryogenesis (%)** (a)	**Intensity** (b)
BM (control)*	8	+
BM + Arginine (60μM)	18	++
BM + Nicotinic acid (40 μM)	15	+
+ Thiamine (15 μM)	10	++
Arginine (60μM) BM + Nicotinic acid (40 μM) Thiamine (15μM)	48	+++

* : Basal embryogenic medium composed of MS salts, supplemented with 43 μM NAA. (a) : Observations were taken 5 weeks after the culture were initiated in darkness using cv, Bonnaire, with 50 - 60 embryos per treatment.(b) : Identical to Table 6.

For the Bonnaire cultivar, using 43 µM NAA in MS medium was sufficient to obtain twelve percent of embryogenic responses (Plate V, figs. 2, 3, 4 and 5). However, using amino acids as well as arginin or vitamins as well as thiamine and nicotinic acid increased slightly the number of positive responses. The best percentage of embryogenic explants, 48 % was achieved with cv, Bonnaire using a combination of arginine, thiamine and nicotinic acid (L1 medium, Table 2). To germinate (Plate V, Fig. 6 and 7), somatic embryos needed to be subcultured on a medium devoid of NAA (L2, Table 2).

CONCLUDING REMARKS :

The organogenic and/or embryogenic pathways were determined by the choice of the explants and the culture media. We have reported earlier, the factors involved in organogenesis in Pea and Sugarbeet tissue cultures (Tétu, 1989), and especially, the influence of TIBA on organogenesis; TIBA, known as an auxin transport inhibitor (Depta and Rubery, 1984; Rubery, 1987) had no influence on pea and sugarbeet calli, but was effective when applied during the initiation of the calli (Tétu *et al*, 1987). Elsewhere, a high frequency regeneration procedure *via* direct bud formation have been pointed out, using a BAP-TIBA medium, for various Pea and Sugarbeet breeding lines and/or cultivars (Tétu, 1989; Detrez *et al*, 1989).

- Somatic embryos developped either from calli or directly from organs without involving any intermediate callus stage. Since the first observation describing that somatic embryos could be obtained directly from somatic cells, repeated attempts have been made to regenerate plants via this pathway, and somatic embryogenesis has now been observed in many plant taxa (Tomes, 1985; Maheswaran and Williams, 1986; Vasil, 1988). Since regeneration of plants from callus cultures often show genetic variability, direct somatic embryogenesis from organs can be more efficiently used for the application of biotechnology in the improvement of crops. Direct somatic embryogenesis also provides a model system for studying the basic developmental morphology of non-zygotic embryos and the influence of hormones and other factors on embryos development. We have observed that both somatic embryos and buds could be obtained in pea and sugarbeet tissue cultures (Tétu *et al*, 1989; Tétu, 1989).

- Cytological observations showed that the early stages of differentiation, in the embryogenic and organogenic sugarbeet calli, started by the organization of meristematic areas (Tétu, 1989). However, the most of the somatic embryos did not developped sponstaneously into complete plants, but needed multiple hormonal sequences to germinate. Moreover, in few cases, incompletions of the somatic embryos led to a subsequent bud-like development, with necrosis of the root axis.

- In Sugarbeet, we have demonstrated that adventitious buds were covered with hair cells, but not the somatic embryos (Tétu *et al*, 1989). At the ultrastructural level, protein bodies were observed in the embryogenic sugarbeet calli (Tétu, 1989) but not in the organogenic ones; it is of interest to note that some proteins for specific embryogenic differentiation patterns in pea (Stirn and Jacobsen, 1987) and carrot (Sung and Okimoto, 1981) have also been reported.

The most limiting factors for using genetic engineering in order to complement the conventional crops breeding programmes are the difficulties in regenerating plants *in vitro*. The various regeneration systems established here, are a prerequisite for the use of *Agrobacteria* -mediated gene transfer. We observed that the frequency of adventitious organogenesis was not correlated to the genotype and/or cultivars. On the contrary, somatic embryogenesis appeared highly controlled by genotypes and/or cultivars. So, the selection of genotypes and/or cultivars that are compatible under the conditions chosen for regeneration, as well as genotypic variability for regeneration must be considered. Moreover, the data concerning the genetic stability or instability of the regenerated plants, via direct organogenesis both in Pea (Rubluo *et al*, 1984; Griga *et al*, 1986; Ahmed *et al*, 1987; Natali and Cavalini, 1987) and in Sugarbeet (Hussey and Hepher, 1978; Freytag *et al*, 1988; Detrez et al, 1989) are discordant.

Using Agrobacterium gene transfer, no success have been reported in Pea and Sugarbeet, despite the recent success with Glycine max (Hinchee *et al*, 1988). The production of transgenic plants in soybean was preceded by investigations on the suceptibility of cultivars to different *Agrobacterium* strains (e.g. Pederson et al, 1983; Byrne et al, 1987). In Pea, (Puonti-Kaerlas *et al*, 1989; Hobbs *et al*, 1989), it has been demonstrated that pea genotypes differed in their susceptibility to Agrobacteria strains. The presence of this genotype/strain interaction complicates the selection of a suitable strain or genotype for transformation experiments; moreover, differences between *in vivo* and *in vitro* genotypes have been found (Hobbs *et al*, 1989).

The above describe system might be useful in obtaining transgenic plants in these important crop species.

BIBLIOGRAPHY

AHMED R., DUTPA S. and GHOSH P.D. 1987. The cytological status of plants regenerated from shoot-meristem culture of *Pisum sativum* L. Plant Breeding, **98** : 306-311.

ARMSTRONG C.L. and GREEN C.E.,. 1985. Establishment and maintenance of friable, embryogenic maize callus and the involvement of L-proline. Planta, **164** :207-214

BARWALE U.B., KERNS H.R. and WIDHOLM J.M. 1986. Plant regeneration from callus cultures of several soybean genotypes via embryogenesis and organogenesis. Planta, **167** : 473-481.

CHRISTIANSON M. L., WARWICK D. A. and CARLSON P. S. 1983. A morphogenetically competent soybean suspension culture. Science. **222** : 632-634.

DEPTA H. and RUBERY P.H. 1984. A comparative study of carrier participation in the transport of 2,3,5-triiodobenzoic acid, indole-3-acetic acid, and 2,4-dichlorophenoxyacetic acid by *Cucurbita pepo* L. hypocotyl segments. J. Plant. Physiol., **115** : 371-387.

DETREZ C., TETU T., SANGWAN R.S. and SANGWAN-NORREEL B.S. 1988. Direct organogenesis from petiole and thin cell layer explants in sugarbeet cultured *in vitro*. J. Exper. Bot., **39** : 917-926.

DETREZ C. 1988. Etude histophysiologique et cytogénétique de l'organogenèse adventive sur pétiole chez *Beta vulgaris* L. Contribution à la lutte contre la rhizomanie. Thèse université de compiègne.

DETREZ C., SANGWAN R. S. and B.S. SANGWAN-NORREEL. 1989. Phenotypic and karyotypic status of Beta vulgaris plants regenerated from direct organogenesis in petiole culture. Theor. Appl. Genet. (In Press)

DUNCAN D.R., WILLIAMS M.E., ZEHR B.E., and WIDHOLM J.M. 1985. The production of callus capabl of plant regeneration from immature embryos of numerous *Zea mays* genotypes. Planta, **165** : 322-332.

FINER J.J 1988. Apical proliferation of embryogenic tissue of soybean (*Glycine max* L. Merrill). Plant Cell Reports, **7** : 238-241.

FREYTAG A.H., ANAND S.C., RAO-ARELLI A.P. and OWENS L.D. 1988. An improved medium for adventitious shoot formation and callus induction in *Beta vulgaris* L. *in vitro* . Plant Cell Reports, **7** : 30-34.

GRIGA M., TEJKLOVA E., NOVAK F.J. and KUBALAKOVA M. 1986. *In vitro* clonal propagation of *Pisum sativum* L. Plant Cell Tissue and Organ Culture, **6** : 95-104.

HAMMATT N. and DAVEY M.R. 1987. Somatic embryogenesis and plant regeneration from cultured zygotic embryos of soybean (*Glycine max* L.). J. Plant. Physiol., **128** : 219-226.

HINCHEE M.A.W., CONNOR-WARD D.V.C.W., NEWELL C.A., MC-DONNELL R.E., SATO S.J., GASSER C.S., FISHHOFF D.A., FRALEY R.T. and HORSCH R.B. 1988. Production of transgenic soybean plants using *Agrobacterium* mediated DNA transfer. Biotechnology, **6** : 915-922.

HUSSEY G. and HEPHER A. 1978. Clonal propagation of sugarbeet plants and the formation of polyploids by tissue culture. Ann. Bot., **42** : 477-479.

JACOBS M.J., BUBGEE W. M. and GABRIELSON D.A. 1985. Enumeration, location and characterization of endophytic bacteria withins sugarbeet roots. Canadian journal of batany. 63 : 1262-1265.

JACOBSEN H.J. and KISELY W. 1984. Induction of somatic embryos in Pea (Pisum sativum L.). Plant. Cell. Tissue. and Organ Culture. **3** : 319-324.

KAMEYA T. et WIDHOLM J. 1981. Plant regeneration from hypocotyl sections of *Glycine* species. Plant. Sci. Letters, **21** : 289-294.

KISELY W., MYERS J.R., LAZZERI P.A., COLLINS G.B. and JACOBSEN H.J. 1987. Plant regeneration via somatic embryogenesis in pea (*Pisum sativum* L.) Plant Cell Reports, **6** : 305-308.

KOMATSUDA T. and OHYAMA K. 1988. Genotypes of high competence for somatic embryogenesis and plant regeneration in soybean (Glycine max. Theor. Appl. Genet. **75** : 695-700.

LAZZERI P.A., HILDEBRAND D.F. and COLLINS G.B. 1985. A procedure for plant regeneration from immature cotyledon tissue of soybean. Plant. Mol. Biol. Rep., **3** : 160-167.

LAZZERI P.A., HILDEBRAND D.F. and COLLINS G.B. 1987. Soybean somatic embryogenesis : effects of hormones and culture manipulations. Plant Cell Tissue and Organ Culture, **10** : 197-208.

LIPPMANN B. and LIPPMANN G. 1984. Induction of somatic embryos in cotyledonary tissue of soybean, *Glycine max* L. Merr. Plant Cell Reports, **3** : 215-218.

MAHESWARAN G. and WILLIAMS E. G. 1984. Direct somatic embryoïd formation on immature embryos of *Trifolium repens*, *T. pratense* and *Medicago sativa* , and rapid clonal propagation of *T. repens*. Annals of Botany, 54 :201-211.

MAHESWARAN G. and WILLIAMS E. G. 1985. Origin and development of somatic embryoids formed directly on immature embryos of *Trifolium repens in vitro*. Annals of Botany, 56 : 619-630.

MAHESWARAN G. and WILLIAMS E. G. 1986b. Direct somatic embryogenesis on immature sexual embryos of *Trifolium repens*, *T. subterraneum* and *T. resupinatum*. Plant. Cell. Reports, **3** : 165-168.

MAHESWARAN G. and WILLIAMS E.G. 1987. Uniformity of plants regenerated by direct somatic embryogenesis from zygotic embryos of *Trifolium repens* . Ann. Bot., **59** : 93-97.

MARGARA J. 1982. Les bases de la multiplication végétative; les méristèmes et l'organogenèse. (Ed. I.N.R.A).

MROGINSKI L.A. and KARTHA K.K. 1981. Regeneration of pea (*Pisum sativum* L. cv. Century) plants by *in vitro* culture of immature leaflets. Plant Cell Reports, **1** : 64-66.

MURASHIGE T. and SKOOG F. 1962. A revised medium for rapid growth and bioassays with tobacco tissue cultures. Physiol Plant, **15** : 473-497.

NATALI. and CAVALLINI A. 1987. Nuclear cytology of callus and plantlets regenerated from pea (*Pisum sativum* L.) meristems. Protoplasma, **141**: 121-125.

PHILLIPS G.C. and COLLINS G.B. 1981. Induction and development of somatic embryos from cell suspension cultures of soybean. Plant Cell Tissue Organ Cult., **1** : 123-129.

RAGAVHAN V. 1986. Embryogenesis in angiosperms. (Ed) Cambridge University Press.

RANCH J.P., OGLESBY L. and ZIELINSKI A.C. 1985. Plant regeneration from embryo-derived tissue cultures of soybeans *in vitro*. Cell. Dev. Biol. **21** : 653-658.

RONCHI V.N., CALIGO, M.A., NOZZOLINI M. and LUCCARINI G. 1984. Stimulation of carrot somatic embryogenesis by proline and serine. Plant Cell Report. **3** : 210-214.

RUBERY P.H 1980. The mechanism of transmembrane auxin transport and its relation to the chemiosmotic hypothesis of the polar transport of auxin. In. : Plant growth substances 1979, pp. 50-60, Skoog, F., Ed. Springer, Berlin Heideberg New York.

RUBERY P.H. 1987. Auxin transport. In : Plant hormones and their role in plant growth and development, DAVIES P.J. (ed.) Martinus NIJHOFF Publishers) : 341-362.

RUBLUO A., MROGINSKI L., and KARTHA K.K. 1982. Morphogenetic responses of pea leaflets cultured *in vitro*. In: Fujiwara A. ed. Plant Tissue Culture 1982. Tokyo: The Japanese Assoc. Plant Tissue Culture, pp. 151-152.

RUBLUO A., KARTHA K.K., MROGINSKI L.A. and DYCK J. 1984. Plant regeneration from Pea leaflets cultured *in vitro* and genetic stability of regenerants. J. Plant Physiol., **117** : 119-130.

SANGWAN R.S. 1983. Effects of exogenous amino acids of in vitro androgenesis of *Datura*. Biochem. Physiol. Pflanzen. 178 : 415-422.

SAUNDERS J.W. 1984. Shoot-regeneration from hormone-autonomous callus fom shoot cultures of several sugarbeet (Beta vulgaris L.) genotypes. Plant. Sci. Let. **34** : 219-223.

SAUNDERS J.W. and DOLEY W.P. 1986. One step shoot regeneration from callus of whole plant leaf explants of sugarbeet lines and a somaclonal variant for *in vitro* behavior. J. Plant Physiol., **124** : 473-479.

SUNG, Z. R. and OKIMOTO R. 1981. Embryogenic proteins in carrot somatic embryos. Proc. Nat. Acad. Sci. USA **78** : 3683-3687

TETU T. 1985. Aptitudes morphogénétiques in vitro de divers explants chez la betterave à sucre. D.E.A. de Biologie et Physiologie Végétales. University of PARIS VI, PARIS, FRANCE.

TETU T., SANGWAN R.S. and SANGWAN-NORREEL B.S. 1987. Hormonal control of organogenesis and somatic embryogenesis in *Beta vulgaris* callus. J. Exp. Bot., **38** : 506-517.

TETU T., SANGWAN-NORREEL B.S. and SANGWAN R.S. 1987. Embryogenèse somatique et régénération *in vitro* chez trois variétés précoces de Soja. C.R. Acad. Sci., Paris, **305** : 613-617.

TETU T.; SANGWAN R.S. and SANGWAN B. 1989. Direct somatic embryogenesis and organogenesis in cultured immature embryos of Pisum sativum L. (In Press).

TETU T. 1989. Régénération de plantes in vitro par organogenèse et embryogenèse somatique chez la betterave sucrière (Beta vulgaris L., le Pois (Pisum sativum L.) et le Soja (Glycine max. L). Etudes physiologiques, morphologiques, anatomiques et ultrastucturales.

TOMES D.T. 1985. Cell culture, somatic embryogenesis and plant regeneration in maize, rice, sorghum and millets. In : Bright, S.W.J. and M.G.K. JONES (Eds), Cereal Tissue and cell culture. pp175-203. Martinus Nijhoff/Dr W. Junk Publ., Dordrecht, Netherlands

TRIGIANO R.N. and CONGER B.V. 1987. Regulation of growth and somatic embryogenesis by proline and serine in suspension culture of *Dactylis glomerata*. J. Plant Physiol. **130** : 49-55

VAN GEYT J.P.C. and JACOBS M. 1985. Suspension culture of sugarbeet (Beta vulgaris L.). Induction and habituation of dedifferentiated and self-regenerating cell lines. Plant Cell Reports, **4** : 66-69.

VASIL I.K. 1988. Progress in the regeneration and genetic manipulation of cereal crops. Bio/Technology, **6** : 397-401.

VON ARNOLD S. 1986. Improved efficiency of somatic embryogenesis in mature embryos of Picea abies L;. Karst. J. Plant Physiol. 128 : 233-244.

WRIGHT M.S., WARD D.V., HINCHEE M.A., CARNES M.G. et KAUFMAN R.J. 1987. Regeneration of soybean (*Glycine max* L. Merr.) from cultured primary leaf tissue. Plant Cell Reports, **6** : 83-89.

IN VITRO CULTURE OF WHEAT AND RICE FOR UNDERSTANDING THE MOLECULAR BASIS OF SOMATIC EMBRYOGENESIS AND FOR TRANSFORMATION

N. MAHESHWARI[1,2], K. RAJYALAKSHMI[1], C.N. CHOWDHRY[2], A. GROVER[2], A.K. TYAGI[2] AND S.C. MAHESHWARI[1,2]
Departments of Botany[1] and Plant Molecular Biology[2]
University of Delhi
Delhi, India

ABSTRACT. The article gives, firstly, a brief overview of the earlier work on *in vitro* regeneration and transformation studies on wheat and rice. This is followed by presentation of results of our work on these crops. In our laboratory, complete plantlets have been regenerated via somatic embryogenesis with high efficiency employing explants from inflorescences, seeds and young leaf bases in two popular Indian varieties, namely, Sonalika in wheat and Basmati in rice. Trials with many other varieties show that there is a marked effect of genotype. A preliminary study with wheat to determine differences in embryogenic and non-embryogenic calli, at protein level, has revealed some interesting quantitative and qualitative changes. Furthermore, attempts to elicit better response and to retain long-term regeneration potential have proved fruitful. Conditions for protoplast isolation and culture have been worked out and experiments have also been initiated on transient expression of genes.

1. INTRODUCTION

Notwithstanding the great achievements of plant breeders in improving wheat and rice in the past decades employing conventional approaches, the fact remains that in many countries the yields have already reached a plateau with little prospect of further gains. It is in this context that recent developments in cell culture and recombinant DNA technology are of great interest since they not only allow unprecedented ease in development of pure lines and generation of variability by induction of novel mutations (e.g. through anther and pollen culture), but they also provide a powerful biotechnological tool to tailor crops according to our requirements by way of genetic engineering at the single cell level. While the possibilities are immense taking a cue from model systems such as tobacco, *Datura*, *Petunia* and other dicots, nonetheless, in reality great difficulties are being experienced in cereals such as wheat and rice. Thus, despite great efforts, many cultivars have proved extremely recalcitrant. Furthermore, these crops are not yet amenable to transformation via *Agrobacterium*, thus necessitating use of alternative strategies for

R.S. Sangwan and B.S. Sangwan-Norreel (eds.), The Impact of Biotechnology in Agriculture, 191–213.

genetic engineering. The real remedy for these problems lies in devising methods of general applicability for culture and efficient regeneration, not only from callus or isolated cells in suspensions but also from protoplasts. It is also important to understand the molecular basis of differentiation so that tissue culture technology becomes less empirical.

Here, we first provide in brief a review of the past work on tissue, cell and protoplast culture of wheat and rice. We then present our findings with respect to several Indian varieties of these crop plants. Additionally, our attempts to understand molecular basis of somatic embryogenesis in wheat and on developing transformation systems are also summarized.

2. REVIEW OF PREVIOUS WORK

2.1. Wheat

Tissue culture studies on wheat employing somatic cells were initiated in the mid-seventies with various explants like mature and immature embryos, inflorescences, leaves and mature seeds. Differentiation of plantlets was reported via the organogenetic mode, but the response was sporadic (<u>see</u> Schaeffer et al. 1984, Maddock 1985, Vasil and Vasil 1986 and references cited therein). It is, therefore, natural that many efforts have been made to optimize conditions for better response -- these include studies on various factors like orientation of explant in relation to medium, choice of carbohydrate source and of growth regulators (Eapen and Rao 1985, Galiba and Erdei 1986, Papenfuss and Carman 1987, Purnhauser et al. 1987, Elena and Ginzo 1988, Brown et al. 1989). More recently, however, somatic embryogenesis has been reported employing as explants immature embryos (Ahloowalia 1982, Ozias-Akins and Vasil 1982, 1983a, b, Maddock et al. 1983, Heyser et al. 1985), inflorescences (Ozias-Akins and Vasil 1982, Maddock et al. 1983) and seeds (Nabors et al. 1983, Heyser et al. 1985). Since the embryogenetic mode of plantlet regeneration gives greater assurance of obtaining plants originating from single cells and avoiding chimeric plantlets, many workers have been trying to optimize conditions for somatic embryogenesis. Such response, from explants comprising immature embryos, has been improved by a judicious selection of hormones as for example by the use of 3,6-dichloro-o-anisic acid (dicamba, Carman et al. 1987b) or abscisic acid (Qureshi et al. 1989) in the culture medium, by incubating cultures in low oxygen atmosphere (Carman 1988), and by adjusting the concentration of the osmoticum (Brown et al. 1989). More such studies are necessary since the precise mode of differentiation, i.e. whether the organogenetic or the embryogenetic mode will dominate, as also the degree of response have been found to vary between various cultivars (Maddock et al. 1983, Wernicke and Milkovits 1984, Galiba et al. 1986, Mathias and Fukui 1986, Mathias and Simpson 1986, Mathias et al. 1986, Carman et al. 1987a, Rajyalakshmi et al. 1988).

In addition to somatic cells, gametic cells can also be made to differentiate *in vitro* and develop into plantlets as first revealed by studies on anther culture by Guha and Maheshwari (1964, 1966). The technique has also been applied to wheat (Research Group 303, 1972, Chu et al. 1973, Ouyang et al. 1973, Picard and De Buyser 1973). With further efforts on optimization of various conditions e.g. stage of pollen at the time of culture, pre-treatments to donor plant and explants, composition of media and temperature during culture, it has become possible to obtain haploids and homozygous plantlets both via direct embryogenesis or an intermediate callus phase (*see* Ouyang 1986 and references cited therein). A strong effect of genotype on culture response has, however, been observed (De Buyser and Henry 1979, Wei 1982, He and Ouyang 1984, Lazar et al. 1984, Henry and De Buyser 1985) pointing to the need of improving cultural conditions further. Employing aneuploid, substitution and translocation lines, it has recently been found that genes for anther culture response and regeneration ability are associated with specific chromosomes (Agache et al. 1989). Although much remains to be done, it is worth mentioning that significant biotechnological applications of the anther culture technique have already come in the form of improved varieties of both spring and winter wheat, as in China (Yin et al. 1978, Hu 1986).

Attempts have also been made to establish suspension cultures which can be employed to regenerate plantlets from individual cells. Unfortunately, such cultures have been difficult to establish in the Gramineae and particularly so in wheat. In early attempts at culture in liquid medium, success was limited to only growing small cell clusters rather than single-celled and rapidly growing cell suspensions (Ahuja et al. 1982, Maddock 1987). Similarly, attempts to culture protoplasts from wheat and other members of the Gramineae have largely been futile (*see* Maheshwari et al. 1986). Nonetheless, following the success reported in pearl millet -- of regenerating protoplasts isolated from embryogenic suspension culture by Vasil's group (Vasil and Vasil 1980) -- similar work has also been attempted in wheat. Recently, a couple of reports on regeneration of plantlets from protoplasts of suspension cultures from anther-derived callus (Harris et al. 1988) and on differentiation of albino shoots (Hayashi and Shimamoto 1988) have appeared.

Finally, a few remarks are appropriate on the state-of-art with regard to genetic engineering studies. The lack of infectivity of *Agrobacterium* and non-availability of reliable protoplast culture systems have been the main reasons for the rather slow progress in transformation of wheat, although a report on stable transformation of its relative, *Triticum monococcum*, appeared a few years ago (Lörz et al. 1985). Therefore, other directions of research have been followed. In recent years wheat protoplasts have been employed for study of transient gene expression and for evaluation of various eukaryotic promoters. Ou-Lee et al. (1986) have found that both the 35 S promoter of the cauliflower mosaic virus and the *copia* long terminal repeat promoter from *Drosophila* work with equal efficiency as assayed by employing a reporter gene -- chloramphenicol acetyltransferase (CAT) -- from bacteria after electroporation of protoplasts isolated from cultured cells. Similar success has been achieved with the CAT or β-glucuronidase (GUS) gene by employing protoplasts from various tissues (Lee et al. 1988, 1989). To increase the level of expression, Lee et al. (1989) and Oard et al. (1989) introduced a DNA segment from alcohol dehydrogenase introns and found a greatly enhanced activity -- this effect might be at the level of RNA processing (splicing). Transient expression of neomycin phosphotransferase II (NPTII) gene has also been demonstrated in dry and viable embryos of wheat by imbibition of DNA solution (Töpfer et al. 1989). These reports indicate that it may not be difficult to express foreign genes in wheat provided a stable transformation and regeneration system is available.

2.2. Rice

Organogenesis and regeneration of plantlets in somatic tissue cultures of rice was reported even before wheat in the late sixties, for which credit is due to Japanese workers. Various explants like seeds (Nishi et al. 1968, Nakano and Maeda 1979, Raghav Ram and Nabors 1984), immature or mature embryos (Maeda 1968, Tamura 1968, Lai and Liu 1982, Paul and Ghosh 1984, Raina et al. 1987), endosperm (Nakano et al. 1975, Bajaj et al. 1980), leaves (Henke et al. 1978, Bhattacharya and Sen 1980, Yan and Zhao 1982) and roots (Kawata and Ishihara 1968, Henke et al. 1978, Abe and Futsuhara 1984) have been employed in these studies. Furthermore, a few cases of regeneration via somatic embryogenesis were also reported by employing various explants e.g. from leaves (Wernicke et al. 1981), immature embryos and seeds (Heyser et al. 1983), roots (Abe and Futsuhara 1985) and inflorescences (Chen et al. 1985).

Despite these efforts, the common problem is of low frequency of plantlet formation. Various media-adjuvants that have been found to enhance the response are tryptophan (Siriwardana and Nabors 1983, Raina et al. 1987) and kinetin in addition to 2,4-D (Raghav Ram and Nabors 1984, Oard and Rutger 1988), but they all seem to act in a cultivar specific manner. Other favourable conditions for improved response are a reasonably high callus/medium volume ratio and pre-conditioning of the medium by growth of embryogenic callus (Raghav Ram and Nabors 1985). Another problem in cereal tissue culture is the decline of response with time; but it has been possible to avoid it in rice, to a large extent, by providing a suitable osmoticum in the culture medium in the form of sorbitol or mannitol in addition to sucrose (Kavi Kishor and Reddy 1986, Kavi Kishor 1987). Despite all this work, as pointed out earlier, the regeneration response has been found to vary greatly between cultivars (Wernicke et al. 1981) with the _japonica_ varieties responding better than _indica_ and _javanica_ varieties (Abe and Futsuhara 1984, 1986). Chen and Luthe (1987) have analysed proteins from embryogenic and non-embryogenic calli and found, interestingly, both qualitative and quantitative differences in one- and two-dimensional gel electrophoretic profiles which may pave the way for understanding differential expression of genes associated with the induction of somatic embryogenesis.

As far as production of haploids from pollen is concerned, rice was the first crop among monocots where success was reported (Niizeki and Oono 1968). Subsequently, several reports have appeared on the optimization of various pre-treatments and culture conditions (such as temperature, hormones, and carbohydrate source) leading to improved response, but success with different genotypes has been variable (Chen 1986a). Nevertheless, the potential of haploid breeding for rice has been demonstrated in many countries including China where more than a dozen improved varieties are being cultivated on more than 1,70,000 ha of land (Chen 1986b).

Although attempts to establish suspension cultures and culture of protoplasts were initiated in the late seventies, success until recently was very limited and, at worst, claims could not be substantiated (see Vasil and Vasil 1986). Recently, however, plantlet formation in protoplast cultures has been reported in japonica varieties from different laboratories. As a source of protoplasts, most workers have employed suspension cultures initiated from calli of various origins e.g. mature seeds or embryos (Fujimura et al. 1985, Abdullah et al. 1986, Yamada et al. 1986, Kyozuka et al. 1987), immature embryos (Fujimura et al. 1985, Kyozuka et al. 1987), leaf bases (Abdullah et al. 1986) or anthers (Toriyama et al. 1986). Others have employed calli originating from young coleoptiles (Coulibaly and Demarley 1986), mature seeds or young embryos (Kyozuka et al. 1987). Interestingly, Abdullah et al. (1986) have reported plantlet regeneration from protoplasts via somatic embryogenesis. With such remarkable progress within only a couple of years, it is hoped that regeneration of protoplasts into plantlets from many other sources will also become possible, particularly in those cases where regeneration of plantlets is already established (Zimny and Lörz 1986, Ozawa and Komamine 1989). This may, however, require employment of various new strategies, like heat shock to protoplasts (Thompson et al. 1987), nurse culture (Kyozuka et al. 1987) or density-gradient purification of protoplasts (Masuda et al. 1989) -- all of which have been found to be beneficial.

In contrast to the success in japonica varieties, until recently considerable difficulties have been encountered in regenerating indica varieties. It is, therefore, a great relief -- to those who wish to extend biotechnological approach to a wider range of cultivars in the South-East Asian sub-continent -- that even indica protoplasts have been regenerated into plantlets (Kyozuka et al. 1988, L. Lee et al. 1989, Wang et al. 1989). The main factor contributing to the success with indica varieties seems to be the use of japonica cells to nurse the cultured protoplasts. Nevertheless, a continuing and critical problem is that many varieties do not produce good suspension cultures and, in these, it may not be easy to establish protoplast culture technology (Kyozuka et al. 1988).

Notwithstanding the difficulties that are still encountered with *indica* varieties, the general success in culturing cells and protoplasts of rice and other related crops has brought new life and interest in the area of production of novel genotypes by somatic hybridization and genetic engineering. Complete plantlets have already been regenerated by fusing protoplasts of not only two different cultivars or of varieties of cultivated and wild rice (Toriyama and Hinata 1988, Hayashi et al. 1988) but also of two different genera such as rice and barnyard grass (Terada et al. 1987). Rice protoplasts have also been employed in transient gene expression assay after electroporation and it has been found that not only hormone and wound inducible plant promoters (Marcotte et al. 1988, Logemann et al. 1989) but also promoters of viral and animal origin (Ou-Lee et al. 1986) can function in rice. Furthermore, transient expression of genes has also been shown following particle bombardment of rice cells (Wang et al. 1988).

To turn to genetic engineering studies, since protoplasts have been finally regenerated (and also efforts to use *Agrobacterium* for genetic transformation have failed), several groups have been studying direct gene transfer using protoplasts. The first report on transformation of rice using kanamycin resistance gene came from Japan (Uchimiya et al. 1986, Morota and Uchimiya 1987), and, subsequently, the same group reported regeneration of transformed plantlets (Toriyama et al. 1988). Similar success has been reported by Cocking's group in U.K. (Yang et al. 1988, Zhang et al. 1988) and by Zhang and Wu (1988) who employed GUS gene for transformation. More interestingly, Shimamoto et al. (1989) have, for the first time, regenerated fertile transformed plants of rice employing hygromycin resistance and GUS genes and the trait has been found to be stably inherited in the progeny.

3. RESULTS AND DISCUSSION

Although our laboratory has long been interested in various aspects of tissue culture since the late fifties (*see* Maheshwari in this volume), it is in 1984 that we initiated work on monocots. Some of the work on *Echinochloa* (Tyagi et al. 1985) and wheat (Rajyalakshmi et al. 1988) has already been published. In the following paragraphs, we provide a summary of our results on wheat and rice (efforts will be made to give some information on the methods employed in the text and legends, but more details of culture conditions can be found in our earlier publications mentioned above).

3.1. Wheat

In wheat, work has been done on regeneration of plantlets from inflorescences, seeds and leaves. Experiments have also been conducted to determine differences in gene expression at the time of initiation of callus and those related with embryogenesis by way of analyzing the steady state and *de novo* profiles of proteins. Attempts have also been

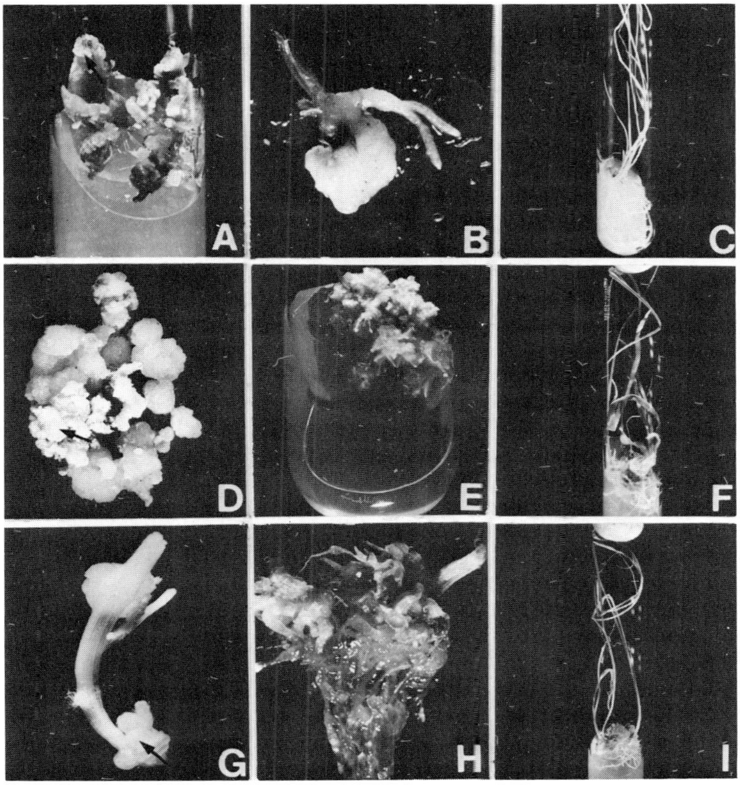

Figure 1. Regeneration of plantlets via somatic embryogenesis in inflorescence, seed and leaf base cultures of wheat. A. Callus from young inflorescence explant on MS + 4.5 x 10^{-5}M 2,4-D + 0.02% CH + 10% CM after 30 days of culture. B. An early stage of plantlet formation after subculture on MS medium. C. A complete plantlet. D. Callus from a seed cultured on MS + 2.5 x 10^{-5}M 2,4-D after 30 days of culture. E. An early stage of plantlet formation on transfer to MS basal medium. F. Complete plantlets. G. Callus from a leaf base explant on MS + 2.5x10^{-5}M 2,4-D after 30 days of culture. H. An early stage of plantlet formation on transfer to MS basal medium. I. Complete plantlets. Inflorescence and seed cultures were initiated in diffuse light, but leaf cultures were maintained in dark for the first 30 days. For plantlet regeneration, cultures were incubated under 3.27 W.m^2 light intensity. Arrows in A, D and G indicate compact embryogenic callus.

made to study transient gene expression in mesophyll derived protoplasts.

3.1.1. Inflorescence culture. Of the twenty Indian varieties tested in inflorescence cultures, Sonalika gave distinctly superior response not only in terms of percentage of cultures which undergo callus induction (ca. 75% in comparison to less than 50% in other varieties), but also in regeneration frequency (66%, whereas in others the response was less than 50%, see Rajyalakshmi et al. 1988). These data clearly indicate the strong effect of genotype on callus induction as well as plantlet regeneration. An examination of callus cultures induced on MS (Murashige and Skoog 1962) medium + 4.5×10^{-5}M 2,4-D + 200 mg/1 casein hydrolysate (CH) + 10% coconut milk (CM) showed the presence of embryogenic as well as non-embryogenic calli (Fig. 1A). On subsequent transfer to basal medium, complete plantlets developed (Fig. 1B, C). The plantlets, obtained via somatic embryogenesis, were transferred successfully onto soil for further growth and have been shown to have the normal chromosome number (2n = 6x = 42).

3.1.2. Seed culture. In cultures initiated from mature seeds of the variety Sonalika on MS + 2.5×10^{-5}M 2,4-D, the embryos proceed to the initial stages of the germination process and cells in the scutellar

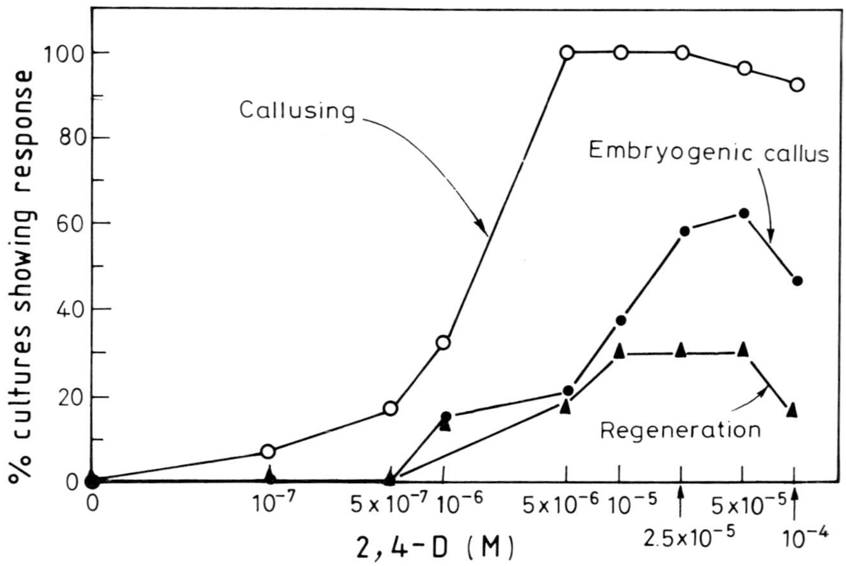

Figure 2. Effect of various concentrations of 2,4-D on callus induction, embryogenic callus formation and plantlet regeneration in seed cultures of wheat. For embryogenesis and plantlet formation, data were taken on 15th and 30th day, respectively, after transferring four-week-old calli, induced at various concentrations of 2,4-D, onto MS medium.

region proliferate giving rise to embryogenic as well as non-embryogenic callus masses (Fig. 1D). From such callus, it is possible to have complete plantlets differentiated (Fig. 1E, F) after subculture on MS medium. This concentration of 2,4-D was selected as optimal on the basis of a detailed study involving closely spaced concentrations. As is clear from Fig. 2, it was possible to obtain callusing even at 10^{-7}M 2,4-D, but such cultures did not produce plantlets. Increasing concentrations of 2,4-D resulted in better response and when these calli were transferred to basal medium, more embryogenic calli as well as plantlets were produced (Fig. 2). The best response in terms of callusing, and then, development of embryogenic callus masses, and finally, plant regeneration was obtained at 2.5×10^{-5}M 2,4-D, the percentage values for cultures proceeding through various stages being 100, 57 and 29, respectively.

3.1.3. Long-term retention of regeneration potential. To study the long-term retention of regeneration potential, the calli from inflorescence as well as seed cultures of the variety Sonalika were maintained on MS medium supplemented with 2.5×10^{-5}M 2,4-D + 200mg/1 CH + 10% CM and the percentage of cultures showing regeneration was periodically assessed (see Rajyalakshmi et al. 1988). The regeneration frequency attenuates to a rather low value in about 12 months -- the decline having already set in by sixth month. The regeneration potential could, however, be restored if the cultures were kept in the dark for one month and subsequently subcultured at fortnightly intervals. Faster subculturing has proved to be of great help in retaining the regenerative capacity for extended periods. The dark treatment, although not obligatory, obviates the problem of browning of tissues and thereby helps in the long-term retention of regeneration potential. By these manipulations, it has been possible to retain regeneration potential up to 21 months of culture.

3.1.4. Leaf base culture. Since in earlier work, regeneration in leaf explants has been reported only via organogenesis, we were interested to see the response of the variety Sonalika and see if an embryogenic culture system could be developed. Plantlets grown on MS medium for 2-3 weeks were used, and the first three leaves were cut into small pieces of about 5 mm and cultured on MS medium with varying concentration of 2,4-D (Fig. 3). After a month, profuse callusing was observed in the two basal segments of the leaves (Fig. 1G), the lowermost being far superior to the others (data not shown). Depending on the origin from various leaves, however, even the response of the lowermost explant varied (Fig. 3). In general, explants from the third (innermost) leaf responded better and up to 75% produced callus masses on medium enriched with $2.5\text{-}5 \times 10^{-5}$M 2,4-D. Of these, as many as 50% cultures were embryogenic. The response of explants from the first (outermost) leaf was clearly poor in comparison to those from the other two leaves. Calli from various explants regenerated into plantlets on transfer to MS medium (Fig. 1H, I). The best response (about 60% cultures showing plantlets) was observed from callus initiated at 5×10^{-6}M 2,4-D from the middle leaf (Fig. 4). This, of course, does not

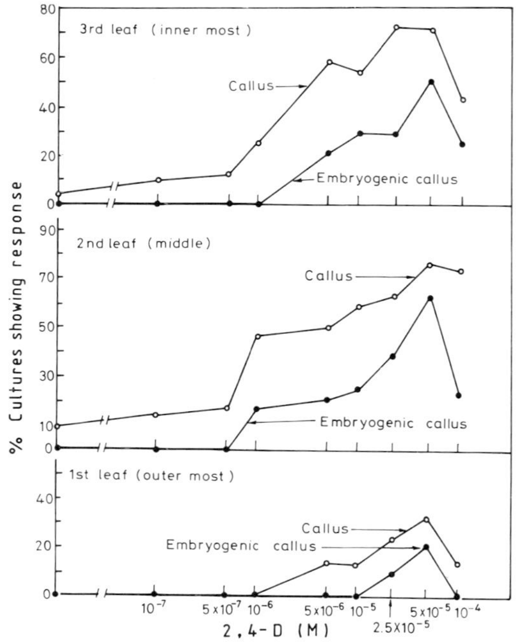

Figure 3. Effect of various concentrations of 2,4-D on callus induction and formation of embryogenic callus from the lowermost leaf base explant of the 1st, 2nd and 3rd leaf of wheat.

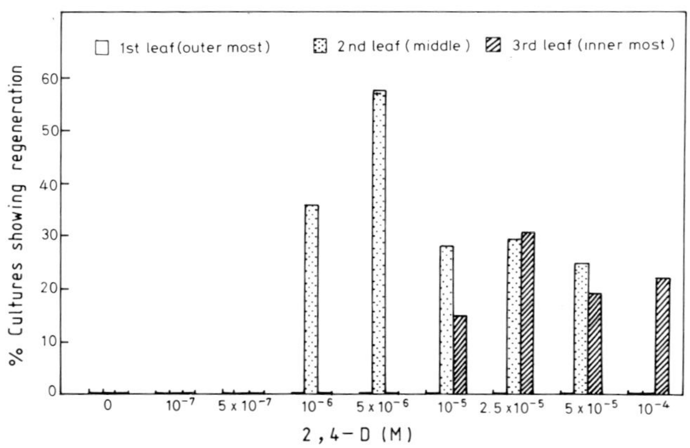

Figure 4. Regeneration of plantlets in wheat from 4-week-old calli from leaf base cultures initiated at various concentrations of 2,4-D (see also Fig. 3) and transferred to MS medium.

represent the maximum yield on per explant basis as the frequency of callusing of leaf bases at this concentration was low (Fig. 3). An overall evaluation of the data, therefore, indicates that the optimal concentration of 2,4-D for the purpose of obtaining the maximum number of plantlets might vary between explants from various leaves.

3.1.5. Protein profiles during embryogenesis. With the embryogenic callus system originating from leaf-base explants, experiments have been initiated to determine differences in gene expression as represented by changes in steady state level (silver-stained) and *de novo* synthesized (^{35}S-methionine-labeled) protein profiles. Our aim is to determine differences that occur at the time of cell division or initiation of the embryogenic pathway and those related with embryogenic and non-embryogenic calli. Preliminary investigations with cultured explants have shown several quantitative changes at different stages after initiation of culture. One interesting change concerns a 26 kDa protein band appearing after few days of culture (data not shown).

To pinpoint differences in protein profiles in embryogenic and non-embryogenic calli, it was first of all necessary to ensure reliability of visual separation of these two types of calli. Towards this end, both types of calli -- after 60 days of initiation of culture and separated visually -- were transferred to basal medium to see their regeneration potential. In such cultures, so far, regeneration of plantlets has been possible only from the embryogenic calli indicating that it is indeed possible to separate two types of calli with a high degree of reliability. From these two types of calli, proteins -- native as well as those synthesized *de novo* -- have been studied and quantitative as well as a few qualitative differences have been observed (Fig. 5). Although as clear, differences are seen by both methods i.e. silver staining and fluorography, they, however, do not appear to be related since the molecular weights of the concerned proteins in the two methods are different. However, further studies are necessary to resolve if there is any functional relationship between these proteins (differences in molecular weight resulting for example due to processing) and whether they are really related with the processes of cell division or differentiation.

3.1.6. Protoplasts and transient gene expression. Finally, coming to our efforts to establish transformation system, it may be mentioned that unfortunately, so far, it has not yet been possible to establish good embryogenic suspension culture in wheat. Therefore, isolation and culture of protoplasts from young leaves has been attempted. Although it is possible to get as many as 4 x 10^6 protoplasts per gram of leaf tissue, only rarely are any divisions seen in them -- in this regard, we share the disappointment of other workers in the field. But progress continues to be made with other systems -- thus Hahne et al. (1989) have been able to regenerate mesophyll protoplasts of *Avena sativa* repeatedly at high frequency and further work with wheat may also prove rewarding. Wheat mesophyll protoplasts have also been used

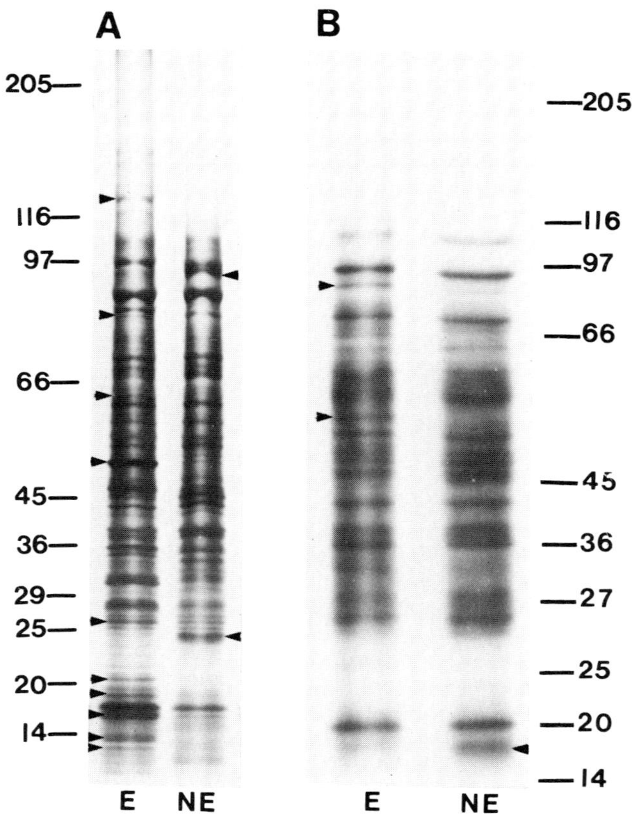

Figure 5. Soluble protein profiles from 60-day-old embryogenic (E) and non-embryogenic (NE) calli of wheat. The samples were extracted in 0.5 mM Tris HCl (pH 6.8) containing 5% β-mercaptoethanol (v/v) and 10% glycerol. The proteins were precipitated with acetone and vacuum dried. The pellet was dissolved and separated on 7.5 - 15% linear gradient gel by the method of Laemmli (1970). A. Five µg protein was loaded in each lane (quantified by the method of Bradford 1976) and stained by the method of Damerval et al. (1987). The molecular weight standards obtained from Sigma were run parallely. Differences between E and NE calli are marked with arrow heads. B. The E and NE tissues were fed with ^{35}S-methionine for 16 h. Conditions for extraction of proteins and running the gel are same as mentioned for A. 50,000 counts were loaded in each lane. The gels were treated with dimethyl sulphoxide (DMSO) and 2,5-diphenyloreazole (PPO) and dried. Fluorographs were obtained by exposing the Kodak X-ray films for 6 days. The differences between E and NE are shown with arrow heads.

by us to study transient expression of GUS gene under control of the ^{35}S promoter employing PEG induced DNA uptake and it has been possible to localize the activity by a histochemical test after 48 h of culture.

3.2. Rice

Four varieties of *indica* rice, namely, Basmati 370, Pusa 169, Ratna, and Improved Sabarmati, were obtained from the Indian Agricultural Research Institute, New Delhi. Immature inflorescences were obtained from field-grown Basmati 370, while the other three varieties served as source material for obtaining callus from seeds or young leaf bases.

3.2.1. *Inflorescence culture*. Callus from about 5 mm pieces of immature inflorescences became visible after two weeks of culture on MS medium containing varying concentrations of 2,4-D and coconut milk. It was initiated first at the cut ends of the pedicel and was noticeable soon after on the surface of spikelets in almost all the explants (Fig. 6A). The callus was compact and nodular in appearance. On transfer to MS medium after 60 days of culture, these calli developed well-differentiated embryos which subsequently germinated to form plantlets (Fig. 6B, C). Of the various combinations of 2,4-D and CM tried, callus initiated at 2.3×10^{-5}M 2,4-D + 10% CM resulted in the maximum production of embryos and plantlets (in approximately 63% of cultures).

3.2.2. *Seed culture*. Variety Pusa 169 was used for initiating callus from the mature seed. . The scutellum of the mature embryo swelled up after 4 days in culture and callusing was noticeable from the seventh day onwards on MS + 10^{-5}M 2,4-D where it was subcultured at monthly intervals. After 60 days, the callus was subcultured on MS + 2.8×10^{-6}M IAA + 2.3×10^{-5}M kinetin + 10% coconut milk where about 33% of cultures produced embryos and plantlets. This response could be increased up to 75% if callus was initiated on the medium enriched with tryptophan, the optimal concentration being 0.24 mM (Fig. 7).

Similar enhancement of response has been observed also on N6 (Chu et al. 1975) medium + 10^{-5}M 2,4-D (Fig. 6D), where the effect of another amino acid, proline, was evaluated. As in case of tryptophan, calli initiated on proline-containing medium have shown enhanced response in terms of production of embryos and plantlets (up to 75% on 12 mM proline in comparison to only 36% without it) as shown in Fig. 7 (*see* also Fig. 6E, F).

3.2.3. *Leaf base culture*. Three varieties, Pusa 169, Ratna and Improved Sabarmati, were used to obtain callus from the leaf base explants. Seeds were grown *in vitro* and from ten-day-old seedlings, all four leaves were excised, cut into 5 mm pieces, and subsequently cultured on MS medium containing 2,4-D (10^{-5}M). Segments from the first (outermost) and fourth (innermost) leaf do not respond and only the lower segments from the second and the third leaf of rice seedlings gave rise to nodular compact callus, similar to embryogenic callus

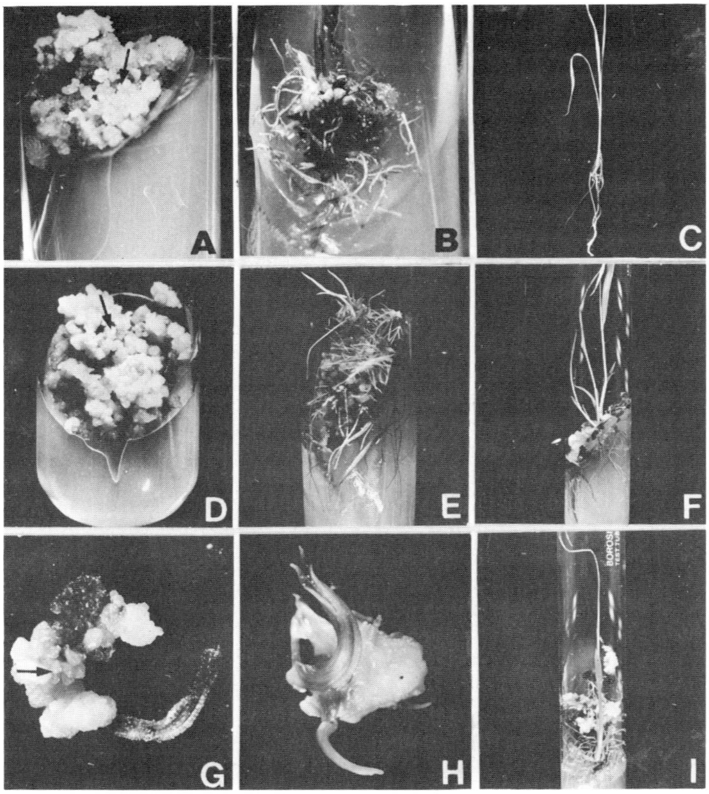

Figure 6. Regeneration of plantlets via somatic embryogenesis in inflorescence, seed and leaf base cultures of rice. A. Callus from a young inflorescence explant on MS + 2.3 x 10^{-5}M 2,4-D and 10% CM after 30 days of culture. B. An early stage of plantlet formation after subculture on MS medium. C. A complete plantlet. D. Callus from a seed cultured on N6 medium + 10^{-5}M 2,4-D + 100 mg/l casein hydrolysate + 12 mM proline after 30 days of culture. E. An early stage of plantlet formation on transfer to MS medium + 2.8 x 10^{-6}M IAA + 2.3 x 10^{-5}M kinetin + 10% coconut milk. F. Complete plantlet. G. Callus from a leaf base explant on MS + 10^{-5}M 2,4-D after 30 days of culture. H. An early stage of plantlet formation on transfer to MS basal medium. I. A complete plantlet. Inflorescence and seed cultures were initiated in diffuse light, but leaf cultures were maintained in dark for first 30 days. For plantlet regeneration, cultures were incubated under 3.27 W.m^2 light intensity. Arrows in A, D and G indicate compact embryogenic callus.

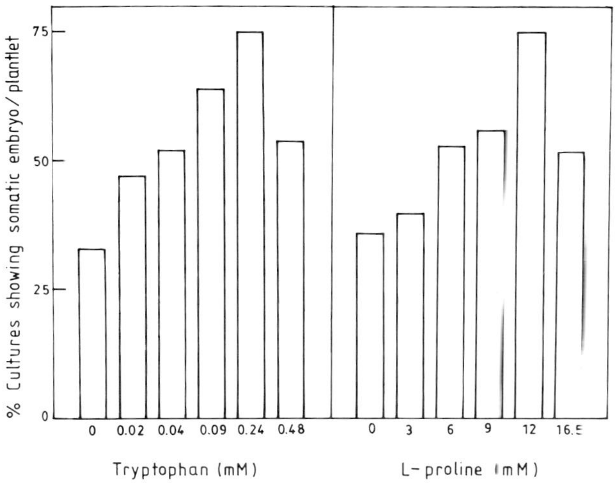

Figure 7. Effect of tryptophan and proline on somatic embryo/plantlet formation in seed cultures of rice. Callus cultures were initiated either on MS + 10^{-5}M 2,4-D alone or with varying concentrations of tryptophan. The effect of varying concentrations of proline added to N6 medium + 10^{-5}M 2,4-D + 100 mg/l casein hydrolysate is shown on the right. After 60 days of culture, the calli were transferred to MS + 2.8 x 10^{-6}M IAA + 2.3 x 10^{-5}M kinetin + 10% coconut milk and data representing per cent cultures showing plantlet formation after 30 days on this medium are given here. The data given are an average of two experiments, each with at least 18 cultures.

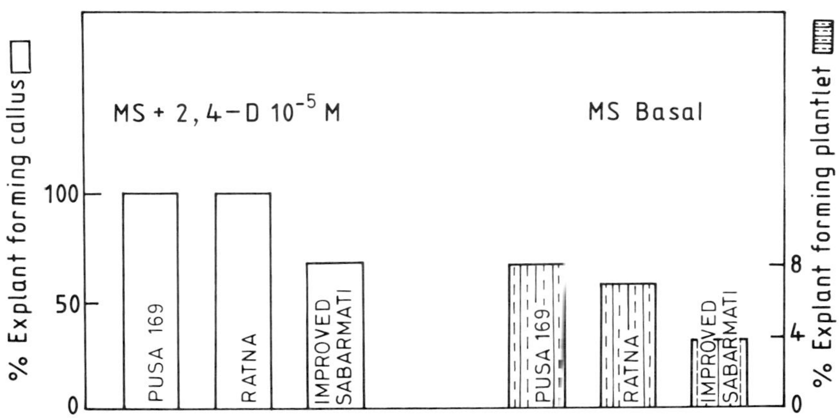

Figure 8. Callus induction and plantlet regeneration in second leaf base cultures of three rice varieties. Callus cultures were initiated on MS + 10^{-5}M 2,4-D and transferred to MS medium after 30 days. Data on regeneration were taken after another 30 days of culture.

(Fig. 6G). Leaf cultures turn brown if kept in light -- the callusing response was also very poor. Of the three varieties tried, callusing was more pronounced in Pusa 169 and Ratna than in Improved Sabarmati (Fig. 8). On transfer to the basal medium, the callus produced green regions. After 15-20 days of subculture, the friable callus showed profuse rooting. The green regions produced plantlets (Fig. 6H, I), the frequency being about 8% (Fig. 8).

3.2.4. Protoplasts and transient gene expression. Our attempts to induce good suspension culture have not yet met with much success and, therefore, protoplasts isolated from mesophyll cells of young leaves were employed to establish transient gene expression system in rice. Both electroporation and PEG-mediated DNA delivery methods have been found to work well (electroporation being better) for GUS gene under control of the 35S promoter, as has recently been shown in our lab (Ashok Chaudhury, unpublished) by histochemical as well as spectrofluorometric methods of assay.

4. CONCLUSION

The above is a summary of a new direction of research in our laboratory aimed towards the application of biotechnology for improvement of wheat and rice. The initial progress is encouraging, but it remains to be seen as to how a more intensive investigation can begin. There are still great difficulties both in respect of wheat and *indica* rice which are to be overcome. Hopefully, with a sustained effort employing both techniques of basic molecular biology and of cell culture, a headway will be made and genetic engineering studies will become a reality in the coming years. Since protoplasts are difficult to regenerate, perhaps the emphasis should be on direct gene transfer methods employing particle-gun technology or other suitable systems of DNA delivery.

5. ACKNOWLEDGEMENTS

We are grateful to Drs J.P. Khurana, P.K. Gharyal and Mr Ashok Chaudhury for help. Financial assistance from DST, UGC and ICAR for the work reported here is also appreciated.

6. REFERENCES

Abdullah, R., Cocking, E.C., and Thompson, J.A. (1986) 'Efficient plant regeneration from rice protoplasts through somatic embryogenesis', Bio/Technology 4, 1087-1090.

Abe, T. and Futsuhara, Y. (1984) 'Varietal difference of plant regeneration from root callus tissues in rice', Jap. J. Breed. 34, 147-155.

Abe, T. and Futsuhara, Y. (1985) 'Efficient plant regeneration by somatic embryogenesis from root callus tissues of rice (*Oryza sativa* L.)', J. Plant Physiol. 121, 111-118.

Abe, T. and Futsuhara, Y. (1986) 'Genotypic variability for callus formation and plant regeneration in rice (Oryza sativa L.)', Theor. Appl. Genet. 72, 3-10.

Agache, S., Bachelier, B., De Buyser, J., Henry, Y., and Snape, J. (1989) 'Genetic analysis of anther culture response in wheat using aneuploid, chromosome substitution and translocation lines', Theor. Appl. Genet. 77, 7-11.

Ahloowalia, B.S. (1982) 'Plant regeneration from callus culture in wheat', Crop Sci. 22, 405-410.

Ahuja, P.S., Pental, D., and Cocking, E.C. (1982) 'Plant regeneration from leaf base callus and cell suspensions of Triticum aestivum', Z. Pflanzenzuchtg 89, 139-144.

Bajaj, Y.P.S., Saini, S.S., and Bidani, M. (1980) 'Production of triploid plants from the immature and mature endosperm cultures of rice', Theor. Appl. Genet. 58, 17-18.

Bhattacharya, P. and Sen, S.K. (1980) 'Potentiality of leaf sheath cells for regeneration of rice (Oryza sativa L.) plants', Theor. Appl. Genet. 58, 87-90.

Bradford, M.M. (1976) 'A rapid and sensitive method for the quantitation of microgram quantities of protein utilizing the principle of protein-dye binding', Anal. Biochem. 72, 248-254.

Brown, C., Brooks, F.J., Pearson, D., and Mathias, R.J. (1989) 'Control of embryogenesis and organogenesis in immature wheat embryo callus using increased medium osmolarity and abscisic acid', J. Plant Physiol. 133, 727-733.

Carman, J.G. (1988) 'Improved somatic embryogenesis in wheat by partial simulation of the in-ovulo oxygen, growth-regulator and desiccation environments', Planta 175, 417-424.

Carman, J.G., Jefferson, N.F., and Campbell, W.F. (1987a) 'Induction of embryogenic Triticum aestivum L. calli. I. Quantification of genotype and culture medium effects', Plant Cell, Tissue and Org. Cult. 10, 101-113.

Carman, J.G., Jefferson, N.E., and Campbell, W.F. (1987b) 'Induction of embryogenic Triticum aestivum L. calli. II. Quantification of organic addenda and other culture variable effects', Plant Cell, Tissue and Org. Cult. 10, 115-128.

Chen, L.-J. and Luthe, D.S. (1987) 'Analysis of proteins from embryogenic and non-embryogenic rice (Oryza sativa L.) calli', Plant Sci. 48, 181-188.

Chen, T.-H., Lam, L., and Chen, S.-C. (1985) 'Somatic embryogenesis and plant regeneration from cultured young inflorescences of Oryza sativa L. (rice)', Plant Cell, Tissue and Org. Cult. 4, 51-54.

Chen, Y. (1986a) 'Anther and pollen culture of rice', in H. Hu and H. Yang (eds), Haploids of Higher Plants in Vitro, China Academic Publishers, Beijing, pp. 3-25.

Chen, Y. (1986b) 'The inheritance of rice pollen plant and its application in crop improvement', in H. Hu and H. Yang (eds), Haploids of Higher Plants in Vitro, China Academic Publishers, Beijing, pp. 118-136.

Chu, C.C., Wang, C.C., Sun, C.S., Hsü, C., Yin, K.C., Chu, C.Y., and Bi, F.Y. (1975) 'Establishment of an efficient medium for anther culture of rice through comparative experiments on the nitrogen sources', Sci. Sin. 18, 659-668.

Chu, C.C., Wang, C.C., Sun, C.S., Chien, N.F., Yin, K.C., and Hsu, C. (1973) 'Investigations on the induction and morphogenesis of wheat (*Triticum vulgare*) pollen plants', Acta Bot. Sin. 15, 1-11.

Coulibaly, M.Y. and Demarly, Y. (1986) 'Regeneration of plantlets from protoplasts of rice, *Oryza sativa* L', Z. Pflanzenzuchtg 96, 79-81.

Damerval, C., Le Guilloux, M., Blaisonneau, J., and de Vienne, D. (1987) 'A simplification of Heukeshoven and Dernick's silver staining of proteins', Electrophoresis 8, 158-159.

De Buyser, J. and Henry, Y. (1979) 'Androgenese sur des Bles tendres en cours de selection. I. L' obtention des plantes in vitro', Z. Pflanzenzuchtg 83, 49-56.

Eapen, S. and Rao, P.S. (1985) 'Factors controlling callus initiation, growth and plant regeneration in bread wheat (*Triticum aestivum* L.)', Proc. Indian Acad. Sci. (Plant Sci.) 94, 33-40.

Elena, E.B. and Ginzo, H.D. (1988) 'Effect of auxin levels on shoot formation with different embryo tissues from a cultivar and a commercial hybrid of wheat (*Triticum aestivum* L.)', J. Plant Physiol. 132, 600-603.

Fujimura, T., Sakurai, M., Akagi, H., Negishi, T., and Hirose, T. (1985) 'Regeneration of rice plants from protoplasts', Plant Tissue Culture Lett. 2, 74-75.

Galiba, G. and Erdei, L. (1986) 'Dependence of wheat callus growth, differentiation and mineral content on carbohydrate supply', Plant Sci. 45, 65-70.

Galiba, G., Kovács, G., and Sukta, J. (1986) 'Substitution analysis of plant regeneration from callus culture in wheat', Plant Breed. 97, 261-263.

Guha, S. and Maheshwari, S.C. (1964) '*In vitro* production of embryos from anthers of *Datura*', Nature 204, 497.

Guha, S. and Maheshwari, S.C. (1966) 'Cell division and differentiation of embryos in the pollen grains of *Datura in vitro*', Nature 212, 97-98.

Hahne, B., Fleck, J., and Hahne, G. (1989) 'Colony formation from mesophyll protoplasts of a cereal, oat', Proc. Natl Acad. Sci. USA 86, 6157-6160.

Harris, R., Wright, M., Byrne, M., Varnum, J., Brightwell, B., and Schubert, K. (1988) 'Callus formation and plantlet regeneration from protoplasts derived from suspension cultures of wheat (*Triticum aestivum* L.)', Plant Cell Rep. 7, 337-340.

Hayashi, Y., Kyozuka, J., and Shimamoto, K. (1988) 'Hybrids of rice (*Oryza sativa* L.) and wild *Oryza* species obtained by cell fusion', Mol. Gen. Genet. 214, 6-10.

Hayashi, Y. and Shimamoto, K. (1988) 'Wheat protoplast culture: embryogenic colony formation from protoplasts', Plant Cell Rep. 7, 414-417.

He, D. and Ouyang, J. (1984) 'Callus and plantlet formation from cultured wheat anthers at different developmental stages', Plant Sci. Lett. 33, 71-79.

Henke, R.R., Mansur, M.A., and Constantin, M.J. (1978) 'Organogenesis and plantlet formation from organ- and seedling-derived calli of rice

(Oryza sativa)', Physiol. Plant. 44, 11-14.
Henry, Y. and De Buyser, J. (1985) 'Effect of the 1B/1R translocation on anther culture ability in wheat (Triticum aestivum L.)', Plant Cell Rep. 4, 307-310.
Heyser, J.W., Dykes, T.A., DeMott, K.J., and Nabors, M.W. (1983) 'High frequency long term regeneration of rice from callus culture', Plant Sci. Lett. 29, 175-182.
Heyser, J.W., Nabors, M.W., MacKinnon, C., Dykes, T.A., DeMott, K.J., Kautzman, D.C., and Mujeeb-Kazi, A. (1985) 'Long-term, high-frequency plant regeneration and the induction of somatic embryogenesis in callus cultures of wheat (Triticum aestivum L.)', Z. Pflanzenzuchtg 94, 218-233.
Hu, D. (1986) 'Jinghua No. 1, a winter wheat variety derived from pollen sporophyte', in H. Hu and H. Yang (eds), Haploids of Higher Plants in Vitro, China Academic Publishers, Beijing, pp. 137-148.
Kavi Kishor, P.B. (1987) 'Energy and osmotic requirement for high frequency regeneration of rice plants from long-term cultures', Plant Sci. 48, 189-194.
Kavi Kishor, P.B. and Reddy, G.M. (1986) 'Regeneration of plants from long-term cultures of Oryza sativa L.', Plant Cell Rep. 5, 391-393.
Kawata, S. and Ishihara, A. (1968) 'The regeneration of rice plant Oryza sativa L. in the callus derived from the seminal root', Proc. Jap. Acad. 44, 549-553.
Kyozuka, J., Hayashi, Y., and Shimamoto, K. (1987) 'High frequency plant regeneration from rice protoplasts by novel nurse culture methods', Mol. Gen. Genet. 206, 408-413.
Kyozuka, J., Otto, O.E., and Shimamoto, K. (1988) 'Plant regeneration from protoplasts of indica rice: genotypic differences in culture response', Theor. Appl. Genet. 76, 887-890.
Laemmli, U.K. (1970) 'Cleavage of structural proteins during the assembly of the head of bacteriophage T4', Nature 227, 680-685.
Lai, K.-L. and Liu, L.-L. (1982) 'Induction and plant regeneration of callus from immature embryos of rice plants (Oryza sativa L.)', Jap. J. Crop Sci. 51, 70-74.
Lazar, M.D., Schaeffer, G.W., and Baenziger, P.S. (1984) 'Cultivar and cultivar x environment effects on the development of callus and polyhaploid plants from anther cultures of wheat', Theor. Appl. Genet. 67, 273-277.
Lee, B., Murdoch, K., Topping, J., Kreis, M., and Jones, M.G.K. (1989) 'Transient gene expression in aleurone protoplasts isolated from developing caryopses of barley and wheat', Plant Mol. Biol. 13, 21-29.
Lee, B.T., Murdoch, K., Topping, J., de Both, M.T.J., Wu, Q.S., Karp, A., Steele, S., Symonds, C., Kreis, M., and Jones, M.G.K. (1988) 'Isolation, culture and morphogenesis from wheat protoplasts, and study of expression of DNA constructs by direct gene transfer', Plant Cell, Tissue Org. Cult. 12, 223-226.
Lee, L., Schroll, R.E., Grimes, H.D., and Hodges, T.K. (1989) 'Plant regeneration from indica rice (Oryza sativa L.) protoplasts', Planta 178, 325-333.
Logemann, J., Lipphardt, S., Lörz, H., Häuser, I., Willmitzer, L., and

Schell, J. (1989) '5' upstream sequences from the wun1 gene are responsible for gene activation by wounding in transgenic plants', The Plant Cell 1, 151-158.
Lörz, H., Baker, B., and Schell, J. (1985) 'Gene transfer to cereal cells mediated by protoplast transformation', Mol. Gen. Genet. 199, 178-182.
Maddock, S.E. (1985) 'Cell culture, somatic embryogenesis and plant regeneration in wheat, barley, oats, rye and triticale', in S.W.J. Bright and M.G.K. Jones (eds), Cereal Tissue and Cell Cultures, Martinus Nijhoff/Dr W. Junk Publishers, Dordrecht, pp. 131- 174.
Maddock, S.E. (1987) 'Suspension and protoplast culture of hexaploid wheat (*Triticum aestivum* L.)', Plant Cell Rep. 6, 23-26.
Maddock, S.E., Lancaster, V.A., Risiott, R., and Franklin, J. (1983) 'Plant regeneration from cultured immature embryos and inflorescences of 25 cultivars of wheat (*Triticum aestivum*)', J. Exp. Bot. 34, 915-926.
Maeda, E. (1968) 'Subculture and organ formation in the callus derived from rice embryos in vitro', Proc. Crop Sci. Soc. Japan 34, 139-147.
Maheshwari, S.C., Gill, R., Maheshwari, N., and Gharyal, P.K. (1986) 'Isolation and regeneration of protoplasts from higher plants', in J. Reinert and H. Binding (eds), Differentiation of Protoplasts and of Transformed Plant Cells, Springer-Verlag, Berlin, pp. 3-36.
Marcotte, W.R., Bayley, C.C., and Quatrano, R.S. (1988) 'Regulation of a wheat promoter by abscisic acid in rice protoplasts', Nature 335, 454-457.
Masuda, K., Kudo-Shiratori, A., and Inoue, M. (1989) 'Callus formation and plant regeneration from rice protoplasts purified by density gradient centrifugation', Plant Sci. 62, 237-246.
Mathias, R.J. and Fukui, K. (1986) 'The effect of specific chromosome and cytoplasm substitutions on the tissue culture response of wheat (*Triticum aestivum*) callus', Theor. Appl. Genet. 71, 797-800.
Mathias, R.J., Fukui, K., and Law, C.N. (1986) 'Cytoplasmic effects on the tissue culture response of wheat (*Triticum aestivum*) callus', Theor. Appl. Genet. 72, 70-75.
Mathias, R.J. and Simpson, E.S. (1986) 'The interaction of genotype and culture medium on the tissue culture responses of wheat (*Triticum aestivum* L. em. thell) callus', Plant Cell, Tissue and Org. Cult. 7, 31-37.
Morota, H. and Uchimiya, H. (1987) 'Stable maintenance of foreign DNA in transformed cell lines of rice (*Oryza sativa* L.)', Jap. J. Genet. 62, 363-368.
Murashige, T. and Skoog, F. (1962) 'A revised medium for rapid growth and bioassays with tobacco tissue cultures', Physiol. Plant. 15, 473-497.
Nabors, M.W., Heyser, J.W., Dykes, T.A., and DeMott, K.J. (1983) 'Long-duration, high-frequency plant regeneration from cereal tissue cultures', Planta 157, 385-391.
Nakano, H. and Maeda, E. (1979) 'Shoot differentiation in callus of *Oryza sativa* L.', Z. Pflanzenphysiol. 93, 449-458.
Nakano, H., Tashiro, T., and Maeda, E. (1975) 'Plant differentiation in

callus tissue induced from immature endosperm of Oryza sativa L.', Z. Pflanzenphysiol. 76, 444-449.
Niizeki, H. and Oono, K. (1968) 'Induction of haploid rice plant from anther culture', Proc. Japan Acad. 44, 554-557.
Nishi, T., Yamada, Y., and Takahashi, E. (1968) 'Organ redifferentiation and plant restoration in rice callus', Nature 219, 508-509.
Oard, J.H., Paige, D., and Dvorak, J. (1989) 'Chimeric gene expression using maize intron in cultured cells of bread wheat', Plant Cell Rep. 8, 156-160.
Oard, J.H. and Rutger, J.N. (1988) 'Callus induction and plant regeneration in elite U.S. rice lines', Crop Sci. 28, 565-567.
Ou-Lee, T.-M., Turgeon, R., and Wu, R. (1986) 'Expression of a foreign gene linked to either a plant-virus or a Drosophila promoter, after electroporation of protoplasts of rice, wheat and sorghum', Proc. Natl Acad. Sci. USA 83, 6815-6819.
Ouyang, J. (1986) 'Induction of pollen plants in Triticum aestivum', in H. Hu and H. Yang (eds), Haploids of Higher Plants in Vitro, China Academic Publishers, Beijing, pp. 26-41.
Ouyang, T.W., Hu, H., Chuang, C.C., and Tseng, C.C. (1973) 'Induction of pollen plants from anthers of Triticum aestivum L. cultured in vitro', Sci. Sin. 16, 79-95.
Ozawa, K. and Komamine, A. (1989) 'Establishment of a system of high-frequency embryogenesis from long-term cell suspension cultures of rice (Oryza sativa L.)', Theor. Appl. Genet. 77, 205-211.
Ozias-Akins, P. and Vasil, I.K. (1982) 'Plant regeneration from cultured immature embryos and inflorescences of Triticum aestivum L. (wheat): evidence for somatic embryogenesis', Protoplasma 110, 95-105.
Ozias-Akins, P. and Vasil, I.K. (1983a) 'Improved efficiency and normalization of somatic embryogenesis in Triticum aestivum (wheat)', Protoplasma 117, 40-44.
Ozias-Akins, P. and Vasil, I.K. (1983b) 'Proliferation of and plant regeneration from the epiblast of Triticum aestivum (wheat; Gramineae) embryos, Amer. J. Bot. 70, 1092-1097.
Papenfuss, J.M. and Carman, J.G. (1987) 'Enhanced regeneration from wheat callus cultures using dicamba and kinetin', Crop Sci. 27, 588-593.
Paul, N.K. and Ghosh, P.D. (1984) 'Regeneration of plantlets of Oryza sativa L. cv. Kiran from scutellar tissues', Proc. Indian Natl Sci. Acad. B50, 332-336.
Picard, E. and De Buyser, J. (1973) 'Obtention de plantules haploides de Triticum aestivum L. à partir de culture d'anthères in vitro', C.R. Acad. Sci. Paris 277, 1463-1466.
Purnhauser, L., Medgyesy, P., Czáko, M., Dix, P.J., and Marton, L. (1987) 'Stimulation of shoot regeneration in Triticum aestivum and Nicotiana plumbaginifolia Viv. tissue cultures using the ethylene inhibitor AgNO3', Plant Cell Rep. 6, 1-4.
Qureshi, J.A., Kartha, K.K., Abrams, S.R., and Steinhauer, L. (1989) 'Modulation of somatic embryogenesis in early and late-stage embryos of wheat (Triticum aestivum L.) under the influence of (±)-abscisic acid and analogs', Plant Cell, Tissue and Org. Cult. 18, 55-69.

Raghav Ram, N.V. and Nabors, M.W. (1984) 'Cytokinin mediated long-term, high frequency plant regeneration in rice tissue cultures', Z. Pflanzenphysiol. 113, 315-323.

Raghav Ram, N.V. and Nabors, M.W. (1985) 'Plant regeneration from tissue cultures of Pokkali rice is promoted by optimizing callus to medium volume ratio and by a medium-conditioning factor produced by embryogenic callus', Plant Cell, Tissue and Org. Cult. 4, 241-248.

Raina, S.K., Sathish, P., and Sarma, K.S. (1987) 'Plant regeneration from in vitro cultures of anthers and mature seeds of rice (Oryza sativa L.) cv. Basmati-370, Plant Cell Rep. 6, 43-45.

Rajyalakshmi, K., Dhir, S.K., Maheshwari, N., and Maheshwari, S.C. (1988) 'Callusing and regeneration of plantlets via somatic embryogenesis from inflorescence cultures of Triticum aestivum L.: Role of genotype and long-term retention of morphogenic potential', Plant Breed. 101, 80-85.

Research Group 303, Institute of Genetics, Acad. Sin. (1972) 'A preliminary report on the induction of pollen plants from rice and wheat', Genet. Commun. (Beijing) 1, 1-4.

Schaeffer, G.W., Lazar, M.D., and Baenziger, P.S. (1984) 'Wheat', in W.R. Sharp, D.A. Evans, P.V. Ammirato, and Y. Yamada (eds), Handbook of Plant Cell Culture, Vol. 2, MacMillan Publishing Company, New York, pp. 108-136.

Shimamoto, K., Terada, R., Izawa, T., and Fujimoto, H. (1989) 'Fertile transgenic rice plants regenerated from transformed protoplasts', Nature 338, 274-276.

Siriwardana, S. and Nabors, M.W. (1983) 'Tryptophan enhancement of somatic embryogenesis in rice', Plant Physiol. 73, 142-146.

Tamura, S. (1968) 'Shoot formation in calli originated from rice embryo', Proc. Japan Acad. 44, 544-548.

Terada, R., Kyozuka, J., Nisibayashi, S., and Shimamoto, K. (1987) 'Plantlet regeneration from somatic hybrids of rice (Oryza sativa L.) and barnyard grass (Echinochloa oryzicola Vasing.)', Mol. Gen. Genet. 210, 39-43.

Thompson, J.A., Abdullah, R., Chen, W.-H., and Gartland, K.M.A. (1987) 'Enhanced protoplast division in rice (Oryza sativa L.) following heat shock treatment', J. Plant Physiol. 127, 367-370.

Töpfer, R., Gronenborn, B., Schell, J., and Steinbiss, H.-H. (1989) 'Uptake and transient expression of chimeric genes in seed-derived embryos', The Plant Cell 1, 133-139.

Toriyama, K., Arimoto, Y., Uchimiya, H., and Hinata, K. (1988) 'Transgenic rice plants after direct gene transfer into protoplasts', Bio/Technology 6, 1072-1074.

Toriyama, K. and Hinata, K. (1988) 'Diploid somatic-hybrid plants regenerated from rice cultivars', Theor. Appl. Genet. 76, 665-668.

Toriyama, K., Hinata, K., and Sasaki, T. (1986) 'Haploid and diploid plant regeneration from protoplasts of anther callus in rice', Theor. Appl. Genet. 73, 16-19.

Tyagi, A.K., Bharal, S., Rashid, A., and Maheshwari, N. (1985) 'Plant regeneration from tissue cultures initiated from immature inflorescences of a grass, Echinochloa colonum (L.) Link', Plant Cell Rep. 4, 115-117.

Uchimiya, H., Fushimi, T., Hashimoto, H., Harada, H., Syono, K., and Sugawara, Y. (1986) 'Expression of a foreign gene in callus derived from DNA-treated protoplasts of rice (Oryza sativa L.)', Mol. Gen. Genet. 204, 204-207.

Vasil, I.K. and Vasil, V. (1986) 'Regeneration in cereal and other grass species', in I.K. Vasil, (eds), Cell Culture and Somatic Cell Genetics of Plants, Vol. 3. Plant Regeneration and Genetic Variability, Academic Press, Orlando, pp. 121-150.

Vasil, V. and Vasil, I.K. (1980) 'Isolation and culture of cereal protoplasts. Part 2: Embryogenesis and plantlet formation from protoplast of Pennisetum americanum, Theor. Appl. Genet. 56, 97-99.

Wang, D., Miller, P.D., and Söndahl, M.R. (1989) 'Plant regeneration from protoplasts of indica type rice and CMS rice', Plant Cell Rep. 8, 329-332.

Wang, Y.-C., Klein, T.M., Fromm, M., Cao, J., Sanford, J.C., and Wu, R. (1988) 'Transient expression of foreign genes in rice, wheat and soybean cells following particle bombardment', Plant Mol. Biol. 11, 433-439.

Wei, Z.M. (1982) 'Pollen callus culture in Triticum aestivum', Theor. Appl. Genet. 63, 71-73.

Wernicke, W., Brettell, R., Wakizuka, T., and Potrykus, I. (1981) 'Adventitious embryoid and root formation from rice leaves', Z. Pflanzenphysiol. 103, 361-365.

Wernicke, W. and Milkovits, L. (1984) 'Developmental gradients in wheat leaves -- response of leaf segments in different genotypes cultured in vitro', J. Plant Physiol. 115, 49-58.

Yamada, Y., Zhi-Qi, Y., and Ding-Tai, T. (1986) 'Plant regeneration from protoplast-derived callus of rice (Oryza sativa L.)', Plant Cell Rep. 5, 85-88.

Yan, C.-J. and Zhao, Q.-H. (1982) 'Callus induction and plantlet regeneration from leaf blade of Oryza sativa L. subsp. indica', Plant Sci. Lett. 25, 187-192.

Yang, H., Zhang, H.M., Davey, M.R., Mulligan, B.J., and Cocking, E.C. (1988) 'Production of kanamycin resistant rice tissues following DNA uptake into protoplasts', Plant Cell Rep. 7, 421-425.

Yin, K.C., Hsu, C., Chu, C.Y., Pi, F.Y., Wang, S.T., Liu, T.Y., Chu, C.C., Wang, C.C., and Sun, C.S. (1978) 'Tan Feng No. 1: by anther culture', in Proc. Symp. on Anther Culture. Science Press, Beijing, p. 283.

Zhang, H.M., Yang, H., Rech, E.L., Golds, T.J., Davis, A.S., Mulligan, B.J., Cocking, E.C., and Davey, M.R. (1988) 'Transgenic rice plants produced by electroporation-mediated plasmid uptake into protoplasts', Plant Cell Rep. 7, 379-384.

Zhang, W. and Wu, R. (1988) 'Efficient regeneration of transgenic plants from rice protoplasts and correctly regulated expression of the foreign gene in the plants', Theor. Appl. Genet. 76, 835-840.

Zimny, J. and Lörz, H. (1986) 'Plant regeneration and initiation of cell suspensions from root-tip derived callus of Oryza sativa L. (rice)', Plant Cell Rep. 5, 89-92.

CALCIUM AND CALMODULIN DURING CARROT SOMATIC EMBRYOGENESIS

A.C.J. TIMMERS
Department of Plant Cytology and Morphology
Agricultural University Wageningen
Arboretumlaan 4
6703 BD Wageningen
The Netherlands

ABSTRACT. Carrot somatic embryogenesis has been used extensively as a model system for the study of control mechanisms of plant embryogenesis. Ca^{2+} and the Ca^{2+} binding protein calmodulin play a fundamental role in the control of plant growth and development. During plant embryogenesis especially important is the role of Ca^{2+} in the regulation of cell polarity, cell growth, mitosis and cytokinesis, cell volume, plant hormone action and distribution, and enzyme activation. Carrot somatic embryos develop in culture on clusters of small cytoplasm rich cells. They progress through the successive stages of globular, heart shaped, and torpedo shaped embryos, comparable with zygotic embryogenesis. The process is influenced by several exogenous factors such as light quality and medium composition.

Calcium and calmodulin both possess a number of chemical and physical properties which makes them very suitable as intracellular messengers in the regulation of embryogenesis. The involvement of Ca^{2+} and calmodulin can be studied with various techniques. Of great importance is the measurement of the cytoplasmic Ca^{2+} concentration and the determination of its distribution. In this paper a summary of methods for the three forms of Ca^{2+} (free cytoplasmic, bound to membrane surfaces or intracellular chelating molecules, and sequestered inside subcellular organelles) which are applicable to the system of carrot somatic embryogenesis, will be presented.

High concentrations of external Ca^{2+} has proven to be promotive for carrot somatic embryogenesis. Embryogenesis increases upon transfer to a medium with a elevated concentration of Ca^{2+}, regardless of the initial concentration. Somatic embryos possess a higher concentration of bound calcium, which is mainly localized in the outer cell layers, as compared with proembryogenic masses. Activated calmodulin is unevenly distributed in all stages of carrot somatic embryogenesis. In the globular and heart stage activated calmodulin is restricted to the basal part of the embryo. In old torpedo shaped stages activated calmodulin is also present in the shoot apex. From these observations it is concluded that Ca^{2+} and calmodulin are important for the initiation of polarity during carrot somatic embryogenesis.

R.S. Sangwan and B.S. Sangwan-Norreel (eds.), The Impact of Biotechnology in Agriculture, 215–234.

1. INTRODUCTION

Carrot somatic embryogenesis forms an ideal model system for the study of the control mechanisms of plant embryogenesis (Nomura & Komamine 1986b). Embryogenesis is a complex process involving coordinated cell division, growth and development. Ca^{2+} ions play an important role in the intricate network of interactions which regulates the growth and development of plants (Hepler & Wayne 1985, Hepler 1988). Many physiological processes in plants are under the control of Ca^{2+}. In most of them also the Ca^{2+} binding protein calmodulin is involved (table 1).

Here especially of interest is the role of Ca^{2+} and calmodulin in the regulation of cell polarity, cell growth, mitosis and cytokinesis, cell volume, plant hormone action and distribution, and enzyme activa-

TABLE 1. Ca^{2+} linked processes in plants

Process	Reference
Cell polarity, polarized growth	Hepler & Wayne 1985, Reiss et al. 1986
Cell growth and proliferation	Poovaiah 1985, Hepler 1988
Exocytosis and endocytosis	Steer 1988
Polar transport and secretion of IAA	Dela Fuente 1984
Mitosis and cytokinesis	Hepler & Wayne 1985
Cytoplasmic streaming	Hepler & Wayne 1985
Phytochrome response	Roux et al. 1986
Gravitropic responses	Poovaiah et al. 1987
Nastic movements	Satter & Galston 1981
Plant hormone action	Hepler & Wayne 1985, Elliot 1986
Secretion and synthesis of α-amylase	Jones et al. 1986
Secretion of peroxidases	Penel & Greppin 1982
Regulation of cell pH	Felle 1989
Volume regulation	Kauss 1987
Phosphorylation of proteins	Poovaiah & Veluthambi 1986
Photosynthesis	Brand & Becker 1984
Senescence	Poovaiah & Leopold 1973
Assembly and disassembly of microtubuli	Kakiuchi & Sobue 1983
Inhibition of respiration in meristematic protoplasts	Owen et al. 1987
Closing of stomata	Smith & Willmer 1988
Gene expression	Guilfoyle 1989

tion. The change of a plant cell from an undifferentiated state to a differentiated state certainly involves changes in metabolic processes which are dependent on the activation or inactivation of the proper enzymes at the right time and place. The activity of a number of plant enzymes seems to be under the control of Ca^{2+} and/or calmodulin (table 2).

TABLE 2. Enzymes activated by Ca^{2+} and calmodulin

Enzyme	Ca^{2+}	Calmodulin	Reference
NAD+ kinase	+	+	Anderson et al. 1980, Dieter 1984
NADH glutamate dehydrogenase	+	o	Das et al. 1989
Ca^{2+} ATPase	+	+	Dieter 1984
H^+ ATPase	+	+	Ranjeva & Boudet 1987
Membrane-bound ATPase	+	+	Dieter 1984
Quinate: NAD^+ oxidoreductase	+	+	Dieter 1984
Protein kinases	+	- or -	Blowers & Trewavas 1989
Protein phosphatases	+	+	Roberts et al. 1986
1,3-β-glucan synthase	+	-	Kauss 1987
Adenylate/guanylate cyclases	+	+	Roberts et al.1986
Glyoxalase-I	+	+	Das et al. 1987
Phospholipase	+	o	Leshem 1987
Isofluoridoside phosphate synthase	+	+	Kauss 1983
α-amylase	+	+	Mitsui et al.1984, Jones et al. 1986

+: activation, -: no effect, o: not known.

Here I will discuss the role of Ca^{2+} and calmodulin during carrot somatic embryogenesis. Before doing this I first will introduce you to the model system, mention some fundamental properties of Ca^{2+} and calmodulin, and summarize the technical approaches to the system. In the last paragraph some conclusions will be presented.

The references given here are selected from a large pool of articles. In most cases only the most recent paper or a reference of a review article is given. Further references can be found therein.

2. SOMATIC EMBRYOGENESIS OF CARROT

The ability of carrot cells to produce embryos *in vitro* was independently discovered in 1958 by Reinert and by Steward et al. Somatic embryos of carrot arise instantaneously by diluting a suspension culture and transferring a sieved fraction to 2,4-D omitted medium. In most cases suspension cultures are derived from callus cultures of hypocotyl sections on solidified medium (e.g. Bhojwani & Razdan 1983). Alternatively, hypocotyl sections can be put directly into 2,4-D containing liquid medium (De Vries et al. 1988b). Embryogenic suspension cultures are composed of small and densely cytoplasmic cells, capable to form embryos, and occurring in clusters or clumps, and of vacuolated free cells and small aggregates of such cells (Street & Withers 1974). The cell clusters which are able to form embryos have been termed proembryogenic masses by Halperin (1966) or embryogenic clumps by McWilliam et al. (1974). In this paper the terminology of Halperin will be used.

The embryogenic potential of a culture is reflected by the number of proembryogenic masses present (De Vries et al. 1988b). The number of them in different embryogenic carrot suspension cultures varies between 0.1% and 5% of the total number of cells. Proembryogenic masses are derived from single cells of a specific type (Nomura & Komamine 1985). They are composed of two distinct types of cells (Bhojwani & Razdan 1983). The central cells have a single large vacuole, a small and compact nucleus with a faintly staining nucleolus, a low population of ribosomes, very few endoplasmic profiles and normal mitochondria, just a few spherosome-like vesicles, low dehydrogenase activity, and a reduced number of amyloplasts. At the periphery of the proembryogenic mass groups of highly meristematic cells are present. In contrast to the central cells, these are characterized by having several small vacuoles, a large diffusely staining nucleus with a single prominently staining nucleolus, a higher density of ribosomes, numerous profiles of rough endoplasmic reticulum, normal mitochondria, spherosome-like vesicles, higher dehydrogenase activity, and prominent amyloplasts. Upon transfer to 2,4-D omitted medium embryos are rapidly initiated from the superficial cells of the proembryogenic masses.

Somatic embryos most probably derive from a single cell (e.g. McWilliam et al. 1974). They progress through the successive stages of globular, heart shaped and torpedo shaped embryos comparable with the development of zygotic embryos (Street & Withers 1974). Only at the early beginning of somatic embryogenesis some differences with the development of zygotic embryos can be observed (McWilliam et al. 1974). Ultrastructural changes accompanying embryogenesis are described in detail by Halperin & Jensen (1967) and Street & Withers (1974). Most obvious are differences in the number of ER profiles, the number and structure of Golgi bodies and mitochondria and the appearance of microtubuli. Besides by cytological changes embryogenesis is also characterized by a number of changes in metabolism (Nomura & Komamine 1986b). During the formation of globular embryos a high rate of DNA synthesis and changes in protein (Sung & Okimoto 1981, Slay et al. 1989) and polyamine (Fienberg et al. 1984) metabolism have been observed. For more information the reader is referred to articles in Terzi et al. (1985).

Somatic embryogenesis is influenced by several exogenous factors. Treatments of a suspension culture with white and blue light are inhibitory to growth and somatic embryogenesis. The highest number of somatic embryos is produced in the dark or after red and green light treatments (Michler & Lineberger 1987). In the culture medium 2,4-D, usually used in a concentration in the range of 0.5-1.0 $mg.l^{-1}$, is thought to be essential for the acquisition of embryogenic potential. However, Smith & Krikorian (1988) demonstrated that somatic embryos can be obtained from certain cells of zygotic embryos in hormone-free medium. Embryogenesis itself is inhibited by exogenously supplied 2,4-D or IAA. Zeatin promotes, other cytokinins inhibit embryogenesis. GA_3 or ABA do not effect the number of embryos formed in the globular and early heart stages but they cause a decrease in the heart and torpedo stages (Fujimura & Komamine 1975). Promoting effects of embryo-conditioned medium on the frequency of somatic embryogenesis have also been reported (Hari 1980, Smith & Sung 1985). This has been found to depend on the presence of correctly glycosylated extracellular proteins in the culture medium (De Vries et al. 1988a). Further, the presence of minimal amounts of endogenous NH_4 is essential for the occurrence of embryogenesis in a carrot suspension culture (Tazawa & Reinert 1969). The importance of the form of the nitrogen source and some other factors in the medium were discussed by Bhojwani & Razdan (1983). The role of exogenously added Ca^{2+} will be described below.

3. FUNDAMENTAL PROPERTIES OF CALCIUM IN LIVING ORGANISMS

Ashley & Campbell (1979) distinguish four principle roles of calcium in living organisms:

a. A structural role of Ca^{2+}, including extracellular calcium precipates, in plants, mainly calcium pectate of the middle lamella; also the binding to phospholipids, proteins and nucleic acids, which is required for the normal function of biological membranes and many intracellular structures.
b. A cofactor for extracellular enzymes (so far mainly found in animals).
c. An electrical role of Ca^{2+}.
d. A role as an intracellular regulator: changes in the concentration of Ca^{2+} are caused by a primary physiological stimulus.

Concerning the last point a number of properties of calcium make it suitable as a second messenger. Important is its low intracellular free concentration (about 0.1 μM) as compared to that outside the cell. At elevated levels Ca^{2+} reacts with phosphate forming an insoluble precipitate which would strongly inhibit the phosphate-based energy metabolism of cells (Kretsinger 1979). A low intracellular concentration of Ca^{2+} is a prerequisite for the normal functioning of a cell. Due to this steep gradient it is possible to raise the Ca^{2+} concentration very quickly which makes it suitable for rapid responses of cells to environmental stimuli. A low internal Ca^{2+} concentration is maintained by actively pumping Ca^{2+} out of the cytoplasm. Ca^{2+} pumps are found in a variety of membranes such as in mitochondria (Dieter & Marme 1980), the tonoplast (Gross 1982), the plasmamembrane (Stosic et al. 1983), and rough ER (Buckhout 1984). Another important property of Ca^{2+} is its binding to a number of proteins in a reversible manner and thereby changing there activity.

Although the concentration of the closely related cation Mg^{2+} is much higher, cellular processes often possess a high selectivity for Ca^{2+}. Important in this respect are the somewhat larger diameter of Ca^{2+}, its more flexible coordination than Mg^{2+}, and its weaker energy of hydration (Levine & Williams 1982). For a detailed discussion of the physical properties which account for Ca^{2+} selectivity the reader is referred to Hepler & Wayne (1985).

Much emphasis has been placed on free cytoplasmic Ca^{2+} as the determinant of physiological triggering. In plants, as in animals, the level of free cytoplasmic Ca^{2+} is thought to be in the range of 10^{-5} to 10^{-6} M (Poovaiah 1985). However, direct measurements of the concentration in the cytosol of plant cells are scarce. Williamson & Ashley (1982) reported a concentration of 0.22 μM for _Chara_ and 1.1 μM for _Nitella_ internodal cells at rest. On electrical of mechanical stimulation this concentration could be raised to 6.7 μM and 43 μM respectively. In _Coscinodiscus concinnus_ and _Guinardia flaccida_, two centric diatoms, the mean cytoplasmic Ca^{2+} concentration was estimated to be 11 μM and 11.5 μM respectively (Brownlee et al. 1987). The Ca^{2+} concentration in rhizoids and thalli of _Riccia fluitans_ was determined with double-barreled microelectrodes to be between 9.4 and 18.7 μM (Felle 1988). The same author found that the concentration in root hairs of _Zea mays_ (14.5-23.1 μM) was higher than in the coleoptiles (8.4-14.3 μM). In comparison with these values the Ca^{2+} concentration in carrot protoplasts (36.1 μM) measured with quin-2 by Gilroy et al. (1989) seems rather high.

Besides free in the cytoplasm Ca^{2+} can be bound to membrane surfaces of intracellular chelating molecules and can also be sequestered inside subcellular organelles where the concentration lies in the mM range (Poovaiah 1985). Estimates of total Ca^{2+} are therefore a sum of the three sources and reflect only indirectly any changes in cytoplasmic Ca^{2+} (Caswell 1979). The distribution of calcium in plant cells is described in detail by Kauss (1987).

4. FUNDAMENTAL PROPERTIES OF CALMODULIN

Calmodulin is a small, acidic, heat-resistent Ca^{2+} binding protein with an isoelectric point of 3.9 and it is present in all eukaryotic cells investigated so far. It has been purified and characterized from a variety of higher plants including peanut (Anderson et al. 1980), zucchini (Marme & Dieter 1983), asparagus, mung bean (Anderson 1983), barley (Schleicher et al. 1983), spinach (Lukas et al. 1984), oats (Biro et al. 1984), white corn (Vantard et al. 1985) and wheat (Toda et al. 1985). It consists of a single polypeptide chain of 148 amino acid residues. The molecular weight of calmodulin from plants is similar but not identical to animal calmodulin. On SDS PAGE it shows a apparent molecular weight of 14,500 in the presence of Ca^{2+} and 17,000-19,000 in the absence of Ca^{2+} (Marme & Dieter 1983). The calmodulin sequence can be divided into four structural domains that are homologous to each other (Watterson et al. 1980). Each domain contains a high affinity Ca^{2+} binding site, known as EF hands (Kretsinger 1980, Van Eldik et al. 1982). Calmodulin seems to bind Ca^{2+} in a sequential and ordered manner. This means that calmodulin in the resting cell presents one site for Ca^{2+} binding. Upon occupancy of this site, a second Ca^{2+} binding site is shaped, a second Ca^{2+} binds and so on. Several Ca^{2+} binding complexes are sequentially present and exhibit different conformations. In a given conformation calmodulin interacts with a given enzyme or set of enzymes (Haiech et al. 1981).

The behaviour of animal and plant calmodulin upon Ca^{2+} binding seems to be very similar (Dieter 1984). In the absence of Ca^{2+}, 40% of it exists in an α-helical configuration. Ca^{2+} binding increases the helicity to more than 50% (Dedman et al. 1977). A comparison of sequences of calmodulins from various sources, including animals and higher plants, reveals a high degree of sequence identity and indicates that calmodulin may be one of the most highly conserved proteins known (Roberts et al. 1986). Spinach calmodulin differs from bovine calmodulin in only 12 amino acid residues. In contrast, calmodulins from phylogenetically earlier plant species differ markedly from higer plant calmodulins (Roberts et al. 1986). Characteristic for all calmodulins is the presence of trimethyllysine at position 115, the lack of tryptophan and the high content of negatively charged amino acids. Antibodies directed against bovine calmodulin cross-react with calmodulin from higher plants (Schleicher et al. 1982). In addition, antisera against spinach calmodulin cross-react with pea, wheat and corn calmodulin but do not react with bovine brain calmodulin (Muto & Miyachi 1984).

The yield of calmodulin from zucchini hypocotyles was found to be about 10 mg per kg of plant tissue, corresponding to a concentration of about 10^{-3}-10^{-2} $mol.m^{-3}$ in the cytoplasm of a plant cell (Marme & Dieter 1983). The calmodulin content in plant tissue is not constant. The level has found to be high in growing tissue and leaves (Muto & Miyachi 1984). In corn root sections calmodulin was mainly detected in the root cap cells. In terminal buds of spinach it is highly concentrated in the apical meristem and leaf primordia (Lin et al. 1986). Most of the calmodulin is localized, free or bound, in the cytoplasm whereas smaller fractions are associated with organelles. In pea seedlings calmodulin seems to be present in vacuoles and amyloplasts (Dauwalder et al. 1986). These immunocytochemical studies do not confirm the presence of calmodulin in the apoplast as found by radioimmunoassays (Biro et al. 1984). Calmodulin is further associated with cytoskeletal elements which is very apparent in mitotic spindles (Vantard et al. 1985). Nucleoli seem to be devoid of calmodulin (Dauwalder et al. 1986). In animal cells a correlation between the calmodulin level and the cell cycle has been found (Means et al. 1982, You et al. 1988). High calmodulin levels are present in late G_1 and early S phase. Similar findings for plant cells have not been reported.

The calmodulin level never acts as a limiting factor in the cell (Poovaiah & Reddy 1987). Therefore the distribution of activated calmodulin is more important. This has been studied in tipgrowing plant cells by Hausser et al. (1984) with fluorescing phenothiazines. These antipsychotic drugs bind in a reversible manner with the Ca^{2+}-calmodulin complex. This binding becomes irreversible after irradiation with UV light (Prozialeck et al. 1981). Young stages of tipgrowing cells possess a high calmodulin level. The localization of calmodulin coincides with that of calcium in Acetabularia mediterranea (Cotton & Van den Driessche 1987).

5. EXPERIMENTAL APPROACHES TO THE SYSTEM

Evidence for Ca^{2+} as intracellular regulator of processes involved in carrot somatic embryogenesis can be obtained by a number of experimental approaches. One is the study of the effect of changes in the Ca^{2+} concentation in the medium or in the organism under investigation. This can be achieved by the use of Ca^{2+} chelating agents such as EGTA, Ca^{2+} channel blockers such as La^{3+}, Ca^{2+} ionophores such as A23187 or by raising the extracellular Ca^{2+} concentration. However, to obtain direct evidence for a link between an extracellular primary signal and the intracellular physiological event it is necessary to measure the intracellular concentration of free Ca^{2+} (Ashley & Campbell 1979). Once it is established that free cytoplasmic Ca^{2+} triggers an event, the source of this Ca^{2+} must be investigated. Therefore it is necessary to measure the total Ca^{2+} concentration and its detailed distribution (Caswell 1979).

Three types of intracellular Ca^{2+} can be distinguished (see also the paragraph on calcium):

a. Cytoplasmic free Ca^{2+}.
b. Ca^{2+} bound to membrane surfaces or intracellular chelating molecules.
c. Ca^{2+} sequestered inside subcellular organelles.

Each type demands a different detection technique. Several reviews on the measurement of intracellular Ca^{2+} are published (e.g. Ashley & Campbell 1979, Caswell 1979, Borle & Snowdowne 1987). Here I will only give the information from these reports which is applicable to the study of the role of Ca^{2+} during somatic embryogenesis.

For calmodulin techniques are available to study its total cellular concentration as well as its intracellular distribution. With the help of phenothiazines it is even possible to distinguish between activated and non-activated calmodulin. A short summary of applications will be given below.

5.1. Total cellular Ca^{2+}

Total amounts of Ca^{2+} can be determined by atomic absorption spectrophotometry (Havelange 1989) which gives an indication of the Ca^{2+} concentration of the whole tissue under investigation. However, more important is the subcellular distribution. This can be visualized by histochemical techniques (McGee-Russel 1958, Slocum & Roux 1982), X-ray microanalysis (Hughes 1986), and autoradiography (Caswell 1979). Especially X-ray microanalysis seems promising for the study of the distribution of intracellular Ca^{2+}. A main problem, however, with this technique is the preparation of the material before study in the electron microscope. The method always requires a fast freezing step, which seems to be not a great problem any more at the moment (Kaesser et al. 1989); but then the biological Ca^{2+} concentrations are very close to the limits of the resolution of the scanning electron microscope. The following procedures include direct observation of frozen-hydrated sections, study of frozen-dried sections or sections of frozen-dried and resin embedded specimens, and the use of freeze substitution. The last method with long substitution times in 20% acrolein in diethyl ether under anhydrous conditions gives the best results (Marshall 1980).

5.2. Free cytoplasmic Ca^{2+}

The three methods of measuring the cytosolic free Ca^{2+} concentration are photoproteins, Ca^{2+} sensitive electrodes and dyes (Thomas 1986). Two photoproteins that have been succesfully used to measure intracellular free Ca^{2+} are aequorin and obelin (Borle & Snowdowne 1987). In the presence of Ca^{2+} these proteins emit light in direct proportion to the Ca^{2+} concentration and to their own concentration. Aequorin is a very sensitive Ca^{2+} indicator with a limit of detection of 2.10^{-7} M (Gilroy et al. 1989). In animal cells aequorin can be incorporated by the use of microinjection (Blinks et al. 1978), scrape-loading (Snowdowne & Borle 1985) of centrifugation-loading (Borle et al. 1986). Until now only microinjection has been performed on plant cells (Williamson & Ashley 1982). No reports are available on the use of photoproteins in large plant tissues because of the difficulty to incorporate them uniformly into the plant cell. Centrifugation loading might be a technique also applicable to plant cells.

The use of Ca^{2+} sensitive electrodes for the measurement of Ca^{2+} in living plant systems is limited by the fact that only the concentration in one cell at a moment is measured. When used in a multicellular system it is very difficult to see in which cell or cell compartment the measurements take place. The only possible application I see for the carrot system is the use of microelectrodes in combination with confocal scanning light microscopy.

A large number of dyes, specific for Ca^{2+}, are nowadays available. These include the metallochromic indicators arsenazo III and antipyralazo III and the fluorescent indicators quin-2 and fura-2. They are all able to indicate low intracellular free Ca^{2+} levels (Ashley & Campbell 1979, Borle & Snowdowne 1987). However, their application to whole plant tissue seems limited because of their cell impermeability and sequestering into intracellular membrane compartments (Hepler & Callaham 1987). The membrane sequestration permeable acetoxymethyl (AM) esters of the recently developed long wavelength Ca^{2+} indicators fluo-3 and rhod-2 (Minta et al. 1987) seem promising for the study of intracellular free Ca^{2+} in large plant tissue especially in combination with the use of confocal scanning light microscopy.

5.3. Membrane-bound calcium

Membrane-bound calcium can very easy be visualized by the use of chlorotetracycline (CTC). Its fluorescence increases upon binding with divalent cations. The selectivity for CTC fluorescence is for Ca^{2+} and Mg^{2+}. A number of properties makes it however suitable as a Ca^{2+} indicator (Caswell 1979). CTC is cell permeable and easy loads into plant cells and it is therefore applicable for the visualization of Ca^{2+} in complete plant tissue.

5.4. Calmodulin

Being a protein, calmodulin can be studied by methods suited for protein analysis. These include radioimmunoassay (a calmodulin ^{125}I RIA kit is commercially available from Amersham, Berks, England), and immunocytochemistry (Wick & Duniec 1986). With these methods total calmodulin levels and its distribution in plant tissue are indicated. The Ca^{2+}-calmodulin complex can be visualized by phenothiazines which bind rather specifically to activated calmodulin (Levin & Weiss 1975, 1977). They can be photooxidized to fluorescent derivatives and then become irreversibly bound to the complex (Prozialeck et al. 1981). The brightest fluorescence is produced by fluphenazine.2HCl (Hausser et al. 1984). This compound is cell permeable and therefore applicable to whole plant tissues.

6. CALCIUM AND CALMODULIN DURING CARROT SOMATIC EMBRYOGENESIS

The role of Ca^{2+} during the development of somatic embryos has not received much attention. Only a few reports on this matter are available. From these, however, it is obvious that calcium is important for the normal development of these embryos and plays a fundamental role in plant embryogenesis. First I will focus on the influence of external Ca^{2+} on the process of somatic embryogenesis. Then I will give some results of localization studies of calcium and calmodulin during the development of the embryos.

6.1. Effects of external Ca^{2+} on somatic embryogenesis

For normal cell proliferation in a carrot suspension culture at least a concentration of 30 µM $CaCl_2$ is essential. Between a Ca^{2+} concentration of 30 µM and 1.0 mM (the regular concentration in the medium) both the amount of somatic embryos as well the cell density in proliferating cultures (cultures with 2,4-D in the medium) increases with the concentration of Ca^{2+}. At concentrations over 1.0 mM there is a clear increase in the amount of somatic embryos which does not coincide with a simultanous increase in cell proliferation (Jansen et al. 1989). High concentrations of Ca^{2+} (higher than 7.5 mM) are also promotive for the initiation of embryogenesis from single cells (Nomura 1987). From this it was concluded that Ca^{2+} influx promotes embryogenesis. Adding the Ca^{2+} ionophore A23187, however, inhibits embryogenesis in the presence of Ca^{2+} concentrations higher than approximately 0.1 mM. Only at low

Ca^{2+} concentrations a promotive effect of an enlarged Ca^{2+} influx could be observed (Jansen et al. 1989). Experiments with embryogenic cells, pregrown at various Ca^{2+} concentrations, indicate that not the actual concentration of Ca^{2+} is responsible for the increased amount of embryos but rather the change in its concentration. The same increase of embryogenesis can be achieved when proembryogenic masses transferred to medium with a ten fold higher concentration of Ca^{2+}, regardless of the initial concentration (Jansen et al. 1989).

Chlorotetracycline, an antibiotic with Ca^{2+} binding properties, inhibits embryogenesis in a concentration of 10^{-4} M. Lower concentrations only slow down the development of the embryos. Transfer to medium without CTC, after CTC treatment, speeds up the development and growth of the embryos just as the transfer to medium with an elevated concentration of Ca^{2+} (Timmers, unpublished results).

6.2. Ca^{2+} distribution during carrot somatic embryogenesis

The data above, together with the observations of polarized DNA synthesis and cell division (Nomura & Komamine 1986a) suggest that a polarized localization of Ca^{2+} is necessary for normal embryogenesis (Nomura 1987). The Ca^{2+} distribution was investigated by the use of fura-2 AM (Nomura 1987), CTC (Timmers et al. 1989) or alizarin red S (Timmers, unpublished results). Nomura (1987) observed a polar distribution of free cytoplasmic Ca^{2+} in cell clusters and developing embryos. This polarity was absent in clusters in medium with 2,4-D. CTC and alizarin red S (method according to Kotenko et al. 1987) reveal that the Ca^{2+} content in developing embryos is higher than in proembryogenic masses. The Ca^{2+} seems to be uniformly distributed over the embryo with a higher concentration in the outer cell layers (H. Kieft, personal communication). If there is a polarized distribution of Ca^{2+} during embryogenesis this is restricted to the free cytoplasmic part of it.

6.3. Calmodulin distribution during carrot somatic embryogenesis

Immunocytological studies on the light microscopical level revealed a uniform diffuse and punctate cytoplasmic distribution of calmodulin during the early phases of carrot somatic embryogenesis. With the electron microscope it was found that short after the initiation of embryogenesis a small gradient in calmodulin distribution from the outside to the center of a proembryogenic mass is present. High amounts of calmodulin were present in mitochondria or plastids, followed by the nucleolus and the nucleoplasm. Cell walls, amyloplasts and vacuoles were negative (Jansen et al. 1989). In torpedo shaped stages calmodulin seems to be mainly present in amyloplasts in the most outer layers of the embryo (Timmers et al. 1989). From this it is concluded that during carrot somatic embryogenesis a redistribution of calmodulin appears.

As mentioned earlier the distribution of activated calmodulin can be visualized with fluphenazine (Hausser et al. 1984). Already before the initiation of embryogenesis a polarized distribution of activated calmodulin proved to be present in proembryogenic masses (Timmers et al. 1989). From the globular to the torpedo shaped stage calmodulin was active mainly in the basal part of the embryo. Only in late torpedo shaped stages activated calmodulin was also present in the shoot apex. In the early developmental stages the signal from the fluorochrome was uniformly distributed over all cell organelles, although the nuclei gave a more faint response. In the later stages, a punctate label could be observed in the organelles surrounding the nuclei, most likely the amyloplasts.

7. CONCLUSIONS

Ca^{2+} is essential for the normal development of carrot somatic embryos. High concentrations of external Ca^{2+} are promotive, while high concentrations of 2,4-D prevent it. At concentrations above 6 mM external Ca^{2+} even in the presence of 2,4-D embryogenesis occurs (Jansen et al. 1989). Therefore Ca^{2+} and 2,4-D clearly are antagonists in this respect. It has been suggested before that auxin alters the Ca^{2+} flux at the plasma membrane and causes a lowered internal concentration of Ca^{2+} (Hepler & Wayne 1985). From these data it seems that the only role of 2,4-D in the medium is the prevention of organized growth which would make it possible to create a proembryogenic mass from which an embryo might develop. The external trigger for embryogenesis to proceed seems to be the dilution of the culture, not the removal of the auxin from the medium. At high densities embryogenesis is inhibited.

Activated calmodulin is already unevenly distributed in proembryogenic masses in medium with 2,4-D. Obviously, polarity is already present in these proembryogenic masses. Therefore, these masses are not just clusters of undifferentiated suspension cells but possess already a high degree of organization. This view is supported by the fact that there is little difference between the protein patterns and the gene-expression programs of proembryogenic masses and those of somatic embryos (Sung & Okimoto 1981, Wilde et al. 1988). A proembryogenic mass can be regarded to be homologous to a single disorganized embryo and those arising from it as adventitious embryos, as was earlier suggested by Kohlenbach (1978).

A proper development of an embryo requires a strictly controlled regulation of the plane of cell division, cell division rate, and cell volume. The microtubular cytoskeleton plays an important role in these events. By regulating the organization of the cytoskeleton in individual cells a developing embryo determines its final morphology. At the onset of embryogenesis, up to the late globular stage, there is no polarity morphologically visible. In the early heart shaped stage polarity is clearly established. Referring to work on other plant systems (Hepler 1988, Marme 1989) these aspects of embryogenesis can be dependent on changes in the free cytoplasmic Ca^{2+} concentration in certain cells or cell compartments, influencing tubulin polymerization or depolymerization. Improvements of the visualization of the distribution of Ca^{2+} and calmodulin during embryogenesis therefore certainly will help us to understand the processes of organized development in plants more profoundly.

ACKNOWLEDGEMENTS. I thank J.H.N. Schel for his helpful criticisms during the preparation of this paper. Further I thank J. Cobben-Molenaar for preparing the typescript.

8. REFERENCES

Anderson, J.M. (1983) Purification of calmodulin, Methods Enzymol. 102, 9-17.

Anderson, J.M., Charbonneau, H., Jones, H.P., McCann, R.O., Cormier, M.J. (1980) Characterization of the plant nicotinamide adenine dinucleotide kinase activator protein and its identification as a calmodulin. Biochemistry 19, 3113-3120.

Ashley, C.C., Campbell, A.K. (1979) Detection and measurement of free Ca^{2+} in cells. Elsevier/North-Holland Biomedical Press, Amsterdam, New York, Oxford.

Bhojwani, S.S., Razdan, M.K. (1983) Plant tissue culture: Theory and practice. Elsevier, Amsterdam, Oxford, New York, Tokyo, pp. 91-112.

Biro, R.L., Daye, S., Serlin, B.S., Terry, M.E., Datta, N., Sopory, S.K., Roue, S.J. (1984) Characterization of oat calmodulin and radio immunoassay of its cellular distribution. Plant Physiol. 75, 382-386.

Blinks, J.R., Mattingly, P.H., Jewell, B.R., Van Leeuwen, M., Harrer, G.C., Allen, D.G. (1978) Practical aspects of the use of aequorin as a calcium indicator: Assay preparation, microinjection and interpretation of signals. Methods Enzymol. LVII, 292-328.

Blowers, D.P., Trewavas, A.J. (1989) Second messengers: Their existence and relationship to protein kinases, in Boss, W.F., Morre, D.J. (eds.) Second messengers in plant growth and development. Alan R. Liss. Inc., New York, pp. 1-28.

Borle, A.B., Freudenrich, C.C., Snowdowne, K.W. (1986) A simple method for incorporating aequorin into mammalian cells. Am. J. Physiol. 251, C323-C326.

Borle, A.B., Snowdowne, K.W. (1987) Methods for the measurement of intracellular ionized calcium ions in mammalian cells: Comparison of four classes of Ca^{2+} indicators, in Cheung, W.Y. (ed.) Calcium and cell function, vol. VII, Academic Press, New York, pp. 159-200.

Brand, J.J., Becker, D.W. (1984) Evidence for direct roles of calcium in photosynthesis. J. Bioenerg. Biomembr. 16, 239-248.

Brownlee, C., Wood, J.W., Briton, D. (1987) Cytoplasmic free calcium in single cells of centric diatoms. The use of fura-2. Protoplasma 140, 118-122.

Buckhout, T.J. (1984) Characterization of Ca^{2+} transport in purified endoplasmic reticulum membrane vesicles from *Lepidium sativum* L. roots. Plant Physiol. 76, 962-967.

Caswell, A.H. (1979) Methods of measuring intracellular calcium. Int. Rev. Cytol. 56, 145-181.

Cotton, G., Vanden Driesche, T. (1987) Identification of calmodulin in *Acetabularia*: Its distribution and physiological significance. J. Cell Sci. 87, 337-347.

Das, R., Bagga, S., Sopory, S.K. (1987) Involvement of phosphoinositides, calmodulin and glyoxlase I in cell proliferetion in callus cultures of *Amaranthus paniculatus*. Plant Sci. 53, 45-51.

Das, R., Sharma, A.K., Sopory, S.K. (1989) Regulation of NADH-glutamatedehydrogenase activity by phytochrome, calcium and calmodulin in *Zea mays*. Plant Cell Physiol. 30, 317-323.

Dauwalder, M., Roux, S.J., Hardison, L. (1986) Distribution of calmodulin in pea seedlings. Immunocytochemical localization in plumules and root apices. Planta 168, 461-470.

Dedman, J.R., Potter, J.D., Jackson, R.L., Johnson, J.D., Means, A.R. (1977) Physicochemical properties of rat testis Ca^{2+}-dependent regulator protein of cyclic phosphodiesterase: Relationship of Ca^{2+}-binding, conformational changes and phosphodiesterase activity. J. Biol. Chem. 252, 8415-8422.

Dela Fuente, R.K. (1984) Role of calcium in the polar secretion of indoleacetic acid. Plant Physiol. 76, 342-346.

De Vries, S.C., Booij, H., Janssens, R., Vogels, R., Saris, L., LoSchiavo, F., Terzi, M., Van Kammen, A. (1988a) Carrot somatic embryogenesis depends on the phytohormone-controlled presence of correctly glycosylated extracellular proteins. Genes & Development 2, 462-476.

De Vries, S.C., Booij, H., Meyerink, P., Huisman, G., Dayton Wilde, H., Thomas, T.L., Van Kammen, A. (1988b). Acquisition of embryogenic potential in carrot cell suspension cultures. Planta 176, 196-204.

Dieter, P. (1984) Calmodulin and calmodulin mediated processes in plants. Plant Cell and Environment 7, 371-380.

Dieter, P., Marme, D. (1980) Ca^{2+} transport in mitochondrial and microsomal fractions from higher plants. Planta 150, 1-8.

Elliot, D.C. (1986) Calcium involvement in plant hormone action, in Trewavas, A.C. (ed.) Molecular and cellular aspects of calcium in plant development. Plenum Press, New York, pp. 285-292.

Felle, H. (1988) Cytoplasmic free calcium in <u>Riccia</u> <u>fluitans</u> L. and <u>Zea mays</u> L.: Interaction of Ca^{2+} and pH? Planta 176, 248-255.

Felle, H. (1989) pH as a second messenger in plants, in Boss, W.F., Morre, D.J. (eds.) Second messengers in plant growth and development. Alan R. Liss. Inc., New York, pp. 57-80.

Fienberg, A.A., Choi, J.H., Lubick, W.P., Sung, Z.R. (1984) Developmental regulation of polyamine metabolisme in growth and differentiation of carrot culture. Planta 162, 532-539.

Fujimura, T., Komamine, A. (1975) Effects of various growth regulators on the embryogenesis in a carrot cell suspension culture. Plant Sc. Lett. 5, 359-364.

Gilroy, S., Hughes, W.A., Trewavas, A.J. (1989) A comparison between quin-2 and aequorin as indicators of cytoplasmic calcium levels in higher plant cell protoplasts. Plant Physiol. 90, 482-491.

Gross, J. (1982) Oxalate-enhanced active calcium-uptake in membrane fractions from zucchini squash, in Marme, D., Marre, E., Hertel, R. (eds.) Plasmalemma and tonoplast: Their functions in the plant cell. Elsevier Biomedical Press, New York, pp. 369-376.

Guilfoyle, T.J. (1989) Second messengers and gene expression, in Boss, W.F., Morre, D.J. (eds.) Second messengers in plant growth and development. Alan R. Liss. Inc., New York, pp. 315-326.

Haiech, J., Klee, C.B., Demaille, J.G. (1981) Effects of cations on affinity of calmodulin for free calcium: Ordered binding of calciumions allow the specific activation of calmodulin-stimulated enzymes. Biochemistry 20, 3890-3897.

Halperin, W. (1966) Alternative morphogenetic events in cell sus-

pensions. Amer. J. Bot. 53, 443-453.
Halperin, W., Jensen, W.A. (1967) Ultrastructural changes during growth and embryogenesis in carrot cell cultures. J. Ultrastruct. Res. 18, 428-443.
Hari, V. (1980) Effect of cell density changes and conditioned media on carrot cell embryogenesis. Z. Pflanzenphysiol. 96, 227-231.
Hausser, I., Herth, W., Reiss, H.D. (1984) Calmodulin in tipgrowing plant cells, visualized by fluorescing calmodulin binding phenothiazines. Planta 162, 33-39.
Havelange, A. (1989) Levels and ultrastructural localization of calcium in Sinapis alba during the floral transition. Plant Cell Physiol. 30, 351-358.
Hepler, P.K. (1988) Calcium and development, in Greuter, W., Zimmer, B. (eds.) Proc. XIV Intern. Bot. Congr. Koeltz, Konigstein/Taunus, pp. 225-246.
Hepler, P.K., Callaham, D.A. (1987) Free calcium increases during anaphase in stamen hair cells of Tradescantia. J. Cell Biol. 105, 2137-2143.
Hepler, P.K., Wayne, R.O. (1985) Calcium and plant development. Ann. Rev. Plant Physiol. 36, 397-439.
Hughes, W.A. (1986) NMR and X-ray micro analysis methods for measurement of calcium in plant cells, in Trewavas, A.J. (ed.) Molecular and cellular aspects of calcium in plant development. Plenum Press, New York, London, pp. 157-164.
Janssen, M.A.K., Kreuger, M., Booij, H., Schel, J.H.N., De Vries, S.C., Van Kammen, A. (1989) The role of calcium and calmodulin in early stages of carrot somatic embryogenesis. Submitted.
Jones, R.L., Deikman, J., Melroy, D. (1986) Role of Ca^{2+} in the regulation of α-amylase synthesis and secretion in barley aleurone, in Trewavas, A.C. (ed.) Molecular and cellular aspects of calcium in plant development. Plenum Press, New York, London, pp. 49-56.
Kaesser, W., Koyro, H.W., Moor, H. (1989) Cryofixation of plant tissue without pretreatment. J. Microsc. 154, 279-288.
Kakiuchi, S., Sobue, K. (1983) Control of cytoskeleton by calmodulin and calmodulin-binding proteins. Trends Biochem. Sci. 8, 59-62.
Kauss, J. (1983) Volume regulation-activation of a membrane associated cryptic enzyme system by detergent-like action of phenothiazine drugs. Plant Sci. Lett. 26, 103-109.
Kauss, H. (1987) Some aspects of calcium dependent regulation in plant metabolism. Ann. Rev. Plant Physiol. 38, 47-72.
Kohlenbach, H.W. (1978) Comparative somatic embryogenesis, in Thorpe, T.A. (ed.) Frontiers of plant tissue culture. Univ. Calgary Press, Canada, pp. 59-66.
Kotenko, J.L., Miller, J.H. Robinson, A.I. (1987) The role of asymmetric cell division in Pteridophyte cell differentiation. I. Localized metal accumulation and differentiation in Vittaria gemmae and Onoclea prothallia. Protoplasma 136, 81-95.
Kretsinger, R.H. (1979) The information role of calcium in the cytosol. Adv. Cyclic Nuc. Res. 11, 1-26.
Kretsinger, R.H. (1980) Structure and evolution of calcium modulated proteins. Crit. Rev. Biochem. 8, 119-174.

Leshem, Y.Y. (1987) Membrane phospholipid catabolisme and Ca^{2+} activity in control senescence. Physiol. Plant. 69, 551-559.

Levin, R.M., Weiss, B. (1975) Mechanism by which psychotropic drugs inhibit cyclic AMP-phosphodiesterase in brain. Mol. Pharmacol. 12, 581-589.

Levin, R.M., Weiss, B. (1977) Binding of trifluperazine to the calcium dependent activator of cyclic AMP-phosphodiesterase. Mol. Pharmacol. 13, 690-697.

Levine, B.A., Williams, R.J.P. (1982) The chemistry of calcium ion and its biological relevance, in Anghileri, L.J., Tuffet-Anghileri, A.M. (eds.) The role of calcium in biological systems 1, CRC Press, Boca Raton, Florida, pp. 3-26.

Lin, C.T., Sun, D., Song, G.X., Wu, J.Y. (1986) Calmodulin: Localization in plant tissues. J. Histochem. Cytochem. 34, 561-567.

Lukas, T.J., Iverson, D.B., Schleicher, M., Watterson, D.M. (1984) Structural characterization of a higher plant calmodulin: Spinacea oleracea. Plant Physiol. 75, 788-795.

Marme, D. (1989) The role of calcium and calmodulin in signal transduction, in Boss, W.F., Morre, D.J. (eds.) Second messengers in plant growth and development. Alan R. Liss. Inc., New York, pp. 57-80.

Marme, D., Dieter, P. (1983) Role of Ca^{2+} and calmodulin in plants, in Cheung, W.Y. (ed.) Calcium and cell function, vol. 4, Academic Press, New York, pp. 263-311.

Marshall, A.T. (1980) Freeze-substitution as a preparation technique for biological X-ray microanalysis. Scan. Elec. Microsc. II, 395-408,

McGee-Russel, S.M. (1958) Histochemical methods for calcium. J. Histochem. Cytochem. 6, 22-42.

McWilliam, A.A., Smith, S.M., Street, H.E. (1974) The origin and development of embryoids in suspension cultures of carrot (Daucus carota). Ann. Bot. 38, 243-250.

Means, A.R., Tash, J.S., Chafouleas, J.G. (1982) Physiological implications of the presence, distribution and regulation of calmodulin in eukaryotic cells. Physiol. Rev. 62, 1-38.

Michler, C.H., Lineberger, R.D. (1987) Effects of light on somatic embryo development and abscisic acid levels in carrot suspension cultures. Plant, Cell, Tissue and Organ Culture 11, 189-207.

Minta, A., Harootunian, A.T., Kao, J.R.Y., Tsien, R.Y. (1987) New fluorescent indicators for intracellular sodium and calcium. J. Cell Biol. 105, 89a.

Mitsui, T., Christeller, J.T., Hara-Nishimura, I., Akazawa, T. (1984) Possible roles of calcium and calmodulin in the biosynthesis and secretion of α-amylase in rice seed scutellar epithelium. Plant Physiol. 75, 21-25.

Muto, S., Miyachi, S. (1984) Production of antibody against spinach calmodulin and its application to radioimmunoassay for plant calmodulin. Z. Pflanzenphsyiol. 114, 421-431.

Nomura, K. (1987) Mechanisms of somatic embryogenesis in carrot suspension cultures. Ph.D. thesis. Univ. Tokyo, Japan.

Nomura, K., Komamine, A. (1985) Identification and isolation of single cells that produce somatic embryos at a high density frequency in carrot suspension cultures. Plant Physiol. 79, 988-991.

Nomura, K., Komamine, A. (1986a) Polarized DNA synthesis and cell division in cell clusters during somatic embryogenesis from single carrot cells. New Phytol. 104, 25-32.

Nomura, K., Komamine, A. (1986b) Molecular mechanisms of somatic embryogenesis, in Miflin, B.J. (ed.) Oxford surveys of Plant Molecular and Cell Biology, vol. 3, Oxford University Press, pp. 456-466.

Owen, J.H., Hetherington, A.M., Wellburn, A.R. (1987) Calcium, calmodulin and the control of respiration in protoplasts isolated from meristematic tissues by abscisic acid. J. Exp. Bot. 38, 1356-1361.

Penel, C., Greppin, H. (1982) Effect of light and phenothiazines on the level of extracellular peroxidases, in Marme, D., Marre, E., Hertel, R. (eds.) Plasmalemma and tonoplast: Their functions in the plant cell, Elsevier Biomedical Press, New York, pp. 53-57.

Poovaiah, B.W. (1985) Role of calcium and calmodulin in plant growth and development. Hort. Science 20, 347-351.

Poovaiah, B.W., McFadden, J.J., Reddy, A.S.N. (1987) The role of calcium ions in gravity signal perception and transduction. Physiol. Plant. 71, 401-407.

Poovaiah, B.W., Leopold, A.C. (1973) Deferral of leaf senescence with calcium. Plant Physiol. 52, 236-239.

Poovaiah, B.W., Reddy, A.S.N. (1987) Calcium messenger systems in plants. CRC-Crit. Rev. Plant Sci. 6, 47-103.

Poovaiah, B.W., Veluthambi, K. (1986) The role of calcium and calmodulin in hormone action in plants: Importance of protein phosphorylation, in Trewavas, A.J. (ed.) Molecular and cellular aspects of calcium in plant development. Plenum Press, New York, pp. 83-90.

Prozialeck, W.C., Amino, M., Weiss, B. (1981) Photoaffinity labeling of calmodulin by phenothiazine psychotics. Mol. Pharmacol. 19, 264-269.

Ranjevo, R., Boudet, A.M. (1987) Phosphorylation of proteins in plants: Regulatory effects and potential involvement in stimulus/response coupling. Ann. Rev. Plant Physiol. 38, 73-93.

Reinert, J. (1985) Morphogenese und ihre Kontrolle an Gewebekulturen aus Carotten. Naturwissenschaften 45, 344-345.

Reis, H.D., Nobiling, R. (1986) Quin-2 fluorescence in lily pollentube: Distribution of free cytoplasmic calcium. Protoplasma 131, 244-246.

Roberts, D.M., Lukas, T.J., Watterson, D.M. (1986) Structure, function, and mechanism of action of calmodulin. Crit. Rev. Plant Sci. 4, 311-339.

Roux, S.J., Wayne, R.O., Datta, N. (1986) Role of calcium ions in phytochrome response: an update. Physiol. Plant 66, 344-348.

Satter, R.L., Galston, A.W. (1981) Mechanisms of control of leaf movements. Ann. Rev. Plant Physiol. 32, 83-110.

Schleicher, M., Iversen, D.B., Van Eldik, L.J., Watterson, D.M. (1982) Calmodulin, in Lloyd, C.W. (ed.) The cytoskeleton in plant growth and development. Academic Press, New York, pp. 85-106.

Schleicher, M., Lukas, T.J., Watterson, D.M. (1983) Futher characterization of calmodulin from the monocotyledon barley (*Hordeum vulgare*). Plant Physiol. 73, 666-670.

Slay, R.M., Grimes, H.D., Hodges, T.K. (1989) Plasma membrane proteins associated with undifferentiated and embryonic *Daucus carota* tissue.

Protoplasma 150, 139-149,
Slocum, R.D., Roux, S.J. (1982) An improved method for the subcellular localization of calcium using a modification of the antimonate precipitation technique. J. Histochem. Cytochem. 30, 617-629.
Smith, D.L., Krikorian, A.D. (1988) Production of somatic embryos from carrot tissues in hormone-free medium. Plant Science 58, 103-110.
Smith, G.N., Willmer, C.M. (1988) Effects of calcium and abscisic acid on volume changes of guard cell protoplasts of Commelina. J. Exp. Bot. 39, 1529-1539.
Smith, J.A., Sung, Z.R. (1985) Increase in regeneration of plant cells by cross feeding with regenerating Daucus carota cells, in Terzi, M., Pitto, L., Sung, Z.R. (eds.) Somatic Embryogenesis. IPRA, Rome, pp. 77-85.
Snowdowne, K.W., Borle, A.B. (1985) Effects of low extracellular sodium on cytosolic ionized calcium. Na^+-Ca^{2+} exchange as a major calcium influx pathway in kidney cells. J. Biol. Chem. 260, 14998-15007.
Steer, M.W. (1988) The role of calcium in exocytosis and endocytosis in plant cells. Physiol. Plant. 72, 213-220.
Steward, F.C., Mapes, M.O., Mears, K. (1958) Growth and organized development of cultured cells II. Organization in cultures grown from freely suspended cells. Am. J. Bot. 45:705-708.
Street, H.E., Withers, L.A. (1974) The anatomy of embryogenesis in culture, in Street, H.E. (ed.) Tissue culture and plant science. Academic Press, London, New York, pp. 71-100.
Stosic, V., Penel, C., Marme, D., Greppin, H. (1983) Distribution of calmodulin-stimulated Ca^{2+} transport into membrane vesicles from green spinach leaves. Plant Physiol. 72, 1136-1138.
Sung, Z.R., Okimoto, R. (1981) Embryogenic proteins in somatic embryos of carrot. Proc. Nat. Acad. Sci. 78, 3683-3687.
Tazawa, M., Reinert, J. (1969) Extracellular and intracellular chemical environment in relation to embryogenesis in vitro. Protoplasma 68, 157-173.
Terzi, M., Pitto, L., Sung, Z.R. (1985) Somatic Embryogenesis. IPRA Roma, Italy.
Thomas, M.V. (1986) The definition and measurement of intracellular free Ca, in Trewavas, A.J. (ed.) Molecular and cellular aspects of calcium in plant development. Plenum Press, New York, London, pp. 141-147.
Timmers, A.C.J., De Vries, S.C., Schel, J.H.N. (1989) Distribution of membrane-bound calcium and activated calmodulin during somatic embryogenesis of carrot (Daucus carota L.). Submitted.
Toda, H., Yazawa, M., Sakiyama, F., Yagi, K. (1985) Amino acid sequence of calmodulin from wheat germ. J. Biochem. 98, 833-842.
Van Eldik, L.J., Zendegui, J.G., Marshall, D.R., Watterson, D.M. (1982) Calcium binding proteins and the molecular basis of calcium action. Int. Rev. Cytol. 77, 1-61.
Vantard, M., Lambert, A.M., De Mey, J. Picquot, P. Van Eldik, L.J. (1985) Characterization and immunocytochemical distribution of calmodulin in higher plant endosperm cells: localization in the mitotic apparatus. J. Cell Biol. 101, 488-499.
Watterson, D.M., Sharief, F., Vanaman, T.C. (1980) The complete amino

acid sequence of the Ca^{2+} dependent modulator protein (calmodulin) of bovine brain. J. Biol. Chem. 255, 962-975.

Wick, S.M., Duniec, J. (1986) Effects of various fixatives on the reactivity of plant cell tubulin and calmodulin in immunofluorescence microscopy. Protoplasma 133, 1-18.

Wilde, H.D., Nelson, W.S., Booij, H., De Vries, S.C., Thomas, T.L. (1988) Gene-expression programs in embryogenic and non-embryogenic carrot cultures. Planta 176, 205-211.

Williamson, R.E., Ashley, C.C. (1982) Free Ca^{2+} and cytoplasmic streaming in the alga Chara. Nature 296, 647-651.

You, J.S., Li, S.W., Wang, D.S., Zhang, Y., Suen, D.Y., Xue, S.B. (1988) The distribution of calmodulin and the calmodulin antagonist TFP in cell cycle of the friend erythro-leukemia cells. Abstr. 4th. Int. Congr. Cell Biol. Montreal, p. 50.

Section 2. Plant genetic engineering

GENOMIC REORGANIZATION INDUCED BY PLANT TISSUE CULTURE

RONALD L. PHILLIPS
Department of Agronomy and Plant Genetics and
Plant Molecular Genetics Institute
University of Minnesota
St. Paul, Minnesota 55108 USA

ABSTRACT. Variation induced by the plant tissue culture process is a common feature of regenerated plants or their progenies. Whether the variation is a result of a random array of disparate genetic phenomena or the end result of a chain of events initiated by some aspect of the culture environment is not clear at this time. Genetic, molecular, and cytological analyses will be presented for oat and maize tissue cultures that lead to an hypothesis accounting for the observed genetic and cytogenetic variation. The data indicate chromosome breakage as the principal cytogenetic event, concomitant increases in single gene mutations, activation of Ac and Spm transposable elements, and quantitative trait variation.

The hypothesis assumes that the tissue culture environment leads to cell cycle disturbances which cause a delay in the replication of normally late replicating DNA of heterochromatic regions. This delay causes the formation of anaphase bridges which yield specific rearrangements depending on the distribution of heterochromatin and the ploidy level. The cell cycle disturbances, either caused by the tissue culture environment or the anaphase bridges, are hypothesized to lead to alterations in DNA methylation. This methylation effect may yield single gene mutations because of the methylation alteration itself or by base substitutions caused by methylation changes. Transposable elements also can be activated by methylation effects on cryptic elements. Mutations, therefore, could in part be the result of transposable element insertions or excisions. A more global methylation effect might alter a large portion of the genome and cause variation in quantitatively inherited traits. Further studies on cell cycles in tissue culture are indeed warranted.

When we first regenerated maize (Zea mays L.) plants from tissue cultures initiated from immature embryos (Green and Phillips, 1975), there were hopes that the procedure would be a means to clonally propagate maize where every regenerant would be genetically identical to the source. We soon realized that this was not the case; many mutants were obvious in the progeny of regenerated plants. The purpose of this paper is to review (1) the recognition of chromosome breakage as the principal cytogenetic event, 2) the extensive qualitative genetic variation among progeny of regenerated plants, 3) the activation of transposable elements, 4) the quantitative trait variation among tissue culture-derived lines, and 5) the hypothesis that these events may all be the result of disturbances in the cell cycle of cultured cells.

R.S. Sangwan and B.S. Sangwan-Norreel (eds.), The Impact of Biotechnology in Agriculture, 237–246.

Chromosome Breakage

Our first extensive data on chromosome breakage arose from analyzing regenerants of oat (Avena sativa L.) tissue cultures (McCoy et al. 1982). Plants were regenerated every four months after culture initiation up to 20 months. Two varieties, Tippecanoe and Lodi, were regenerated under the same conditions and analyzed at meiosis for alterations in chromosome number or structure by analyzing microsporocytes. At 4 months of culture, approximately 11% of the plants had an obvious cytogenetic alteration; and the frequency was found to increase until at 20 months, approximately 49% of the plants were cytologically abnormal. Most of these changes appeared to be rather simple chromosomal structural alterations. Plants regenerated from cultures of the other variety, Lodi, exhibited a 49% abnormality rate after 4 months of culture which increased to about 88% among regenerants from the 20-month-old cultures. These results indicated an age effect as well as a genotype (Tippecanoe versus Lodi) effect. Although these results were interesting, examining the actual types of chromosome changes proved to be even more informative. First, no polyploid regenerants were obtained out of 321 Tippecanoe and 478 Lodi plants examined at meiosis. Of course, cultivated oat is an allohexaploid and higher ploidy levels would probably not be expected. Second, the most common chromosomal types present were telocentrics, now suspected of being near-telocentrics (Johnson et al. 1987) where a break occurred near the centromere in heterochromatin. Regenerated plants with a telocentric chromosome occurred about 20% of the time (20.6% in Tippecanoe, and 23.8% in Lodi). This is indeed an unusual result since telocentrics in plants grown from seed are very rare; 6000 seedlings of the variety Garry were analyzed and none with telocentrics were observed (McGinnis, 1966). The existence of such a high frequency of regenerants with a near-telocentric chromosome is unexpected and signals that unusual events are occurring in tissue culture. The second most common chromosomal alteration was that leading to a chromosome interchange. The third most common change was monosomy, which was about five times higher than trisomy; this implies that nondisjunction is not involved.

The above results with oat led us to hypothesize that heterochromatin often surrounds centromeres, that it is late replicating, and that under certain circumstances breakage events occur in chromosomes with heterochromatic blocks. Rhoades and Dempsey (1971, 1973) showed that knobbed arms in maize would occasionally break in the presence of two or more B chromosomes. The hypothesis is that the B chromosomes produce a gene product(s) that cause certain heterochromatic blocks to replicate abnormally late leading to the inability of these blocks to separate in anaphase of a specific division. Depending on the chromosomal location of the heterochromatin, chromosome bridges form and lead to a breakage event. The near-telocentrics in oat could have been caused by such a mechanism; perhaps the tissue culture environment also causes such an effect relative to late-replicating DNA. If two nonhomologous chromosomes are involved, an interchange could result from the various breakages and reunions. The monosomics might have occurred due to somatic instability of the telocentrics or near-telocentrics resulting in loss of that chromosome at some point in development. Johnson et al. (1987) subsequently showed the existence of large heterochromatic blocks around oat centromeres and that these chromosomal regions are late-replicating in seedling root tips, as expected for heterochromatin.

The hypothesis that tissue culture-induced delayed replication of late-replicating DNA leads to the high frequency of chromosome breakage among regenerants predicts that the types of cytogenetic alterations observed should reflect the distribution of heterochromatin in the species. With that in mind, it is interesting to review the results we have obtained with maize regenerants. From 1036 regenerated maize plants meiotically analyzed (Armstrong and Phillips, 1988; Benzion and Phillips, 1988; Lee and Phillips, 1987a; Rhodes et al. 1986), changes in chromosome number were found infrequently (6%), and many of these arose from one culture which had apparently a doubled chromosome number. However, regenerants with interchanges and deletions or duplications were common (20%). Inversions, which require two breaks within a single chromosome, were very rare (0.2%); inversions also were rare (0.2%) in oats (McCoy et al. 1982). The distribution of breaks in the maize regenerants indicated that most breaks occurred in chromosomes with knobs, in the same arm as the heterochromatic knob, and in between the knob and the centromere. These cytological results fit the hypothesis forwarded based on oat. The lack of near-telocentrics in the maize regenerants would be expected since the heterochromatic blocks are not next to the centromere, except for chromosome 7, but usually distal in the chromosome arm. Thus, the two species, oat and maize, both produce a high frequency of regenerants with chromosome breaks. Although the distribution of breaks is quite different in the two species, the distribution in each case is predicted by the common hypothesis.

Single Gene Mutations

Some single-gene recessive mutations have been reported among progenies of oat regenerants (Cummings et al. 1976). Because oat is an allohexaploid, we would not expect to see a high frequency of recessives. This appears to be the case. First and second self-pollinated generation progenies of maize regenerants, however, often segregate for new mutations not present in the source material. Lee and Phillips (1987b) report that among plants regenerated after 3 to 4 months of culture, approximately 38% segregate for a new mutation in the first or second self generation. After 8 to 9 months of culture, 77% of the progenies segregated. Furthermore, 10% of the regenerants segregated for more than one new mutation among the 3 to 4 month regenerants whereas 40% segregated for more than one new mutation among the 8 to 9 month regenerants. The progeny of one plant segregated for five new mutations. At 3 to 4 months of culture age, progeny of the 71 regenerants segregated for 37 variants - a frequency of 0.5 variants per regenerated plant. At 8 to 9 months of culture age, progeny of 177 regenerants segregated for 237 variants - a frequency of 1.3 variants per regenerated plant. This maize tissue culture system, therefore, appears to be highly mutagenic.

Cultures that regenerate by forming somatic embryos conceivably could result in fewer mutations among progenies of regenerated plants because of the precise developmental process that occurs in forming a somatic embryo with a bipolar axis (Vasil, 1983). Armstrong and Phillips (1988) initiated maize tissue cultures and generated from the common source both compact, organogenic (Type I) and friable, embryogenic (Type II) cultures. Regenerating plants after 4 months of culture resulted in a variant segregation frequency among progenies of 22 and 19% for the organogenic and embryogenic cultures, respectively. After 8 months of culture, variant frequencies of 38 and 42% for the organogenic and embryogenic cultures, respectively, were obtained. Thus, the two culture types did not appear to differ in genic mutation frequencies. Considering all measures of variability, i.e. cytological aberrations, pollen sterility, and variant segregation, the embryogenic cultures gave rise to significantly more variability.

Transposable Elements

The above discussion indicates a high frequency of chromosomal and genic changes among progeny of regenerants. One should also realize that we are able to detect only a portion, probably a small portion, of the variation present. Clearly, many cryptic aberrations occur that are undetectable by standard light microscopy of microsporocytes. Similarly, only a portion of the genic changes are detected since we only scored for visible kernel, seedling, or mature plant characteristics. We did not score for isozymic variants, RFLPs, etc. Probably every regenerated plant has been altered via the tissue culture process.

Because of the high frequencies of chromosome breaks and mutations, we decided to test for the possible presence of active transposable elements. Earlier reports with maize indicated that chromosome breakage, no matter how it is generated, can lead to the activation of transposable elements (McClintock, 1950, 1951; Neuffer, 1966; Bianchi et al., 1969; Doerschug, 1973). Since transposable elements can lead to mutations, we thought that the coincidently high frequencies of chromosome breakage and genic mutations might be related to the activation of transposable elements. Peschke et al. (1987) made approximately 1200 tests for Ac (Activator) activity with progenies from 301 regenerated plants; 54 of these tests were positive. The 54 positive tests represented 11 regenerated plants from cultures initiated from three different immature embryos. Various lines of evidence indicated that the Ac activation occurred in culture and that the embryos used to initiate the cultures did not possess an active Ac. More recently, about 700 tests were made to detect Spm (Suppressor-Mutator) activity among progenies of regenerated plants (Peschke et al., 1989). Two tests were positive tracing back to one regenerated plant out of the 120 tested. Evola and coworkers (1984) also have reported activation of Ac and Spm in a smaller population of regenerated plants.

Again, the results reported here do not represent the actual magnitude of transposable element activation. More than a dozen transposable element systems have been described for maize; we tested for activation of only two of these systems. How do these active transposable elements arise? They likely arise by alterations in silent Ac and Spm sequences already present in the genome (Fedoroff et al. 1983; Cone et al. 1986). Peschke (1989) demonstrated the presence of six to ten Ac-homologous DNA sequences in the materials used in the above mentioned studies. She has identified a 10 kb Bgl II band, detected by hybridization with the Hind III internal Ac fragment, which cosegregates with Ac activity based on 38 seedlings. We are now trying to clone this fragment to determine whether it contains a complete Ac sequence and if that sequence is different from the two published Ac sequences.

DNA Methylation

Methylation alterations of Ac-homologous sequences can be dramatic in certain circumstances. Culley (1986) reported extensive hypomethylation of all of the Ac-homologous sequences in maize endosperm tissue cultures, independent of Ac-activity. Brown and Lorz (1986) and Brown (1989) report the methylation of DNA of regenerated maize plants and their progenies. Kaeppler (Univ. Minnesota, personal communication) also has observed methylation differences among certain regenerated plants using RFLP probes and methylation sensitive restriction enzymes. Correlations have been reported between transposable element activity and methylation of specific sites within the Ac element (Schwartz and Dennis, 1986; Chandler and Walbot, 1986; Bennetzen

1987; Chomet et al., 1987; and Schwartz, 1989). Therefore, the possibility exists that the transposable element activity we observed could have been the result of methylation alterations in a silent sequence that is a regular component of the genome of these maize lines. Peschke (1989) examined this question and found that the relationship of methylation and Ac activity is complex in these materials. Plants with more extensive hypomethylation were always plants with Ac activity, but most of the plants showed some degree of hypomethylation independent of whether or not they possessed an active Ac. Peschke (1989) suggests that the hypomethylation may have occurred as the result of Ac activity rather than be the cause. Other researchers have reported that the degree of methylation of Ac can change over time (Schwartz and Dennis 1986; Kunze et al. 1988). Mu-homologous sequences also are occasionally modified by methylation in callus cultures (James and Stadler, 1988). Active Mu lines in culture usually retained their low degree of methylation but some sub-lines had a much higher degree of methylation. Transposition of Mu in cultured cells was demonstrated by the appearance of novel restriction bands in certain sub-lines with unmodified Mu elements. Further tests are needed to determine the role of DNA methylation in transposable element activation.

Quantitative Trait Variation

A striking result of tissue culture is the generation of variation in agronomic traits among lines tested several generations removed from the original regenerated plants. Results for both oat and maize demonstrate the point.

Dahleen (1989) evaluated 56 R4 and R5 lines derived from the oat variety Lodi and 213 R4 and 147 R5 lines from Tippecanoe. These lines traced to those studied by McCoy et al. (1982) and were derived from plants they determined to be cytologically normal with good seed set. Dahleen and coworkers evaluated the lines with six replications, two years, and two locations for changes in height, heading data, 100 seed weight, seed number, percent seed protein, flag leaf area, grain yield, and bundle weight. Nearly all of the families (82% of Lodi, 91% of Tippecanoe) showed significant alterations in at least one of the eight traits. Quantitative trait variation also has been reported in barley (Dunwell et al., 1986) and wheat (Ryan et al., 1987; Chen et al., 1987; Galiba et al., 1985).

Quantitative trait variation was evaluated in maize using 305 tissue culture derived lines from 4 and 8 month-old cultures and 48 control lines that traced to seed from the same ears from which the immature embryos were taken to initiate the cultures (Lee et al., 1988). These materials were evaluated as lines per se (S_2) and in a testcross at three locations in one year. The lines and testcrosses generally were shown to have lower grain yield and moisture. Grain yield and plant height tended to decrease with culture age. The highest yielding line per se in 3 of the 6 trials and the top ranked line in 5 of the 6 trials for yield and moisture were derived from tissue culture. Zehr et al. (1987) also demonstrated quantitative trait variation among maize tissue culture-derived materials.

The results to date indicate that most quantitative traits can be affected by passage through tissue culture and that these variations are heritable.

Hypothesis

The hypothesis proposed is based on the contention that the tissue culture process is causing rather global changes in the cultured cells. The changes observed could be the result of cell cycle disturbances caused by the artificial environment of in vitro grown materials (Fig. 1) Disturbances in the cell cycle could cause a replication delay of late-replicating DNA. This delay could lead to anaphase bridges formed in a manner consistent with the genomic distribution of such late replicating DNA (heterochromatin). These bridges would lead to breakage events and specific kinds of rearrangements depending on the heterochromatin distribution and ploidy level. Another result of the anaphase bridges might be to cause further cell cycle disturbances.

A major effect of cell cycle disturbances is hypothesized to be DNA methylation alterations; these could be specific or global. Specific methylation alterations could lead to single gene mutations either by changes in expression modulated by methylation or specific base substitutions since it is known that deamination of methylcytosine leads to thymine. Such methylation alterations in a specific DNA sequence also could lead to the activation of transposable elements. The newly activated transposable elements could lead to mutations by several routes including insertions. Excisions of the transposable elements also would be expected to lead to mutations (Chen et al., 1986; Donner and Nelson, 1987, 1979; Echt and Schwartz, 1981). The excision process may not be precise and could result in deletions or duplications. Even precise excision leaves footprints which may or may not be repaired (Sachs et al., 1983; Pohlman et al., 1984; Schwarz-Sommer et al., 1985; Saedler and Nevers, 1985; Chen et al., 1986). Also possible is that some of the chromosome breaks could be the result of transposable elements since certain forms of Ds (Dissassociation) and dSpm (Defective Spm) are known to break chromosomes.

Transposable elements should perhaps be viewed more as "reporter" genes indicative of DNA alterations in the cultured cells rather than a major cause of the increased mutation frequency. Brettell et al. (1986) and Dennis et al. (1987) sequenced two maize alcohol dehydrogenase mutants found segregating in progeny of regenerants. Each mutant was shown to be due to a single base change. In addition, most mutants deriving from tissue culture are stable. Lack of mutability is consistent with transposable element-induced mutations, yet the paucity of unstable mutants is striking. Bingham et al. (1988) documented the activation of a mutable allele arising from alfalfa tissue cultures. In maize, Armstrong (1986) observed a green sectoring albino mutant and Woodman and Kramer (1986) reported a mutable cob color phenotype. Such reported examples are rare.

Global methylation alterations could lead to the quantitative trait variation. Alterations in the quantitative traits studied presumably require changes in expression of a large number of genes, especially to achieve the observed degree of variation. This quantitative trait variation coupled with the high frequency of genic mutations, including transposable element activation, implies a very high mutation frequency indeed.

The hypothesis proposed here (Fig. 1) indicates some of the potential wide-ranging effects of perturbing the cell cycles of cultured plant materials. On this basis, expanded studies would be useful on the changes of cell cycle parameters upon initiation of the cultures, during subculturing and extended growth periods, as well as under varying culture conditions. Since we know that DNA methylation alterations can occur in culture, further analyses to determine the extent of such DNA modifications of cultured cells seem appropriate. Additional tests on the role of late-replicating DNA in the chromosome breakage events also would be useful.

In conclusion, the regeneration of plants from callus cultures is not a means to generate genetically identical individuals. Fortunately, plants and lines can be derived which are quite uniform and high performing. At the same time, variability is relatively easy to generate and lends the tissue culture process to in vitro selection schemes. The possibility that today's methods of callus-culturing and regeneration of plant materials may represent a mutagenesis system with a unique underlying mechanism (e.g. hypo- and hypermethylation) should not be dismissed. Perhaps the modification of important agronomic traits can be readily achieved in this manner. Although the average performance of tissue culture derived lines appears to be depressed, that significant positive changes in individual agronomic traits are readily apparent dictates that experimentation should not diminish until more information is available.

References

Armstrong, C.L. (1986) 'Genetic and cytogenetic stability of maize tissue cultures: A comparative study of organogenic and embryogenic cultures', Ph.D. thesis, University of Minnesota.

Armstrong, C.L. and Phillips, R.L. (1988) 'Genetic and cytogenetic variation in plants regenerated from organogenic and friable, embryogenic tissue cultures of maize', Crop Sci., 28, 363-369.

Bennetzen, J.L. (1987) 'Covalent DNA modification and the regulation of Mutator element transposition in maize', Mol. Gen. Genet., 208, 45-51.

Benzion, G. and Phillips, R.L. (1988) 'Cytogenetic stability of maize tissue cultures": A cell line pedigree analysis', Genome, 30, 318-325.

Bianchi, A., Salamini, F., and Parlavecchio, R. (1969) 'On the origin of controlling elements in maize', Genet. Agrar., 22, 335-344.

Bingham, E.T., Groose, R.W., and Ray, I.M. (1988) 'Activation of a mutable allele in alfalfa tissue culture', in O. Nelson (ed.), Plant Transposable Elements, Plenum Press, N.Y., pp. 325-338.

Brettell, R.I.S., Dennis, E.S., Scowcroft, W.R., and Peacock, W J. (1986) 'Molecular analysis of a somaclonal mutant of maize alcohol dehydrogenase', Mol. Gen. Genet., 202, 235-239.

Brown, P.T.H. (1989) 'DNA methylation in plants and its role in tissue culture', Genome, (in press).

Brown, P.T.H. and Lorz, H. (1986) 'Molecular changes and possible origins of somaclonal variation', in J. Semal (ed.), Somaclonal Variation and Crop Improvement, Martinus Nijhoff Publ., Dordrecht, pp. 148-159.

Chandler, V.L. and Walbot, V. (1986) 'DNA modification of a maize transposable element correlates with loss of activity', Proc. Natl. Acad. Sci. USA, 83, 1767-1771.

Chen, T.H., Lazar, M.D., Scoles, G.J., Gustav, L.V., and Kartha, K.K. (1987) 'Somaclonal variation in a population of winter wheat', J. Plant Physiol. 130, 27-36.

Chen, C.H., Freeling, M.L., and Merckelbach, A. (1986) 'Enzymatic and morphological consequences of Ds excisions from maize Adh1', Maydica, 31, 93-108.

Chomet, P.S., Wessler, S., and Dellaporta, S.L. (1987) 'Inactivation of the maize transposable element Activator (Ac) is associated with its DNA modification', EMBO J., 6, 295-302.

Cone, K.C., Burr, F.A., and Burr, B. (1986) 'Molecular analysis of the maize anthocyanin regulatory locus C1', Proc. Natl. Acad. Sci. USA, 83, 9631-9635.

Culley, D.E. (1986) 'Evidence for the activation of a cryptic transposable element Ac in maize endosperm cultures', VI International Congress on Plant Tissue and Cell Culture, Minneapolis, Minnesota, Abstracts, p. 220.

Cummings, D.P., Green, C.E., and Stuthman, D.D. (1976) 'Callus induction and plant regeneration in oats', Crop Sci. 16, 465-470.

Dahleen, L.S. (1989) 'Somaclonal variation in oat (Avena sativa L.) lines derived from tissue culture', Ph.D. Thesis, Univ. of Minnesota.

Dennis, E.S., Brettell, R.I.S., and Peacock, W.J. (1987) 'A tissue culture induced Adh1 null mutant of maize results from a single base change', Mol. Gen. Genet., 210, 181-183.

Doerschug, E.B. (1973) 'Studies of Dotted, a regulatory element in maize, I. Induction of Dotted by chromosome breaks, II. Phase variation of Dotted', Theor. Appl. Genet., 43, 182-189.

Dooner, H.K. and Nelson, O.E., Jr. (1979) 'Heterogeneous flavonoid glucosyltransferases in purple derivatives from a controlling element-suppressed bronze mutant in maize', Proc. Natl. Acad. Sci. USA, 76, 2369-2371.

Dooner, H.K. and Nelson, O.E., Jr. (1977) 'Controlling element-induced alterations in UDPglucose:flavonoid glucosyltransferase, the enzyme specified by the bronze locus in maize', Proc. Natl. Acad. Sci. USA, 74, 5623-5627.

Dunwell, J.M., Cornish, M., Powell, W., and Borrion, E.M. (1986) 'An evaluation of the field performance of the progeny of plants regenerated from embryos of Hordeum vulgare cv. Golden Promise', J. Agric. Sci. Camb., 107, 561-564.

Echt, C.S. and Schwartz, D. (1981) 'Evidence for the inclusion of controlling elements within the structural gene at the waxy locus in maize', Genetics, 99, 275-284.

Fedoroff, N., Wessler, S., and Shure, M. (1983) 'Isolation of the transposable maize controlling elements Ac and Ds', Cell, 35, 235-242.

Galiba, G., Kertesz, Z., Stuka, J., and Sagi, L. (1985) 'Differences in somaclonal variation in three winter wheat (Triticum aestivum) varieties', Cereal Res. Comm., 13, 343-350.

Green, C.E. and Phillips, R.L. (1975) 'Plant regeneration from tissue cultures of maize', Crop Sci., 15, 294-304.

James, M.G. and Stadler, J. (1989) 'Molecular characterization of Mutator systems in maize embryogenic callus cultures indicates Mu activity in vitro', Theor. Appl. Genet., 77, 383-393.

Johnson, S.S., Phillips, R.L., and Rines. H.W. (1987) 'Possible role of heterochromatin in chromosome breakage induced by tissue culture in oats (Avena sativa L.)', Genome, 29, 439-446.

Kunze, R., Starlinger, P., and Schwartz, D. (1988) 'DNA methylation of the maize transposable element Ac interferes with its transcription', Mol. Gen. Genet., 214, 325-327.

Lee, M., Geadelmann, J.L., and Phillips, R.L. (1988) 'Agronomic evaluation of inbred lines derived from tissue cultures of maize', Theor. Appl. Genet., 75, 841-849.

Lee, M. and Phillips, R.L. (1987a) 'Genomic rearrangements in maize induced by tissue culture', Genome, 29, 122-128.

Lee, M. and Phillips, R.L. (1987b) 'Genetic variants in progeny of regenerated maize plants', Genome 29, 834-838.

McClintock, B. (1951a) 'Chromosome organization and genic expression', Cold Spring Harbor Symp. Quant. Biol., 16, 13-47.

McClintock, B. (1950) 'The origin and behavior of mutable loci in maize', Proc. Natl. Acad. Sci. USA, 36, 344-355.

McCoy, T.J., Phillips, R.L., and Rines, H.W. (1982) 'Cytogenetic analysis of plants regenerated from oat (Avena sativa) tissue cultures; high frequency of partial chromosome loss'. Can. J. Genet. Cytol., 24, 37-50.

McGinnis, R.C. (1966) 'Establishing a monosomic series in Avena sativa L.', in R. Riley and K.R. Lewis (eds.), Chromosome Manipulations and Plant Genetics, Oliver and Boyd, Edinburgh and London, pp. 86-97.

Neuffer, M.G. (1966) 'Stability of the suppressor system in two mutator systems of the A1 locus in maize', Genetics, 53, 541-549.

Peschke, V.M. (1989) 'Tissue culture-induced variability in maize: genetic and molecular analysis of tissue culture-derived transposable elements', Ph.D. thesis, University of Minnesota.

Peschke, V.M., Phillips, R.L., and Gengenbach, B.G. (1987) 'Discovery of transposable element activity among progeny of tissue culture-derived maize plants', Science 238, 804-807.

Peschke, V.M., Phillips, R.L., and Pritchard, L. (1989) 'Activiation of the Spm transposable element in a tissue culture-derived plant', Maize Genet. Coop. News Lett., 63, 103.

Pohlman, R.F., Fedoroff, N.V., and Messing, J. (1984) 'The nucleotide sequence of the maize transposable element Activator', Cell, 37, 635-643; Cell, 39, 417.

Rhoades, M.M. and Dempsey, E. (1971) 'On the mechanism of chromatin loss induced by the ß chromosome of maize', Genetics, 71, 73-96.

Rhoades, M.M. and Dempsey, E. (1973) 'Chromatin elimination induced by the ß chromosome of maize', J. Hered., 64, 12-18.

Rhodes, C.A., Phillips, R.L., and Green, C.E. (1986) 'Cytogenetic stability of aneuploid maize tissue cultures', Can. J. Genet. Cytol., 28, 374-384.

Ryan, S.A., Larkin, P.J., and Ellison, F.W. (1987) 'Somaclonal variation in some agronomic and quality characters in wheat', Theor. Appl. Genet., 74, 77-82.

Sachs, M.M., Peacock, W.J., Dennis, E.S., and Gerlach, W.L. (1983) 'Maize Ac/Ds controlling elements--a molecular viewpoint', Maydica, 28, 289-301.

Saedler, H. and Nevers, W. (1985) 'Transposition in plants: a molecular model', EMBO J., 4, 585-590.

Schwartz, D. and Dennis, E. (1986) 'Transposase activity of the Ac controlling element in maize is regulated by its degree of methylation', Mol. Gen. Genet., 205, 476-482.

Schwartz, D. (1989) 'Gene-controlled cytosine demethylation in the promoter region of the Ac transposable element in maize', Proc. Natl. Acad. Sci. USA, 86, 2789-2793.

Schwarz-Sommer, Zs., Gierl, A., Berndtgen, R., and Saedler, H (1985) 'The Spm (En) transposable element controls the excision of a 2kb DNA insert at the wx-m8 allele of Zea mays', EMBO J., 3, 1021-1028.

Vasil, I.K. (1983) 'Regeneration of plants from single cells of cereals and grasses', in P.F. Lurquin and A. Kleinhofs (eds.), Genetic Engineering in Eukaryotes, Plenum Press, New York, pp. 233-252.

Woodman, J.C. and Kramer, D.A. (1986) 'The recovery of somaclonal variants from tissue cultures of B73, an elite inbred line of maize', VI International Congress on Plant Tissue and Cell Culture, Minneapolis, Minnesota, Abstracts, p. 215.

Zehr, B.E., Williams, M.E., Duncan, D.R., and Widholm, J.M. (1987) 'Somaclonal variation in the progeny of plants regenerated from callus cultures of seven inbred lines of maize', Can. J. Bot., 65, 491-499.

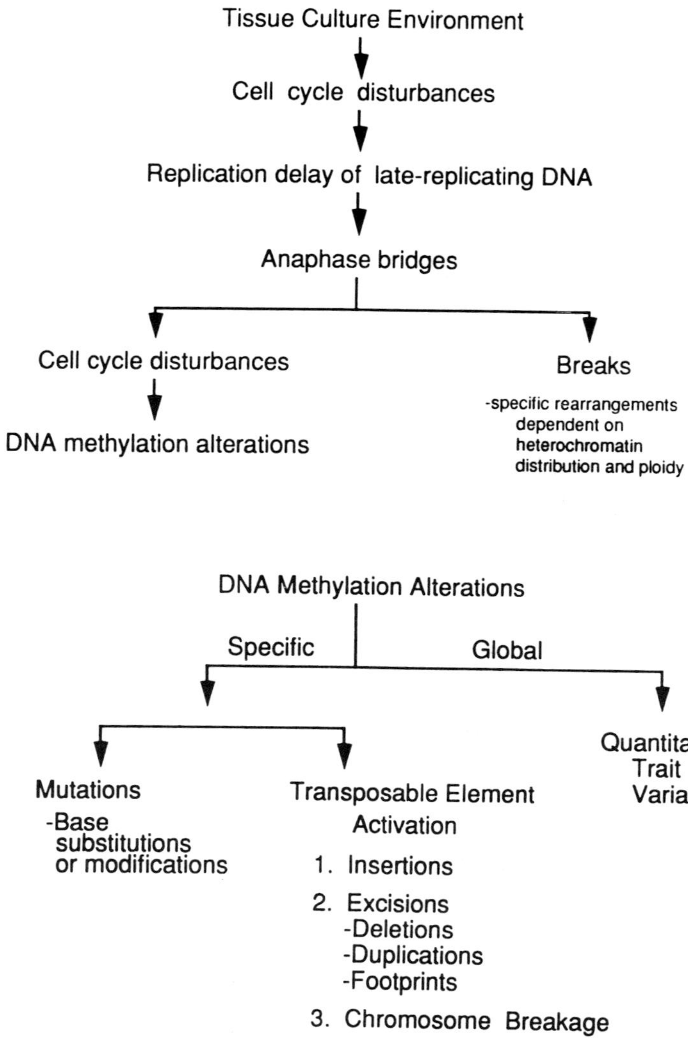

Figure 1. Hypothesis indicating possible mechanisms related to tissue culture-induced variation.

MUTANTS IN THE BIOSYNTHESIS OF AMINO ACIDS

M. Jacobs, V. Frankard and M. Ghislain
Laboratory of Plant Genetics
Vrije Universiteit Brussel
65, Paardenstraat
B-1640 Sint-Genesius Rode
Belgium

ABSTRACT. Mutagenesis-selection procedures used in our laboratory resulted in the obtention of plants with modified regulatory mechanisms leading to a higher production of specific free amino acids. In particular, two mutants of the aspartate pathway in *Nicotiana sylvestris* were isolated following selection on growth inhibitory concentrations of lysine plus threonine, and on a lysine analog, S-(2-aminoethyl)-L-cysteine (AEC), respectively named RLT 70 and RAEC-1.The resistance property in both cases was inherited as a monogenic dominant nuclear character. In RLT 70, up to 70% of the pool of free amino acids was made up of threonine, against 6% in the wild type plant. A completely desensitized aspartate kinase (AK) isoenzyme normally feedback inhibited by lysine was found to be the molecular basis of this overproduction. In RAEC-1, up to 25% of lysine could be found in the pool of free amino acids, compared to 1.5% in the wild type plant. The enzyme implicated in this mutant was dihydrodipicolinate synthase (DHDPS) totally desensitized to normal feedback inhibition by lysine. Evolution of the expression of both mutations during plant development revealed a peak of accumulation of the concerned amino acid, just before the elongation phase. In calli, overproduction was elevated and constant throughout the subcultures. Purification to homogeneity of the DHDPS subunit was achieved through a 2D polyacrylamide gel electrophoresis with a partially purified enzyme fraction. It was found to be a tetramer of four identical subunits, each of 39 kDa, and to be localized in the chloroplast. The purified protein was then microsequenced and the first 11 amino acids of the NH_2-end were determined (collaboration with Dr.J.C. Guillemot, ELF Sanofi Bioresearch). Moreover the eluted protein was injected into rabbits to induce the production of anti-DHDPS antibodies. These polyclonal antibodies were used as probe to screen a cDNA library built in lambda gt11 from total poly A+ RNA fractions of *N. sylvestris*. Several positive clones were isolated and are now being characterized.

1. Introduction

Amino acid biosynthesis is an essential process for plant growth and development, yet little is known about the genetic regulation of specific enzymatic steps in most of the biosynthetic pathways. The accumulation of free amino acids and their incorporation into proteins affect the nutritional value of crops for non-ruminant food or feed and as such, can be an important factor in crop-breeding.

For example, cereal crops are mostly deficient in lysine and threonine and legume crops are generally deficient in methionine and cysteine. Increasing the amount of the desired essential amino acids in the soluble amino acid fractions of the plant represents a possible approach for improving crop quality.

R.S. Sangwan and B.S. Sangwan-Norreel (eds.), The Impact of Biotechnology in Agriculture, 247–258.

2. Aspartate-derived amino acid biosynthesis

The nutritionally essential amino acids, lysine, threonine, methionine and isoleucine derive all from aspartate via a branched biosynthetic pathway (1). The major way of regulation of this pathway is feedback inhibition at the level of the key enzymes, such as aspartate kinase (AK) (2-4), the first enzyme of the pathway, dihydrodipicolinate synthase (DHDPS) (5), the first enzyme of the lysine branch, and homoserine dehydrogenase (HDH) (6), the first enzyme of the branch leading to threonine and methionine (see figure 1). Generally AK is inhibited by both lysine and threonine, while DHDPS is strongly feed-back inhibited by lysine.

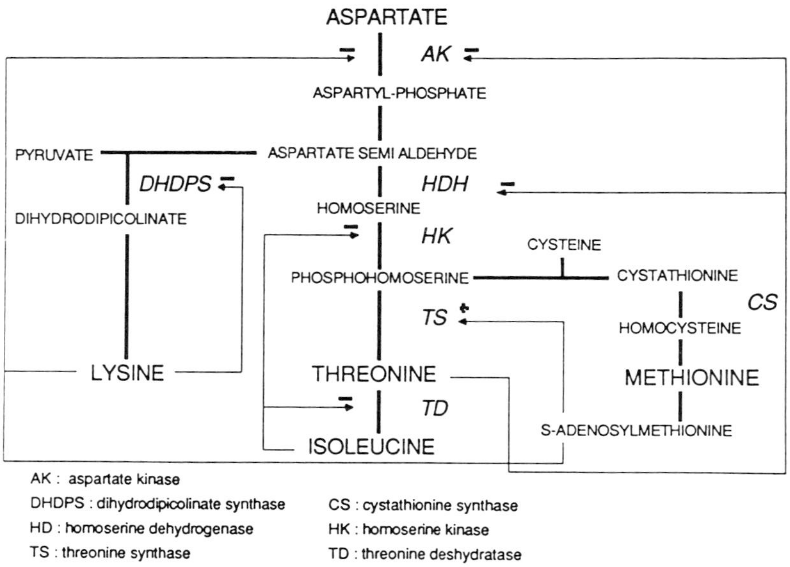

Figure 1 : Regulation of the aspartate-derived biosynthesis

3. Selection of mutants in this biosynthetic family

3.1 GENERAL PRINCIPLES

Approaches to improve the nutritional quality of crops with regard to their amino acid content have been mainly based on the possibility to modify by mutation the protein spectrum of the storage proteins and in particular to decrease the amount of prolamine, such as zein in corn (opaque-2-mutation), or hordein in barley (high lysine mutation). However, reduced yield and protein content characterize such types of mutants. Another approach consists in modifying the amino acid composition of specific storage proteins by genetical engineering. Most of the genes involved in the synthesis of such proteins unfortunately belong to multiple gene families so that the expected effects will be diluted.

Our approach consisted in increasing the amount of the desired amino acid in the soluble fraction by mutagenesis-selection procedures with the goal to develop plants with modified regulatory mechanisms leading to a higher production of specific amino acids.

The selection methods used are based on the growth inhibition caused by a combination of lysine plus threonine present in the culture medium or by the addition of a lysine analog such as S-(2-aminoethyl)-L-cysteine (AEC). This growth inhibition is considered to be the result of feedback inhibition of, respectively, aspartate kinase leading to methionine starvation, and dihydrodipicolinate synthase affecting the lysine content (7).

3.2 SELECTION APPLIED TO VARIOUS SPECIES

A series of plant species were considered : *Nicotiana sylvestris*, *Nicotiana plumbaginifolia*, *Daucus carota*, *Arabidopsis thaliana*, and *Hordeum sativum*. Selection at embryo level (barley) or seed level (Arabidopsis) offers guarantees in obtaining fully developed mutant plants. Embryogenic cell suspensions of carrot lead to plantlets and leaf protoplast cultures of the model species *Nicotiana sylvestris* regenerate readily.

This mutagenesis-selection procedure led for each species to the obtention of a certain number of potentially interesting mutants which are presently being characterized in our laboratory (see table 1) (8-10). Ultimately, the goal should be to clone the implicated genes in order to transfer them to crops.

Table 1 : Mutants of the aspartate pathway obtained by applying the mutagenesis-selection scheme as described in the text.

SPECIES AND EXPERIMENTAL SYSTEM	MUTAGEN	SELECTIVE AGENT (mM)	Nb	Nb of RESISTANT	ANALYZED MUTANTS
Hordeum sativum - embryos	$NaNO_3$	lys + thr 2.5 / 2.5	60.000 M_2 embryos	3	MD, thr+ileu OP AK-LysII feedback-insensitive
Arabidopsis thaliana - seeds	EMS	lys+thr+arg 1 / 1 / 0.5 AEC + arg 0.15 / 0.5	600.000 M_2 seeds 200.000	52 31	D and R, thr OP, lys OP, impaired lys uptake
Daucus carota - embryoids	MNG	lys + thr 2 / 2	70.000 embryoids	1	thr OP AK-lys max 50% inhibition
Nicotiana sylvestris - protoplasts	U.V.	AEC 0.04	$4.1.10^7$ cells	2	MD, lys OP DHPS feed-back insensitive
		lys + thr 0.5 / 0.25	$3.3.10^7$	2	MD, thr OP AK-lys feedback insensitive
Sorghum bicolor - grains	EMS	lys + thr 2.5 / 2.5	17.000 M_2 seeds	8	MD, thr OP

M = monogenic D = dominant R = recessive OP = overproducer

4. Properties of some of these mutants

We shall emphasize the properties of the mutants selected from diploid protoplast cultures of *Nicotiana sylvestris*. As a matter of fact, the AEC-resistant mutant in this species represents the only example of lysine overproduction related to the insensitivity of DHDPS, the main regulatory enzyme of the lysine pathway.

4.1 THE MUTANT RLT-70

Mutagenised diploid leaf protoplasts of *N. sylvestris* were submitted to selection for resistance to growth inhibitory concentrations of lysine plus threonine (9). Genetical analysis of the isolated mutant, RLT 70, revealed that its resistance property was inherited as a monogenic nuclear character, and heterozygous in RLT 70.

After biochemical analysis were performed, an alteration of the regulatory properties of the lysine-sensitive aspartate kinase activity was determined.

In wild type *Nicotiana sylvestris* plants, approximately 80% of the aspartate kinase activity is sensitive to feedback inhibition by lysine, whereas the remaining 20% are inhibited by threonine. An additive inhibition is observed when both amino acids are added.

In the homozygote mutant, the enzyme is completely insensitive to the lysine feedback inhibition, while the heterozygote displays an intermediate inhibition pattern (see table 2).

Table 2 : Inhibition patterns of aspartate kinase in wild type and RLT 70 mutant as a % of the total enzymatic activity, in presence of lysine and/or threonine at a final concentration of 10 mM.

	Lysine sensitive	Threonine sensitive	Insensitive
Wild type	70-80	20-30	0-10
RLT 70			
hêterozygote	35-40	20-30	35-40
homozygote	0-10	20-30	70-80

Analysis of the free amino acid content of leaves of the regenerated plants reveal up to 70% of free threonine against 5% in the wild type. Parallel increase in the soluble content of isoleucine (up to 10x) and to a lesser extent, of lysine (3x) are observed.

When the expression of the mutation was studied during the development of RLT 70, a pattern with a peak before elongation was established (see figure 2 and table 3).

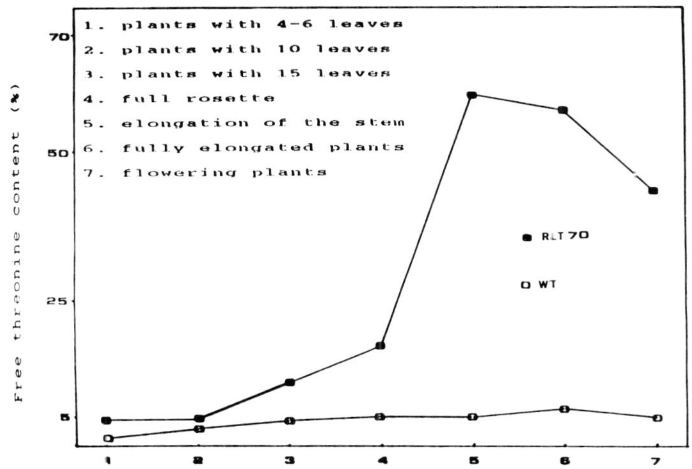

Figure 2 : Relative variation of free threonine during the plant cycle (expressed as % of total free amino acids)

Table 3 : Content of some free amino acids in leaves and calli of Nicotiana mutant RLT 70 and wild type (wt) (expressed as % of the total free amino acids pool in nmoles/gr fresh weight).

Amino Acid	Leaves		Calli	
	WT	RLT 70	WT	RLT 70
Asp	15.1	6.4	13.0	18.1
Thr	6.3	67.1	1.4	24.6
Glu	19.4	5.8	29.2	16.6
Ile	1.7	2.7	2.0	3.4
Lys	1.9	2.6	0.6	0.9
Arg	1.2	1.6	2.3	3.1
Total	7228	25971	1060	6098

Seed-derived calli also present high amounts of free threonine. In both cases, the increase in total free amino acid content of the mutant compared to the wild type, is essentially due to the higher threonine content (see table 3).

It is particularly interesting to mention that immature and mature seeds also display an increased threonine amount, respectively reaching 40% and 30%, compared to 10% and 5% in the wild type seeds (see table 4).

Table 4 : Content of some free amino acids in immature and mature seeds of RLT 70 and wild type *Nicotiana sylvestris* (expressed as % of the total free amino acid pool, in nmoles/gr fresh weight)

Amino Acid	Immature seeds		Mature seeds	
	WT	RLT 70	WT	RLT 70
Asp	17.8	15.0	11.8	19.8
Thr	11.3	41.5	5.2	29.7
Glu	25.3	21.6	16.7	12.2
Ile	1.2	0.8	1.5	1.0
Lys	1.0	1.5	1.3	1.6
Arg	1.1	4.3	1.9	3.4
Total	30019	63783	3413	38155

4.2 THE MUTANT RAEC-1

Diploid leaf protoplasts of *Nicotiana sylvestris* were U.V.- mutagenised and consequently submitted to selection for resistance to AEC (lysine analog) (11). After regeneration to whole plants, the resistant line RAEC-1 was crossed several times with the wild type to improve its fertility and phenotype.

The analysis of the offsprings of these regenerated plants showed that the gene responsible for AEC resistance behaved in a dominant mendelian fashion, and revealed that RAEC-1 was isolated as a heterozygote for this character.

It was established that the molecular basis of this resistance was due to a less sensitive DHDPS to feedback inhibition by lysine (see figure 3).

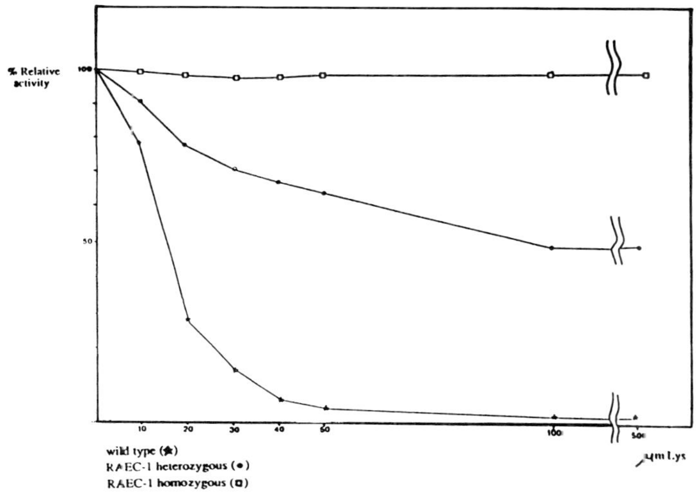

Figure 3 : Feedback inhibition of DHDPS by lysine in the wild type *Nicotiana sylvestris*, and in RAEC-1 heterozygous and homozygous plants.

Recent studies of the evolution of lysine overproduction during the development of homozygous RAEC-1 plants, have established wide variations of the lysine content in the pool of free amino acids. Overproduction itself seems expressed at a preferential phase preceeding elongation, but is not expressed at the same level throughout the mutant population (see table 5 and figure 4).

Table 5 : Content of some free amino acids in leaves of wild type *Nicotiana sylvestris*, and of two RAEC-1 plants at the same developmental stage, but presenting different lysine contents (expressed as % of the total free amino acid pool, in nmoles/ gr fresh weight).

Amino Acid	WT	RAEC-1	
Asp	13.9	15.6	12.3
Thr	5.8	6.7	5.0
Glu	20.3	20.1	19.7
Ile	1.1	1.0	1.1
Lys	1.3	22.5	0.9
Arg	2.4	1.5	0.8
Total	5550	6321	4308

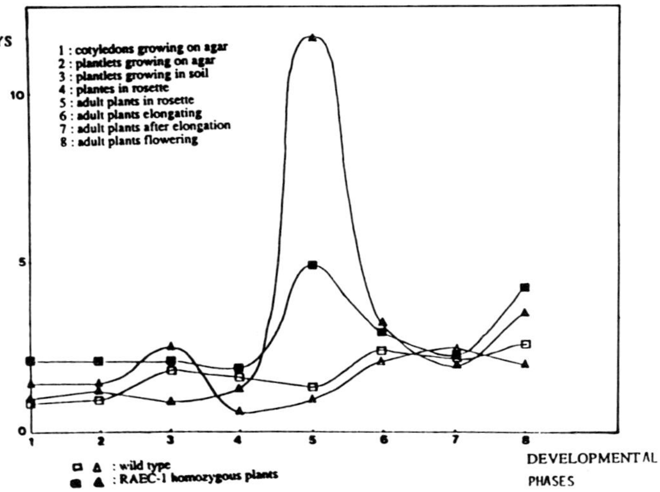

Figure 4 : Evolution of lysine content (expressed as a percentage of the total pool of free amino acids) during the growth of *Nicotiana sylvestris* wild type and RAEC-1 homozygous mutant plants.

It should be noted that the developmental stage at which lysine overproduction eventually occurs in RAEC-1, is the same one at which maximum threonine overproduction is observed in RLT 70, the first mutant.

On the other hand, offsprings of a non-overproducing RAEC-1 homozygous plant may exhibit high amounts of lysine during their development. Thus, the capacity to overproduce lysine seems maintained in all homozygous plants but is more or less expressed.

The fact that lysine overproduction is observed in all calli issued from homozygous RAEC-1 seeds corroborates this hypothesis (see table 6 and figure 5).

Table 6 : Content of some free amino acids in calli of *Nicotiana sylvestris* wild type and RAEC-1 (expressed as % of the total free amino acid pool, in nmoles/ gr fresh weight).

Amino Acid	WT	RAEC-1
Asp	15.0	30.0
Thr	0.6	1.8
Glu	34.9	30.3
Ile	0.4	1.2
Lys	0.8	10.1
Arg	1.1	3.8
Total	13827	17980

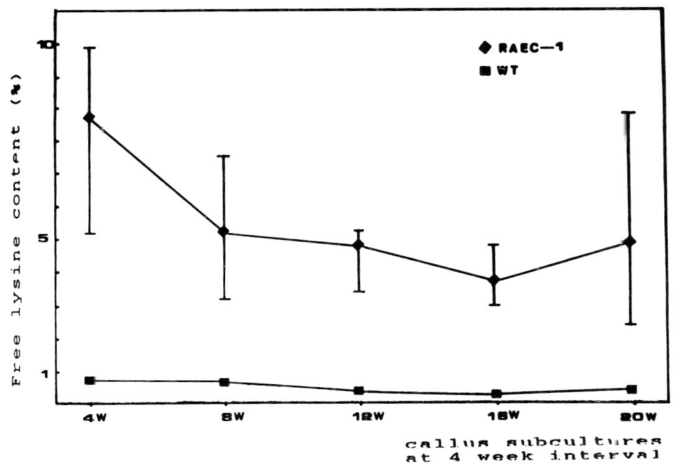

Figure 5 : Evolution of free lysine content during subcultures of *Nicotiana sylvestris* seed derived calli (expressed as % of the total free amino acid content) (W= week).

Among a population of homozygous mutant plants is also observed at a frequency of 1:30, a peculiar phenotype which, when analysed for its free amino acid content, is characterized by high amounts of soluble lysine. On the contrary, elevated lysine content does not necessarily involve an altered phenotype.

In relation to this observation, transgenic *Nicotiana tabacum* plants transformed with the dap A gene from *E. coli* also presented an aberrant phenotype when the gene was overexpressed and consequent high lysine levels were reached (13).

These observations tend to indicate that a complex regulation of the concerned gene exists, susceptible of widely modifying its expression. The study of this expression at the gene level should open the way to a better understanding of the mecanisms involved. The steps leading to the cloning of the DHDPS gene are mentioned in the following part.

5. Dihydrodipicolinate synthase : a cloning approach

5.1 PROTEIN PURIFICATION AND ENZYME CHARACTERIZATION

The study of the key enzyme of the lysine biosynthetic branch of the aspartate pathway, dihydrodipicolinate synthase (DHDPS), was achieved with partially purified enzyme to characterize mainly four aspects : a two-substrates kinetic, the inhibitory properties, the subunit composition and its subcellular localization.

DHDPS enzyme was extracted from young ribless leaves of *N. sylvestris* Speggazini and Comes, fractionated with ammonium sulfate and heat-treated at 65°C in the presence of pyruvate, one of the two DHDPS substrates. DHDPS fractions were further fractionated on an anion exchange chromatography. Gel filtration was the last step of the purification of the native form of DHDPS. Purification to homogeneity of the DHDPS subunit was achieved through a 2D polyacrylamide gel electrophoresis with partially purified enzyme fractions (5).

DHDPS kinetic was proposed to be a Ping Pong system where pyruvate binds first, activates the enzyme by a Schiff base formation. Water is released, then follows the binding of aspartate semi-aldehyde, the second substrate. The synthesis of dihydrodipicolinate involves condensation, dehydration and cyclisation steps in an unknown sequence. Its synthesis is accompagnied by the release of water (12).

One of the most characterized properties of the plant DHDPS is the inhibitory effect of lysine, the naturally occuring effector. It exerts a strong feedback inhibition with an $I_{0.5}$ between 10 and 50 µM. AEC, the lysine analog used to isolate the mutant, can also feedback inhibit DHDPS, but concentrations 10 times higher are needed to reach the $I_{0.5}$.

DHDPS was found to be a tetramer of identical subunits ranging from 32 kDa for the wheat enzyme, to 39 kDa for the *N. sylvestris* enzyme. Four identical subunits are also the rule for bacterial DHDPS (14).

DHDPS, as the majority of the other enzymes of the lysine biosynthesis, is localized in the chloroplast. Indeed, the *N. sylvestris* enzyme was found in the stroma, by using the enzymatic test and the immune detection assay in sub-chloroplastic fractions. However, the enzyme itself is synthetised in the cytoplasm and then translocated to the organelles (5,11).

5.2 ATTEMPTS TO ISOLATE cDNA CLONES OF *NICOTIANA SYLVESTRIS*

Purification to homogeneity of DHDPS allowed us to develop two approaches to clone the gene of *Nicotiana sylvestris*. Proteins resolved by 2D polyacrylamide gels were electroblotted on polyvinylidene difluoride membranes (Immobylon-PVDF). The DHDPS spot, previously identified through a specific labelling, was isolated and the NH_2-end of the protein sequenced (in collaboration with Dr. Guillemot, Sanofi ELF Bioresearch). The first 11 amino acids have been identified using 100 pmole of protein. A 96 fold degenerated oligonucleotide mixture has been deduced from this sequence and can now be used to screen a cDNA library.

The purified protein, eluted from the 2D polyacrylamide gel, was injected into rabbits to induce the production of anti-DHDPS antibodies. The best immunoreacting serum on Western blotting analysis was used to screen a gene expression library. The Eco RI site of the lambda gt11 expression vector was used to clone cDNA from leaf poly A+ RNA fractions of wild type *Nicotiana sylvestris*. About 10^6 clones of this expression library were screened using the polyclonal antibodies as a probe.

Screening with both immune and non-immune sera allowed us to isolate several clones. One of these remained positive when tested with purified anti-DHDPS antibodies. We are currently analysing its sequence for homology to the 11 amino acid sequence of DHDPS.

6.Conclusions and perspectives

The mutagenesis-selection procedures used in our laboratory led to the obtention of potentially interesting mutants, which are now being characterized at biochemical and molecular level.

The availability of the genes carrying these mutations will open the way to different areas of research :

- the increase of our knowledge about genetic control, molecular regulation of expression, physiological and developmental effects and agronomic performance of the mutant and transformed plants, which will allow future applications based also on specific modifications in basic plant biochemical pathways;
- the devising and comparison of methods to alter biochemical processes, namely mutagenesis-selection versus gene isolation, modification and transfer in the genetic program of crops;
- the obtention of crops and specially maize, with added lysine and threonine content in grain by a procedure avoiding drawbacks involved by modifying storage proteins and ensilage maize to take full profit of the high proportion of soluble amino acids in such plant material;
- the opening of new avenues for herbicide resistance : regulatory enzymes such as DHDPS or AK can be suitable targets for herbicide action (high-sensitivity of the lysine binding site), as already demonstrated for key enzymes of the valine or shikimate pathway.

Acknowledgements

This research was partly supported by grants to V.F. and M.G. from the "Institut pour l'encouragement de la Recherche Scientifique dans l'Industrie et l'Agriculture", by a Concerted Research Action (Belgian State) and by an E.C.C. contract.

References

(1) Bryan J K (1980),"Synthesis of the aspartate family and branched-chain amino acids", in Miflin B J (ed.), The Biochemistry of Plants, Vol 5, Amino acids and derivatives, Academic Press, New York, pp 404-452.

(2) Yamada Y, Kumpaisal R, Hashimoto T, Sugimoto Y, Suzuka Y (1986),"Growth and aspartate kinase activity in wheat cell suspension culture: effects of lysine analogs and aspartate-derived amino acids", Plant Cell Physiol. 27: pp 607-617.

(3) Giovanelli J, Mudd S H, Datko A (1989),"Aspartokinase of *Lemna paucicostata* Hegelm. 6746", Plant Physiol. 90: pp 1577-1583.

(4) Dotson S B, Somers D A, Gengenbach B G (1990),"Purification and characterization of lysine-sensitive aspartate kinase from maize cell cultures", Planta (submitted).

(5) Ghislain M, Frankard V, Jacobs M (1990),"Purification and characterization of dihydrodipicolinate synthase of *Nicotiana sylvestris* (Spegg. and Comes)", Planta (in press).

(6) Sainis, Mayne, Wallsgrove R, Lea P, Miflin B (1981),"Localisation and characterisation of homoserine deshydrogenase isolated from barley and pea leaves", Planta 152: pp 491-496.

(7) Green C E, Phillips R L (1974)," Potential selection system for mutants with increased lysine, threonine, and methionine in cereal crops", Crop Science 14: pp 827-830.

(8) Cattoir-Reynaerts A, Degryse E, Jacobs M (1981),"Selection and analysis of mutants overproducing amino acids of the aspartate family in barley, Arabidopsis, and carrot", in "Induced mutations as a tool for crop plant improvement", IAEA-SM-251/28, Vienna, pp 58-63.

(9) Negrutiu I, Jacobs M, Cattoir-Reynaerts A (1984),"Progress in cellular engineering of plants : biochemical and genetic assessment of selectable markers from cultured cells", Plant Molec. Biol. 3: pp 289-302.

(10) Jacobs M, Negrutiu I, Dirks R, Cammaerts D (1987),"Selection programmes for isolation and analysis of mutants in plant celle cultures", in : Green C E, Somers D A, Hackett W P, Biesboer D D, (eds), Plant Biology, vol 3, Plant tissue and cell culture, Alan R Liss, New York, pp. 243-264.

(11) Negrutiu I., Cattoir-Reynaerts A., Verbruggen I., Jacobs M. (1984)," Lysine overproducer mutants with an altered dihydrodipicolinate synthase from protoplast culture of *Nicotiana sylvestris* (Spegazzini and Comes)", Theor. Appl. Genet. 68, 11-20

(12) Kumpaisal R., Hashimoto T., Yamada Y. (1987)," Purification and characterization of dihydrodipicolinate synthase from wheat suspension cultures", Plant Physiol. 85, 145-151

(13) Glassman, Barnes, and Ernst (1989)," Elevation of free lysine in plants by the introduction of a bacterial gene", Dechema European Workshop, Bad Soden, Germany (Poster).

(14) Hallings S.M., Stahly D.P. (1976) "Dihydrodipicolinic acid synthase of *Bacillus licheniformis* , quaternary structure, kinetics, and stability in the presence of sodium chloride and substrates", B. B. A. 452, 580-596

GENETIC ENGINEERING OF RICE

M.R. DAVEY and P.T. LYNCH
Plant Genetic Manipulation Group,
Department of Botany,
University of Nottingham,
University Park, Nottingham, NG7 2RD UK.

ABSTRACT. Somatic cell techniques, involving somaclonal variation, transformation and somatic hybridisation, can be used to generate novel material for the plant breeder. Plant regeneration from japonica cultivars of rice is now established and is being extended to indica varieties. Such protoplast-to-plant systems are essential for somatic hybridisation. Electrofusion of protoplasts has resulted in both intra- and inter-specific somatic hybrid plants of rice. Isolated protoplasts are also preferable as starting material for the generation of somaclonal variants and as recipients of foreign genes. Direct uptake of DNA into rice protoplasts mediated by chemical methods and/or electroporation has resulted in transient expression of CAT and GUS reporter genes and in the production of stably transformed plants expressing GUS and antibiotic resistance. Other methods of gene delivery, including particle bombardment and injection, are discussed and their merits and limitations evaluated.

1. INTRODUCTION.

Rice is the staple food for more people worldwide than any other crop, with approximately 350 million tonnes being grown annually on 150 million hectares. However, compared with production, only small amounts of rice are traded (Greenland, 1984). Rice belongs to the genus *Oryza*, two species of which are cultivated. *O. sativa* L., the common rice, is grown in tropical and subtropical regions. *O. glaberrima* is grown in West Africa, but is being displaced by the former species. *O. sativa* probably originated from *O. rufipogea* in Asia, while *O. glaberrima* may be derived from *O. breviliguliata* in Africa (Chang, 1976; Oka, 1988).

There are three dominant ecographic races of *O. sativa*, previously thought to be subspecies, namely indica, javanica and japonica (or sinica), of different geographical origin and environmental adaptation. Indica varieties are tall, heavy tillering with pendant leaves and heavy, flat grains and are adapted to tropical conditions. Japonica varieties have darker green, erect leaves with moderate tillering, short rounded grains and are more tolerant to cool conditions. Javanica or bulu rices are tall, low tillering and have large grains borne on long panicles.

R.S. Sangwan and B.S. Sangwan-Norreel (eds.), The Impact of Biotechnology in Agriculture, 259–280.

The differentiation of rice into the two main indica and japonica varietal types by morphological and serological characters, is still used by rice breeders and scientists, but is a rather empirical and over-simplified distinction. Glaszmann (1987) subdivided the indica type varieties, by isozyme polymorphism, into varietal groups 1 to 5. Varietal group 1 is considered to be the true or "typical" indicas. Japonica and javanica rices are classed in varietal group 6.

Conventional breeding methods involving pedigree selection, backcrossing and mutation breeding have been successful in improving characteristics such as the productivity, yield stability, and disease and insect resistance in rice. A range of molecular and somatic cell techniques are now available to supplement conventional plant breeding. Some of these approaches are comparatively simple and rely upon relatively basic tissue culture. Others are dependent upon sophisticated molecular methods, including DNA cloning, sequencing, vector construction and the introduction of such engineered DNA into recipient plant cells.

Whilst monocotyledons have presented problems at the culture level, considerable progress has been made recently. Indeed, the development of protoplast-to-plant systems for several grasses and the major cereals, such as maize and rice, has enabled transformation and somatic hybridisation to become realistic methods of genetic modification in these crops. The relevance of these technologies to cereal improvement has already been discussed in previous reviews (Cocking and Davey, 1987; Vasil, 1988).

2. TISSUE CULTURE OF RICE AND THE PRODUCTION OF SOMACLONAL VARIANTS FROM CALLUS.

The ability to induce explants to produce shoots directly or with minimum tissue proliferation, is a basic requirement of a micro-propagation system. Although not used extensively in rice, micro-propagation has been employed, on a small scale, to multiply important germplasm such as that carrying cytoplasmic male sterility (CMS; Kamari et al, 1988).

In cases where shoot regeneration occurs through an intervening callus phase, the regenerated plants may exhibit a lack of uniformity, including variation in morphology, yield, disease resistance and other traits. Although a disadvantage to clonal plant production, it provides a source of variation which, if stable, can be exploited by the plant breeder. Indeed, the exposure of somaclonal variation (Larkin and Scowcroft, 1981) probably represents the simplest form of plant genetic manipulation. Somaclonal variation may result from gross karyotic changes, and chromosome rearrangements; it probably results, in other instances, from more subtle changes at the DNA level (Scowcroft and Larkin, 1988). Such changes have been reported in several crop plants, including potato (Bidney and Shepard, 1981), maize (McCoy and Phillips, 1982) and *Triticale* (Nakamura and Keller, 1982).

Variation in phenotypic features of primary regenerant rice plants, such as the number of fertile tillers per plant, panicle length, plant height and number of seeds, have been reported by several workers (Nishi et al, 1968; Henke et al, 1978; Nakano and Maeda, 1979). Albino and polyploid regenerant rice plants have also been produced (Niizek and Oono, 1971; Nishi and Mitsuoka, 1969). Some of the variation has been shown to be heritable in the progeny of somaclones. Ogura (1981) observed that 70% of the third generation progeny from somaclones differed from the parental material. Similarly, third generation progeny of somaclonal variants derived from haploid cultures of the japonica variety Calrose 76 had smaller stature, leaf size and seed weight than parental material (Schaeffer, 1982). Other workers have reported variation of agronomic potential in rice, including tolerance to aluminium (Yashida and Ogawa, 1983), copper (Tobita et al, 1988) and salt (Brar et al, 1985), and increased lysine and protein content (Zapata, 1984). However, problems of mutation stability have been encountered in the progeny of rice somaclones, Oono (1984) reporting the gradual loss of salt tolerance with successive generations in the cultivar Norin 8.

3. THE DEVELOPMENT OF PROTOPLAST-TO-PLANT SYSTEMS FOR RICE.

The current success in regenerating plants from protoplasts of members of the Gramineae relates to the use of embryogenic cell suspension cultures as a source of protoplasts (Vasil, 1987). Using this approach, protoplast-to-plant systems have been established for several grasses (Vasil, 1987; Creemers-Molenaar et al, 1988; Dalton, 1988; Horn et al, 1988; van der Valk and Zaal, 1988), sugarcane (Srinivasan and Vasil, 1986; Chen et al, 1987), and for the cereals maize (Rhodes et al, 1988), including elite lines (Prioli and Söndahl, 1989; Shillito et al, 1989), millet (Vasil and Vasil, 1980; Heyser, 1984) and rice. To date, protoplasts isolated directly from leaves, leaf bases or roots of grasses and cereals, have failed to undergo sustained cell division.

Embryogenic cell suspension cultures suitable for protoplast isolation are normally initiated from embryogenic callus derived from a range of explants, including anthers, immature and mature embryos and leaf base tissue (Thompson and Cocking, 1986). Several reports have been made of somatic embryogenesis in rice (Wernicke et al, 1981; Ling et al, 1983; Chen et al, 1985; Abe and Futsuhara, 1985), and the histology of the process described (Jones and Rost, 1989). Embryogenic callus, vital for intitiating cell suspensions from which protoplasts can be isolated, tend to be sporadic and transient, particularly from indica varieties.

Suspensions of rice suitable for protoplast isolation were described by Wakasa et al (1984) and Toriyama and Hinata (1985) by culturing cells in amino-acid based media. The protoplast-derived tissues reported by Toriyama and Hinata (1985) developed chlorophyll and produced roots. Subsequently, Fujimura et al (1985), Coulibaly and Demarly (1986) and Yamada et al (1986) reported plant regeneration from rice protoplasts isolated from cultured cells. Abdullah et al (1986) described a procedure for reproducible regeneration of plants from protoplasts of Taipei 309. Mature embryos or immature basal portions of leaves excised from 7 day old dark-grown seedlings produced nodular embryogenic callus on LS (Linsmaier and Skoog, 1965) medium containing 2.5 mg/L 2,4-D, 1.0 mg/L thiamine hydrochloride, 3.0% w/v sucrose and 0.8% w/v agarose (Sigma Type I). Transfer of callus to an amino-acid based medium (AA-2) with 2.0 mg/L 2,4-D resulted in fast growing, finely divided cell suspensions composed of densely cytoplasmic, isodiameteric, thin walled cells, which could be maintained in exponential growth by subculture every 7 days. Such cultures released large numbers of protoplasts when 6 to 8 months old, after incubation in an enzyme mixture consisting of Cellulase RS (1.0% w/v) and Pectolyase Y23 (0.1% w/v). Freshly isolated protoplasts were heat shocked (45°C for 5 minutes, followed by 10 seconds on ice), since this procedure was known to increase the number of protoplasts of rice dividing after 10 to 15 days of culture to 45% and increased the plating efficiency after 30 days to 1.1% (Thompson et al, 1987). Protoplasts were plated at 3.0 X 10^5/ml in KPR medium solidified with 1.2% w/v Sea Plaque LMT agarose and containing 0.5 mg/L 2,4-D with 10% w/v glucose.

Although protoplasts of Taipei 309 and Fujisaka 5 divided in both liquid and solidified medium, the importance of agarose as a gelling agent had been demonstrated previously by Thompson et al (1986). Sea Plaque agarose used at the normal concentration of 0.4 to 0.6% w/v sustained protoplast division, but a higher concentration of 1.2% w/v improved protoplast survival, giving division and plating efficiencies of 26% and 0.5% respectively for protoplasts of Fujisaka 5 and values of 18% and 0.3% for the same parameters for those of Taipei 309. Embedding the protoplasts in agarose droplets or in agarose sectors and bathing these in the same liquid medium was equally effective. Subsequently, after 4 to 6 weeks, 15 to 20% of the protoplast-derived cell colonies each produced 3 or 4 somatic embryos within 5 days of transfer of the tissues to hormone-free N6 regeneration medium containing 8.0% w/v sucrose and soldified with 0.8% w/v agar (designated N60). Such somatic embryos germinated within 14 days of tissues being placed on the N60 regeneration medium. Transfer of regenerating cultures to the light at the time of coleoptile elongation was accompanied by chlorophyll production. Green plantlets could be excised from the parent tissue, rooted on hormone-free MS (Murashige and Skoog, 1962) based medium, transferred to compost and grown to maturity.

More recently, additional refinements have been made to the procedure described by Abdullah et al (1986) for regenerating fertile plants from protoplasts of cell suspensions of Taipei 309 (Finch et al, 1989). Careful pipetting was carried out to ensure that the correct 3:1 volume-to-volume ratio of fresh medium to spent medium and cells was maintained at each subculture. The use of disposable plastic pipettes also facilitated selection of fine cell colonies, reducing the need to sieve the cultures. Other workers have also modified the procedure of Abdullah et al (1986). Instead of transferring protoplast-derived tissues to hormone-free medium for plant regeneration, Zhang and Wu (1988) sub-cultured cells of Taipei 309 and Pi-4 to MS based medium containing 0.8% w/v agarose, 2.0 mg/L kinetin and 0.5 mg/L NAA. The addition of growth regulators at this stage was essential for plant regeneration from the variety Pi-4.

To date, the procedures described for several japonica rice genotypes rely on embryogenic cell suspension cultures as the protoplast source, with the exception of the method of Coulibaly and Demarly (1986) which employed callus. The use of protoplast nurse cultures has been found to be important by some research groups to achieve sustained protoplast division (Kyozuka et al, 1987). Most of the procedures result in comparable plant regeneration efficiencies of 10 to 20%, although Kyozuka et al (1987) reported that up to 50% of protoplast-derived tissues were capable of plant regeneration. Recently, Masuda et al (1989) showed that cytoplasmically-rich rice protoplasts could be selected by density gradient centrifugation and that these protoplasts had a greater plating efficiency during subsequent culture than protoplasts in unselected populations.

Since indica rice varieties are a major food source in most tropical regions of the world, it is also essential to develop reproducible plant regeneration systems for these cultivars. Lee et al (1989) isolated protoplasts from cell suspensions, derived from immature embryo callus, of the cultivar IR54 and cultured the protoplasts in a modified Kao (1977) medium on a Millipore filter overlaying a nurse culture of suspension cells of the japonica rice, Calrose 76. Protoplast plating efficiencies ranged from 0.5 to 3.0%; protoplasts failed to divide in the absence of the nurse culture. Whether cells of Calrose 76 could be replaced by cells of other varieties was not determined. Transfer of the Millipore filter with the adhering protoplast-derived colonies to a LS based medium with 2.0% w/v sucrose. 0.5 mg/L 2,4-D and 0.4% w/v agarose (BRL), but lacking nurse cells, resulted in the development of embryo-like structures on the protoplast-derived tissues. Subsequent transfer of selected tissues with their embryo-like structures to N6 or MS based regeneration medium supplemented with 3.0% w/v sucrose, 0.4% w/v agarose and 0.03 μM NAA with BAP or kinetin at 9.0 μM, induced shoot emergence. BAP was most effective in inducing shoot regeneration. It remains to be seen whether the method reported by Lee et al (1989) is reproducible in other laboratories, and applicable to other indica varieties. The efficiency of plant regeneration in this system also needs to be assessed to determine whether it is high enough for transformation and somatic hybridisation studies. The centrifugation procedure to concentrate cytoplasamically-rich protoplasts described by Masuda et al (1989) may be particularly useful in increasing the throughput of protoplast-derived colonies of indica rice.

4. SOMACLONAL VARIATION IN PROTOPLAST-DERIVED JAPONICA AND INDICA RICE PLANTS.

Ogura et al (1987) examined the performance and cytology of protoplast-derived plants of the four japonica cultivars Fujisaka 5, Iwaimochi, Nipponbare and Norin 14, and observed high yield and a low degree of variation in these regenerated (R0) plants. Lee et al (1989) also reported a preliminary analysis of protoplast-derived plants of IR54. The height of 76 plants grown to maturity ranged from 61 to 102 cm, with an average of 83.3±8.5 cm. In comparison, plants grown from seed under the same conditions averaged 106.6±11.6 cm. Sixty-two of the regenerated plants flowered, but their fertility was less than that of seed-grown plants, with 30-80% of the caryopses being fertile compared to 80-95% for seed-grown plants.

The long-term value and genetic basis of such variation in protoplast-derived plants (protoclonal variation) can only be determined by examination of sexual progeny. In collaboration with the International Rice Research Centre, Abdullah et al (1989) grew seed progeny (R1) of protoplast-derived plants of Taipei 309 in a randomised complete block design field experiment during the 1987 dry season. The R1 protoclones produced more tillers/plant, had wider but shorter flag leaves resulting in a significant reduction in the flag leaf length/width ratio, and required longer to flower than control plants. At flowering, the R1 plants had shorter panicles, but increased numbers of primary and secondary branches and spikelets on each panicle. Grain length and width were also increased, but grain weight was reduced. The authors concluded that protoclonal variation results in a shift away from the direction of the previous selection history of the plant variety. Ogura et al (1989) also examined the protoclonal variation of the seed progeny (R1) of protoplast-derived plants from the japonica cultivars Fujisaka 5, Iwaimochi and Nipponbare. Compared with the controls, the R1 plants tended to exhibit a slight increase in the number of panicles, a small decrease in the number of spikelets per panicle, but a similar grain yield. Nipponbare and Iwaimochi were phenotypically more stable than Fujisaka 5. Material from this study with favourable characters, such as short culm length, has now been established as breeding lines. In the longer term, it will be interesting to see whether similar variation is seen in R1 plants of indica rice varieties.

The progeny of protoplast-derived plants have also been analysed at the molecular level. Several restriction fragment length polymorphisms (RFLPs) were detected in cultivars of Asian (*O. sativa*) and African (*O. glaberrima*) rice using a human minisatellite DNA probe (Dallas, 1988). Specific fragments were inherited in a Mendelian fashion and were thought to represent unlinked loci. The hybridisation patterns were cultivar specific. The RFLPs of plants produced by selfing individuals regenerated from protoplasts of *O. sativa* cv Taipei 309 were similar to those of seed-derived plants, indicating that the hybridisation patterns were largely unchanged following regeneration of plants from cultured protoplasts.

5. TRANSIENT GENE EXPRESSION STUDIES BY DIRECT DNA UPTAKE INTO RICE PROTOPLASTS.

Rice, like the other major cereals and members of the Gramineae, is not amenable to the normal methods of *Agrobacterium* transformation, which has been used to introduce foreign genes into a range of dicotyledonous crop plants. Nevertheless, it is well established, particularly from early virus studies (see Davey and Kumar, 1983) and investigations of endocytosis in isolated soybean protoplasts (Tanchak et al, 1984), that isolated plant protoplasts are capable of taking up a range of macromolecules. Pioneering studies using isolated *Agrobacterium* Ti plasmid confirmed that this DNA could be introduced into *Petunia* protoplasts (Davey et al, 1980) using poly-L-ornithine and into tobacco protoplasts by polyethylene glycol (PEG)-mediated DNA delivery (Krens et al, 1982). Although the T-DNA of introduced Ti plasmids showed a range of integration patterns in the genome of the recipient protoplasts (Davey et al, 1989), these experiments demonstrated that foreign DNA could be introduced into isolated protoplasts and that expression of T-DNA specific genes resulted in transformation of *Petunia* and tobacco protoplasts causing tumourous growth. Subsequent work, and undoubtedly the most significant advance in this area of research, demonstrated that T-DNA sequences were unnecessary for stable transformation and expression of foreign DNA in plant cells (Paszkowski et al, 1984). This methodology has also become routine for transient studies, where gene expression is monitored within hours of the DNA uptake event. Indeed, this technique is used extensively to assess plasmid constructs and gene promoters, since the latter can be linked to readily assable reporter genes, such as those for chloramphenicol acetyltransferase (CAT) and β-glucuronidase (GUS) (Jefferson, 1987). Transient gene expression studies avoid the complication of positional effects at the whole plant level and can also be used to optimise the DNA delivery conditions for use in subsequent stable transformation studies.

To date, the number of transient gene expression studies in rice is limited. Using PEG-mediated plasmid uptake, Junker et al (1987) studied the expression of the neomycin phosphotransferase (NPTII) gene from Tn5 fused to four different promoters. Specifically, the CaMV 35S promoter, the 1'-2' promoter and the nopaline synthase promoter of the T-DNA of *Agrobacterium tumefaciens*, and the sucrose synthase promoter with part of exon 1 from maize. Protoplasts were isolated from cell suspensions and leaf mesophyll of Taipei 309. Optimum NPTII expression was obtained 2 days after uptake of circular plasmid, with NPTII activity being dependent on the amount of DNA used for transformation, this increasing with DNA concentrations of 5 to 100 µg/1.0 X 10^6 protoplasts. Saturation was not achieved at the highest concentration of 100 µg/ml of DNA. The level of enzyme activity depended on the promoter fused to the NPTII gene, the CaMV35S being the most efficient followed by the 1'-2' promter, the nopaline synthase promoter and the sucrose synthase promoter respectively.

Em is a major protein of the mature wheat embryo that begins to accumulate in embryos by 21 days post anthesis. Using PEG-mediated plasmid uptake, chimaeric genes composed of a translational fusion between 5' flanking fragments of an *Em* genomic clone and GUS, linked to a 3' fragment from the CaMV 35S gene, were introduced into rice cell suspension protoplasts (Marcotte et al, 1988). The *Em* promoter was inducible by abscisic acid, responding within 60 minutes of treatment. Only 650 bp of the promoter was required for GUS expression. These studies provided direct evidence for phytohormone acting to regulate the initiation of gene transcription.

6. PRODUCTION OF TRANSGENIC RICE PLANTS BY DNA UPTAKE INTO PROTOPLASTS.

The development of reproducible protoplast-to-plant systems for rice has provided the basis for the production of transgenic plants. Using kanamycin at 100 µg/ml for selection, Uchimiya et al (1986) obtained protoplast-derived colonies of *O. sativa* C5924 following PEG-induced uptake of pCYT2T3 into cell suspension protoplasts. This 18.3 kb plasmid carried a chimaeric gene consisting of the nopaline synthase promoter, the NPTII [APH(3')II] structural gene and a terminator of CaMV DNA. The construct conferred resistance to aminoglycoside antibiotics on transformed plant cells. Transformation frequencies of 2 to 3% were recorded in several experiments. In a continuation of these experiments by the same research group, but using electroporation as the plasmid delivery method, Toriyama et al (1988) selected transformed rice cells on G418-containing medium following uptake of pCN (5.4 kb) carrying the CaMV 35S promoter, the nopaline synthase and CaMV terminators and the NPTII gene. Resistance to this antibiotic was encoded by the same NPTII gene as used for kanamycin selection. Five rice plants were regenerated from 19 G418 resistant colonies. Molecular analysis of these plants confirmed NPTII activity in their leaves and the presence of the foreign gene.

In a simultaneous series of experiments, Yang et al (1988a) assessed the conditions for transformation of cell suspension protoplasts of Taipei 309 by pCaMVNEO, the 4.4 kb plasmid previously used to transform maize cell suspension protoplasts to kanamycin resistance (Fromm et al, 1986). Using electroporation with 3 successive pulses separated by 10 second intervals, Yang et al (1988a) employed a range of capacitances from 20 to 50 nF and voltages from 500 to 2500 V. Protoplasts were electroporated with plasmid DNA and sheared calf thymus DNA as carrier in a medium modified from the formulation of Fromm et al (1986). Subsequently, the protoplasts were incubated on ice for 10 minutes after electroporation. Other experiments compared electroporation with PEG 6000 and PEG in combination with electroporation as methods of plasmid delivery. Electroporation was the most efficient method of gene delivery into cell suspension protoplasts of Taipei 309, with transformed colonies selected following exposure of protoplast-derived cell colonies to 100 μg/ml of kanamycin from days 14 to 42 of culture. Expression of transformation in absolute terms gave frequencies of 19.9 X 10^{-5} and relative values of 26%, the latter being approximately 10 fold higher than those reported by Uchimiya et al (1986). NPTII enzyme activity was demonstrated in six randomly chosen kanamycin resistant clones, while DNA hybridisation confirmed the presence of the 1.0 kb fragment containing the NPTII gene in plant DNA. Of 20 clones analysed by DNA hybridisation, one contained an apparently unmodified form of the NPTII gene. Additional bands present in the other clones and hybridising to the NPTII gene probe, were probably the result of rearrangement of plasmid sequences in the transformed cells, since some bands were common to certain clones.

Although these preliminary experiments by Yang et al (1988a) utilised a non-regenerating protoplast system, application of these plasmid uptake conditions to protoplasts from a totipotent suspension culture resulted in transgenic plants of Taipei 309 (Zhang et al, 1988). Kanamycin resistant protoplast-derived cell colonies underwent somatic embryogenesis following transfer to hormone-free N60 regeneration medium containing 8.0% w/v sucrose and solidified with 0.4% w/v Sigma agarose type 1. Of 400 kanamycin resistant colonies selected, 6 produced 12 green plants, while 2 colonies gave rise to albino shoots, although the latter failed to survive. An important feature of the regeneration procedure was the omission of kanamycin from the N60 regeneration medium, since the presence of the antibiotic inhibited somatic embryogenesis. Six regenerated plants chosen randomly for analysis contained the NPTII gene as shown by DNA hybridisation, but only 2 expressed NPTII enzyme activity. The experiments reported by Zhang et al (1988) confirmed that heat shocking of rice protoplasts prior to plasmid uptake maximised transformation, and that carrier DNA was unnecessary during plasmid uptake. Future experiments will determine the stability and expression of the NPTII gene in seed progeny of the protoplast-derived transgenic plants, and the extent to which parameters such as tiller number, leaf size and shape, flowering characteristics and grain yield are affected by the presence of a foreign gene.

Hygromycin B has also been used to select transformed protoplast-derived cells of rice. Thus, Shimamoto et al (1989) incorporated this antibiotic at 20 µg/ml into the protoplast culture medium 5 weeks after electroporation of *O. sativa* cv Nipponbare cell suspension protoplasts with pGL2 carrying the bacterial *hph* gene. Fertile transgenic plants were regenerated from the selected protoplast-derived colonies. As in the experiments of Zhang et al (1988), the antibiotic used for selection was omitted from the plant regeneration medium. DNA hybridisation showed the presence of multiple chromosomal integration sites in some of the transgenic plants, with 2 to 10 copies of the integrated *hph* gene per diploid cell. Transmission of the hygromycin resistance phenotype to seed progeny was confirmed by germinating seeds in the presence of the antibiotic and by Southern hybridisation. Such experiments also involved co-transformation, in which protoplasts were electroporated with pGL2 mixed with pBI221 carrying the GUS gene, and calf thymus DNA as carrier. Seeds of hygromycin resistant transgenic plants also expressed GUS in their embryos and endosperm, demonstrating co-transformation with pBI221 carrying a non-selectable reporter gene.

Other workers have employed GUS to confirm transformation of rice protoplasts. Zhang and Wu (1988) used PEG to introduce pAI$_1$GusN, linearised with *Eco*R1, into cell suspension protoplasts of japonica rice varieties Pi-4 and Taipei 309. Up to 0.6% of the protoplasts showed transient GUS expression 3 days after gene transfer, as detected histochemically by the presence of blue cells. Linear plasmid gave at least 10 times more blue cells than circular plasmid. In the absence of a selection system, protoplasts were grown to cell colonies and randomly chosen colonies also screened for GUS expression. Of 378 colonies assayed, 61 exhibited foreign gene expression as shown by the presence of blue cells in the tissues. At least 86 regenerated plants of 378 screened were transgenic following confirmation of the presence of the foreign gene by DNA hybridisation. The plasmid used in these experiments contained the maize Adh1 gene promoter, which could be induced by anaerobic conditions. Indeed, GUS activity was 5 to 6 fold higher in the roots of some of the transgenic plants under anaerobic conditions.

The fact that transformation, including co-transformation, of rice protoplasts has been demonstrated by independent research groups now makes possible the introduction of agronomically important characteristics, such herbicide and insect resistance, into this cereal. These characteristics have already been introduced into dicotyledons, including tobacco, tomato and potato. In the longer term, it should be possible to introduce into rice other agronomically desirable characters, such as improvement in yield and stress tolerance, once gene sequences for these traits have been identified through RFLP mapping prior to their isolation and cloning by recombinant DNA techniques.

7. OTHER METHODS OF INTRODUCING DNA INTO RICE.

The difficulties inherent in developing efficient protoplast-to-plant systems for cereals, including rice, have stimulated workers to assess methods of transformation other than direct DNA uptake into protoplasts. Currently, transfer of genes directly into plant cells and tissues by particle bombardment is receiving much attention. Wang et al (1988) assessed this technique for suspension cells of Taipei 309. The cells were absorbed onto three layers of Whatman No 4 filter paper and placed in a plastic Petri dish before being bombarded with tungsten particles coated with pAI₁GusN carrying the GUS gene, or with pCaMVICN carrying the CAT gene. Following bombardment with the GUS construct, blue cells were detected with an *in situ* enzyme assay. The number of blue cells was used to monitor the effect of a range of experimental parameters, including the type of particle stopping plate, particle size and DNA concentration. Particles of 1.2 µm in diameter gave the largest number of blue cells. Excess DNA in the DNA-particle mixture decreased the efficiency of gene transfer, probably as a result of increased aggregation of the particles. The optimum conditions resulting in expression of GUS also resulted in successful delivery of pCaMVI₁CN into rice cells, with expression of CAT activity when recipient cells were assayed 48 hours after plasmid delivery. Such experiments as those described by Wang et al (1988) hold promise for the future in obtaining transgenic rice plants by particle bombardment, especially if the recipient tissues are capable of undergoing plant regeneration by somatic embryogenesis from individual cells.

Other methods of plasmid delivery to rice have been reported. Duan and Chen (1985) used the "pollen tube pathway" which was reported more extensively in a later publication (Luo and Wu, 1988). The stigmas of pollinated flowers were excised and DNA applied to the cut surface of the styles. Transformation was thought to occur by DNA passing down the pollen tubes to the ovules, with up to 20% of the seed progeny from rice plants treated with plasmid carrying the NPTII gene containing and expressing the foreign gene. However, it remains to be seen whether other workers are able to repeat this method of transformation.

Using NPTII as a reporter gene in a transient expression system, Töpfer et al (1989) assessed the ability of viable, dry, isolated rice embryos to take up DNA when imbibed in plasmid solution. Gene expression in these experiments resulting from contaminating micro-organisms was discounted. No attempt was made to determine whether DNA-treated embryos developed into transgenic plants. Again, it remains to be seen whether these results are reproducible, since the experiments are similar in their approach to those of Ledoux et al

(1974) which raised controversy as to the ability of intact seeds and tissues to take up and express exogenously supplied DNA.

In addition to the approaches already described for rice, injection of DNA into cells and tissues has been used to produce transgenic plants in other crops. Thus, injection of relatively large volumes of DNA into floral tillers has been used to transform rye (de la Pẽna et al, 1987), microinjection of microspore-derived somatic embryos has been reported for *Brassica napus* (Neuhaus et al, 1987), while micro-injection of protoplasts (Crossway et al, 1986; Reich et al, 1986) is also established. The applicability of these techniques to rice still needs to be determined, but the ability to regenerate plants from cultured rice microspores (Cho and Zapata, 1988) should enable a comparison to be made with the *Brassica* system. Indeed, it may be more advantageous to use microspores than somatic cells for micro-injection, since it should be possible to avoid deletion of the inserted gene in future generations.

8. SOMATIC HYBRIDISATION OF RICE.

Khush (1987) noted that wide hybridisation may be useful for transferring genes for disease resistance and tolerance to adverse environmental stress from wild *Oryza* species into rice cultivars. To this end, pre- and post-fertilisation barriers have been overcome, to some extent, by the use of growth regulators (Mujeebkazi, 1985) and by embryo rescue (Jena and Khush, 1984). Barriers to recombination have been circumvented by, for example, induced homologous pairing (Riley et al, 1968) and the production of alien addition (Jena and Khush, 1986) and substitution lines (Shepherd and Islam, 1981). Protoplast fusion also provides a means of transferring both nuclear and cytoplasmic genes between sexually isolated rice species.

Somatic hybrid tissues of rice have been obtained following electrical or chemical fusion of protoplasts. Although fusion can be achieved between protoplasts isolated from rice cell suspension cultures with protoplasts isolated directly from the plant, as yet, hybrid callus and plants have been reported only after fusion of protoplasts from cultured cells.

The first reports of somatic hybrid rice callus were both after fusion of rice protoplasts with those from distantly related species. Niizeki and Kita (1981) reported the formation of hybrid cell colonies of rice and soybean after PEG/high pH-high Ca^{2+} induced fusion. Hybrid cell clusters were distinguished by their nuclear and cytoplasmic features. Subsequently, Niizeki et al (1985) described the formation of hybrid rice-soybean callus, such tissue being identified phenotypically since it exhibited both the blackish-purple flavanoid pigmentation of rice callus and the softness and pale green colouration characteristic of soybean callus. Sala et al (1985) reported the formation of hybrid rice-carrot callus after PEG-induced protoplast fusion. Hybrid tissue had the capacity of carrot cells to multiply, combined with the resistance to azetidine-2-carboxylic acid of rice cells. However, both research groups demonstrated rapid elimination of most of the rice genome from the hybrid tissues by changes in esterase and peroxidase isoenzyme banding patterns (Niizeki et al, 1985) and nuclear DNA restriction analysis (Sala et al, 1985).

Recent advances in both protoplast fusion and tissue culture techniques have facilitated the production of somatic hybrid rice callus, plantlets and mature fertile plants. Terada et al (1987) reported the electrofusion of suspension culture derived protoplasts of rice and barnyard grass (*Echinochla oryzicola* Vazing). Hybrid callus was selected by iodoacetamide inactivation of the rice protoplasts; barnyard grass protoplasts did not divide under the culture conditions which favoured rice protoplast growth (Kyozuka et al, 1987). One hundred and sixty six calli were identified as hybrid by isozyme and chromosome analysis. The hybrid callus, described as highly morphogenic, gave rise to 44 shoots, most of which were abnormal. However, 9 of the shoots grew into plantlets with a morphology distinct from that of either parent.

Fertile diploid somatic hybrid rice plants have been produced between sexually compatible japonica (sinica) varieties (Toriyama and Hinata, 1988). Protoplasts were isolated from anther-derived haploid suspension cultures of both varieties, electrofused, and cultured without selection using the methods of Toriyama et al (1986). Hybrid plants were identified morphologically. One parent, Murasakidaikoku, was a mutant rice with dominant purple colouration of the whole plant and recessive dwarfism. The other parent, Yamahaushi, had normal colouration and stature. The somatic hybrid plants, like the sexual hybrids, had purple colouration and normal stature. Both diploid and triploid somatic hybrid plants were produced. No significant difference occurred, at the 5% level, in the segregation of dwarfism and purple colouration between the selfed progenies of the sexual hybrids and somatic hybrids, demonstrating that fertile diploid hybrids of rice can be obtained by somatic hybridisation of haploid cells.

One of the potentials of rice somatic hybridisation is that of introducing agronomically important traits into the crop from sexually isolated wild *Oryza* species and other members of the Gramineae. Hayashi et al (1988) obtained somatic hybrid plants by electrofusion of protoplasts from cell suspension cultures of several japonica rice varieties with protoplasts isolated from suspension cultures of the four rice subspecies *O. officinalis, O. eichingeri, O. brachyantha and O. perrieri*. Hybrid material was selected by iodoacetamide inactivation as described by Terada et al (1987). The hybridity of regenerated plants was confirmed by their intermediate morphology compared to the parents, by fraction I protein analysis and by chromosome number. The viabilty of pollen varied between the somatic hybrids. Those of *O. sativa* + *O. brachyantha* did not produce pollen. The hybrids between *O. sativa* and *O. eichingeri* had 60% pollen viability, but seed set and development were poor. Seeds were obtained only from hybrids with *O. eichingeri* and *O. officinalis*. Currently, the hybrids are being backcrossed to *O. sativa* and embryo rescue incorporated into the experimental procedure.

The transfer of CMS into breeding lines of rice is important for the efficient production of hybrid rice seed. Several types of CMS have been identified in rice (Shinjo, 1984), and protoplast fusion could reduce the numerous backcrosses required by conventional methods to transfer this trait. Somatic hybrid rice callus has been obtained by Yang et al (1988b) from the electrofusion of ^{60}Co-irradiated protoplasts of the CMS japonica line A-58 CMS with iodoacetamide treated protoplasts of the fertile japonica variety Fujiminori. The hybrid tissue exhibited the peroxidase isozyme patterns of the fertile parent and 4 plasmid-like mitochondrial DNAs from the sterile parent. Yang et al (1989) have since reported the regeneration of rice cybrids, the morphology of the regenerated plants being nearly identical to that of the fertile parent. The regenerants had Fujiminori peroxidase isozyme banding patterns and the diploid chromosome number (2n=24), indicating that the regenerants had only the nuclear genome of the Fujiminori parent. The mitochondrial DNA restriction fragment patterns of the regenerants differed from those of the parents, with a stable heterogenous mitochondrial population in the cybrids. Thus, the regenerants had 2 of the 4 A-58 CMS *Pst*I restriction bands, 6 of the 8 Fujiminori *Pst*I bands and the 4 bands attributable to the 4 plasmid-like DNAs of A-58 CMS. Similar rice cybrids have been produced by asymmetric protoplast fusion between protoplasts of the CMS line MTC-9A and the cultivar Norin 8 (Akaqi et al, 1989).

Apart from the use of protoplast fusion in relation to the transfer of cytoplasmic traits such as CMS and herbicide resistance, there exists the opportunity to produce unique nuclear-cytoplasmic combinations, particularly between sexually incompatible species, as discussed by Kumar and Cocking (1987). The implication for rice improvement in relation to the possibility of additional heterosis arising from these interactions and novel combinations could be significant.

Protoplast fusion techniques require adequate methods of selecting heterokaryons and hybrid material. Iodoacetamide selection has already been used to isolate rice somatic hybrids, while selection based on dominant markers, such as kanamycin resistance (Pental et al, 1988), offers another possibility. An alternative to chemical selection of fused protoplasts is fluorescence activated cell sorting (Glimelius et al, 1986), a technique which may also be useful for heterokaryon selection following fusion of rice protoplasts.

The recent reports of somatic hybrid and cybrid rice plant production indicate the potential of protoplast fusion to transfer agronomically importanttraits. However, the products of somatic hybridisation must be fertile to allow integration into conventional breeding programmes. Poor somatic hybrid fertility may be overcome by the fusion of somatic cell protoplasts with gametic cell (pollen tetrad) protoplasts, resulting in triploid hybrid plants. Using such an approach, fertile interspecific gametosomatic hybrids have been produced in the Solanaceae between *Nicotiana glutinosa* and *N. tabacum* (Pirrie and Power, 1986) and between *N. rustica* and *N. tabacum* (Pental et al, 1988). This approach should also be applicable to rice.

9. THE INTERACTION OF RICE ROOT HAIRS WITH *RHIZOBIUM* AND *AGROBACTERIUM* - A NOVEL APPROACH TO RICE GENETIC ENGINEERING.

Whilst somaclonal variation, transformation and somatic hybridisation are established methods of generating material of interest to the breeder, other methods of plant modification are also being assessed. The symbiotic association of *Rhizobium* with legume roots has been an area of intensive research, with the possibility of extending this association to non-legumes. Following the demonstration that enzymatic treatment of the roots of seedlings of a range of crop plants removes the cell wall at the tips of root hairs (Cocking, 1985), Al-Mallah et al (1987) used this technique to eliminate the barrier to *Rhizobium*-host plant specificity in the legume *Trifolium repens*. Similarly, in a series of experiments involving treatment of the roots of 2 day-old rice seedlings with an enzyme mixture consisting of 1.0% w/v Cellulase YC, 0.1% w/v Pectolyase Y23 and 8% w/v mannitol, followed by incubation with *Rhizobium* in the presence of PEG, elongate and spherical nodular structures developed on the roots of seedlings of the cultivars Taipei 309 and Taipei 177 (Al-Mallah et al, 1989a). Light and electron microscopy confirmed the presence of *Rhizobium* both within and between cells of the nodular structures. Bacteria re-isolated from these structures retained their ability to nodulate their normal legume host plants. Nitrogenase activity, as measured by acetylene reduction, was negligible in these studies, but future experiments will be directed towards attempting to establish a nitrogen fixing symbiosis between *Rhizobium* and rice roots.

Studies similar to those described for *Rhizobium* have been carried out on the interaction of *Agrobacterium rhizogenes* with rice. Roots of 2 and 3 day-old seedlings treated with a Cellulase-Pectolyase mixture were inoculated with the supervirulent strain R160-1 of *A. rhizogenes* in the presence of PEG 6000 and 50 mM $CaCl_2$. About 20% of the treated seedlings developed tumour-like outgrowths on their roots (Al-Mallah et al, 1989b). It remains to be seen whether such structures contain and express the NPTII gene carried by the Ri plasmid of strain R160-1, and whether such outgrowths can be excised from the parent plants, maintained in culture, and induced to undergo plant regeneration.

10. CONCLUSION.

The last five years have witnessed significant advances in the genetic engineering of rice through somatic cell techniques. Fundamental to these studies has been the use of embryogenic cell suspensions as source material to enable protoplast-to-plant systems to be developed for japonica varieties. Current improvements in culture methods, including the use of nurse cells, are being applied to indica rice, since there is a need to regenerate plants reproducibly from protoplasts of a range of commercially important varieties and breeding lines. To date, protoplasts isolated directly from the plant have remained unresponsive to culture.

Somaclonal variants expressing agronomically useful characters have been reported, while protoplast fusion has resulted in somatic hybrid and cybrid plants. Currently, emphasis is on the production of fertile somatic hybrids, posssibly through gametosomatic hybridisation, which can be incorporated into breeding programmes. A reproducible protoplast-to-plant system has also been essential for the production of transgenic plants, expressing antibiotic resistance and readily assayable reporter genes, following direct DNA uptake into protoplasts. This transformation technology now provides the basis and stimulus for future attempts to insert agronomically important genes into this crop plant.

11. REFERENCES.

Abdullah,R., Cocking,E.C. and Thompson,J.A. (1986) Efficient plant regeneration from rice protoplasts through somatic embryogenesis, Bio/Technology 4, 1087-1090.

Abdullah,R., Thompson,J.A., Khush,G.S., Kaushik,R.P. and Cocking,E.C. (1989) Protoclonal variation in the seed progeny of plants regenerated from rice protoplasts, Plant Sci. (in press).

Abe,T. and Futsuhara,Y. (1985) Efficient plant regeneration by somatic embryogenesis from root callus tissues of rice (*Oryza sativa*), J. Plant Physiol. 121, 111-118.

Akagi,H., Sakamoto,M., Negishi,T. and Fujimura,T. (1989) Construction of rice cybrid plants, Molec. Gen. Genet. 215, 501-506.

Al-Mallah,M.K., Davey,M.R. ansd Cocking,E.C. (1987) Enzymatic treatment of clover root hairs removes a barrier to *Rhizobium*-host specificity. Bio/Technology 5, 1319-1322.

Al-Mallah,M.K., Davey,M.R. and Cocking,E.C. (1989a) Formation of nodular structures on rice seedlings by rhizobia, J. Exp. Bot. 40, 473-478.

Al-Mallah,M.K., Davey,M.R. and Cocking,E.C. (1989b) A new approach to the nodulation of non-legumes by rhizobia and the transformation of cereals by agrobacteria using enzymatic treatment of root hairs, Internat. J. Plant Genet. Manipulat. (in press).

Bidney,L.D. and Shepard,J.F. (1981) Phenotypic variation in plants regenerated from protoplasts : the potato system. Biotechnol. Bioeng. 23, 2691-2701.

Brar,D.S., Ling,D.H. and Yoshida,S. (1985) Plant regeneration from somatic cell cultures of some IR varieties of rice, in Biotechnology in International Agricultural Research, IRRI, Manila, pp. 169-177.

Chang,T.T. (1976) Rice, in N.W.Simmonds (ed.) Evolution of Crop Plants, Longman, London and New York, pp. 98-104.

Chen,T.H., Lam,L. and Chen,S.H. (1985) Somatic embryogenesis and plant regeneration from cultured young inflorescences of *Oryza sativa* L. (rice), Plant Cell Tissue and Organ Cult. 4, 51-54.

Chen,W.H., Davey,M.R., Power,J.B. and Cocking,E.C. (1988) Sugarcane protoplasts : factors affecting division and plant regeneration, Plant Cell Rep. 7, 344-347.

Cho,M.S. and Zapata,F.J. (1988) Callus formation and plant regeneration in isolated pollen culture of rice (*Oryza sativa* cv Taipei 309), Plant Sci. 58, 239-244.

Cocking,E.C. (1985) Protoplasts from root hairs of crop plants, Bio/Technology 3, 1104-1106.

Cocking,E.C. and Davey,M.R. (1987) Gene transfer in cereals, Science 236, 1259-1262.

Coulibaly,M.Y. and Demarly,Y. (1986) regeneration of plantlets from protoplasts of rice *Oryza sativa*, Z. Pflanzenphysiol. 96, 79-81.

Creemers-Molenaar,T., Loeffen,J.P.M. and Zaal,M.A.C.M. (1988) Isolation, culture and regeneration of *Lolium perenne* and *Lolium multiflorum* protoplasts, in K.J.Puite, J.J.M.Dons, H.J.Huizing, A.J.Kool. M.Koornneef and F.A.Krens (eds.), Progress in Plant Protoplast Research, Kluwer Academic Publishers, Dordrecht, The Netherlands, pp. 53-54.

Crossway,A., Oakes,J.V., Irvine,J.M., Ward,B., Knauf,V.C. and Shewmaker,C.K. (1986) Integration of foreign DNA following microinjection of tobacco mesophyll protoplasts, Mol. Gen. Genet. 20, 179-185.

Dallas,J.F. (1988) Detection of DNA "fingerprints" of cultivated rice by hybridization with a human minisatellite DNA probe, Proc. Natl. Acad. Sci. USA. 85, 6831-6835.

Dalton,S.J. (1988) Plant regeneration from cell suspension protoplasts of *Festuca arundinacea* Schreb. (tall fescue) and *Lolium perenne* L. (perennial ryegrass), J. Plant Physiol. 132, 170-175.

Davey,M.R., Cocking,E.C., Freeman,J., Pearce,N. and Tudor,I. (1980) Transformation of *Petunia* protoplasts by isolated *Agrobacterium* plasmids, Plant Sci. Lett. 18, 307-313.

Davey,M.R. and Kumar,A. (1983) Higher plant protoplasts - retrospect and prospect, Internatl. Rev. Cytol. Suppl. 16, 219-299.

Davey,M.R., Rech,E.L. and Mulligan,B.J. (1989) Direct DNA transfer to plant cells, Plant Molec. Biol. (in press).

Duan,X. and Chen,S. (1985) Variation of the characters in rice (*Oryza sativa*) induced by foreign DNA uptake, China Agric. Sci. 3, 6-9.

Finch,R.P., Lynch,P.T., Jotham,J.P. and Cocking,E.C. (1989) Isolation, culture and genetic manipulation of rice protoplasts, in Y.P.S.Bajaj (ed.), Biotechnology in Agriculture and Forestry Vol 14, Springer-Verlag, Amsterdam, (in press).

Fromm,M.E., Taylor,L.P. and Walbot,V. (1986) Stable transformation of maize after gene transfer by electroporation, Nature 319, 791-793.

Fujimura, T., Sakurai, M., Akagi, H., Negishi, T. and Hirose, A. (1985) Regeneration of rice plants from protoplasts, Plant Tissue Cult. Lett. 2, 74-75.

Glaszmann, J.C. (1987) Isozymes and classification of Asian rice varieties, Theor. Appl. Genet. 74, 21-30.

Glimelius, K., Djupsjobacka, M. and Fellner-Feldegg, H. (1986) Selection and enrichment of plant protoplast heterokaryons of Brassicaceae by flow sorting, Plant Sci. 45, 133-141.

Greenland, D.J. (1984) Rice, Biologist 31, 219-225.

Hayashi, Y., Kyozuka, J. and Shimamoto, K. (1988) Hybrids of rice (*Oryza sativa* L.) and wild *Oryza* species obtained by cell fusion, Mol. Gen. Genet. 214, 6-10.

Henke, R.R., Mansur, M.A. and Constantin, M.J. (1978) Organogenesis and plantlet formation from organ and seedling-derived calli of rice (*Oryza sativa*), Physiol. Plant. 44, 11-14.

Heyser, J.W. (1984) Callus and shoot regeneration from protoplasts of Proso millet (*Panicum miliaceum* L.), Z. Pflanzenphysiol. 113, 293-299.

Horne, M.E., Conger, B.V. and Harms, C.T. (1988) Plant regeneration from protoplasts of embryogenic suspension cultures of orchardgrass (*Dactylis glomerata* L.), Plant Cell Rep. 7, 371-374.

Jefferson, R.A. (1987) Assaying chimaeric genes in plants : the GUS gene fusion system, Plant Mol. Biol. Reporter 5, 387-405

Jena, K.K. and Khush, G.S. (1984) Embryo rescue and interspecific hybrids and its scope in rice improvement, Internatl. Rice Genet. Newsletter 1, 133-134.

Jena, K.K. and Khush, G.S. (1986) Production of monosomic alien addition lines of *O. sativa* having a single chromosome of *O. officinalis*, Rice Genet. 1, 199-208.

Jones, T.J. and Rost, T.L. (1989) The developmental anatomy and ultrastructure of somatic embryos from rice (*Oryza sativa* L.) scutellum epithelial cells, Bot. Gaz. 150, 41-49.

Junker, B., Zimny, J., Lührs, R. and Lörz, H. (1987) Transient expression of chimaeric genes in dividing and non-dividing cereal protoplasts after PEG-induced DNA uptake, Plant Cell Rep. 6, 329-332

Kao, K.N. (1977) Chromosomal behaviour in somatic hybrids of soybean-*Nicotiana glauca*, Molec. Gen. Genet. 150, 225-230.

Khush, G.S. (1987) Innovative approaches to rice improvement, in, International Symposium on Rice Farming Systems : New Directions, Sakha, Egypt, pp. 1-20.

Krens,F.A., Molendijk,L., Wullems,G.J. and Schilperoort,R.A. (1982) *In vitro* transformation of plant protoplasts with Ti-plasmid DNA, Nature 296, 72-74.

Kumar,A. and Cocking,E.C. (1987) Protoplast fusion : a novel approach to organelle genetics in higher plants, Amer. J. Bot. 74, 1289-1303.

Kumari,D.S., Sarma,N.P. and Rao,G.J.N. (1988) Micropropagation of cytosterile stocks, Internatl. Rice Res. Newslett. 13, 5-6.

Kyozaka,J., Hayashi,Y. and Shimamoto,K. (1987) High frequency plant regeneration from rice protoplasts by novel nurse culture methods, Molec. Gen. Genet. 206, 408-413.

Larkin,P.J. and Scowcroft,W.R. (1981) Somaclonal variation - a novel source of variability from cell cultures for plant improvement, Theor. Appl. Genet. 60, 197-214.

Ledoux,L., Huart,R. and Jacobs,M. (1974) DNA-mediated genetic correction of thiaminless *Arabidopsis thaliana*, Nature 249, 17-21.

Lee,L., Schroll,R.E., Grimes,H.D. and Hodges,T.K. (1989) Plant regeneration from indica rice (*Oryza sativa* L.) protoplasts, Planta 178, 325-333.

Ling,D.H., Chen,W.Y. and Ma,Z.R. (1983) Somatic embryogenesis and plant regeneration in an interspecific hybrid of *Oryza*, Plant Cell Rep. 2, 169-171.

Linsmaier,E.M. and Skoog,F. (1965) Organic growth factor requirements of tobacco tissue cultures, Physiol. Plant. 8, 100-127.

Luo,Z-x. and Wu,R. (1988) A simple method for the transformation of rice via the pollen-tube pathway, Plant Mol. Biol. Reporter 6, 165-174.

Marcotte,W.R., Bayley,C.C. and Quatrano,R.S. (1988) Regulation of a wheat promoter by abscisic acid in rice protoplasts, Nature 335, 454-457.

Masuda,K., Kudo-Shiratori,A. and Inoue,M. (1989) Callus formation and plant regeneration from rice protoplasts purified by density gradient centrifugation, Plant Sci. 62, 237-243.

McCoy,T.J. and Phillips,R.L. (1982) Chromosome stability in maize (*Zea mays*) tissue cultures and sectoring in some regenerated plants, Can. J. Genet. Cytol. 24, 559-565.

Mujeeb-Kazi,A. (1985) Cytogenetics of *Hordeum vulgare* X *Elymus patagonicus* hybrid (n=4x=28), Theor. Appl. Genet. 69, 475-479.

Murashige,T. and Skoog,F. (1962) A revised medium for rapid growth and bioassays with tobacco tissue culture, Physiol. Plant. 15,473-497.

Nakamura,C. and Keller,W.A. (1982) Plant regeneration from inflorescence cultures of hexaploid *Triticale*, Plant Sci. Lett. 24, 275-280.

Nakano,H. and Maeda,E. (1979) Shoot differentiation in callus of *Oryza sativa* L., Z. Pflanzenphysiol. 93, 449-458.

Neuhaus,G., Spangenberg,G., Mittelsten Scheid,O. and Schweiger,H-G. (1987) Transgenic rapeseed plants obtained by the microinjection of DNA into microspore-derived embryoids, Theor. Appl. Genet. 7, 30-36.

Niizeki,H. and Oono,K. (1971) Rice plants obtained by anther culture, Colloq. Internat. Centre National de la Recherche Scientifique 193, 251-257.

Niizeki,M. and Kita,F. (1981) Cell division of rice and soybean and their fused protoplasts, Japan. J. Breed. 31, 161-167.

Niizeki,M., Tanaka,M. Akada,S., Hirai,A. and Saito,K. (1985) Callus formation of somatic hybrid of rice and soybean and characteristics of the hybrid callus, Japan. J. Genet. 60, 81-92.

Nishi,T., Yamada,Y. and Takahashi,E. (1968) Organ differentiation and plant restoration in rice callus, Nature 219, 508-509.

Nishi,T. and Mitsuoka,S. (1969) Occurrence of various ploidy plants from anther and ovary culture of rice plants, Japan. J. Genet. 44, 341-346.

Ogura,H., Kyozuka,J., Hayashi,Y., Koba,T. and Shimamoto,K. (1987) Field performance and cytology of protoplast-derived rice (*Oryza sativa*) : high yield and low degree of variation of four japonica cultivars, Theor. Appl. Genet. 74, 670-676.

Ogura,H., Kyozuka,J., Hayashi,Y. and Shimamoto,K. (1989) Yielding ability and phenotypic traits in the selfed progeny of protoplast-derived rice plants, Japan. J. Breed. 39, 47-56.

Oka,H.I. (1988) Origin of Cultivated Rice, Elsevier, Amsterdam.

Paszkowski,J., Shillito,R.D., Saul,M., Mandak,V., Hohn,T., Hohn,B. and Potrykus,I. (1984) Direct gene transfer to plants, EMBO J. 3, 2717-2722.

Pental,D., Mukhopadhyay,A., Grover,A. and Pradhan,A.K. (1988) A selection method for the synthesis of triploid hybrids (3n) by fusion of microspore protoplasts (n) with somatic cell protoplasts (2n), Theor. Appl. Genet. 76, 237-243.

Pirrie,A. and Power,J.B. (1986) The production of fertile, triploid somatic hybrid plants (*Nicotiana glutinosa* (n) + *N. tabacum* 2n) via gametic : somatic protoplast fusion, Theor. Appl. Genet. 72, 48-52.

Prioli,L. and Söndahl,M.R. (1989) Plant regeneration and recovery of fertile plants from protoplasts of maize (*Zea mays* L.), Bio/Technology 7, 589-594.

Reich,T.J., Iyer,V.N. and Miki,B.L. (1986) Efficient transformation of alfalfa protoplasts by the intranuclear microinjection of Ti plasmids, Bio/Technology 4, 1001-1004.

Rhodes,C., Lowe,K. and Ruby,K. (1988) Plant regeneration from protoplasts isolated from embryogenic maize cell cultures, Bio/Technology 6, 56-60.

Riley,R., Chapman,V. and Johnson,R. (1968) Introduction of yellow rust resistance of *Aegliops comosa* into wheat by genetic induced homologous recombination, Nature 217, 383-384.

Sala,C., Biasini,M.G., Morandi,C., Nielsen,E., Parisi,B. and Sala,F. (1985) Selection and nuclear DNA analysis of cell hybrids between *Daucus carota* and *Oryza sativa*, J. Plant Physiol. 118, 409-419.

Schaeffer,G.W. (1982) Recovery of heritable variation in anther-derived doubled haploid rice, Crop Sci. 22, 1160-1164.

Scowcroft,W.K. and Larkin,P.J. (1988) Somaclonal variation, in, G.Block and J.Marsh (eds.), Applications of Plant Cell and Tissue Culture, Ciba Foundation Symposium 137, Wiley, Chichester, pp. 21-35.

Shepherd,K.W. and Islam,A.K.M.R. (1981) Wheat : barley hybrids - the 1st 80 years, in, L.T.Evans and W.J.Peacock (eds.), Wheat Science Today and Tomorrow, Cambridge University Press, Cambridge, pp. 107-128.

Shillito,R.D., Carswill,G.K., Johnson,C.M., DiMaio,J.J., and Harms,C.T. (1989) Regeneration of fertile plants from protoplasts of elite inbred maize, Bio/Technology 7, 581-587.

Shimamoto,K., Terada,R., Izawa,T. and Fujimoto,H. (1989) Fertile transgenic rice plants regenerated from transformed protoplasts, Nature 338, 274-276.

Shinjo,C. (1984) Cytoplasmic male sterility and fertility restoration in rice having genome A, in, S.Tsunoda and N.Takahashi (eds.), Biology of Rice, Japan. Sci. Soc., Tokyo, Elsevier, Amsterdam, pp. 321-338.

Srinivasan,C. and Vasil,I.K. (1986) Plant regeneration from protoplasts of sugarcane (*Saccharum officinarum* L.), J. Plant Physiol. 126, 41-48.

Tanchak,M.A., Griffing,C.R., Mersey,B.G. and Fowke,L.C. (1984) Endocytosis of cationized ferritin by coated vesicles of soybean protoplasts, Planta 162, 481-489.

Terada,R., Kyazuka,J., Nishibayashi,S. and Shimamoto,J. (1987) Plantlet regeneration from somatic hybrids of rice (*Oryza sativa* L.) and barnyard grass (*Echinochloa oryzicola* Vasing.), Mol. Gen. Genet. 210, 39-43.

Thompson,J.A. and Cocking,E.C. (1986) Rice protoplast and cell culture, in, Rice Genetics, IRRI, Manila, pp. 791-798.

Thompson, J.A., Abdullah, R. and Cocking, E.C. (1986) Protoplast culture of rice (*Oryza sativa* L.) using media solidified with agarose, Plant Sci. 47, 123-133.

Thompson, J.A., Abdullah, R., Chen, W-H. and Gartland, K.M.A. (1987) Enhanced protoplast division in rice (*Oryza sativa* L.) following heat shock treatment, J. Plant Physiol. 127, 367-370.

Tobita, S., Takahashi, H., Miyake, H. and Totsuka, T. (1988) Selection and partial characterization of copper resistant line of rice (*Oryza sativa*) callus culture, J. Plant Physiol. 133, 545-549.

Töpfer, R., Gronenborn, B., Schell, J. and Steinbiss, H.H. (1989) Uptake and transient expression of chimeric genes in seed-derived embryos, The Plant Cell 1, 133-139.

Toriyama, K., Arimoto, Y., Uchimiya, H. and Hinata, K. (1988) Transgenic rice plants after direct gene transfer into protoplasts, Bio/Technology 6, 1072-1074.

Toriyama, K. and Hinata, K. (1985) Cell suspension and protoplast culture in rice, Plant Sci. 41, 179-183.

Toriyama, K. and Hinata, K. (1988) Diploid somatic-hybrid plants regenerated from rice cultivars, Theor. Appl. Genet. 76, 665-668.

Toriyama, K., Hinata, K. and Sasaki, T. (1986) Haploid and diploid plant regeneration from protoplasts of anther callus in rice, Theor. Appl. Genet. 73, 16-19.

Uchimiya, H., Fushimi, T., Hashimoto, H., Harada, H, Syōno, K. and Sugawara, Y. (1986) Expression of a foreign gene in callus derived from DNA-treated protoplasts of rice (*Oryza sativa* L.), Mol. Gen. Genet. 206, 204-207.

van der Valk, P. and Zaal, M.A.C.M. (1988) regeneration of plantlets from protoplasts of *Poa pratensis* L. (Kentucky bluegrass), in, K.J. Puite, J.J.M. Dons, H.J. Huizing, A.J. Kool, M. Koornneef and F.A. Krens (eds.), Progress in Plant Protoplast Research, Kluwer Academic Publishers, Dordrecht, The Netherlands, pp. 59-60.

Vasil, I.K. (1987) Developing cell and tissue culture systems for the improvement of cereal and grass crops, J. Plant Physiol, 128, 193-218.

Vasil, I.K. (1988) Progress in the regeneration and genetic manipulation of cereal crops, Bio/Technology 6, 397-402.

Vasil, V. and Vasil, I.K. (1980) Isolation and culture of cereal protoplasts. II. Embryogenesis and plantlet formation from protoplasts of *Pennisetum americanum*. Theor. Appl. Genet. 56, 97-99.

Wakasa, K., Kobayashi, M. and Kamada, H. (1984) Colony formation from protoplasts from nitrate reductase deficient rice cell lines, J. Plant Physiol. 117, 223-228.

Wang,Y-C., Klein,T.M., Fromm,M., Cao,J., Sanford,J.C. and Wu,R. (1988) Transient expression of foreign genes in rice, wheat and soybean cells following particle bombardment, Plant Mol. Biol. 11, 433-439.

Wernicke,W., Brettell,R., Wakizuka,T. and Potrykus,I. (1981) Adventitious embryoid and root formation from rice leaves, Z. Pflanzenphysiol. 103, 361-365.

Yamada,Y., Qi,Y.Z. and Tai,T.D. (1965) Plant regeneration from protoplast-derived callus of rice (*Oryza sativa*), Plant Cell Rep. 5, 85-88.

Yang,H., Zhang,H.M., Davey,M.R., Mulligan,B.J. and Cocking,E.C. (1988a) Production of kanamycin resistant rice tissues following DNA uptake into protoplasts, Plant Cell Rep. 7, 421-425.

Yang,Z-Q., Shikana,T. and Yamada,Y. (1988b) Asymmetric hybridisation between cytoplasmic male-sterile (CMS) and fertile rice (*Oryza sativa* L.) protoplasts, Theor. Appl. Genet. 76, 801-808.

Yang,Z-Q., Shikanai,T., Mori,K. and Yamada,Y. (1989) Plant regeneration from cytoplasmic hybrids of rice (*Oryza sativa* L.), Theor. Appl. Genet. 77, 305-310.

Yoshida,S. and Ogawa,M. (1983) The application of tissue culture-induced mutagenesis to crop improvement, Food Fert. Technol. Cent. Tech. Bull. 73, 1-20.

Zapata,F.J. (1984) Rice anther culture at IRRI, in, Biotechnology in International Agricultural Research, IRRI, Manila, pp. 85-95.

Zhang,H.M., Yang,H., Rech,E.L., Golds,T.J., Davis,A.S., Mulligan,B.J., Cocking,E.C. and Davey,M.R. (1988) Transgenic rice plants produced by electroporation-mediated plasmid uptake into protoplasts, Plant Cell Rep. 7, 379-384.

Zhang,W. and Wu,R. (1988) Efficient regeneration of transgenic plants from rice protoplasts and correctly regulated expression of the foreign gene in the plants, Theor. Appl. Genet, 76, 835-840.

MICROSCOPIC OBSERVATIONS OF FUSION PROCESS OF RICE AND LETTUCE PROTOPLASTS

T. Taniguchi, T. Sato, K. Maeda and E. Maeda
Faculty of Agriculture, Nagoya University, Chikusa, Nagoya, 464-01, Japan

ABSTRACT. Protoplasts were isolated from rice callus and lettuce cotyledons. Their fusion process was studied by light and electron microscopy. Lettuce protoplast fusion could be easily produced by either of polyethylene glycol and electric methods. The transformation of fusion bodies during the fusion process was more easily observed by using the electric method than chemical one. After electric pulse application, the fusion of lettuce protoplasts was finished within about one hr, judging from their externals. On the other hand, the fusion of rice protoplasts occurred in a very short time after electric pulse application. Fusion was performed between lettuce protoplasts and rice protoplasts. Rice protoplasts contain no chloroplasts and lettuce protoplasts contain them. Therefore, both protoplasts could be easily distinguished under a light microscope. When the fusion between rice and lettuce protoplasts was induced by electric pulse application, chloroplasts in lettuce protoplasts moved into rice protoplasts within about one hr. Electron micrographs showed that the components from both protoplasts was located separately in the fusion bodies. Big rice fusion bodies produced by fusion of rice protoplasts contained some round-shaped particles. Their size was almost the same as that of the original protoplasts. These results indicate that it take a long time until the cytoplasm of rice protoplasts mix with the cytoplasm of lettuce protoplasts in the fusion body.

1. Introduction

In 1909, Küster already reported the fusion of plant protoplasts which had been isolated mechanically from plant tissues. However, there was no extensive progress in studies on the phenomenon. About 50 years later, the methods of producing a large amount of protoplasts were developed by Cocking (1960), and Takebe et al. (1968) and Otsuki and Takebe (1969) by employing crude fungal enzymes. Soon Takebe et al. (1971) and Nagata and Takebe (1971) reported that tobacco plants were regenerated from the isolated protoplasts. Since then, many papers described the regeneration of plants from fusion bodies of protoplasts

R.S. Sangwan and B.S. Sangwan-Norreel (eds.), The Impact of Biotechnology in Agriculture, 281–298.

of intra- and interspecies and intergenus. Almost all the studies have given weight on the production of new plants. However, there are not so many reports describing the process of the fusion of protoplasts and the following processes in detail (reviewed by Evans 1983, Fowke 1980, and Schieder and Vasil 1980).

The authors think that the process of fusion of protoplasts and the early processes of the morphogenesis of the fusion bodies are very important. The organogenesis and regeneration of plants must proceed via the earlier processes, namely cell wall production, cell division and so on. In other words, the later processes will be greatly affected by the activity of fusion bodies in the earlier processes. Therefore, to study the fusion and the following processes is thought to be very important. In the previous paper (Taniguchi et al. 1989), we described light microscopic observations of the fusion of protoplasts from rice callus and lettuce cotyledons. This paper describes microscopic observations of the fusion process of both the protoplasts.

2. Materials and Methods

2.1. Plants

2.1.1. Rice. Rice (*Oryza sativa*) var. Nipponbare seeds (a gift from Aichi-Ken Agricultural Research Center) were husked and surface sterilized with 70%(v/v) ethanol for 3 min and then with 1% sodium hypochlorite for 25 min.

MS basal medium supplemented with 10^{-5}M 2,4-D, 200mg/l myo-inositol, 1.0mg/l thiamine-HCl, 3g/l casamino acids, 3% sucrose and 0.8% agar adjusted pH 5.8 was used for callus induction. Seeds were placed on the medium and incubated at 25°C under a constant illumination (about 3000 lux). About 30-50mg (fresh weight) callus was transferred to the fresh medium every one month (Taniguchi et al. 1987).

2.1.2. Lettuce. Lettuce (*Lactuca sativa*) seeds var. Kaiza were sowed on 0.8% agar and incubated 24°C for 5 days under a constant illumination (about 3000 lux). The developed cotyledons were cut at the basal part and used for isolating protoplasts.

2.2. Protoplast isolation

Rice callus protoplasts were isolated by using the enzyme solution shown in Table 1. The callus was cut into small pieces and treated with the enzyme solution for about 2-5 hour at 25°C. Protoplasts were isolated by filtration through nylon filter (200 mesh) and washed three times in a centrifuge at 100x*g* for 2-5 min. Washing solution was consisted of 2.5mM $CaCl_2$ and 0.5M mannitol (Taniguchi et al. 1989).

Lettuce cotyledon protoplasts were isolated by using the enzyme solution in Table 1. The procedure was almost the same as that of rice.

Table 1. Enzyme solutions used for protoplast isolation

	Rice callus[a]	Lettuce cotyledon[b]
Meicelase P-1	4%	
Cellulase R-10		1%
Macerozyme R-10	0.2%	0.5%
Mannitol	0.5M	
Sucrose		0.3M
pH	5.6	5.7

(a:Shahin and Shepard 1980; b:Sumardi et al. 1987)

Meicelase P-1 was purchased from Sanko Junyaku, Takyo, Japan. Cellulase "Onozuka" R-10 and Macerozyme R-10 were purchased from Yakult Honsha, Tokyo, Japan.

2.3. Staining protoplasts

2.3.1. Neural red (NR) and brilliant cresyl blue (BCB). Soon after protoplasts were isolated from lettuce cotyledons or rice callus, the protoplasts were mixed with 1/10 volume of 0.1% NR or BCB for 20 min at room temperature. The stained protoplasts were washed with the washing solution three times (Gahan 1984). Unstained and stained protoplasts were observed with a differential interference microscope (Nikon), unless otherwise stated.

2.3.2. Rhodamine 123 (R-123)(Sigma Chemical Co. U.S.A.) Protoplasts were stained with 100μg/ml R 123 for one hour (Chen 1989, Taniguchi and Maeda 1988, and Wu 1987). They were washed with the washing solution three times and examined by a Nikon fluorescence microscope. Filter module B (excitation: 410-485nm; suppression: 515nm) was used to examine the auto-fluorescence of chloroplasts (red) and the fluorescence of dyes (yellow or yellow green).

2.3.3. Fluorescein isothiocyanate (FITC)(Sigma Chemical Co. U.S.A.). At the beginning of enzyme incubation of plant tissues, FITC in acetone (5 mg/ml) was added to cell-wall degrading enzyme solution according to the method of Thomas and Rose (1988). The protoplasts were isolated and examined by fluorescent microscopy as described previously.

2.4. Viability of protoplasts

The viability of protoplasts was tested with Evans blue (Kanai and Edwards 1973). Based on the staining reaction, more than 80% of rice and lettuce protoplasts was viable immediately after their isolation.

2.4. Electrofusion apparatus

Electric fields were generated by Shimadzu Somatic Hybridizer SSH-1 (Kyoto, Japan), according to the operation manual and the method of Togawa et al. (1987) with slight modification. Protoplasts were aligned in a 1.0 MHz, 200 volt/cm electric field. The distance between electrodes was 0.2 mm. After 1 to 2 min, protoplasts were fused applying one or two 320 volt/cm fusion pulses of 50 μs duration, with a 1 sec period between pulses.

2.5. Protoplast immobilization

Alginate solution was made as follow: 1.5g of sodium alginate (Wako Pure Chemical Industries, Tokyo, Japan) was suspended in 100ml of a 10mM MES buffer (pH 5.8) containing glucose (0.5M). The suspension was autoclaved at 120° for 15 min. This treatment reduced the viscosity of the solution.

Protoplasts were immobilized in alginate gel beads by the method of Draget et al. (1988) with some modification.

Fused and untreated protoplasts were washed with washing solution containing 0.5M manitiol and 0.1M MES buffer (pH 5.8) three times, to remove calcium ions. After centrifugation at 100xg for 2 min, the supernatant was removed. Alginate solution was gently added to the protoplasts and suspended in the solution. This alginate-protoplast mixture was dipped from a Pasteur pitette into a gelling solution 50mM with respect to $CaCl_2$ and 0.4M with respect to glucose in 10mM MES buffer (pH 5.8). The alginate beads were left in a calcium solution for 45 min at room temperature to complete gelling.

2.5. Electron microscopy

Fixation and embedding were performed by two methods.

2.5.1. Alginate bead method (Draget et al. 1988). The alginate beads were transferred directly from the gelling medium to 3% glutaraldehyde in 0.1M MES buffer (pH 5.8) 0.3M with respect to glucose and left at room temperature for 6 hr. The alginate beads were then washed three times with 0.1M MES buffer-0.5M glucose and postfixed in 2% osmium teroixe in the same glucose-containing MES-buffer for 3 hr. The beads were washed three times with 0.1M MES buffer (pH 5.8) and dehydrated in a graded alcohol series. Finally, they were treated with absolute alcohol for 30 min three times and propylene oxide was introduced.

2.5.2. Direct method. Protoplasts and fusion products were fixed in a mixture of 2.8% glutaraldehyde and 0.35M mannitol in MES buffer (pH 5.8) for 3 hr. They were then centrifuged and immobilized in calcium aliginate described above. The alginate beads were rinsed with 0.5 M mannitol overnight. The materials were post-fixed in 2% osmium tetroide in 0.05 M MES buffer (pH 5.8) for 3 hr, washed in distilled water, dehydrated in graded alcohol series and replaced with propylene oxide.

2.5.3. Embedding. They were embedded in TAAB Epon 812 resin.

Ultrathin sections were cut with a diamond knife on a Porter Blum MT2-B ultramicrotome. The section-mounted grids were stained with aqueous uranyl acetate followed by lead citrate. The sections were examined in a Hitachi H600 transmission electron microscope (TEM)(Maeda and Maeda, 1987).

3. Results

3.1. Fusion by polyethylene glycol (PEG)

PEG (Wako Pure Chemical Industries, Osaka Japan) is an efficient agglutinating-inducing agent. Protoplast fusion occurred mainly after its dilution in the incubation. The fusion inducing activity of PEG 6000 was stronger than PEG 1500.

One of the main purposes of this study is to show the morphological change of fusion process. It was not easy to do time-course experiments using PEG system. To determine the start point of the fusion reaction was usually difficult. Therefore, electrofusion was mainly used for the study hereafter.

3.2. Electrofusion of lettuce and rice protoplasts

Lettuce protoplasts and rice protoplasts were mixed and fusion was induced by electric pulse application. Fig. 1 shows a mixture of both the protoplasts before pulse application. A few protoplasts already started to fuse with the others. Protoplasts containing chloroplasts were isolated from lettuce cotyledons and color-less ones were isolated from rice callus. Therefore, both the porotoplats could be easily distinguished from each other. Fig. 2a shows the morphological changes of the fusion bodies of lettuce and rice protoplasts about 10 sec after applying electric pulse. Fig. 2b shows the fusion product 2 min after applying pulse. Chloroplasts in lettuce are moving into rice protoplasts (white area). Fig. 2c shows the fusion 5 min after pulse application. The rice protoplast (center cell) starts to fuse with the other side of lettuce protoplast. Ten min later(Fig. 2d), the color-less area of rice protoplast almost disappears. However, the shape of the fusion body has not yet been round completely.

3.3. Fusion of protoplats stained with NR and BCB

Suspension of isolated rice protoplasts was divided into two parts. One was stained with NR and the other was with BCB for about 20 min. It is known that both the dyes stain vacuoles in plant cells (Gahan 1984). After washing, both the stained protoplasts were mixed and given electric pulse.

Fig. 3a shows the protoplasts in AC field. The protoplasts form pearl chains. The protoplsats having red spots are stained with NR and those having blue spots are stained with BCB. Fig. 3b shows the fusion product 15 sec after electric pulse application. Two

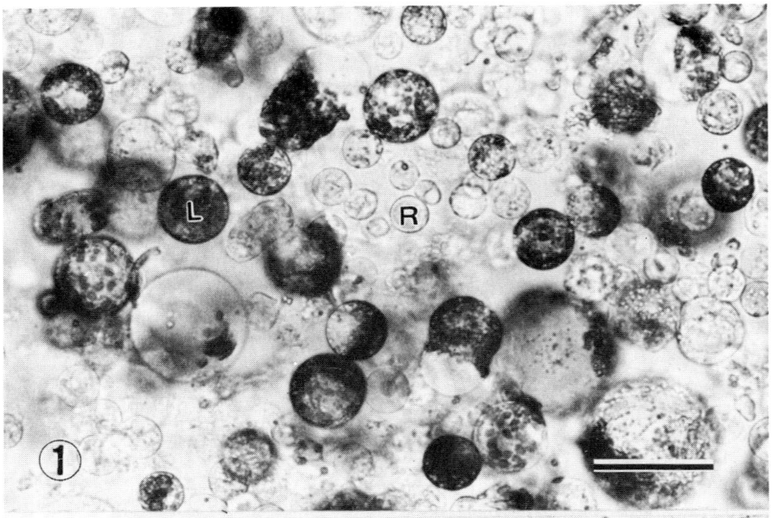

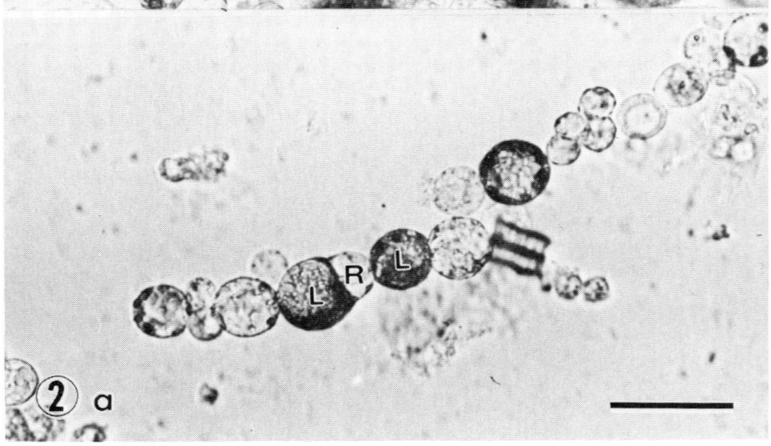

Fig. 1. Mixed protoplast preparation of rice callus and lettuce cotyledon protoplasts before electric pulse application. R: rice callus protoplasts, L: lettuce cotyledon protoplasts. Bar indicates 50 μm.

Fig. 2. Electrofusion between rice callus protoplasts (R) and lettuce cotyledon protoplasts (L). a: 10 sec, b: 2 min, c: 5 min, d: 10 min after electric pulse application, respectively. Bar indicates 50 μm.

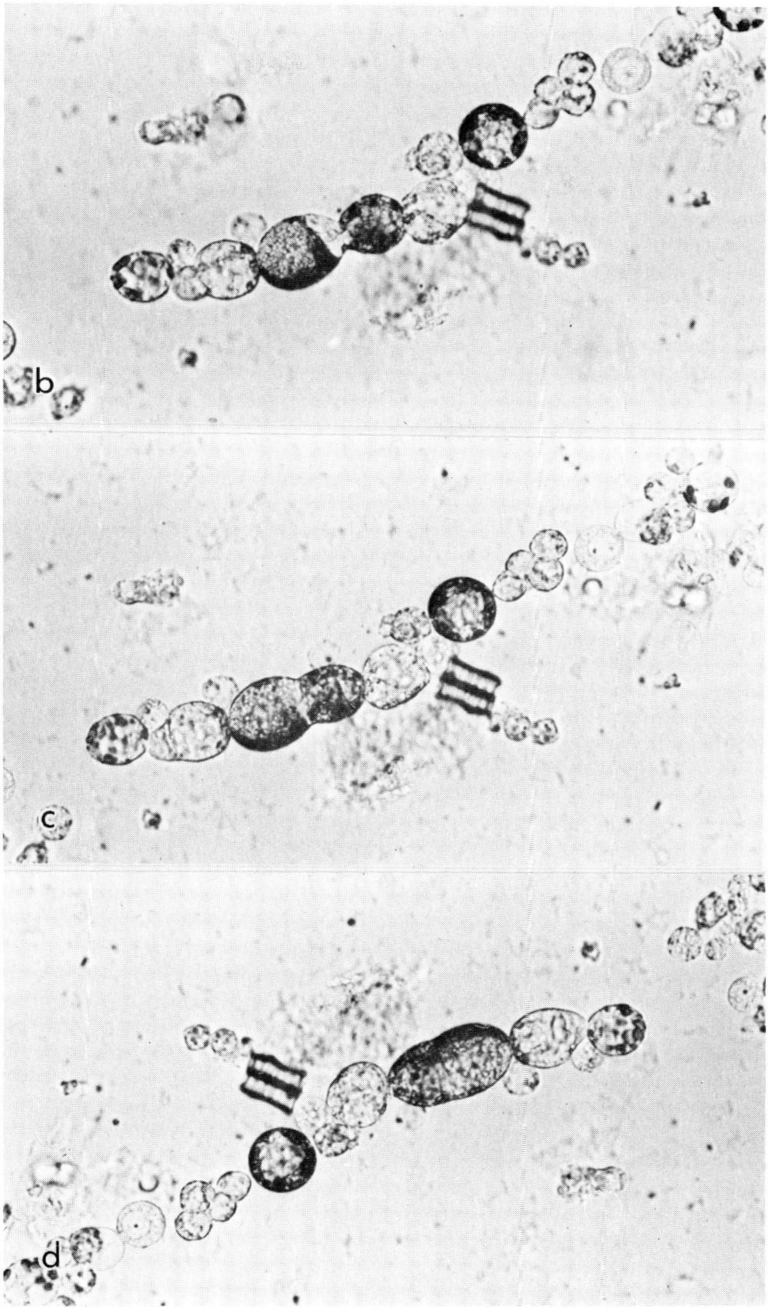
b
c
d

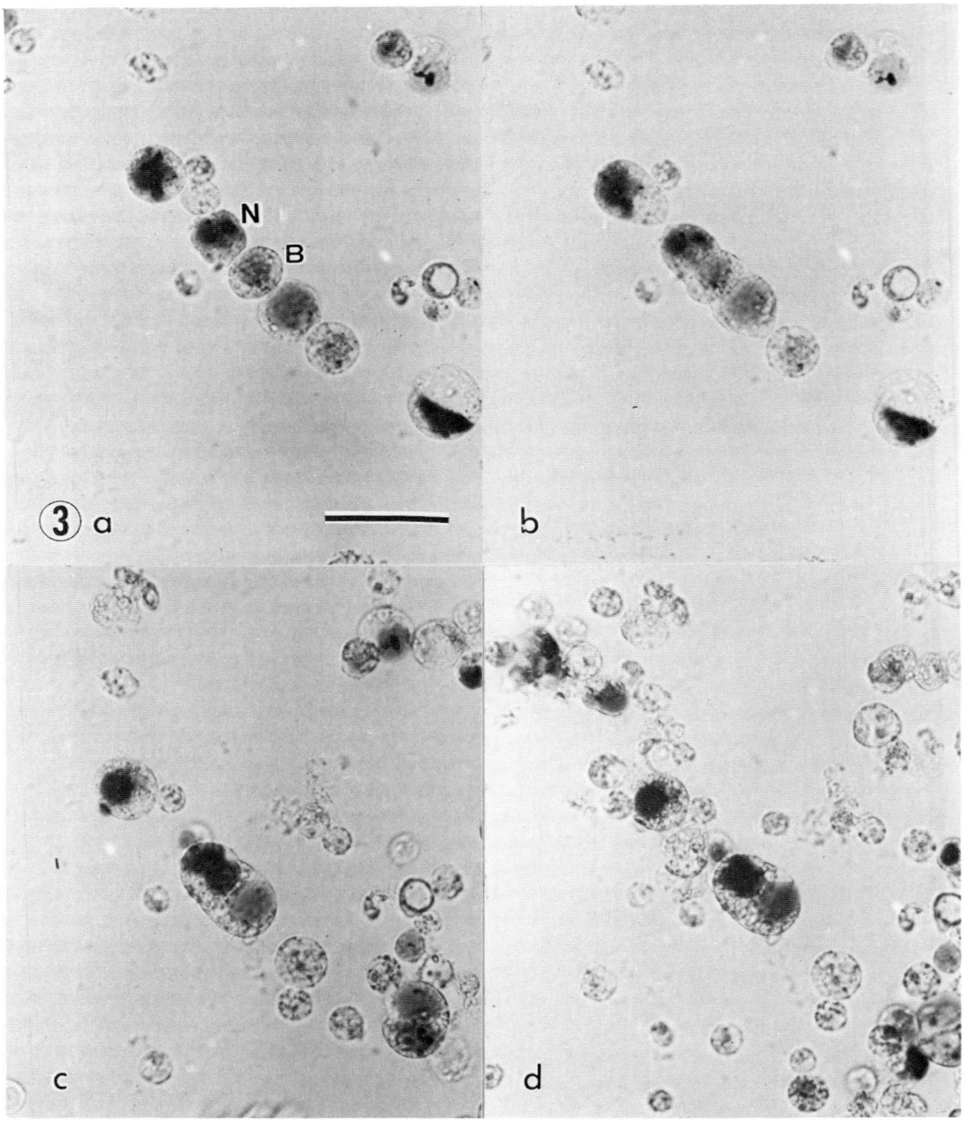

Fig. 3. Electrofusion between neutral red (N) stained rice callus protoplasts and brilliant crecyl blue (B) stained rice callus protoplasts. a: immediately before, b: 15 sec after, c: 7 min after, d: 23 min after electric pulse application, respectively. Bar indicates 50 μm.

protoplasts stained with NR starts to fuse with a protoplast stained with BCB. Two to 7 min later, the stained area approach each other (Figs. 3c). Fig. 3d shows that both the stained areas fused each other. This result may indicate that vacuoles can fuse each other.

3.4. Electrofusion of rice protoplasts stained with R-123 or FITC

Rice protoplast suspensions were divided into two parts. One suspension was stained with R-123 and the other was not stained. After washing with the washing solution, the stained protoplasts were mixed with unstained protoplasts and applied an electric pulse. The fusion bodies were observed by fluorescent microscopy (Fig. 4). The inside of the fusion body is not homogeneous. Some round yellow green fluorescent spots are observed. The size of the spots is approximately the same as that of non-fused protoplasts.

In the process of the isolation of protoplasts, protoplasts were stained with FITC. After isolating the protoplasts, the same experiments were done as that of R-123 (Fig. 5). The round shaped fluorescent spots were observed in the fusion bodies.

Moreover, in the fusion bodies induced from R-123- or FITC-stained protoplasts, some non-stained spots were also observed.

3.5. Ultrastructures of fusion bodies

3.5.1. Fusion of rice callus protoplasts and lettuce cotyledon protoplasts. Rice and lettuce protoplasts were mixed and fusion was induced. Figs. 6 and 7 show the ultrastructure of rice and lettuce protoplasts before the induction of fusion. Fig. 8 shows that the ultrastructure of the fusion product between both the protoplasts about one hr after applying electric pulse. In the round-shaped product, structurally different two areas are observed. One area (arrow) contains many chloroplasts, which may originate from lettuce protoplasts. The other area (arrow head) does not contain chloroplasts, which may originate from rice protoplasts. The shapes of nuclei and other small organelles in one area are different from those of the other area. Fig. 9 shows the magnification of the border area of rice and lettuce protoplast origins shown in Fig. 8.

This result may show that after fusion of both protoplasts, the big cell still contains the original structure of protoplasts.

3.5.2. Fusion bodies of lettuce protoplasts. The speed of the fusion of lettuce protoplasts was slow compared with that of rice protoplasts. Therefore, we could observe the earlier step of the fusion of lettuce protoplasts easier.

Fig. 10 shows one example of the fusion bodies. There are still cell membranes in some areas (arrow). However, the other areas already fuse each other. The membrane is lost. Fig. 11 shows the high magnification of the fusion area shown in Fig. 10. This micrograph clearly shows that the fused area already completely loses the cell membranes.

These result indicates that fusion will start at narrow areas of

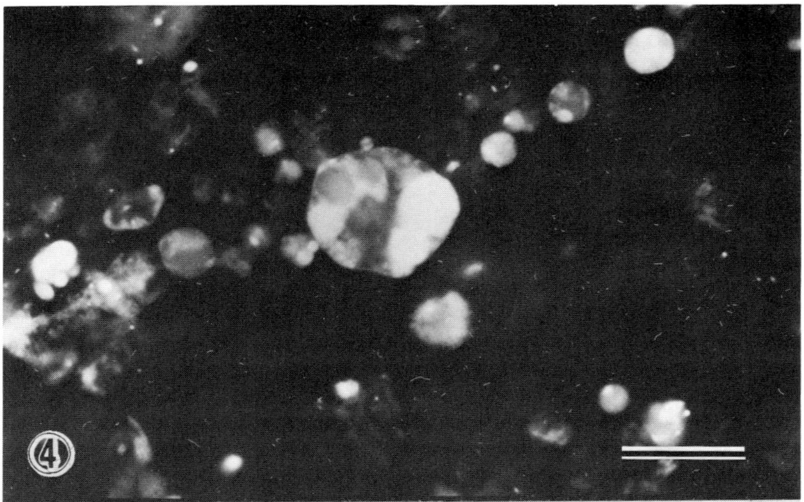

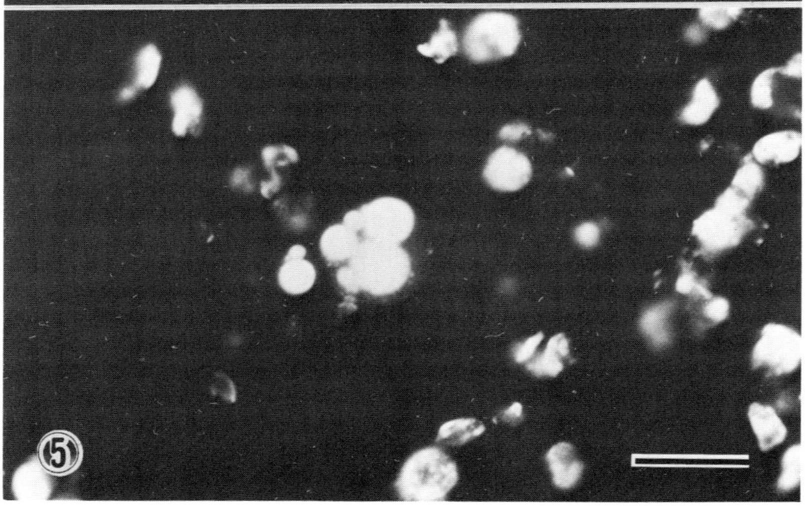

Fig. 4. Fusion body between rhodamine 123 stained and non-stained rice callus protoplasts. Bar indicates 50 μm.

Fig. 5. Fusion body between FITC stained and non-stained rice callus protoplasts. Bar indicates 50 μm.

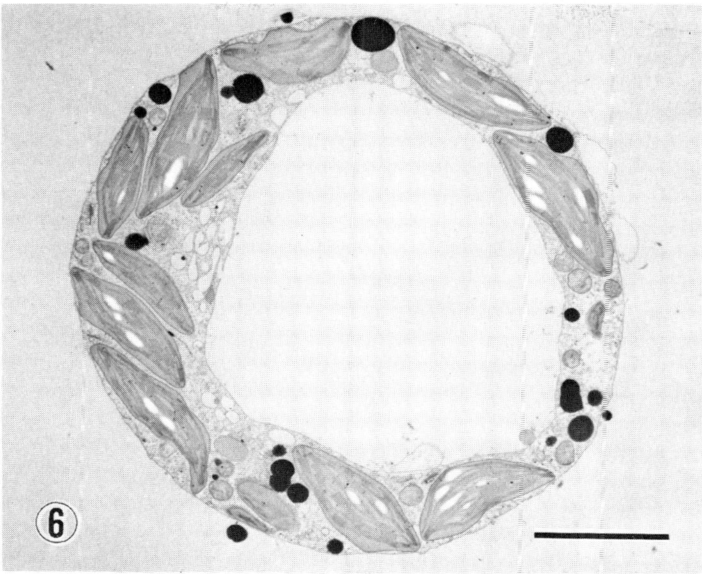

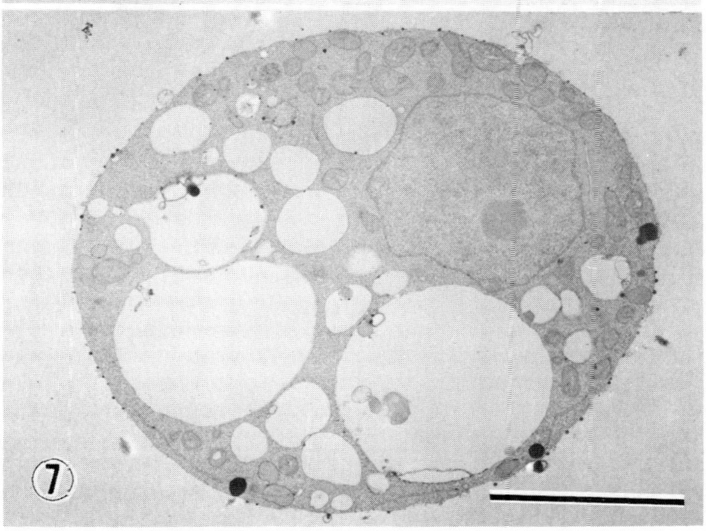

Fig. 6. Electron micrograph of a lettuce cotyledon protoplast. Bar indicates 4 μm.

Fig. 7. Electron micrograph of a rice callus protoplast. Bar indicates 4 μm.

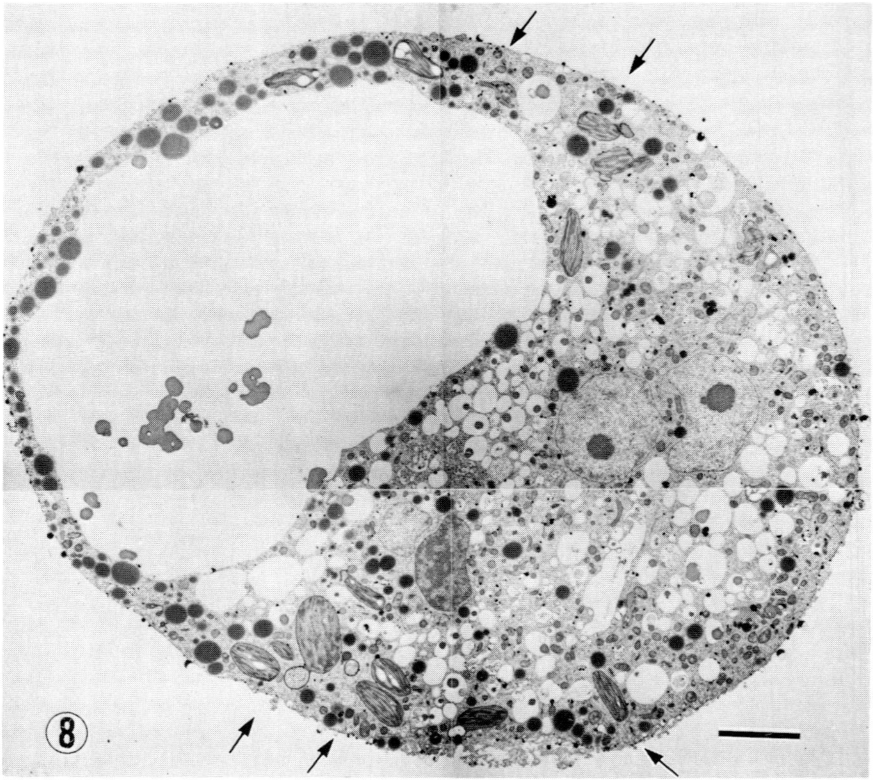

Fig. 8. Electron micrograph of a fusion body between lettuce cotyledon and rice callus protoplasts. Bar indicates 5 μm.

Fig. 9. Electron micrograph of magnification of lettuce protoplast area of a fusion body shown in Fig. 8. Bar indicates 1 μm.

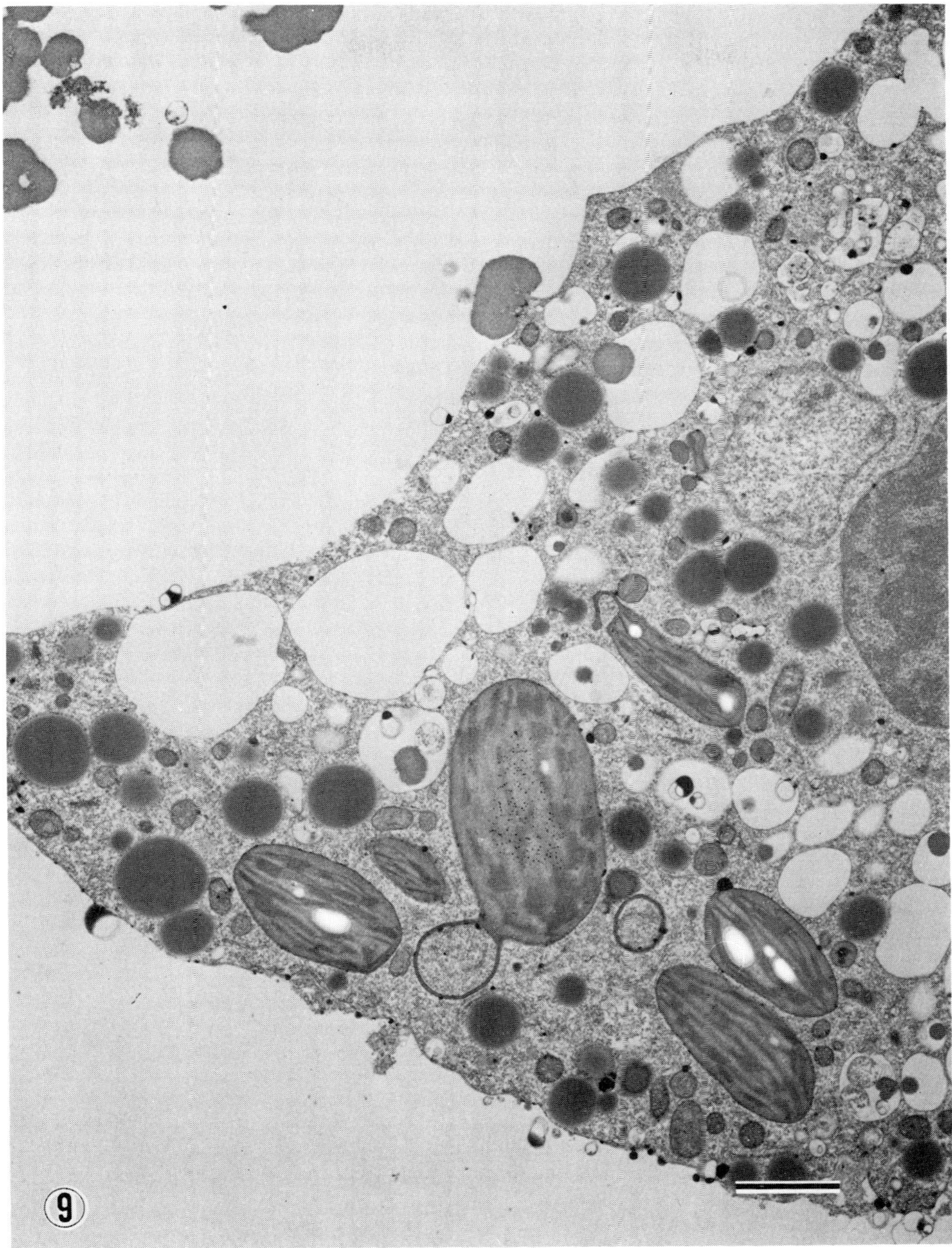

9

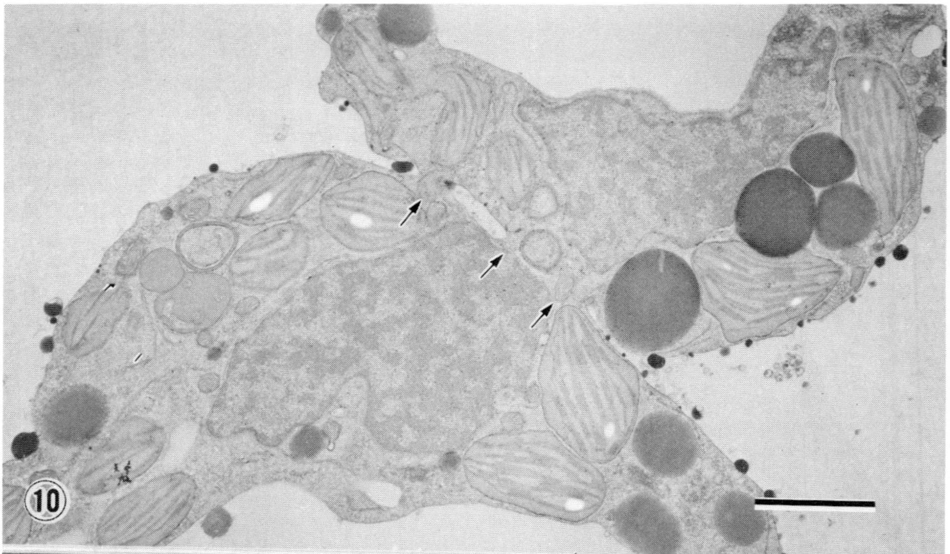

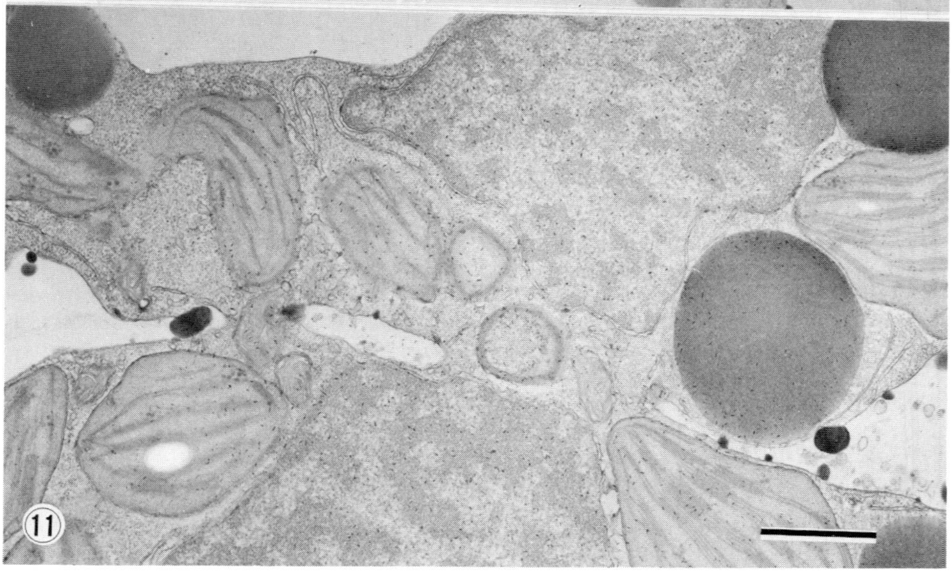

Fig. 10. Electron micrograph of the early step of fusion of lettuce cotyledon protoplasts. Small areas are already fused each other (arrow). Bar indicates 2 μm.

Fig. 11. Electron micrograph of magnification of fusion areas of lettuce fusion body shown in Fig. 10. Bar indicates 1 μm.

the surface of protoplasts after contact.

4. Discussion

One of the main purposes of this study is to examine the morphological changes of the fusion process of protoplasts. PEG is widely used for inducing fusion of plant protoplasts (reviewed by Evans et al. 1983, and Schieder and Vasil 1980).

When this chemical is added to protoplast suspensions, the aggregation of protoplasts occur. The fusion of protoplasts mainly occurs when the PEG-protoplast mixture is diluted by adding some solution not containing PEG. Therefore, it is not easy to do time-course experiments using PEG. Because it is difficult to determine the start point of the fusion reaction.

On the other hand, a number of other treatments have been examined for protoplast fusion by many scientists. Of these treatments, most attention has been addressed to electric stimulation. The influence of electric fields on plant protoplast fusion has been investigated in some laboratories. Senda et al. (1979) found that two glass capillary microelectrodes could be used to fuse adhering protoplasts. Zimmermann and Scheurich (1981) introduced a technique for protoplast fusion which is based on the action of alternating electric fields. After that many reports were described about electrofusion, reviewed by Zachrisson and Bornman (1987).

Electrofusion is superier than the chemical method for determine the start point of the fusion process and monitoring the process microscopically (Lynch et al. 1989). When protoplasts are placed in the AC field, they had dipole characteristics and form pear-chains structures. After that, when electric pulses are applied, the fusion between contacting protoplasts start. Therefore, it is possible to examine the morphological change of the fused bodies (Zachrisson and Bornman 1986).

The fusion between lettuce protoplasts could be easily induced by electric pulses. The speed is not high. It usually takes 0.5 to one hr until two fused protoplasts form a round-shape. However, rice protoplasts fused rapidly with rice protoplasts after pulse application.

When rice protoplasts were fused with lettuce protoplasts (Fig. 2a-2d), the remarkable change in the fused products was the movement of chloroplasts from lettuce protoplasts. As soon as electric pulses were given to the mixture of both the protoplasts (Fig. 2a), the chloroplasts in the lettuce cells started to move into rice protoplasts. The speed of the fusion of lettuce and rice protoplasts was not so rapid. Therefore, the speed of the fusion may be controlled by the lettuce protoplasts. During moving of chloroplasts, some materials in rice protoplasts may also move into lettuce protoplasts. However, this movement can not be observed by light microscopy.

Rice protoplasts stained with NR were fused with those stained with BCB (Fig. 3a-3d). Both NR and BCB stained areas finally fused each other. It is known that both the dyes stain vacuoles (Gahan 1984). Therefore, this result may indicate that vacuoles in the fused cells can fuse each other. The fusion speed of NR and BCB stained protoplasts was not so high. This phenomenon might be due to the effect of staining on the activity of the fusion bodies.

Rice protoplasts fuse rapidly each other and produce sometimes big or giant cells. We are interested in the changes of the cell structure in fusion bodies. When the rice protoplasts stained with R-123 or FITC fused with nonstained protoplasts, some fluorescence spots were observed in the fusion bodies (Figs. 4 and 5). The size of the spots were almost the same as that of non-fused cells. Therefore, we conclude that fusion bodies may conserved the original structure of source protoplasts for a rather long time and finally mix each other.

The studies on the ultrastructures of fusion bodies of some plant protoplasts have been reported by Burgess and Fleming 1974, Davey et al. 1980, and Rennnie et al. 1980, and also discussed by Binding et al. 1986. However, the morphological changes of the earlier step of fusion has not yet been elucidated in detail. We also examined the ultrastructure of fusion bodies. About one hr after electric pulses application to the mixture of lettuce and rice protoplasts, two structurally different areas were observed in the fusion bodies (Figs. 8 and 9). One area contained chloroplasts and the other did not. The feature of nuclei and other organelles were different from each other. This result indicates that it takes a long time until cytoplasm from both the fused cells mix thoroughly. This result consists with that obtained with R-123 or FITC-stained cells.

The early process of fusion of lettuce was also examined by electron microscopy. Small areas of surfaces of two contacting protoplasts were already fused each other (Figs. 10 and 11). If this step was observed by a light microscope, we may judge only adsorption between two protoplasts. Therefore, at the stage of pearl-chain form in the AC field, some of protoplasts might already fused each other.

Finally it is concluded that the fusion reaction occurs already at the earlier step, namely the aggregation of protoplasts and that it takes a long time until cell constituents from source protoplasts mix thoroughly.

5. Acknowledgments

This research was supported in part by a grant from the Ministry of Education, Science and Culture of Japan.

6. References

Binding, H., Krumbiegel-Schroeren, G., and Nehls, R. (1986) 'Protoplast fusion and early development of fusants', in J. Reinert and H. Binding (eds), Results and Problems in Cell Differentiation 12, Differentiation of Protoplasts and Transformed Plant Cell, Springer-Verlag Berling Heidelberg, pp. 37-66.

Burgess, J., and Fleming, E.N. (1974) 'Ultrastructural studies of the aggregation and fusion of plant protoplasts', Planta 118, 183-193.

Chen, L. B. (1989) 'Fluorescent labeling of mitochondria' in Y.-L. Wang, and D. Taylor (eds) Method in Cell Biology vol. 29, Fluorescence Microscopy of Living Cells in Culture, Part A, Fluorescent Analogs, Labeling Cells, and Basic Microscopy, Academic Press, San Diego, pp. 103-123.

Cocking , E. C. (1960) 'A method for the isolation of plant protoplasts and vacuoles', Nature 187, 962-963.

Davey, M. R., Pearce, N., and Cocking, E. C. (1980) 'Fusion of legume root nodule protoplasts with non-legume protoplasts: Ultrastructural evidence for the functional activity of *Rhizobium* bacteroids in a heterokaryotic cytoplasm', Z. Pflanzenphysiol. 99, 435-447.

Draget, K. I., Myhre, S., Evjen, K., and Østgaard, K. (1988) 'Plant protoplast immobilized in calcium alginate; a simple method of preparing fragile cell for transmission electron microscopy', Stain Tech. 63, 159-164.

Evans, D. A., Bravo, J. E., and Gleba, Y. Y. (1983) 'Somatic hybridization: Fusion methods, recovery of hybrids, and genetic analysis', Intern. Rev. Cytol. Suppl. 16, 143-159.

Fowke, L. C. (1980) 'Applications of protoplasts to the study of plant cells', Intern. Rev. Cytol. 68, 9-51.

Gahan, P. B. (1984) 'Plant Histochemistry and Cytochemistry', Academic Press, London.

Kanai, R., and Edwards, G. E. (1973) Purification of enzymatically isolated mesophyll protoplasts from C3, C4 and crassulacean acid metabolism plnts using a aqueous dextran poly-ethylene glycol two-phase system. Plant Physiol. 52, 484-490.

Küster, E. (1909) 'Über die Verschmelzung nackter Protoplasten', Ber. Dtsch. Bot. Ges. 27, 589-598.

Lynch, P. T., Isaac S., and Collin, H. A. (1989) 'Electrofusion of protoplasts from celery (*Apium graveolens* L.) with protoplasts from the filamentous fungus *Aspergillus nidulans*', Planta 178, 207-214.

Maeda, E, and Maeda, K. (1987) 'Ultrastructural studies of leaf hydathodes. I. Wheat (*Triticum aestivum*) leaf tips', Japan. Jour. Crop Sci. 56, 641-651.

Nagata, T., and Takebe, I. (1971) 'Plating of isolated tobacco mesophyll protoplasts on agar medium', Planta 99, 12-20.

Otsuki, Y., and Takebe, I. (1969) 'Isolation of intact mesophyll cells and their protoplasts from higher plants', Plant Cell Physiol. 10, 917-921.

Rennie, P. J., Weber, G., Constabel,, F., and Fowke, L. C. (1980) 'Dedifferentiation of chloroplasts in interspecific and homospecific protoplast fusion products', Protoplasma 103, 253-262.

Schieder, O., and Vasil, I. K. (1980) 'Protoplast fusion and somatic hybridization', Intern. Rev. Cytol. Suppl. 11B, 21-46.

Senda, M., Takada, J., Abe, S., and Nakamura, T. (1979) 'Induction of cell fusion of plant protoplasts by electrical stimulation', Plant Cell Physiol. 20, 1441-1443.

Shahin, E. A., and Shepard, J. F. (1980) 'Cassava mesophyll

protoplsts; isolation, prolifiration, and shoot formation', Plant Sci. Lett. 17, 459-465.
Sumardi, I., Taniguchi, T., and Maeda, E. (1987) 'Effect of enzyme solutions on the isolation of protoplasts from rice, lettuce, sugarcane and coffee leaves', Rep. Tokai Br. Crop Sci. Soc. Japan 104, 7-13.
Takebe, I., Labib, G., and Melchers, G. (1971) 'Regeneration whole plants from isolated mesophyll protoplasts of tobacco', Naturwissenschften 58, 318-320.
Takebe, I., Otsuki, Y., and Aoki, S. (1968) 'Isolation of tobacco mesophyll cells in intact and active state', Plant Cell Physiol. 9, 115-124.
Taniguchi, T., and Maeda, E. (1988) 'Fluorescence microscopic observations of chloroplasts in plant protoplasts', Rep. Tokai Br. Crop Sci. Soc. Japan 106, 15-16.
Taniguchi, T., Sanada, Y., and Maeda, E. (1987) 'Effect of coconut water and coconut cream on organ development in calli of rice and lettuce', Rep. Tokai Br. Crop Sci. Soc. Japan 103, 25-29.
Taniguchi, T., Sato, T., and Maeda, E. (1989) 'Observation of plant plant protoplasts by various light-microscopes', Rep. Tokai Br. Crop Soc. Japan 107, 33-38.
Thomas, M. R., and Rose, R. J. (1988) 'Enrichment for *Nicotiana* heterokaryons after protoplast fusion and subsequent growth in agarose microdrops', Planta 175, 396-402.
Togawa, Y., Toda, K., Takayama, S., Miura, Y., Mochisuki, T., and Iwasaki, I. (1987) 'Cell fusion by electric fields', Shmadzu Rev. 44, 17-28.
Wu, F. S. (1987) 'Localization of mitochondria in plant cells by vital staining with rhodamine 123', Planta 171, 346-357.
Zachrisson, A., and Bornman, C. H. (1986) 'Electromanipulation of plant protoplasts', Physiol. Plant. 67, 507-516.
Zimmermann, U., and Scheurich, P. (1981) 'High frequency fusion of plant protoplasts by electrical fields', Planta 151, 26-32.

GENETIC TRANSFORMATION AND PLANT IMPROVEMENT

R.S. SANGWAN and B.S. SANGWAN-NORREEL
UNIVERSITE DE PICARDIE
Faculté des Sciences, Androgenèse et Biotechnologie
33, rue Saint-Leu
80039 AMIENS Cédex France

1 - INTRODUCTION

During the past three decades "green revolution" has resulted in considerable increases in crop productivity in the developing countries. The increased crop yields have been achieved through a combination of genetic improvements of cultivars and advances in agricultural technology and management. Although, genetic improvement of crops began with the domestication of plants, it was not till 1886, when Mendelian laws of inheritance provided a scientific basis of crop improvement. The procedures of sexual gene transfer and recombination, a prerequisite for improving cultivars were developed several years after the rediscovery of Mendelian laws of heredity.

Most of our modern cultivated varieties have originated from gene transfer within and between species. Nearly all the well known cultivars of wheat, maize, rice, potato, oat, sugarbeet, tomato etc. have been obtained by transfer of gene and traits through sexual hybridations. Such gene transfer by sexual methods in a broad sense can be termed "Genetic engineering", although in a strict sense this term is restricted to genetic manipulation through recombinant DNA technology. It is now evident that further crop improvements would result through new genetic approaches, e.g., gene transfer by non-sexual methods which require an increasing degree of technological sophistication. Indeed, the past decade has seen much progress in the new technology, e.g., the development and refinement of methodology for plant tissue culture and recombinant DNA (Plant biotechnology). Plant genetic engineering gained a momentum when it was shown that T-DNA could be introduced into the plant genome stabily, and normal fertile plants were obtained. This is because one intends to create whole plants with new characteristics by genetic manipulation. A schematical representation of the advances in plant genetic engineering during this century is given in Figure 1. This figure will probably help to understand developments in gene transfer and

R.S. Sangwan and B.S. Sangwan-Norreel (eds.), The Impact of Biotechnology in Agriculture, 299–337.

Figure 1 : The evolution of gene transfer technology in crop plants in the 20 th century

Before 1960	1960-1980	After 1980
Gene transfer by sexual methods	**Gene Transfer by non sexual methods**	
Phase I : Since 1900	Phase II : Since 1960	Phase III : Since 1980
Production of hybrids by sexual crosses Early, and Mid 20th century 1900-Mendel's laws of inheritance rediscovered 1900- De Vries studies mutation in plants Begining of plant breeding 1919- Intraspecific and Interspecific hybrids 1929- intergeneric gene transfer 1940-70- development of superior commercial hybrids in Maize, Wheat, Barley, Rice, Sugarbeet, Potato, Brassica, Tomato, Cotton etc...	Somatic hybrids by cell fusion - In vitro techniques - Totipotency of single cell demonstrated - Plant culture media, growth hormones -Protoplast, organ, embryo, meristem, anther, ovary cultures - Plant regeneration via direct organogenesis and somatic embryogenesis - Haploidy -Protoplast fusion - In vitro multiplication	Prokaryote-Eukaryote hybrids or trans Kingdom hybrids: Obtained by Bacteria plant conjugal mating Use of recombinant DNA technology Transfer of plasmid DNA to plants via Agrobacterium or via direct DNA transfer : - Electroporation - Microinjection - Particle bombardment system -Gene inactivation through anti-sense DNA and m RNA insertion

discoveries in plant biotechnology. However, the long history of gene transfer by plant breeders at interspecific and intraspecific levels is not reviewed here, since extensive reviews are already available (Allard 1969, Simmonds 1979). It should, also, be pointed out that conjugal mating and cross fertilization between plant species is prevented by natural barriers such as cross-incompatibility, although conjugation is common among prokaryotes (Zambryski, 1988; Stachel and Zambryski, 1989). Thus, genetic manipulation in higher plants is restricted by natural incompatibility mechanisms in pollination and fertilization. Although, some of these limitations such as incompatibility barriers can be over come by *in vitro* fertilization (for example in *Brassica,* interspecific hybrids can be obtained by putting the pollen directly into the ovules), only closely related species can be crossed in this way.

During the sixties and seventees, *in vitro* technique of protoplast fusion was successfully used for combining genetic material from unrelated species. Other plant tissue culture techniques such as mass multiplication, haploid production, somatic embryogenesis, artificial seed production, embryo rescue, somaclonal variation etc. have been reviewed in detail in this volume. There has been a renewed interest in conjunction with plant tissue culture recombinant DNA technology for gene transfer. This later technique makes possible genetic manipulations that were outside the repertoire of conventional breeding and cell fusion techniques. At present, genes can be taken from completely unrelated organisms : plants, animals, bacteria, viruses, and can be introduced into a desired crop plant. Thus, now it is possible to transfer specific and well-characterized traits across the broadest evolutionary boundaries. However, the application of recombinant DNA technology for crop improvement is dependent on the availability of efficient systems for the transfer of foreign genetic material into cells capable of generating fertile plants. In recent years, a range of techniques have been developed to transform plant cell with specific pieces of DNA. As discussed below, these techniques are based on two variations : of *Agrobacterium* mediated gene transfer, and direct DNA transfer.

Numerous reviews on plant genetic engineering particularly on recombinant DNA technology have appeared in recent years (Goodman et al. 1987; Schell, 1987; Klee et al. 1987; Cocking and Davey, 1987; Tempé and Schell 1987; Weising et al. 1988; Zambryski, 1988; Gasser and Fraley, 1989). In this article, we examine only the state of the art of gene transfer particularly the techniques used for gene transfer in higher plants. We donot review *Agrobacterium* biology except where they touch on gene transfer function. The biochemistry of isolation, cloning and directed mutagenesis also falls outside the scope of this paper, which have been extensively reviewed elsewhere (Fraley et al. 1986; Schell 1987; Kuhlemeir et al. 1987; Zambryski, 1988).

2 - AGROBACTERIUM-MEDIATED GENE TRANSFER

Agrobacterium tumefaciens and *A. rhizogenes*, are gram-negative soil borne, plant pathogens and have a wide range of hosts among mainly broad leaved dicots but also a few monocots (De Cleene and De lay, 1976; Bytebier et al.1987; Weising et al. 1988). These bacteria are natural genetic engineers of plants. During the course of infection *A. tumefaciens* transfers a portion of its extrachromosal tumour inducing (Ti) plasmid into the cell of the host plant. The plamid DNA is stably integrated into the host genome and causes transformation of the cells for hormone independence, resulting in the formation of a tumerous growth called crown gall, which provides the *bacterium* with specialized nutrients. A similar infection by *A. rhizogenes* generally induces adventitious roots formation from the wounded tissue (Chilton et al. 1982; Tepfer, 1984). Accordingly, the tumour inducing plasmid of *A. tumefaciens* is designated pTi, and root inducing plasmid of *A. rhizogenes* as pRi and the transferred DNA as T-DNA. Apparently, *A. tumefaciens* and *A. rhizogenes* transform plant cells in a similar manner for T-DNA transfer, and production of opines in the host tissues (Tempé et al. 1977). It is now possible to insert DNA sequences of interest first into T-DNA and then into the plant genome, though the integration site of T-DNA appears to be random. The ability to cause crown gall disease can be removed by deletion of genes in the T-DNA without loss of functions for DNA transfer and its integration. Such strains of agrobacteria are called "disarmed strains". Such disarmed bacterial strains are now been routinely used as gene transfer vectors for plant genetic engineering in many laboratories, and transgenic plants has been obtained in more than 30 species (Table 1). The list of transgenic plants is increasing day by day. The most outstanding research on *Agrobacterium* mediated gene transfer in higher plants has been carried out in laboratories of Schell/Van Montagu (Gent/Koln, Belgium, Germany) Nester (Seattle, Washington, USA) Chilton (CIBA-GEIGY, North Carolina, USA) Fraley/Horsch (Monsanto, St Louis, USA) Tempé (Orsay, Gif, France). Recently, binary vectors and non-oncogenic *A. tumefaciens* strains have been developed to eliminate the transfer of oncogenes into plants for obtaining transgenic plants with normal phenotypes (Hockema et al. 1983; De Framond et al. 1983; Fraley et al 1986; Schell 1987;

Table 1. Species in which genetic transformation has been performed and transgenic plants have been obtained.

Species	Techniques of DNA transfer	Explant used	Mendelian inheritance	Reference
Dicotyledons				
Nicotiana tabacum	Ti,	P, LD	+	27, 63
	DDT,	P	+	52, 96, 99
	Li,	P	+	15, 37
	Ep,	P	+	103
	Mi	P	+, -	23
N. plumbaginifolia	Ti,	P, LD	+	62
	DDT	P	-	98
Arabidopsis thaliana	Ti	P, LD, C, R, ZE	+	5, 80, 111, 127
Lycopersicum esculentum	Ti,	LD	+	20, 63
	Ri,	S	-	114, 119
	DDT	P	-	70
Petunia hybrida	Ti,	LD	+	63
	DDT	P	-	25
Solanum tuberosum	Ti,	P, LD	+	4, 30
	Ri		-	93
Brassica napus	Ti	S	+	48

Medicago sativa	Ri,	S	-	114
	DDT(Mi)	P	-	101
Medicago varia	Ri	P, LD		26
Trifolium repens	Ti	P, LD	-	133
Glycine max	Ti,	C	+	6, 39
	Ri,	S	-	114
	Ep,	P	-	19, 53
	Pb	P, C	-	87
Vigna aconitifolia	Ep	P	-	75
Lotus corniculatus	Ri	S	-	67
Daucus carota	Ep,	P	-	46, 95
	Ti	P, C, S	-	107
Linum usitatissimum	Ti	LD, S	-	7
Gossypium hirsutum	Ti	S	-	125
Kalanchoe	Ti	LD	-	110
Helianthus annuus	Ti	LD		89
Populus	Ti	S, LD	-	43
Sugarbeet	Ti	-	-	unpublished

Apple	Ti	LD	-	64
Celery	Ti	LD	-	16
Datura	Ti	LD, E	+	unpublished (Sangwan)
Monocotyledon				
Oryza sativa	DDT,	P	-	71, 124
	Ep	P	-	112
Triticum	DDT,	P	-	81
	Ep	P	-	94
Secale cereale	DDT,	P	-	71
	Fti	Ft	-	34
Zea mays	DDT,	P	-	71
	Pb	P	-	109
	Ep	P	-	57
Asparagus	Ti	P	-	18, 70
Lolium	DDT	P	-	92
Sorghum	EP	P	-	94
Hordeum vulgare	DDT	P	-	71

Saccharum officinarum	DDT	P	-	17
Pennisetum	EP	P	-	53
Panicum maximan	EP	P	-	53

Abbreviations :

P - Protoplast, LD - Leaf disc, S - Stem or other plant organs, ZE - Zygotic embryo, E - Embryo, R - Root, C - Cell/Callus, Ft - Floraltiller, DDT - direct DNA transfer, Li - liposomes, Fti - floraltiller injection

Ep - electroporation, Pb - particle bombardment, Mi - microinjection, Ti - Tumour inducing plasmid : *A. tumefaciens* mediated transformation, Ri - Root inducing plasmid : *A. rhizogenese* mediated transformation, + : presence of Mendelian inheritance, - : not determined.

Weising et al. 1988). To-date, *Agrobacterium*-mediated transformation is one of the best methods available for DNA transfer into the host tissue. A schematic representation of the introduction of desired foreign DNA via the Ti plasmid of *Agrobacterium* to the plant cell is given in Figure 2. However, this method depends on the susceptibility of the target plant to *Agrobacterium* and availability of regeneration procedures in the host plant.

The mechanism involved in T-DNA transfer from the bacterium to the host plant cell is unknown. However, genetic complementation studies and molecular analyses, have permitted to find the essential elements of this transfer process. These are : I) the borders sequences, i.e. the 25 base pairs imperfect direct repeats, flanking the T-DNA (Wang et al. 1984) and II) vir region (Zambryski et al. 1980) the virulence genes of the Ti plasmid which encode the functional products that mediate T-DNA transfer (Stachel and Nester, 1986). It was also observed that the "oncogenes" (genes responsible for hormone independent tumerous growth) are not required for T-DNA transfer or integration. Thus, the T-DNA element is delimited by border sequences, and its transfer is mediated by the vir functions. It is interesting to note that although the vir-region is required for virulence, yet it has never been detected in the crown galls. Moreover, the vir-region is expressed in the bacteria for their virulence, but is weakly or not expressed in bacteria in the absence of plant cells. A compound excreted by plant roots, even in the absence of the Agrobacteria, has been found to activate the expression of genes in the vir-region. The activation of vir results in the formation of a unipolar single stranded T-DNA complexed with several proteins (Zambryski 1988; Stachel and Zambryski 1989). The structure and mode of synthesis of this intermediate T-DNA, together with the requirement for direct bacterial plant cell contact, suggests that the T-DNA is transferred to plant cells by a mechanism analogous to bacterial conjugation. (Lichtenstein 1987; Stachel and Zambryski, 1989).

Gene transfer by means of Agrobacterium was further extended by advances in plant regeneration from protoplast, leaf, stem, root and recently zygotic embryo culture systems, in which selectable markers are used to identify and favor the growth of transformed cells. Chimeric constructs, consisting of a plant promoter sequence and a bacterial antibiotic resistance gene, can be used as efficient marker genes to select plant cells that have acquired an integrated T-DNA. The most commonly used selectable marker has been the neomycin phosphotransferase type II (NPTII) enzyme, which was originally isolated from the prokaryotic transposon Tn5 (Bevan et al. 1983; Herrera-Estrella et al. 1983; De Block et al. 1984). This enzyme detoxifies aminoglycoside

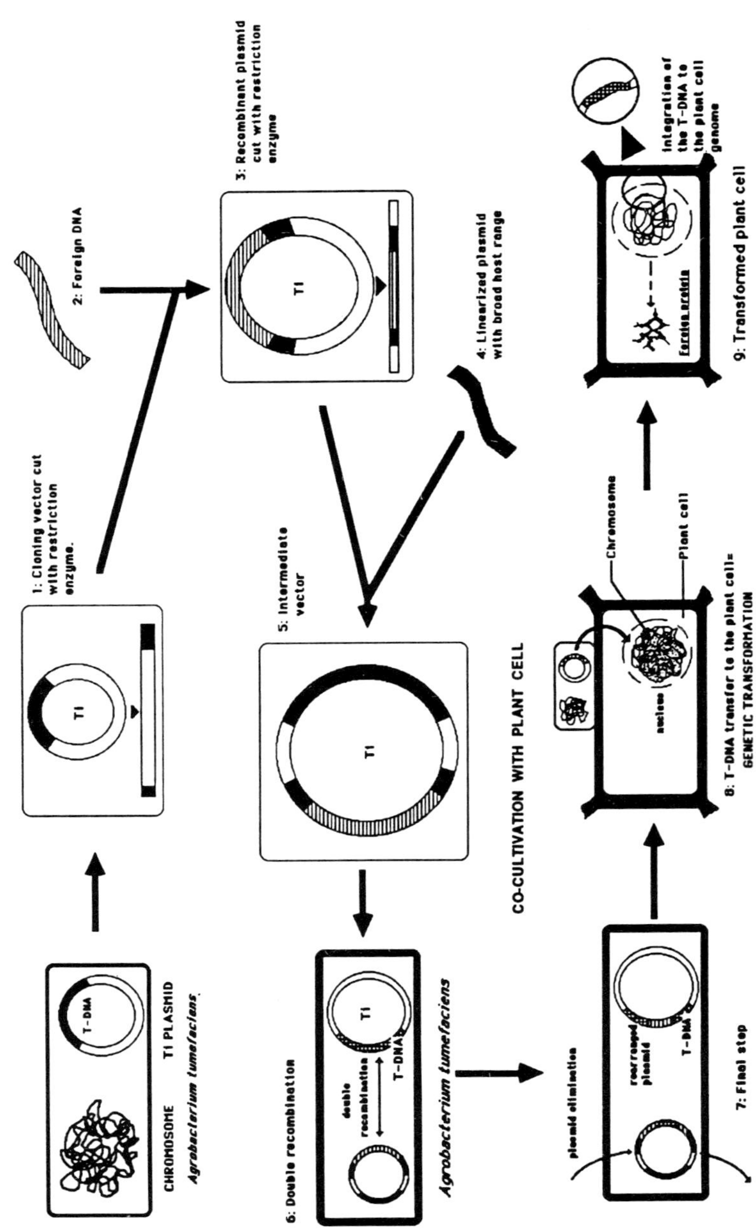

Fig.2 A SCHEMATIC REPRESENTATION OF THE INTRODUCTION OF DESIRED FOREIGN DNA IN THE TI PLASMID OF Agrobacterium tumefaciens AS USED FOR GENETIC ENGINEERING OF PLANT CELL

compounds such as kanamycin and G418 by phosphorylation. Other selectable markers used are hygromycin-B resistance, dehydrofolate reductase which confers methotrexate resistance. Several such selectable markers are listed in Table 2.

These genes fused to constitutive plant transcriptional promoters have been successfully used to transform a large number of plant species and have been incorporated into numerous plant transformation vectors. The introduction of foreign genes in plants via *Agrobacterium* has been mainly observed into the plant chromosomes and rarely in the organelles such as chloroplasts or mitochondria. However, chloramphenicol acetyltransferase, which confers resistance to chloramphenicol, has been inserted into the chloroplast genome (De Block et al. 1985).

Thus, new marker genes can be inserted into genetic constructs which produce multiple transformations of selected lines, and optimize the selection process in the different species. Among these marker genes, B-glucuronidase (GUS gene) from *E. coli* (Jefferson et al. 1987) and luciferase from fireflies (Ow et al. 1986) are the most attractive. Recently, B-glucuronidase has been used for identifying transformants in tissue explants, and for analyzing gene expression and heritability of foreign DNA inserts. Using the GUS gene as a reporter gene, we have studied the early transformation events particularly the mating of the bacteria to the plant cells and then the T-DNA transfer system through the plant cell wall. It was found that in *Arabidopsis* the actively dividing and hormone treated cells are more susceptible for bacterial conjugation and transformation than the non-dividing resting cells (Sangwan, unpublished).

The use of regenerative explants such as leaf disc, stem, root or cotyledons instead of protoplasts, enables researchers to best utilize *Agrobacteria* for transforming intact plant tissues. Limited but encouraging success with engineered plants have been obtained in crops such as potato, tomato, sunflower, sugarbeet, tobacco, rice etc. (Table 1). In addition, the Ti-plasmid has enhanced the genetic variability in some crops (e.g., potato, Ooms et al. 1987) by inducing mutations in gene(s) for desirable agronomic traits. So far, the goal of gene transfer technology has been to engineer bacterial or plant genes which encode for tolerance to broad-spectrum but environmentally safe herbicides, and enhanced resistance to insect pests, viruses, and fungi. Among these, the in-built genetic resistance to insect pests, in view of the production losses world-wide and heavy costs of pesticides, is the most important for economic and ecological reasons.

Table 2. Selectable and scoreable marker genes used in recombinant DNA technology.

Marker gene	Origin	Resistance against	Reference
<u>**Selectable**</u>			
1. Neomycin phosphotransferase (NPT) Type I	Tn 601	Neomycin Kanamycin	44
Type II	Tn 5	"	9,27,56
2. Hygromycin phosphotransferase (hpt)	*E. coli*	Hygromycin B	80, 128
3. Bacterial dihydrofolate reductase (DH FR)	Plasmid R67	Methotrexate	12,27,56
4. Mammalian dihydrofolate reductase	Mouse	Methotrexate	38
5. Streptomycin phosphotransferase	Tn5	Streptomycin	69
6. EPSP synthase (aro A gene)	*Salmonella typhimurium*	Glyphosate	22, 42
7. Phosphinothricin acetyl transferase (bar gene)	*Streptomyces hygroscopicus*	Phosphinothricin bialaphos	29, 123
8. Bleomycin resistance gene	Tn5	bleomycin	59

Scoreable marker	Origin	Resistance against	Reference
Nopaline synthase (nos)	T-DNA	-	68, 135
Octopine synthase (ocs)	T-DNA	Toxic opine precurssor analogues i.e aminoethycystein	24, 33, 61
Beta galactosidase	*E. coli*	-	54
Beta glucuronidase (GUS)	*E. coli*	-	65, 66
Firefly luciferase	*Photinus pyralis*	-	95
Bacterial luciferase	*Vibrio harveyi*	-	76
Chloramphenicol acetyltransferase (Cat)	Tn9	chloramphenicol	27

Abbreviations :

Tn - Transposon, EPSP - 5,enolpruvylshikimate-3-phosphate synthetase

Recently transgenic plants with resistance to herbicides, Basta® and Round- up®, have been reported (Comai et al. 1985; Shah et al. 1986; De Block et al. 1987). It has been reported that the gene encoding cowpea trypsin inhibitor, which gives some degree of field resistance to insects, when transferred to tobacco conferred enhanced resistance (Hilder et al. 1987). Another example of insect resistance has been reported by Plant Genetic Systems, Gent (Belgium) , who have introduced endotoxin genes of *Bacillus thuringiensis* into tobacco (Vaeck et al. 1987). The bacterial toxin present in the transgenic plants kills the insect which feed on its leaves. Moreover, the toxicity of the protein is very specific. It is not toxic to mammals, plants and other insects. Interesting results about resistance to virual infection have also been reported; however, the mechanism of resistance is not well understood (Sequeira 1984 ; Powell Abel et al. 1986).

Thus, the experiences summarized here clearly indicate that the stable introduction of foreign genes via Ti plasmid into plant cells is now possible without affecting the morphogenic potential of the cells. This is an important prerequisite for the successful genetic engineering of plants.

3 - METHODS OF TRANSFORMATIONS

3.1 *Agrobacterium* based transformation systems : If the *Agrobacterium* mediated gene transfer technology is to have an impact on agricultural productivity, then it is necessary to develop efficient methods of transformation and regeneration. The commonly used systems and their variants are :

a. Protoplast system

b. Leaf disc system

a. Protoplast system

The first and early transformation experiments used only co-cultivation of *A. tumefasiens* with tobacco protoplasts (Marton et al., 1979). Later on this technique was improved and transgenic plants were obtained (De Block et al., 1984, Depicker et al., 1985). It is based on the isolation of leaf/callus cell protoplasts and their subsequent infection with a specific *Agrobacterium* strain. The technique of isolation, culture and subsequent regeneration of plants from the protoplast has been well described (Abdullah et al. 1986; Cocking and Davey, 1987). The main point to note is that the co-cultivation with *Agrobacterium* should be carried out only after the regeneration of the protoplast cell walls. Since only a limited number of plants can be regenerated from the protoplasts, this system is not suitable for obtaining transgenic plants in the 'recalcitrant' species. In addition to being technically laborious, it is time consuming and requires, even in the most responsive species such as *Nicotiana*, approximately three months to obtain small transgenic shoots.

b. Leaf-disc system

Figure 3 summerize the protocol for leaf-disc transformation procedure. Technically this is far more easier then the protoplast protocol and plant regeneration is rapid and of short duration. For example, it takes only 5 to 7 weeks in *Arabidopsis* to get complete plants. Although used for model plants e.g *Arabidopsis, Nicotiana, Datura* etc, this technique, described below, is the modified version of Horsch et al., 1984, 1985, De Block et al., 1985, Van Lijsebettens et al., 1989 and has also been successfully used for other species (Table 1). This leaf-disc transformation system permits efficient gene transfer, selection and regeneration in a simple procedure.

Important steps in *Arabidopsis* are :

1) Collect leaves from the axenic plants grown in low light and long day conditions

2) Isolate leaf-discs (1cm in *Nicotiana, Datura*), in *Arabidopsis* ; cut leaves into halves. Avoid excessive wounding during isolation of leaf-discs.

3) Put 12-15 leaf discs in a petri dish (9cm) containg organogenic medium (for *Arabidopsis* MS medium + supplemented with 1 mg/l BA+0.5mg/l NAA) and pre-culture for 4 days. No pre-culture is required for solanaceous plants.

4) Grow Agrobacteria strain (carrying vector of choice) over-night in 5 to 10 ml of LB medium without antibiotic at 28°C, on a shaker at 200 rpm.

5) Infect leaf disc by adding 25-50µl of the bacterial suspension in 20ml of MS medium for 2-5 min. by shaking.

6) Blot dry leaf discs on sterile filter paper, and co-culture for 2 to 4 days on the organogenic medium (e.g., for *Arabidopsis* MS+1mg/l BA+ 0.5mg/l NAA)

7) Wash leaf discs after co-culture with medium containing an antibiotic (e.g., MS (liquid)+750mg/l cefotaxim or vancomycin) to remove and kill the excess of bacteria.

8) Transfer blot-dried leaf-discs to a selective medium (e.g., for *Arabidopsis* MS+BA/NAA+Kanamycin 50mg/l or hygromycin 25mg/l + vancomycin/cefotaxim 500mg/l).

Fig.3 **AN EXPERIMENTAL PROTOCOL FOR GENETIC TRANSFORMATION IN HIGHER PLANTS**

"LEAF DISC TRANSFORMATION PROTOCOL"

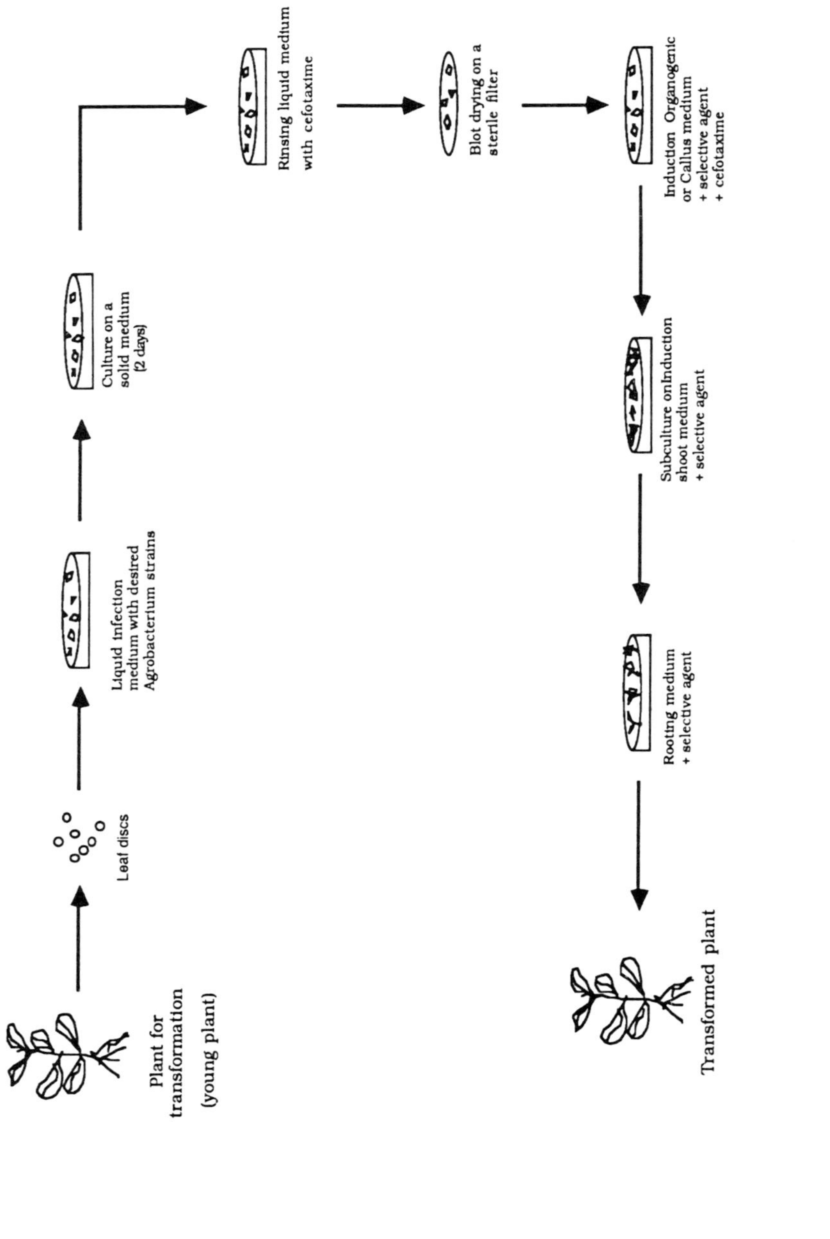

9) Subculture leaf-discs on fresh medium every 2 to 3 weeks. Green organogenic calli or buds appears after 2 to 4 weeks.

10) Isolate organogenic calli from the dead leaf-discs. Remove shoots from the calli.

11) Transfer shoots to the rooting medium (e.g., for *Arabidopsis* MS medium without growth regulators with or without selection pressure).

12) After rooting, transfer the "putative" transformed plants to a greenhouse.

13) In order to confirm the transgenic nature of the plants, following analyses are usually performed :

Recallusing assay

Culture, leaf-discs taken from the leaves of putative transformants and control plants on a callus inducing (selective) medium, e.g., 50mg/l of kanamycin in case of *Arabidopsis*. Leaf-discs of transformants give calli on kanamycin medium while no callus are observed from those of control plants.

NPTII assay

Neomycin phosphotransferase II activity can be determined by the dot blot precedure described by McDonnel et al., 1987.

Genomic hybridization analysis

Isolation of plant DNA and Southern blotting

Transformed plants are examined for the presence of T-DNA by Southern blot analysis. This DNA-DNA blotting method is called "Southern blotting" after the name of its inventor. For example, DNA is obtained by Dellaporta et al., (1983) technique, cut into pieces with specific restriction enzymes and subsequently hybridized (Southern 1975, Maniatis et al., 1982, Feldmann and Marks 1987; De Block et al., 1987, 1989).

Northern and western Blotting

Other techniques for assaying foreign gene expression in transgenic plants are also described in the literature. For example, tests for the expression of the protein and RNA, western and northern blotting should be performed. The details of the techniques have been well described elsewhere (Southern, 1975; Maniatis et al. 1982).

It seems that to date the leaf-disc system of transformation is perhaps the best available technique for transformation of plants which can regenerate from the leaf explants. Since many economically important crops cannot be regenerated from leaf explants, but only from cotyledons (e.g., soybean) or from the immature/mature zygotic embryos, this protocol has a limited value. However with slight modification, i.e., hormonal treatment before Agrobacterial infection, this technique has been successfully applied to transform other explants, particularly the zygotic embryos. Recently we have developed a technique for embryo transformation (Sangwan unpublished) in the model plant *Arabidopsis*, and also in the leguminous crop species such as pea, and soybean. It was observed that hormonal treatment or activation of cell division during the pre-culture period for 4 to 7 days was necessary before co-culture with *Agrobacterium* (Sangwan, unpublished).

Another promising approach that does not require tissue culture techniques is *Agrobacterium* mediated seed transformation (Feldman and Marks, 1987). However, the frequency of transformation was low, and contradictory results were reported. Moreover, only chimeric plants were obtained by this method.

3.2 Direct DNA transfer systems

The observation of Lurquin and Kado (1977) that DNA can pass plasma membrane of the plant cells/protoplasts which are totipotent, provides an efficient system for gene transfer in higher plants. Thus, transformation of plant protoplasts with naked DNA either by direct uptake or microinjection avoids the complex technology of plasmid engineering required in *Agrobacteria* mediated transformation. Since DNA uptake is not host specific, it can potentially be applied to a wide variety of plant species. The prospects to produce transgenic plants by direct DNA uptake, though low in frequency, are encouraging in the crops. Treatments such as addition of polyethylene glycol (PEG), and application of intense electric fields (electroporation) which increase the permeability of membranes by forming large pores, can considerably increase transformation frequency. Under the most favorable conditions of electroporation, DNA was introduced in about 25% of the protoplasts. Most of the DNA was eventually lost; however, if the cells were allowed to proliferate under selective conditions, transformed calli could be obtained at a frequency of approximately 10^{-4} - 10^{-3} (Fromm et al., 1986). It is also possible to increase the transformation frequency by refining.

a. Electroporation

Electroporation apparatus in design and circuitry. Waldron et al., (1985) by modifing the protocol developed by Krens et al., (1982) obtained an average transformation frequency of about 1%. Recently, Shimamoto et al (1989) reported the production of fertile transgenic rice plants. They electroporated the bacterial hph gene, encoding hygromycin B resistance, into protoplasts of *Oryza sativa*. However, the transformation frequencies were in the range of 0.1 to 0.6%, and were low as compared to those from electroporation of maïze protoplasts (Fromm et al., 1986, Rhodes et al., 1988) and those from non-embryogenic rice protoplasts transformed with PEG (Uchimiya, et al., 1986).

b. Particle bombardment system

Although many methods of gene transfer in plant tissues are now available, the gene transfer by bombardment with DNA-coated microprojectile is a rapid and simple method for transforming intact regenerable cells and tissues (Sanford et al., 1987, Klein et al., 1988). This technique has been successfully used for DNA delivery to epidermal tissue of *Allium cepa* (Klein et al., 1987), scutellar tissue of *Zea mays* (Klein et al., 1988), leaf tissues and cell suspension cultures of several species (Klein et al., 1987). The plant materials and the techniques of particle gun device (bombardment) have been described in detail by Klein et al., (1987, 88). Thus, this transformation technique is simple and promising in plants which are difficult to regenerate e.g., cereals, pea, soybean (Mc Cabe et al. 1988) etc., and also those are not susceptible to *Agrobacterium* mediated transformation.

c. Micro-injection system

Micromanipulation techniques have been frequently used and were initially developed for microinjection in animal cells. In plants, foreign DNA has been successfully injected into protoplasts (Crossway et al., 1986), and more recently in the organized structures such as pollen embryos (Neuhaus et al. 1987). To be effective, this method should deliver DNA directly in the nucleus without damaging the host cell (Aly and Owens, 1987). In a study (Reich et al., 1986) with alfalfa protoplasts, transformation frequencies of upto 26% were obtained. In addition to the protoplasts, embryogenic and meristematic tissues which are highly regenerative are also seen as potent recipients for micro-injection of foreign genes. This technique is not limited by any host range. Moreover, using micro-injection, transfer of organelles, nuclei and individual chromosomes is possible, and thus could extend genetic engineering beyond the transfer of small number of nuclear genes. In certain cases, this technique would reduce the time required for the transfer of herbicide resistance, or disease/stress resistance etc.

3.3 Viruses as vector system

Although viral based vectors are well developed in animals, there are only a few reports on viruses for efficient gene transfer into plants. (Gronenborn et al. 1981; Brisson et al. 1984; Brisson and Hohn, 1986) have shown that cauliflower mosaic virus can be used successfully to replicate and express a foreign gene in plants. Thus, this virus can be of some use as a plant gene vector and hence, this system can be effective in the plants (particularly in graminae family) which are not amenable for *Agrobacterium* mediated transformation. Other viruses tested for gene transfer are : maize streak virus (MSV), gemini viruses, wheat dwarf virus etc. However, the fact that gemini viruses are not seed-transmitted and also that integration of viral genes into plant chromosomes has not been detected, limits their use in transformation. *Agrobacterium* mediated virus infection and their suitability for plant genetic engineering has also been reported. However, their contribution as efficient gene transfer vectors in crop plants is not very clear at present.

3.4 Pollen mediated transformation system

Hess (1980) reported that transformation could be obtained by incubating pollen with DNA, followed by pollination and seed set. Recently using a variation of this method, transgenic plants have been obtained in rice (Luo and Wu, 1988) wheat and Maize (Picard, unpublished) thus showing the potential of pollen as a vector for transformation. However, very controversial results have been obtained and serious doubts have been made on the validity of this technique for transformation. Moreover, precise and clear cut molecular analyses were not carried out to evaluate the effectiveness of this otherwise highly attractive gene transfer system. If successful, this technique would have a major implication in crop improvement for recalcitrant species since this method does not require tissue culture.

3.5 Other systems of DNA delivery into plant cells

Following techniques although not frequently used for genetic transformation of plants, have also been described in the litterature.

a. Liposomes and bacterial spheroplasts systems

Caboche and Deshayes (1986) have reported transformation of tobacco protoplasts by encapsulated DNA in artificial lipid vesicles (liposomes). Similarly, bacterial spheroplasts were used for protoplast transformation (Cocking and Davey, 1987).

b. Another interesting approach, as described by De la Pena et al., (1987) is the direct injection of chimeric plasmid DNA into rye floral tillers. Although transformation frequency was very low (about 0.07%) this is a highly promising but unconfirmed technique since it does not involve tissue culture system and also is not subject to host range restrictions.

Recently it has been reported that isolated mature zygotic embryos of wheat were able to take up DNA by imbibition of DNA solution, and express a chimeric Npt II gene transiently (Schell, 1987). These embryos later develop into normal fertile plants. However, details of the system and efficiency of transformation and its limitations if any, have not yet been ascertained.

4 - ONTOGENY OF TRANSFORMATION IN *ARABIDOPSIS*

Transgenic *Arabidopsis* plants were readily obtained with the leaf disc transformation-regeneration system as described above. Adaptation of this system to the plants such as *Arabidopsis, Nicotiana* or Petunia required a selectable marker e.g kanamycin or hygromycin. In *Arabidopsis* both kanamycin and hygromycin can be efficiently used for the transformation experiments. For example, in our experiments, leaf pieces co-cultivated with *A. tumefaciens* strain PGS Gluc 1 (supplied by J. Leemans PGS Gent, Belgium) carrying a scoreable marker GUS gene in the plasmid began to form organogenic calli after 2-3 weeks, while the control explants died on kanamycin (50mg/l) medium. Shoots emerged from these calli within 3-4 weeks after inoculation, and transfer to rooting medium was required for efficient root development. The rooted plants were transferred to a greenhouse. Presently, about 50% of the leaf pieces give rise to calli and shoots. We also observed that the frequency of transformation was independent of the selective agent and of the promoter used. This agrees with the results reported by Van Lijsebettens et al., (1989).This is in contrast to the results of Lloyd et al., (1986) who obtained transformation only with hygromycin as a selectable marker.

Since GUS gene was present, the transgenic nature of these plants could be shown by cytological techniques (i.e. presence of blue coloration in the transformed cells), resistance to kanamycin (50mg/l) in the leaf callus assay, Npt II test and also by DNA blot hybridization. Genetic analysis and southern blot analysis showed that transgenic plants of *Arabidopsis* contained one to many copies of T-DNA. In most cases T-DNA were integrated at a single locus, however multiple insertions into 2 or more different sites on different chromosomes were also reported (De Block et al. 1984; Lloyd et al. 1986 ; Feldmann and Marks, 1987). Moreover, stable Mendelian inheritance and segregation of one or more dominant traits was oberved (Lloyd et al. 1986). Furthermore, the progeny also showed coinheritance of the selectable marker (kanamycin/hygromycin resistance). The coinheritance of an unselected marker provides formal proof that the hygromycin/kanamycin resistance was caused by *Agrobacterium* mediated transformation rather then by an endogenous mutation. In *Arabidopsis*, the genetic and molecular analyses of the transformants and their progenies have been detailed elsewhere (Horsch et al. 1984, 85; Lloyd et al. 1986; Feldmann and Marks, 1987).

5 - EARLY EVENTS OF TRANSFORMATION

The precise mechanism of plasmid transfer from the *Agrobacterium* to the plant cell is still unknown, although it is evident that crown gall tumors result from the extracellular infection at a woundsite by *A. tumefaciens* . However, in order to enter the plant cell the plasmid DNA must be transferred across both the bacterial plasma membrane and plant cell wall. The first step is therefore the attachment of the bacteria to the host cells. This attachment is a complex process with probably several intermediate steps. For examples : 1) Agrobacteria induce syntheses of cellulose fibrils or extracellular polysaccharide threads which bind the bacteria to the plant cell surface. The term exopolysaccharide (EPS) has been widely used to cover polysaccharides found externally on the surface of the microbial cell, and has the advantage of including polymers of different physical types (and also of very diverse chemical composition). This term will be used in the present review to embrace all polysaccharides. 2) The bacteria also possess enzymes capable of digesting the pectin portion of the plant cell wall. 3) Although EPS attachment to the host cell are widely recognized, the role of EPS in the infection and transfer of T-DNA is unknown. However, this attachment and pectin digestion may facilitate the intimate contact of the bacterial membrane with the plasmalemma of the plant cell. In addition, some areas of plant plasma membrane may become exposed during the formation of the wound. Thus, an intimate contact of the bacterial plasma membrane with the plant plasma membrane may occur by either of two routes. Plant cells apparently do not play an active role in the binding of *Agrobacterium* since Agrobacteria can bind to dead plant cells with little or no change in strain specificity or in the kinetics of binding (Matthysse et al. 1982; Sangwan, unpublished). Moreover, the binding does not dependent on the presence of plant growth regulators (auxins/cytokinins) in the culture media, and on the nature of plasmid in the bacteria. Thus, initial receptor for bacterial attachment may be a compound of the plant cell plasma membrane. Matthysse et al. (1982) using carrot protoplasts observed that *A. tumefaciens* can bind specifically to the plasma membrane of the plant cells. The bacteria attached both individually and in clusters to the surface of the carrot protoplasts soon (1-2 hours) after the addition to the protoplasts. Using cotyledons, zygotic embryos, leaf and root explants of *Arabidopsis,* we observed that the *Agrobacterium* developed EPS threads within a short period (7-8 hours), and were attached singly or in groups to all the cells of the explants (Sangwan, unpublished). This suggests that the attachment of bacteria to protoplasts or to intact cells is very similar, and probably involves the same receptor for binding to both type of cells. Perhaps, the receptor is present on both the cell wall and plasma membrane or either localized in some region of the cell membrane which is accessible to the bacterium even in the presence of the plant cell wall. However, it is important to remember that in natural infection bacteria bind only to wound sites. Thus, the presence of receptors for bacterial attachment on the plant cell wall or plasma membrane and the resulting binding of the bacterium to the plasmalemma would enable the bacterium to transfer plasmid DNA across the plant plasmalemma resulting in the formation a crown gall tumor cell. For better understanding of the mechanism of transfer of T-DNA to the plant, we have given in Figure 4, a schematic presentation of the early events of transformation. But how and why the T-DNA goes to the nuclear genome and not to the plastid or mitochondrial genome remains an intriguing question. According to Stachel and Zambryski (1989) T-DNA is transferred to plant cells by a mechanism analogous to bacterial conjugation although there is no experimental proofs for such a hypothesis.

It has also been observed that rapidly growing, and actively dividing cells were more prone to transformation than the non dividing resting cells in *Nicotiana* (An, 1985). Moreover, transformation of tobacco protoplast was strongly dependent on cell cycle, with the a highest efficiency at metaphase (Nagata et al., 1986). This is in accordance with the observations that

INFECTION PROCESS

1. Physical contact between bacteria and wounded/cultured plant cells

2. Plant recognition and defense system initiation

3. Multiplication, migration and colonization by the bacteria

4. Induction of cellulose (infection) threads and attachment to the plant cell wall or plasma membrane

5. Small molecules produced by plant cells bind to receptors on the bacterial surface

6. Induction of enzyme necessary to cleave the plasmid DNA

7. Activation of vir gene and the formation of single-strand (SS) DNA-protein complex

8. Plant cell wall degradation/pore formation at the receptor/attachment site

INTEGRATION

9. Passage of SS T-DNA-protein complex from the bacteria to the plant cell wall/membranes and to the cytoplasm

10. Transfer of T-DNA "probably aided by reticulum endoplasmic" to the nucleus

11. Integration and expression of T-DNA in the plant genome

Fig. 4 . Schematic presentation of possible stage-sequences during the transfer of T-DNA to the plant cell (whether these stages can occur sequentially remains to be investigated) only the metabolically active plant cells produce molecules which activate specifically the vir region (Bolton et al., 1986, Stachel and Zambryski, 1986) of the plasmid. In our experiments with zygotic embryo/cotyledon transformation in *Arabidopsis*, a drastic increase in transformation frequency was obtained after a preculture period of 4 to 6 days in the hormone-supplied medium before infection with *Agrobacterium tumefaciens* (Sangwan, unpublished). Similar observations were reported in tobacco pith cells with *A. rhizogenesis* (Bercetche et al., 1987). This suggests that wounding of the explants alone does not cause enough plant cell activation in order to obtain an efficient transformation. It also appears that for efficient mating of *Agrobacterium* to the plant cell, a certain degree of activation or division is also required in the plant cell which is provided by *in vitro* culture conditions. During cell division not only physiological changes but also the changes in cell wall structures occur. e.g synthesis and formation of new cell wall. It is possible that during these events there is a direct attachment of the bacterium to the receptor site of plasma-membrane of the plant cell instead of cell wall. This may facilitate the transfer of T-DNA to the cell cytoplasm and then to the nucleus. So, for efficient transformation of a species one has to optimize not only the regeneration system but also the activation of cell division in the explant cells before or during the co-cultivation period. We also observed that *Agrobacteria* were present not only on the wounded surface but in the conducting tissues and in certain intercellular spaces of the (primary) explants (Sangwan, unpublished). This suggests that bacteria can migrate through the conducting tissues of the explants, and can reach and eventually transform the actively dividing cells e.g., the cambium in the root explant of *Arabidopsis*, even far away from the wounded and cut surface. Transformed buds and roots have been obtained from these transformed cambial cells. It is of interest to note that when *in vitro* direct organogenesis occurrs from leaf, or root explants it is mainly multicellular and rarely unicellular in origin. However, the transformed roots/shoots mainly originate from single transformed cells which developed into micro callus before forming roots/shoots.

6 - CONCLUSIONS AND FUTURE PROSPECTS

The transgenic plant systems based on recombinant DNA technology, as is evident from this and other reviews, will retain their place as a new method in plant improvement and developmental genetics. In the future, experiments similar to those already conducted will provide more crucial informations on the intricate mechanisms of plant development. Therefore, the future role of transgenic systems is the continuation and broadening of the type of experiments on crop plants that have been previously or hitherto conducted with model plants. However, some new areas which we think of particular importance are 1) basic investigations on plant development 2) applied studies for producing improved varieties.

6.1 Basic plant development studies:

a. As more and more transgenic plants of new species have been investigated, new and unexplained phenomena have emerged. Although, at present it is difficult to understand these events or even devise strategies to analyse them, these phenomena may provide clues to the future of gene regulation in plants of which we currently understand very little. Moreover, it is important to understand the factors which determine the pattern of expression of recombinant DNA molecules, and thereby to predict specificity of the tissue of expression. Such knowledge might lead to vector designs that limit expression of foreign genes in the specific cell types. The introduction of genes which express only at a particular stage of plant development would be very desirable.

b. Directed integration of foreign DNA is an important goal. Although efficient site-directed integration by homologous recombination appears difficult to accomplish from experiments conducted, so far, it remains an important goal to predetermine the integration site of inserted DNA.

c. Since, there is a random integration of T-DNA and since, the T-DNA is inserted as a single locus, *Agrobacterium* mediated gene transfer can be successfully used for the isolation of plant genes by gene tagging. In order, to study the gene tagging, some specific Ti plasmid vectors have already been designed.

d. The use of antisense constructs to dampen the expression of genes has potential for creating phenocopy mutations and for basic studies on plant and trait development.

6.2 Applied studies

The successful genetic manipulation of plants by recombinant DNA technology leading to routine production of fertile plants will have a number of practical applications in plant improvment. This technology offers the potential to precisely add characterized genes to the existing genomes. For example, a superior cultivar developed through conventional plant breeding would provide the genetic background for the introduction of new genes developed in the laboratory. Recently, isogenic lines are being used by biotechnologists to isolate genes related to specific traits such as disease resistance. In particular, the new gene transfer technique would reduce dependence on hazardous chemicals in agricultural production systems, and promote long-term sustainability of highly productive agriculture. As we have discussed in the introduction, transfer of single gene traits such as those for herbicide resistance, insect resistance, as already accomplished in model plants, appears to us the most promising and technically feasible step in crop plants. However, most agronomically desired improvements e.g., in yield, pest and pathogen resistance, stress tolerance, and photosynthetic efficiency involve many genes of unknown identity. To achieve this the use of transposable elements would be of great help (Fedoroff 1983) . Moreover, a better understanding of the control of fungal pathogenicity, photosynthesis, solute uptake etc. is also beginning to open up new possibilities for gene transfer. For example, the observation that plant chitinases are important antifungal agents in plants (Schlumbaum et al., 1986) indicates a good target for transformation. So far genes have been transferred into plant chromosomes and expressed as nuclear coded genes, as single genetic traits. Similar procedures, however, have not yet been developed to transfer and express foreign genes (genetic matrice) specially into organelles. Since the cereals are difficult to regenerate *in vitro* and are not susceptible to *Agrobacterium*, it would be difficult to transfer marker gene in these crops. However, if significant advances are to be made in the use of these new techniques for crop improvement overall gene transfer in cereals will require close integration of cell culture and molecular approaches. For example, it may be possible to transfer, both nuclear and cytoplasmic encoded genes, by protoplast fusion in rice. In future, somatic hybrids could be obtained between the cultivated and wild species of rice, wheat, potato etc. combining highly desirable traits such as resistance to fungal or virus diseases and salt tolerance, with high yielding ability. In order to introduce a desirable trait, it is essential to choose the correct genes. This choice depends on an understanding of the biochemical basis of the processes underlying the trait. In general, the biochemistry of plant traits of agronomic importance is not well known, and therefore, should be a major area for future research.

Finally, the unusual power of recombinant DNA technology, uncertainty over the behavior of modified plant/microbe have led to violent manifestations, and generated a great concern in the ecological movements in the world; hence, we should introduced these techniques cautiously.

7 - ACKNOWLEDGEMENTS

The senior author is indebted to Prof. M. VAN MONTAGU of the state university of Gent, Belgium for support of his research during his stay (in 1988) in his laboratory. Dr. R.S. SANGWAN was recipient of a senior fellowship from European Economic Community. We thank our colleagues (Drs. D. INZE, A. KAPLAN, D. VALVEKENS, M. VAN LIJSEBETTENS, C. BOWLER, T. TETU, C. VAN DUN, A. De LAAT) both in Gent and in Amiens for valuable discussions. We are also grateful to N. PAWLICKI and F. FLANDRE for their help in planning and drawing figures 2, 3 Dr. B.S. AHLOOWALIA and E. ZYPRIAN for critically reading the manuscript, and G. VASSEUR and M. POIRET for their very efficient help in preparing the manuscript. Special thanks to all other members of our group at Amiens.

8 - LITERATURE CITED

1. ABDULLAH, R., COCKING, E.C., THOMPSON, J.A. 1986. Efficient plant regeneration from rice protoplasts through somatic embryogenesis. Biotechnology, **4** : 1087-1090.

2. ALLARD, R.W. 1969. Principles of plant breeding. Wiley, New York.

3. ALY, M.A.M., OWENS, L.D. 1987. A simple system for plant cell microinjection and culture. Plant Cell Tissue Organ Cult., **10** : 159-174.

4. AN, G. 1985. High efficiency transformation of cultured tobacco cells. Plant Physiol., **79** : 568-570.

5. AN, G., WATSON, B.D., CHIANG, C.C. 1986. Transformation of tobacco, potato and *Arabidopsis thaliana* using a binary Ti vector system. Plant Physiol. **81** : 301-305.

6. BALDES, R., MOOS, M., GEIDER, K. 1987. Transformation of soybean protoplasts from permanent suspension cultures by cocultivation with cells of *Agrobacterium tumefaciens*. Plant Mol. Biol., **9** : 135-145.

7. BASIRAN, N., ARMITAGE, P., SCOTT, R.J. DRAPER, J. 1987. Genetic transformation of flax (*Linum usitatissimum*) by *Agrobacterium tumefaciens* : regeneration of transformed shoots via a callus phase. Plant Cell Rep., **6** : 396-399.

8. BERCETCHE, J., CHRIQUI, D., ADAM, S., DAVID. C. 1987. Morphogenetic and cellular reorientations induced by *Agrobacterium rhizogenes* on carrot, pea tobacco. Plant Science, **52** : 195-210.

9. BEVAN, M. 1984. Binary *Agrobacterium* vectors for plant transformation. Nucleic Acids Res., **12** : 8711-8721.

10. BEVAN, M., FLAVELL, R.B., CHILTON, M.D. 1983. A chimaeric antibiotic resistance gene as a selectable marker for plant cell transformation. Nature, **304** : 185-187.

11. BOLTON, G.W., NESTER, E.W., GORDON, M.P. 1986. Plant phenolic compounds induce expression of the *Agrobacterium tumefaciens* loci needed for virulence. Science, **232** . 983-985.

12. BRISSON, N., PASZKOWSKI, J., PENSWICK, J.R., GRONENBORN, B., POTRYKUS, I., HOHN, T. 1984. Expression of a bacterial gene in plants by using a viral vector. Nature, **310** : 511-514.

13. BRISSON, N., HOHN, T. 1986. Plant virus vectors : cauliflower mosaic virus. Methods Enzymol., **118** : 659-668.

14. BYTEBIER, B., DEBOECK, F., DE GREVE, H., VAN MONTAGU, M., HERNALSTEENS, J.P. 1987. T-DNA organization in tumor cultures and transgenic plants of the monocotyledon *Asparagus officinalis* . Proc. Natl. Acad. Sci., USA, **84** : 5345-5349.

15. CABOCHE, M., DESHAYES, A. 1986. Utilisation de liposomes pour la transformation de protoplastes de mésophylle de Tabac par plasmide recombinant de *E. coli* leur conférant la résistance à la kanamycine. C.R. Acad. Sci., **299** : 663-666.

16. CATLIN, D., OCHOA, O., Mc CORMICK, S., QUITOS, C.F. 1988. Celery transformation by *Agrobacterium tumefaciens* : cytological and genetic analysis of transgenic plants. Plant Cell Rep., **7** : 100-103.

17. CHEN, W.H., GARTLAND, K.M.A., DAVEY, M.R., SOTAK, R., GARTLAND, J.S. et al. 1987. Transformation of sugarcane protoplasts by direct uptake of a selectable chimaeric gene. Plant Cell Rep., **6** : 297-301.

18. CHILTON, M.-D., TEPFER, D.A., PETIT D.C., CASSE-DELBART, F., TEMPE, J. 1982. *Agrobacterium rhizogenes* inserts T-DNA into the genomes of the hostplant cells. Nature, **295** : 432-434.

19. CHRISTOU, P., MURPHY, J.E., SWAIN, W.F. 1987. Stable transformation of soybean by electroporation and root formation from transformed callus. Proc. Natl. Acad. Sci., USA, **84** : 3962-3966.

20. CHYI, Y.-S., PHILLIPS, G.C. 1987. High efficiency *Agrobacterium*-mediated transformation of *Lycopersicon* based on conditions favorables for regeneration. Plant Cell Rep., **6** : 105-108.

21. COCKING, E.C., DAVEY, M.R. 1987. Gene transfer in cereals. Science, **236** : 1259-1262.

22. COMAI, L., FACCIOTTI, D., HIATT, W.R., THOMPSON, G., ROSE, R.E., STALKER, D.M. 1985. Expression in plants of a mutant *aroA* gene from *Salmonella typhimurium* confers tolerance to glyphosate. Nature, **317** : 741-744.

23. CROSSWAY, A., OAKES, J.V., IRVINE, J.M., WARD, B., KNAUF, V.C., SHEWMAKER, L.K. 1986. Integration of foreign DNA following microinjection of tobacco mesophyll protoplasts. Mol. gen. Genet., **202** : 179-185.

24. DAHL, G.A., TEMPE, J. 1983. Studies on the use of toxic precursor analogues of opines to select transformed plant cells. Theor. Appl. Genet., **66** : 233-239.

25. DAVEY, M.R., COCKING, E.C., FREEMAN, J., PEARCE, N., TUDOR, I. 1980. Transformation of petunia protoplasts by isolated *Agrobacterium* plasmid. Plant Sci. Lett., **18** : 307-313.

26. DEAK, M., KISS, G.B., KONCZ, C., DUDITS, D. 1986. Transformation of Medicago by *Agrobacterium*-mediated gene transfer. Plant Cell Rep., **5** : 97-100.

27. DE BLOCK, M., HERRERA-ESTRELLA, L., VAN MONTAGU, M., SCHELL, J., ZAMBRYSKI, P. 1984. Expression of foreign genes in regenerated plants and in their progeny. EMBO J., **3** : 1681-1689.

28. DE BLOCK, M., SCHELL, J., VAN MONTAGU, M. 1985. Chloroplast transformation by *Agrobacterium tumefaciens*. EMBO J., **4** : 1367-1372.

29. DE BLOCK, M., BOTTERMANN, J., VANDEWIELE, M., DOCKX, J., THOEN, C. et al. 1987. Engineering herbicide resistance in plants by expression of a detoxifying enzyme. EMBO J., **6** : 2513-2518.

30. DE BLOCK, M. 1989. Genotype-independent leaf disc transformation of potato (*Solanum tuberosum*) using *Agrobacterium tumefaciens*. Theor. Appl. Genet., **76** : 767-774.

31. DE CLEENE, M., DE LEY, J. 1976. The host range of crown gall. Bot. rev., **42** : 389-466.

32. DE FRAMOND, A.J., BARTON, K.A., CHILTON, M.D. 1983. Mini-Ti : A new vector strategy for plant genetic engineering. Biotechnology, **1** : 262-269.

33. DE GREVE, H., LEEMANS, J., HERNALSTEENS, J.-P., THIA-TOONG, L., DEBEUCKELEER, M. et al. 1982. Regeneration of normal and fertile plants that express octopine synthase, from tobacco crown galls after deletion of tumor controlling functions. Nature, **300** : 752-755.

34. DE LA PENA, A., LORZ, H., SCHELL, J. 1987. Transgenic rye plants obtained by injecting DNA into young floral tillers. Nature, **325** : 274-276.

35. DELLAPORTA, S.L., WOOD, J., HICKS, J.B. 1983. A plant DNA mini-preparation : version II. Plant Mol. Biol. Rep., **1** : 19-21.

36. DEPICKER, A., HERMAN, L., JACOBS, A., SCHELL, J., VAN MONTAGU, M. 1985. Frequencies of simultaneous transformation with different T-DNAs and their relevance to the *Agrobacterium*/plant cell interaction. Mol. Gen. Genet., **201** : 477-484.

37. DESHAYES, A., HERRERA-ESTRELLA, L., CABOCHE, M. 1985. Liposome-mediated transformation of tobacco mesophyll protoplasts by an *Escherichia coli* plasmid. EMBO J., **4** : 2731-2737.

38. EICHHOLTZ, D.A., ROGERS, S.G., HORSCH, R.B., KLEE, H.J., HAYFORD, M., et al. 1987. Expression of mouse dihydrofolate reductase gene confers methotrexate resistance in transgenic petunia plants. Somat. Cell Mol. Genet., **13** : 67-76.

39. FACCIOTTI, D., O'NEAL, J.K., LEE, S., SHEWMAKER, C.K. 1985. Light-inducible expression of a chimeric gene in soybean tissue transformed with *Agrobacterium*. Biotechnology, **3** : 241-246.

40. FEDOROFF, N. 1983. In mobile genetic elements. J.A. Shapiro Ed. (Academic Press, New York), pp 1-63.

41. FELDMANN, K.A., MARKS, M.D. 1987. *Agrobacterium*-mediated transformation of germinating seeds of *Arabidopsis thaliana* : a non-tissue culture approach. Mol. Gen. Genet., **208** : 1-9.

42. FILLATTI, J.J., KISER, J., ROSE, R., COMAI, L. 1987. Efficient transfer of a glyphosate tolerance gene into tomato using a binary *Agrobacterium tumefaciens* vector. Biotechnology, **5** : 726-730.

43. FILLATTI, J.J., SELLMER, J., MCCOWN, B., HAISSIG, B., COMAI, L. 1987. *Agrobacterium*-mediated transformation and regeneration of *Populus*. Mol., Gen. Genet., **206** : 192-199.

44. FRALEY, R.T., ROGERS, S.G., HORSCH, R.B., SANDERS, P.R., FLICK, J.S., et al. 1983. Expression of bacterial genes in plant cells. Proc. Natl. Acad. Sci., USA, **80** : 4803-4807.

45. FRALEY, R.T., ROGERS, S.G., HORSCH, R.B. 1986. Genetic transformation in higher plants. CRC Critical Rev. Plant Sci., **4** : 1-46.

46. FROMM, M., TAYLOR, L.P., WALBOT, V. 1985. Expression of genes transferred into monocot and dicot plant cells by electroporation. Proc. Natl. Acad. Sci. USA, **82** : 5824-5828.

47. FROMM, M., TAYLOR, L.P., WALBOT, V. 1986. Stable transformation of maize after gene transfer by electroporation. Nature, **319** : 791-793.

48. FRY, J., BARNASON, A., HORSCH, R.B. 1987. Transformation of *Brassica napus* with *Agrobacterium tumefaciens* based vectors. Plant Cell Rep., **6** : 321-325.

49. GASSER, C.S., FRALEY, R.T. 1989. Genetically Engineering plants for crop improvement. Science, **244** : 1293-1299.

50. GOODMAN, R.M., HAUPTLI, H., CROSSWAY, A., KNAUF, V.C. 1987. Gene transfer in crop improvement. Science, **236** : 48-64.

51. GRONENBORN, B., GARDNER, R.C., SCHAEFER, S., SHEPHERD, R.J. 1981. Propagation of foreign DNA in plants using cauliflower mosaic virus as a vector. Nature, **294** : 773-776.

52. HAIN, R., STABEL, P., CZERNILOFSKY, A.P., STEINBISS, H.H., HERRERA-ESTRELLA, L., SCHELL, J. 1985. Uptake, integration, expression and genetic transmission of a selectable chimeric gene by plant protoplasts. Mol. Gen. Genet., **199** : 161-168.

53. HAUPTMANN, R.M., OZIAS-AKINS, P., VASIL, V., TABAEIZADEH, Z., ROGERS, S.S. et al. 1987. Transient expression of electroporated DNA in monocotyledonous and dicotyledonous species. Plant Cell Rep., **6** : 265-270.

54. HELMER, G., CASADABAN, M., BEVAN, M., KAYES, L., CHILTON, M.D. 1984. A new chimeric gene as a marker for plant transformation : the expression of *Escherichia coli* *ß*-galactosidase in sunflower and tobacco cells. Biotechnology, **2** : 520-527.

55. HERNALSTEENS, J.P., THIA-TOONG, L., SCHELL, J., VAN MONTAGU, M. 1984. An *agrobacterium*-transformed cell culture from the monocot *Asparagus officinalis*. EMBO J., **3** : 3039-3041.

56. HERRERA-ESTRELLA, L., DE BLOCK, M., MESSENS, E., HERNALSTEENS, J.P., VAN MONTAGU, M;, SCHELL, J. 1983. Chimeric genes as dominant selectable markers in plant cells. EMBO J., **2** : 987-995.

57. HESS, D. 1980. Investigations on the intra-and interspecific transfer of anthocyanin genes using pollen as vectors. Z. Pflanzenphysiol., **98** : 321-337.

58. HILDER, V.A., GATEHOUSE, A.M.R., SHEERMAN, S.E., BARKER, R.F., BOULTER, D. 1987. A novel mechanism of insect resistance engineered into tobacco. Nature, **330** : 160-163.

59. HILLE, J., VERHEGGEN, F., ROELVINK, P., FRANSSEN, H., VAN KAMMEN, A., ZABEL, P. 1986. Bleomycin resistance : a new dominant selectable marker for plant cell transformation. Plant Mol. Biol., **7** : 171-176.

60. HOEKEMA, A., HIRSCH, P.R., HOOYKAAS, P.J.J., SCHILPEROOT, R.A. 1983. A binary plant vector stategy based on separation of *vir-* and T-region of the *Agrobacterium tumefaciens* Ti plasmid. Nature, **303** : 179-180.

61. HOEKEMA, A., VAN HAAREN, M., FELLINGER, A., HOOYKAAS, P.J.J., SCHILPEROOT, R.A. 1985. Non-oncogenic plant vectors for use in the *Agrobacterium* binary system. Plant Mol. Biol., **5** : 85-89.

62. HORSCH, R.B., FRALEY, R.T., ROGERS, S.G., SANDERS, P.R., LLOYD, A., HOFFMANN, N. 1984. Inheritance of functional foreign genes in plants. Science, **223** : 496-498.

63. HORSCH, R.B., FRY, J.E., HOFFMANN, N.L., EICHHOLTZ, D., ROGERS, S.G., FRALEY, R.T. 1985. A simple and general method for transferring genes into plants. Science, **227** : 1229-31.

64. JAMES, D.J., PASSEY, A.J., BARBARA, D.J., BEVAN, M. 1989. Genetic transformation of apple (*Malus pumila* Mill) using a disarmed Ti- binary vector. Plant Cell Report, **7** : 658-661.

65. JEFFERSON, R.A., KAVANAGH, T.A., BEVAN, M.W. 1987. GUS fusions : β-glucuronidase as a sensitive and versatile gene fusion marker in higher plants. EMBO J., **6** : 3901-3907.

66. JEFFERSON, R.A. 1987. Assaying chimeric genes in plants : the GUS gene fusion system. Plant Mol. Biol. Reporter, **5** : 387-405.

67. JENSEN, J.S., MARCKER, K.A., OTTEN, L., SCHELL, J. 1986. Nodule-specific expression of a chimaeric soybean leghemoglobin gene in transgenic *Lotus corniculatus*. Nature, **321** : 669-674.

68. JONES, J.D.G., DUNSMUIR, P., BEDBROOK, J. 1985. High level of expression of introduced chimaeric genes in regenerated transformed plants. EMBO J., **4** : 2411-2418.

69. JONES, J.D.G., SVAB, Z., HARPER, E.C., HURWITZ, C.D., MALIGA, P. 1987. A dominant nuclear streptomycin resistance marker for plant cell transformation. Mol. Gen. Genet., **210** : 86-91.

70. JONGSMA, M., KORNNEEF, M., ZABEL, P., HILLE, J. 1987. Tomato protoplast DNA transformation : physical linkage and recombination of exogenous DNA sequences. Plant Mol. Biol., **8** : 383-394.

71. JUNKER, B., ZIMNY, J., LUHRS, R., LORZ, H. 1987. Transient expression of chimaeric genes in dividing and non-dividing cereal protoplasts after PEG-induced DNA uptake. Plant Cell Rep., **6** : 329-332.

72. KLEE, H., HORSCH, R., ROGERS, S. 1987. *Agrobacterium*-mediated plant transformation and its further applications to plant biology. Annu. Rev. Plant Physiol., **38** : 467-486.

73. KLEIN, T.M., WOLF, E.D., WU, R., SANFORD, J.C. 1987. High-velocity microprojectiles for delivering nucleic acids into living cells. Nature, **327** : 70-73.

74. KLEIN, T.M., FROMM, M.E., WEISSINGER, A., TOMES, D., SCHAAF, S., SLETTEN, M., SANFORD, J.C. 1988. Transfer of foreign genes into intact maize cells using high-velocity microprojectiles. Proc. Natl. Acad. Sci. USA, In press.

75. KOHLER, F., GOLZ, C., EAPEN, S., KOHN, H., SCHIEDER, O. 1987. Stable transformation of moth bean *Vigna aconitifolia* via direct gene transfer. Plant Cell Rep., **6** : 313-317.

76. KONCZ, C., OLSSON, O., LANGRIDGE, W.H.R., SCHELL, J., SZALAY, A.A. 1987. Expression and functional assembly of bacterial luciferase in plants. Proc. Natl. Acad. Sci. USA, **84** : 131-135.

77. KRENS, F.A., MOLENDIJK, L., WULLEMS, G.J., SCHILPEROORT, R.A. 1982. *In vitro* transformation of plant protoplasts with Ti-plasmid DNA. Nature, **296** : 72-74.

78. KUHLEMEIR, C., GREEN, P.J., CHUA, N.H. 1987. Regulation of gene expression in higher plants. Ann. Rev. Plant Physiol., **38** : 221-257.

79. LICHTENSTEIN, C. 1987. Bacteria conjugate with plants. Nature, **328** : 108-109.

80. LLOYD, A.M., BARNASON, A.R., ROGERS, S.G., BYRNE, M.C., FRALEY, R.T., HORSCH, R.B. 1986. Transformation of *Arabidopsis thaliana* with *Agrobacterium tumefaciens*. Science, **234** : 464-466.

81. LORZ, H., BAKER, B., SCHELL, J. 1985. Gene transfer to cereal cells mediated by protoplast transformation. Mol. Gen. Genet., **199** : 178-182.

82. LUO ZHONG-XUN, WU, R. 1988. A simple method for the transformation of rice *via* the pollen tube pathway. Plant Molecular Biol. Reporter., **6** : 165-174

83. LURQUIN P.F., KADO, C.I. 1977. Mol. Gen. Genet., **154**.

84. MANIATIS, T., FRISCH, E.F., SAMBROOK, J. 1982. Molecular cloning : A laboratory manual. (Cold Spring Harbor, N.Y : cold spring Harbor Laboratory).

85. MARTON, L., WULLEMS, G.J., MOLENDIJK, L., SCHILPEROORT, R.A. 1979. *In vitro* transformation of cultured cells from *Nicotiana tabacum* by *Agrobacterium tumefaciens*. Nature, **277** : 129-131.

86. MATTHYSSE, A.G., HOLMES, K.V., GURLITZ, R.H.G. 1982. Binding of *Agrobacterium tumefaciens* to carrot protoplasts. Physiol. Plant Pathol., **20** : 27-33.

87. Mc CABE, D.E., SWAIN, W.F., MARTINELL, B.J., CHRISTOU, P. 1988. Stable transformation of Soybean (*Glycine max*) by particle acceleration. Bio/Technology, **6** : 923-926.

88. Mc DONNEL, R.E., CLARK, R.D., SMITH, W.A., HINCHEE, M.A. 1987. A simplified method for the detection of neomycin phosphotransferase II activity in transformed plant tissue. Plant Mol. Biol. Report., **5** : 380-386.

89. MURAI, N., SUTTON, D.W., MURRAY, M.G., SLIGHTOM, J.L., MERLO, D.J. et al. 1983. Phaseolin gene from bean is expressed after transfer to sunflower via tumor-inducing plasmid vectors. Science, **222** : 476-482.

90. NAGATA, I., OKADA, K., TAKEBE, I. 1986. Fallen leaf taked conference on *Agrobacterium* and crown gall. Sept. 11-14. Abstract page 19.

91. NEUHAUS, G., SPANGENBERG, G., MITTELSTEN SCHEID, O., SCHWEIGER, H.G. 1987. transgenic rapeseed plants obtained by the microinjection of DNA into microspore-derived embryoids. Theor. Appl. Genet., **75** : 30-36.

92. OHTA, Y. 1986. High efficiency genetic transformation of maize by a mixture of pollen and exogenous DNA. Proc. Natl. Acad. Sci. USA, **83** : 715-719.

93. OOMS, G., BURRELL, M.M., KARP, A., BEVAN, M., HILLE, J. 1987. Genetic transformation in two potato cultivars with t-DNA from disarmed *Agrobacterium*. Theor. Appl. Genet., **73** : 744-750.

94. OU-LEE, T.-M., TURGEON, R., WU, R. 1986. Expression of a foreign gene linked to either a plant virus or a *Drosophila* promoter, afterelectroporation of protoplasts of rice, wheat and sorghum. Proc. Natl. Acad. Sci. USA, **83** : 6815-6819.

95. OW, D.W., WOOD, K.V., DELUCA, M., DEWET, J.R., HELINSKI, D.R., HOWELL, S.H. 1986. Transient and stable expression of the firefly luciferase gene in plant cells and transgenic plants. Science, **234** : 856-859.

96. PASZKOWSKI, J., SHILLITO, R.D., SAUL, M., MANDAK, V., HOHN, T., et al. 1984. Direct gene transfer to plants. EMBO J., **3** : 2717-2722.

97. POTRYKUS, I., SAUL, M.W., PETRUSKA, P., PASZKOWSKI, P., SHILLITO, R.D. 1985. direct gene transfer to cells of a graminaceous monocot. Mol. Gen. Genet., **199** : 183-188.

98. POTRYKUS, I., PASZKOWSKI, J.P., SAUL, M.W., PETRUSKA, P., SHILLITO, R.D. 1985. Molecular and general genetics of a hybrid foreign gene introduced into tobacco by direct gene transfer. Mol. Gen. Genet., 199 : 169-177.

99. POTRYKUS, I., SHILLITO, R.D., SAUL, M.W., PASZKOWSKI, J.P. 1985. Direct gene transfer. State of the art and furure potential. Plant Mol. Biol. Rep., **3** 117-128.

100. POWELL ABEL, P., NELSON, R.S., DE, B., HOFFMAN, N., ROGERS, S.G., FRALEY, R.T., BEACHY, R.N. 1986. Delay of disease development in transgenic plants that express the tobacco mosaic virus coat protein. Science, 232 : 738-743.

101. REICH, T.J., IYER, V.N., MIKI, B.L. 1986. Efficient transformation of alfalfa protoplasts by the intranuclear microinjection of Ti plasmids. Biotechnology, **4** : 1001-1004.

102. RHODES, C.A., PIERCE, D.A., METLER, I.J., MASCARENHAS, D., DETMER, J.J. 1988. Genetically transformed maize plants from protoplasts. Sciences, **240** : 204-207.

103. RIGGS, C.D., BATES, G.W. 1986. Stable transformation of tobacco by elctroporation : evidence for plasmid concatenation. Proc. Natl. Acad. Sci. USA, **83** : 5602-5606.

104. SANFORD, J.C., KLEIN, T.M., WOLF, E.D., ALLEN, N. 1987. Particle science Technology, **5** : 27-37.

105. SCHELL, J. 1987. Transgenic plants as tools to study the molecular organization of plant genes. Science, **237** : 1176-1183.

106. SCHLUMBAUM, A., MAUCH, F., VOGELI, U., BOLLER, T. 1986. Plant chitinases are potent inhibitors of fungal growth. Nature, **324** : 365-367.

107. SCOTT, R.J., DRAPER, J. 1987. Transformation of carrot tissues derived from proembryogenic suspension cells: a useful model system for gene expression studies in plants. Plant Mol. Biol., **8** : 265-274.

108. SEQUEIRA, L. 1984. Cross protection and induced resistance. Their potential for plant disease control. Trends Biotechnol., **2** : 25-29.

109. SHAH, D.M., HORSCH, R.B., KLEE, H.J., KISHORE, G.M., WINTER, J.A. et al. 1986. Engineering herbicide tolerance in transgenic plants. Science, **233** : 478-481.

110. SHAW, C.H., SANDERS, D.M., BATES, M.R., SHAW, C.H. 1986. Light regulation of a ssRubisco-nos chimaeric gene: photoregulatory control sequences from a C3 plant function in cells of a CAM plant. Nucleic Acids Res., **14** : 6603-6612.

111. SHEIKOLESLAM, S.N., WEEKS, D.P. 1987. Acetosyringone promotes high efficiency transformation of *Arabidopsis thaliana* explants by *Agrobacterium tumefaciens*. Plant Mol. Biol., **8** : 291-298.

112. SHIMAMOTO, K., TERADA, R., IZAWA, T., FUJIMOTO, H. 1989. Fertile transgenic rice plants regenerated from transformed protoplasts. Nature. **338**, 274-276.

113. SIMMONDS, N.W. 1979. Principles of crop improvement. Longman, London, 339pp.

114. SIMPSON, R.B., SPIELMANN, A., MARGOSSIAN, L. MCKNIGHT, T.D. 1986. A disarmed binary vector from *Agrobacterium tumefaciens* function in *Agrobacterium rhizogenes*. Frequent co-transformation of two distinct T-DNAs. Plant Mol. Biol., **6** : 403-415.

115. SOUTHERN, E.M. 1975. Detection of specific sequences among DNA fragments separated by gel electrophresis. J. Mol. Biol., **98** : 503-517.

116. STACHEL, S.E., ZAMBRYSKI, P.C. 1986. *Agrobacterium tumefaciens* and the susceptible plant cell: a novel adaptation of extracellular recognition and DNA conjugation. Cell **47** : 155-157.

117. STACHEL, S.E., NESTER, E.W. 1986. The genetic end transcriptional organization of the vir region of the A6 Ti plasmid of *Agrobacterium tumefaciens* . EMBO, J., **7** : 27-37.

118. STACHEL, S.E., ZAMBRYSKI, P.C. 1989. Generic trans-kingdom Sex. Nature, **340** : 190-191.

119. SUKHAPINDA, K., SPIVEY, R., SIMPSON, R.B., SHANIN, E.A. 1987. Transgenic tomato (*Lycopersicon esculentum* L.) transformed with a binary vector in *Agrobacterium rhizogenes* : nonchimeric origin of callus clone and low copy numbers of integrated vector T-DNA. Mol. Gen. Genet., **206** : 491-497.

120. TEMPE, J., PETIT, A., HOLSTERS, M. VAN MONTAGU, M., SCHELL, J. 1977. Ther mosensitive step associated with transfer of the Ti-plasmid during conjugation : possible relation to transformation in crown-gall. Proc. Nat. Acad. Sci., **74** : 2848-2849.

121. TEMPE, J., SCHELL, J. 1987. La manipulation des plantes. La Recherche, **188** : 696-709.

122. TEPFER, D. 1984. Transformation of several species of higher plants by *Agrobacterium rhizogenes*: sexual transmission of the transformed genotype and phenotype. Cell **37** : 959-967.

123. THOMPSON, C.J., RAO MOVVA, N., TIZARD, R., CRAMERI, R., DAVIES, J.E. et al. 1987. Characterization of the herbicide resistance gene bar from Streptomyces hygroscopicus. EMBO J., **6** : 2519-2523.

124. UCHIMIYA, H., FUSHIMI, T., HASHIMOTO, H., HARADA, H., SYONO, Y., SUGAWARA, Y. 1986. Expression of a foreign gene in callus derived from DNA-treated protoplasts of rice (*Oryza sativa* L.). Mol. Gen. Genet., **204** : 204-207.

125. UMBECK, P., JOHNSON, G., BARTON, K., SWAIN, W. 1987. Genetically transformed cotton (*Gossypium hirsutum* L.) plants. Biotechnology, **5** : 263-266.

126. VAECK, M., REYNAERTS, A., HOFTE, H., JANSENS, S., DE BEUCKELEER, M. et al. 1987. Transgenic plants protected from insect attack. Nature, **328** : 33-37.

127. VALVEKENS, D., VAN MONTAGU, M., VAN LIJSEBETTENS, M. 1988. *Agrobacterium tumefaciens*-mediated transformation of *Arabidopsis thaliana* root explants by using kanamycin selection. Proc. Natl. Acad. Sci. U.S.A. **85** : 5536-5540.

128. VAN DEN ELZEN, P.J.M., TOWNSEND, J., LEE, K.Y., BEDBROOK, J.R. 1985. A chimeric hygromycin resistance gene as a selectable marker in plant cells. Plant Mol. Biol., **5** : 299-302.

129. VAN LIJSEBETTENS, M., VALVEKENS, D., VANDERHAEGEN, R., VAN MONTAGU, M. 1989. Highly efficient *Agrobacterium* mediated transformation of *Arabidopsis thaliana* : A genetic and molecular evaluation. In Proceedings genetic Manipulation in plant breeding, Plenum Press, Eds. J. Jensen and C.N. Law.

130. WALDRON, C., MALCOLM, S.K., MURPHY, E.B., ROBERTS, I.L. 1985. Methos for high frequency DNA-mediated transformation of plant protoplasts. Plant Mol. Biol. Rep., **3** : 169-173.

131. WANG, K., HERRERA-ESTRELLA, L., VAN MONTAGU, M., ZAMBRYSKI, P. 1984. Right 25bp terminus sequence of the nopaline T-DNA is essential for and determines direction of DNA transfer from *Agrobacterium* to the plant genome. Cell, **38** :45-62.

132. WEISING, K., SCHELL, J., KAHL, G. 1988. Foreign genes in plants : transfer, structure, expression, and application. Annu. Rev. Genet., **22** : 421-477.

133. WHITE, D.W.R., GREENWOOD, D. 1987. Transformation of the forage legume *Trifolium repens* L. using binary *Agrobacterium* vectors. Plant Mol. Biol., **8** :461-469.

134. ZAMBRYSKI, P., JOOS, H., GENETELLO, C., LEEMANS, J., VAN MONTAGU, M., GOODMAN, H. 1980. Tumor DNA structure in plant cells transformed by *A. tumefaciens*. Science, **209** : 1385-1391.

135. ZAMBRYSKI, P., JOOS, H., GENETELLO, C., LEEMANS, J., VAN MONTAGU, M., SCHELL, J. 1983. Ti plasmid vector for the introduction of DNA into plant cells without alteration of their normal regenerative capacity. EMBO, J., **2** : 2143-2150.

136. ZAMBRYSKI, P. 1988. Basic process underlyings *Agrobacterium*-mediated DNA transfer to plant cells. Annu. Rev. Genet., **22** :1-30.

POTENTIAL TRANSFORMATION SYSTEMS IN *DACTYLIS GLOMERATA*

B. V. CONGER
Department of Plant and Soil Science
University of Tennessee
Knoxville, TN 37901-1071
USA

ABSTRACT. *In vitro* systems for *Dactylis glomerata* L. include the production of somatic embryos directly from mesophyll cells in cultured leaf explants and the complete development of embryos (to a germinable stage) in a single liquid medium. Regeneration has also been achieved from suspension-derived protoplasts and the plants grown to maturity. These systems offer attractive possibilities for gene transfer. The following experiments have been attempted: (1) abrasion of leaf surfaces and application of *Agrobacterium* vectors containing a kanamycin resistance gene (NPTII), (2) direct uptake of a DNA plasmid containing a hygromycin resistance gene (APHIV) by suspension-derived protoplasts, and (3) treatment of leaf segments and plated suspensions with microprojectiles coated with DNA containing both the NPTII and beta-glucuronidase (GUS) genes. A few false positives were obtained with *Agrobacterium* and transient expression of the GUS gene was obtained with microprojectiles. The only stable transformants were produced through direct uptake of DNA by protoplasts. Transformation was confirmed in hygromycin resistant calli and regenerated plants by Southern blot hybridization.

1. Introduction

Major components required in genetic engineering or gene transfer in higher plants include: (a) identifying, locating and isolating the gene or DNA sequence of interest (b) transferring the gene to cells of a chosen recipient by means of a selected delivery system and (c) regenerating and recovering normal fertile plants from those transformed cells. The development of efficient and repeatable regeneration systems remains a major problem for many of our important crop species, including the cereals and grasses.

Regeneration systems developed for *Dactylis glomerata* (orchardgrass or cocksfoot) have very high regeneration capacity through somatic embryogenesis

R.S. Sangwan and B.S. Sangwan-Norreel (eds.), The Impact of Biotechnology in Agriculture, 339–344.

and are among the most advanced for the Gramineae. These include complete embryo development from mesophyll cells in cultured leaf segments (Conger et al., 1983) and the full maturation of embryos (to a germinable stage) in a single liquid medium (Gray et al., 1984). Histological studies provide strong evidence for a single cell origin of somatic embryos from both leaf (Trigiano et al., 1989) and suspension (Conger et al., 1989) cultures. More recently, regeneration has also been achieved from suspension-derived protoplasts and the plants grown to maturity (Horn et al., 1988a).

These systems offer attractive possibilities for gene transfer and the objective of this communication is to describe the following experiments: (a) abrasion of leaf surfaces and application of various *Agrobacterium* vectors containing a selectable marker, neomycin phosphotransferase (NPTII), for kanamycin resistance, (b) direct uptake by suspension-derived protoplasts of a plasmid containing the aminoglycoside phosphotransferase type IV gene (APHIV) which confers hygromycin resistance and (c) bombardment of both leaf segments and plated suspensions with microprojectiles coated with DNA containing NPTII and a reporter gene, beta-glucuronidase (GUS).

2. Experiments

2.1. *AGROBACTERIUM* VECTORS

These experiments were initiated in early 1985 well before development of particle gun technology and before plants had been established from protoplasts in any cereal or grass species. The rationale was based on the leaf disc transformation technique developed for *Nicotiana* and *Petunia* (Horsch et al., 1985). Although *Agrobacterium*, under natural conditions, does not infect the Gramineae (DeCleene, 1985) limited reports at the time indicated low level infection and physical presence of T-DNA after application to some monocot species (Hernalsteens et al., 1984; Hooykaas-Van Slogteren et al., 1984). More recently *Agrobacterium* mediated delivery of maize streak virus DNA has been demonstrated in *Zea mays* L. (Grimsley et al., 1987) and other Gramineae species (Boulton et al., 1989). The leaf culture system in orchardgrass was considered to provide an exciting opportunity to test whether *Agrobacterium* could be utilized as a gene vector system for cereals and grasses. The experiments were conducted as a scientific collaboration between The University of Tennessee and CIBA-GEIGY Corporation, Research Triangle Park, North Carolina.

Leaf surfaces were abraded with carborundum and various strains of *A. tumefaciens* and *A. rhizogenes* containing the NPTII gene were applied directly to the wounded surfaces. Preliminary experiments showed that the abrasion did not affect embryogenic capacity. Protocols for leaf culture were as described previously (Conger et al., 1983) and segments were placed on selection medium at various times after treatment. Kanamycin at 5 μg/ml inhibited somatic embryogenesis in control leaf segments. A few putative resistant calli and shoots were recovered. Upon further testing, calli proved to be not resistant and chlorophyll loss accompanied by anthocyanin accumulation occurred in escape seedlings. The experiments were discontinued after several months.

2.2. DIRECT UPTAKE OF DNA BY PROTOPLASTS

Results of these experiments have been described in two papers (Horn et al., 1988a,b) and will only be summarized here. The reader is referred to the original sources for experimental details.

Successful accomplishment of transformation by this approach is based on high frequency regeneration from protoplasts. This was achieved and reported by Horn et al. (1988a) making orchardgrass one of the few grass or cereal species in which protoplast regeneration has been attained. These protoplast cultures were used for the transformation experiments (Horn et al., 1988b). Protoplasts were treated immediately after isolation with a plasmid containing the APHIV gene flanked by the expression signals from the 35S transcript of cauliflower mosaic virus (see Horn et al., 1988b for details of the construct). Uptake of the plasmid was enhanced by heat shock (45°C for 5 min) plus electroporation and/or treatment with polyethylene glycol (PEG). Resistant colonies were able to grow in medium supplemented with 20 μg/ml hygromycin.

Plants were regenerated from selected calli on SH (Schenk and Hildebrandt, 1972) medium without hormones or hygromycin. A few colonies grew from control protoplasts under selective conditions. However, growth of the escapes was usually slower than that of the transgenic colonies and were fewer in number if the selection was started at 10 rather than 14 days after plating. Several hygromycin resistant calli which did not contain the foreign gene, as shown by Southern blot hybridization, were obtained. A total of 90 plants were regenerated from 11 of 21 resistant callus lines. Transgenic plants were confirmed by Southern analysis of DNA from harvested leaves. Although the plants are green and develop with apparently normal morphology, they are slower growing and have not been induced to flower. Therefore, sexual transmission of the foreign gene has not yet been accomplished. Nevertheless, this example of genetic transformation is currently one of the very few among the Gramineae.

2.3. MICROPROJECTILES

Compared to bacterial or viral vectors or direct uptake of DNA by protoplasts, development and use of the particle gun represents a more recent technology for transferring genes to plant cells (Klein et al., 1987; Sanford et al., 1987). This technique offers exciting possibilities and potential for genetic transformation of crop species, such as cereals and grasses, for which suitable vectors have not been discovered or developed and/or from which plants are difficult to regenerate and establish from protoplasts.

The high regeneration capacity of both leaf and suspension cultures make the orchardgrass system an attractive candidate for gene transfer experiments with microprojectiles. In a collaboration with Agracetus, Middleton, Wisconsin, such experiments were initiated in early 1988. The approach was to treat leaf segments and plated suspension cultures with gold microprojectiles coated with DNA containing both the NPTII and GUS genes. The particle delivery apparatus utilizing an electrical discharge mechanism and the preparation of DNA-coated gold particles have been described (Christou et al., 1988; McCabe et al., 1988).

Analyses for the GUS gene indicated transient expression in both tissues. Culture of either tissue on selection (kanamycin) medium did not produce any stable transformants. Only two major experiments were conducted and it is possible that further "fine tuning" of the methodology might produce positive results.

Microprojectile experiments have been reinitiated at The University of Tennessee utilizing a recently constructed gun which also propels the particles by an electrical discharge. Results of a preliminary experiment with plated suspension cultures again indicate transient expression of the GUS gene.

3. Conclusions

Currently, orchardgrass is only one of three cereal or grass species (the others are corn and rice) in which there are published reports of obtaining transgenic plants through a cellular approach (see recent review by Gasser and Fraley, 1989). This was achieved for hygromycin resistance through direct uptake of DNA by protoplasts. Infection, and hence gene transfer, by *Agrobacterium* vectors was very little or none. Transient expression for the GUS gene was obtained after treatment of both leaf segments and plated suspension cultures with microprojectiles. The highly efficient and repeatable somatic embryogenesis systems developed for this species offer attractive possibilities for future gene transfer experiments.

4. References

Boulton, M.I., Bucholz, W.G., Marks, M.S., Markham, P.G. and Davies, J.W. (1989) Specificity of *Agrobacterium*-mediated delivery of maize streak virus DNA to members of Gramineae. Plant Molec. Biol. 12, 31-40.

Christou, P., McCabe, D.E. and Swain, W.F. (1988) Stable transformation of soybean callus by DNA-coated gold particles. Plant Physiol. 87, 671-674.

Conger, B.V., Hanning, G.E., Gray, D.J. and McDaniel, J.K. (1983) Direct embryogenesis from mesophyll cells of orchardgrass. Science 221, 850-851.

Conger, B.V., Hovanesian, J.C., Trigiano, R.N. and Gray, D.J. (1989) Somatic embryo ontogeny in suspension cultures of orchardgrass. Crop Sci. 29, 448-452.

DeCleene, M. (1985) The susceptibility of monocotyledons to *Agrobacterium tumefaciens*. Phytopath. Z. 113, 81-89.

Gasser, C.S. and Fraley, R.T. (1989) Genetically engineering plants for crop improvement. Science 244, 1293-1299.

Gray, D.J., Conger, B.V. and Hanning, G.E. (1984) Somatic embryogenesis in suspension and suspension-derived callus cultures of *Dactylis glomerata*. Protoplasma 122, 196-202.

Grimsley, N., Hohn, T., Davies, J.W. and Hohn, B. (1987) *Agrobacterium* mediated delivery of infectious maize streak virus into maize plants. Nature 325, 177-179.

Hernalsteens, J.P., Thia-Toong, L., Schell, J. and Van Montagu, M. (1984) An *Agrobacterium* - transformed cell culture from the monocot *Asparagus officinalis*. EMBO J. 3, 3039-3042.

Hooykaas-Van Slogteren, G.M.S., Hooykaas, P.J.J. and Schilperoort, R.A. (1984) Expression of Ti plasmid genes in monocotyledonous plants infected with *Agrobacterium tumefaciens*. Nature 311, 763-764.

Horn, M.E., Conger, B.V. and Harms, C.T. (1988a) Plant regeneration from protoplasts of embryogenic suspension cultures of orchardgrass *(Dactylis glomerata* L.) Plant Cell Rep. 7, 371-374.

Horn, M.E., Shillito, R.D., Conger, B.V. and Harms, C.T. (1988b) Transgenic plants of orchardgrass *(Dactylis glomerata* L.) from protoplasts. Plant Cell Rep. 7, 469-472.

Horsch, R.B., Fry, J.E., Hoffman, N.L., Eicholz, D., Rogers, S.G. and Fraley, R.T. (1985) A simple and general method for transferring genes into plants. Science 227, 1229-1231.

Klein, T.M., Wolf, E.D., Wu, R. and Sanford, J.C. (1987) High velocity microprojectiles for delivering nucleic acids into living cells. Nature 327, 70-73.

McCabe, D.E., Swain, W.F., Martinell, B.J and Christou, P. (1988) Stable transformation of soybean *(Glycine max*) by particle acceleration. Bio/Technology 6, 923-926.

Sanford, J.C., Klein, T.M., Wolf, E.D. and Allen, N. (1987) Delivery of substances into cells and tissues using a particle bombardment process. Particulate Sci. Technol. 5, 27-37.

Schenk, R.U. and Hildebrandt, A.C. (1972) Medium and techniques for induction and growth of monocotyledonous and dicotyledonous plant cell cultures. Can. J. Bot. 50, 199-204.

Trigiano, R.N., Gray, D.J., Conger, B.V. and McDaniel, J.K. (1989) Origin of direct somatic embryos from cultured leaf segments of *Dactylis glomerata*. Bot. Gaz. 150, 72-77.

SUNFLOWER TRANSFORMATION: A STUDY OF SELECTABLE MARKERS

A. E. ESCANDON and G. HAHNE
Institut de Biologie Moléculaire des Plantes du CNRS, Université Louis Pasteur,
12, Rue du Général Zimmer
67084 Strasbourg Cédex
France

ABSTRACT. The overall utility of a transformation protocol depends to a large extent on the efficiency of the selection step. In the work presented here, we evaluated three selectable markers (kanamycin, paromomycin, and phosphinotricin) for their performance in sunflower transformation experiments, using hypocotyl explants of three inbred lines and one hybrid. The assays were performed on several culture media currently in use for sunflower tissues.

The different sunflower genotypes showed a differential sensitivity depending on marker and medium used. Kanamycin did not show a significant inhibitory effect, while paromomycin and phosphinotricin proved to be efficient selective agents. More detailed studies revealed that the sensitivity towards a selectable marker is strongly influenced by factors such as the N-sources contained in the medium, as well as its hormonal balance.

1. Introduction

Sunflower (*Helianthus annuus* L.) has proven difficult in tissue culture, and the protocols developed for plant regeneration often appear limited to certain cultivars (Hartman et al. (1988), Wilcox McCann et al. (1988), Paterson and Everet (1985)). Although the infection with *Agrobacterium tumefaciens* is easily possible, the combination of the two techniques, i.e., *Agrobacterium* mediated transformation and subsequent plant regeneration, to date seems difficult to achieve. Transformed sunflower plants have only been reported once (Everett et al. (1987)), and they reported serious problems with the selectable marker used, kanamycin, which inhibited morphogenesis.

In order to optimize conditions for sunflower transformation, we have investigated the response of hypocotyl explants of four sunflower lines towards three selectable markers, under various culture conditions. Resistance towards kanamycin and paromomycin, both aminoglycosides, is conferred by the enzyme, neomycin phospho transferase (Hain et al. (1985)), while the herbicide phosphinotricin (Basta) is inactivated by the enzyme, phosphinotricin acetyl transferase (De Block et al. (1987)). The utility of a selectable marker not only depends on the level of resistance conferred to a tissue by the introduction of a foreign gene, but is critically influenced by the level of spontaneous resistance of this tissue towards the utilized toxin. We observed in our study that this level of resistance was strongly influenced by the composition of the medium on which the selection was performed. Auxin, cytokinin, and gibberellic acid concentrations, as well as the type of N-source, all were identified as causes for variation of spontaneous resistance of sunflower hypocotyl slices towards phosphinotricin. The response towards paromomycin was much less dependant on the N-source, but still strongly modulated by the hormonal conditions.

2. Materials and Methods

Hypocotyl slices of *in vitro* grown sunflower plants (inbred lines HA 401B, HA 300 B, RHA 274, and hybrid line Mirasol) were cocultured with *Agrobacteria* according to

R.S. Sangwan and B.S. Sangwan-Norreel (eds.), The Impact of Biotechnology in Agriculture, 345–353.

Media	Nitrate (g/l)	Caseine-hydrolysate (g/l)	Naphthalene-acetic acid (NAA) (mg/l)	Benzyl-adenine (BAP) (g/l)	Gibberellic acid (GA_3) (mg/l)	Ratio NAA/BAP
PER	6.9	0.5	1.0	1.0	0.1	1
HaR	1.9	1.0	2.0	0.2	0	10
H 1	1.9	1.0	5.0	0.02	0	75
H 2	1.9	1.0	5.0	0.1	0	50
H 3	1.9	1.0	5.0	0.5	0/ 0.1	10
H 4	1.9	1.0	1.0	0.5	0	2
H 5	1.9	1.0	1.0	1.0	0/ 0.1	1
H 6	1.9	1.0	0.1	1.0	0	0.1
H 7	1.9	1.0	0.1	5.0	0/ 0.1	0.02
H 8	1.9	1.0	0.02	5.0	0	0.013
N 1	6.9	0.50	5.0	0.5	0	10
N 2	4.4	0.50	5.0	0.5	0	10
N 3	1.9	0.50	5.0	0.5	0	10
N 4	1.9	0.75	5.0	0.5	0	10
N 5	1.9	1.00	5.0	0.5	0	10

Table 1
Composition of the media used in this study. Both PER (Paterson and Everett 1985) and HaR (Hahne unpublished) are modified MS media (Murashige and Skoog 1962). Medium variations N 1 - N 5 served to investigate the dependance of phosphinotricin - resistance on the nitrogen source, while H 1 - H 8 were used to study the influence of the hormonal balance.

Horsch et al. (1985), and cultured on the respective medium (table 1), containing 500 µg/ml Cefotaxim to control growth of *Agrobacteria*. The strains of *Agrobacterium tumefaciens* used in this study were GV 3101 (pGV 3850) - negative control; GV 3101 (pGV 3850::1103) - *neo* gene under control of the *nos* promoter; C58 C1 Rif^r (pGSFR 1280) - *bar* gene under control of the Cauliflower Mosaic Virus 35 S promoter, as well as the *neo/nos* gene. The explants were evaluated after 10-12 days of culture under selective conditions. The difference between green explants showing developing callus and brown, dying explants was obvious. This visual analysis corresponded well with vital staining assays (Towill and Mazur (1975)). For the detection of NPT II activity we used the protocol described by McDonnell et al. (1987), and GUS assays were performed according to the protocol provided with the GUS-kit (Clontech Labs., Palo Alto, California, USA). All experiments were repeated at least three times, with twenty explants for each point.

3. Results

The goal of this study was to define conditions for efficient and reliable selection of transformed sunflower tissues. We therefore studied the response of hypocotyl slices (transformed and *Agrobacterium*-treated, but non-transformed) from four lines on two different media currently in use for callus induction in our laboratory. Since the response

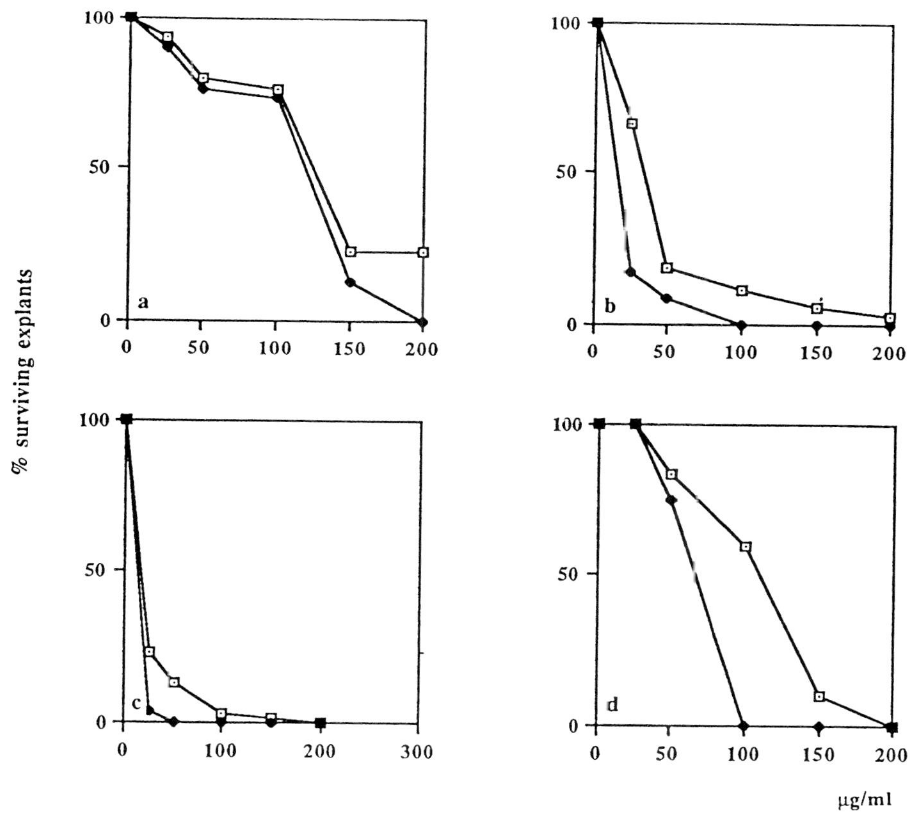

Fig. 1
Response of hypocotyl slices of four genotypes (RHA 274: **a**; HA401B: **b**; Mirasol: **c**; HA300B: **d**) towards selection on phosphinotricin on PER medium (see table 1). Open squares: transformed explants, solid squares: untransformed controls.

was found to be different on these two media, key media components were subsequently tested for their effect on the selection efficiency, using one representative genotype.

3.1. RESISTANCE TO SELECTABLE AGENTS: GENOTYPIC EFFECTS

We tested the four lines for their resistance (spontaneous and induced by transformation) towards the three selective agents: kanamycin, paromomycin, and phosphinotricin. The level of spontaneous resistance to all three selective agents varied strongly in function of the genotype. The response curve of the transformed tissues followed relatively closely that of the controls (tissues treated with *Agrobacteria* not carrying the resistance gene), leaving only a relatively small window for selection of transformed material. An example of this differential response is depicted in fig. 1, showing the differential behaviour of the four genotypes towards selection on phosphinotricin. Similar differences in response were obtained using the two aminoglycoside antibiotics (data not shown).

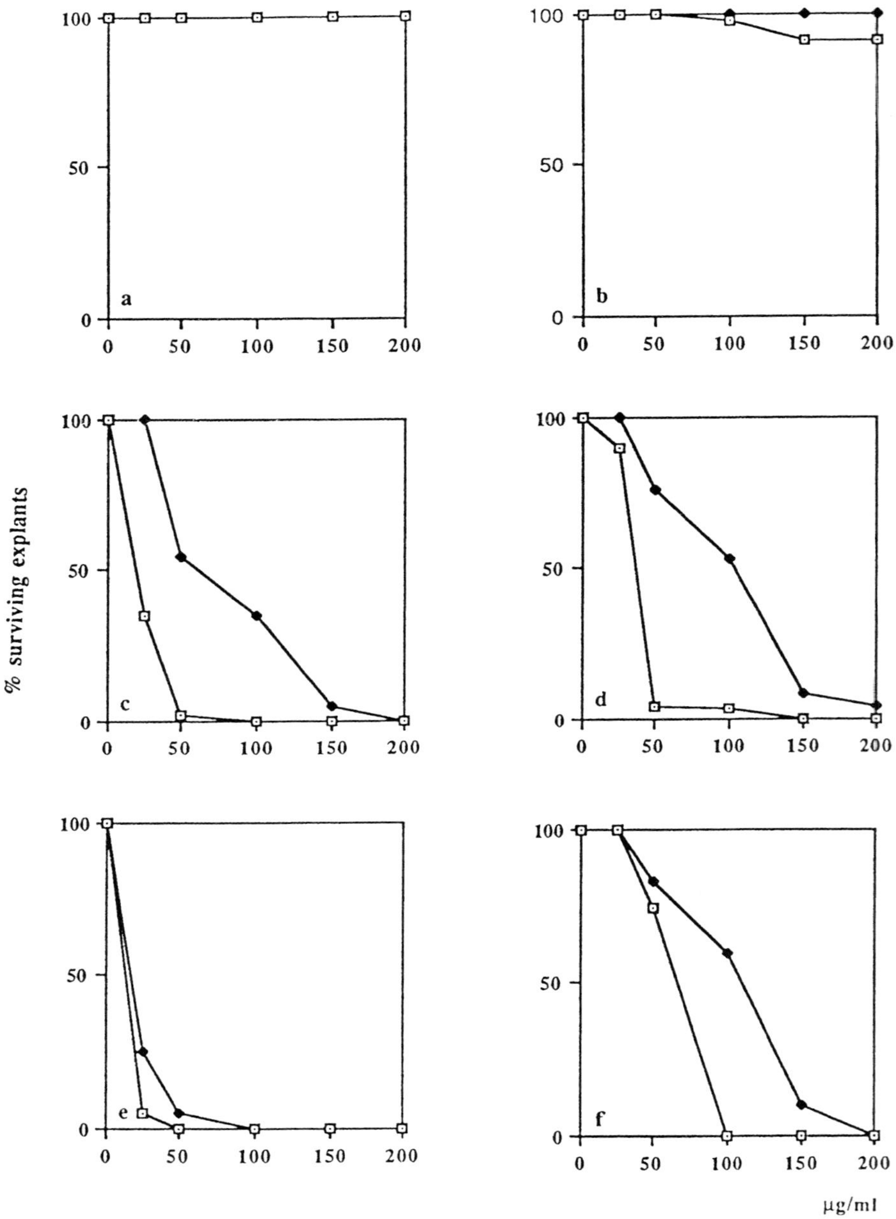

Fig. 2
Response of hypocotyl slices (HA300B) towards selection on kanamycin (**a, b**), paromomycin (**c, d**), and phosphinotricin (**e, f**) on 2 different media: HaR (**a, c, e**) and PER (**b, d, f**); see table 1 for compositions. Solid squares: transformed explants, open squares: untransformed controls.

3.2. RESISTANCE TO DIFFERENT SELECTABLE AGENTS: MEDIA EFFECTS

The three markers differed markedly in their utility for selection of transformed tissues. Kanamycin did not show a significant inhibitory effect on untransformed tissue even at concentrations of up to 200 µg/ml. Higher concentrations did decrease viability, but to the same extent in transformed and non-transformed tissue. Paromomycin, however, was inhibitory in all four lines at concentrations between 50 and 150 µg/ml. Transformed tissue showed a significantly higher resistance towards this antibiotic. A similar observation was made when phosphinitricin was used as selective marker. Toxicity towards untransformed tissues was observed at even lower concentrations, and transformed tissues showed a good level of resistance towards phosphinotricin. Using this selective agent, the difference between transformed and untransformed tissue was easier to assess visually than using paromomycin, because tissue browning was more pronounced. However, both are useful as selectable markers. Representative results obtained with one genotype are displayed in fig. 2. The other genotypes showed similar responses.

Apart from the variation of resistance in function of genotype and marker used it became evident that the media composition also has a considerable influence on the level of spontaneous and induced resistance (fig. 2). This effect was observed for all three markers in all four genotypes. Depending on the medium used, paromomycin and phosphinotricin are both useful markers in combination with all four lines, while kanamycin in our hands is not a successful marker. The calli visually selected as being "transformed" were subjected to an assay of NPT II activity. For both paromomycin and phosphinotricin, NPT II activity was always found, while the controls were negative.

In order to identify the factors important for this medium effect, we performed a detailed investigation of the influence of key media components on the response of transformed and control hypocotyl explants of one genotype, HA 300 B, towards the herbicide, phosphinotricin. We concentrated on the factors where the media used in the previous study differed: the hormonal composition and the nitrogen source (table 1).

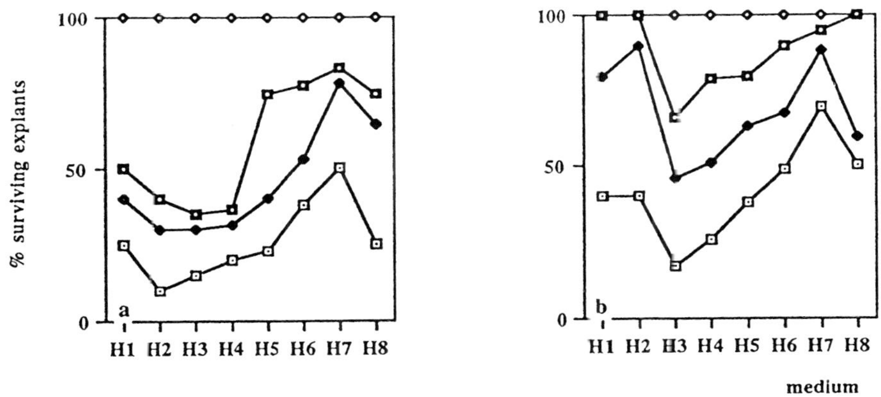

Fig. 3
Response of hypocotyl slices (HA300B) towards selection on phosphinotricin on HaR medium containing a variable ratio of naphthyl acetic acid and benzyl adenine (see table 1). The curves correspond to (from top to bottom): 0, 25, 50, 100 µg/ml phosphinotricin. **a**: untransformed control, **b**: transformed explants.

3.3. HORMONAL BALANCE

Keeping the basal composition of the medium constant, we varied the ratio between auxin (naphthalene acetic acid, NAA) and cytokinin (benzyladenine, BAP) from 75:1 to 1:75. Both control and transformed cultures exhibit a strong variation of their resistance in function of the hormonal composition (fig. 3).

Gibberellic acid was found to have a strong modulating influence on the response of the hypocotyl explants towards the herbicide (data not shown). While both transformed tissue and untransformed control showed the usual strong variation in function of the auxin/ cytokinin ratio in the absence of gibberellic acid towards phosphinotricin at 50 mg/l, the response was levelled for both explant types in the presence of gibberellic acid. For all tested auxin/ cytokinin values, the control explants showed a low survival when cultured on medium containing 0.1 mg/l gibberellic acid. In contrast, the survival rate of transformed tissue was uniformly high under these conditions.

3.4. NITROGEN SOURCE

The balance between reduced and oxidized nitrogen sources is the second major difference between the two media utilized in the previous study. Both contain the same amount of ammonium ions, but differ significantly in their content of nitrate, and both are supplemented with different amounts of casein hydrolysate. Therefore we used a range of media, all of which contained the three nitrogen sources, but at varying proportions. Ammonium was left constant at the level of the basic MS medium. Nitrate was stepwise decreased from the very elevated level of the PER medium to the value of MS medium, and casein hydrolysate was increased stepwise from the level of PER to that of HaR. In this way, the ratio between oxidized inorganic (nitrate) and reduced organic (casein hydrolysate) nitrogen supply was constantly lowered.

The transformed tissues showed no clear dependance on the nitrogen source in the selective medium at any of the three phosphinotricine concentrations (fig. 4). The

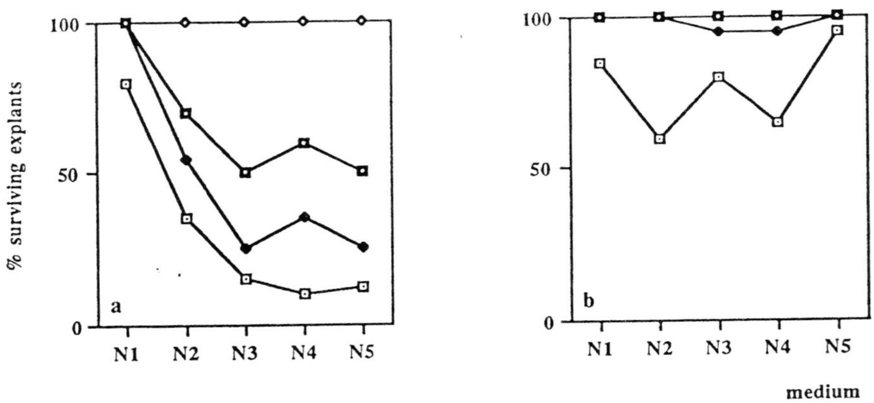

Fig. 4
Response of hypocotyl slices (HA300B) towards selection on phosphinotricin on medium containing varying ratios of nitrate/ casein hydrolysate (see table 1). The curves correspond to (from top to bottom): 0, 25, 50, 100 μg/ml phosphinotricin. **a**: untransformed control, **b**: transformed explants.

untransformed tissues, however, were clearly more resistant towards the herbicide in the presence of high levels of nitrate, even up to 100 μg/ ml phosphinotricin. Decreasing nitrate concentrations corresponded to decreasing spontaneous resistance (fig. 4). Further decrease of the nitrate/ casein hydrolysate ratio by increasing the casein hydrolysate level at a constant nitrate concentration did not produce a further clear corresponding variation (fig. 4).

From this experiment it is clear that the nitrogen supply contained in the selective medium has a strong influence on the behaviour of untransformed sunflower hypocotyl explants towards phosphinotricine, an effect that is not found with transformed tissue. High degrees of spontaneous resistance are apparently caused by a high nitrate concentration. The herbicide, phosphinotricin, acts as inhibitor of the enzyme, glutamine synthetase, thus increasing intracellular ammonium concentrations (Thompson et al. (1987), De Block et al. (1987)). Since this enzyme is also involved in the nitrate assimilation pathway, this effect of nitrate to increase the spontaneous resistance might possibly be confined to phosphinotricin action. To test this possibility, a second series of experiments was conducted, monitoring the resistance of hypocotyl slices towards the aminoglycoside, paromomycin. When selecting on paromomycin containing media, no nitrogen dependance of the resistance was observed in transformed, nor in untransformed tissue (data not shown). A high nitrate level, a high casein hydrolysate level, as well as an intermediate level all produced a uniform response towards the two paromomycin concentrations utilized (data not shown). The nitrate effect on spontaneous resistance does therefore not seem to be a universal phenomenon, but rather limited to phosphinotricin selection.

3.5. TRANSFORMATION EFFICIENCY

Of all cells present in a hypocotyl explant, only a relatively small proportion will be actually transformed by *Agrobacterium*. We conducted a series of experiments to determine the extent of transformation. Sunflower (HA300 B) hypocotyl explants were treated with *Agrobacteria* carrying the GUS gene. The histological reaction with x-gluc allows to visualize and localize those cells that express the enzyme activity.

Surprisingly few cells were found to be transformed, the number of blue spots detected per explant depending, to a certain degree, on the culture medium employed. A ratio of auxin/cytokinin of 10 consistently only yielded 1-5 transformed cells per hypocotyl slice, while with a ratio of 0.1, 10-30 transformed cells were observed per explant (data not shown). In both cases, the negative controls were absolutely free of blue spots. Since the transformation itself was performed on an identical medium in both cases (hormone-free basal MS medium), the medium effect must consist in a differential stimulation of the transformed cells.

The few transformed cells are apparently sufficient to confer to the whole tissue slice resistance towards the selectable marker, at least considering survival and green color. However, formation of callus under selection pressure proceeds from transformed cells, as GUS-transformed explants analyzed at a later time mainly contained callus with blue-staining cells.

The cells responding under our experimental conditions are all found in the central region of the hypocotyl, and not clearly associated with a particular tissue type. The cell type responding with callus formation, however, depends on the hormonal composition of the medium. This fact, in combination with the small number of transformed cells, may explain the variation of the transformation efficiency between the two media.

4. Discussion

Selectable markers have been evaluated for several species (Negrutiu et al. (1984), Hauptmann et al. (1988)), and one ideal marker suitable for all cases does not yet exist. We have studied the interaction between sunflower genotypes and three markers with two different mechanisms of action (inhibition of translation and inhibition of glutamine synthetase). A practically useful selectable marker can be found in both categories, provided the culture conditions are chosen correctly.

The complex media used in plant tissue culture provoke a multitude of interconnected physiological changes in the cultured tissues, some of which may have profound influence on their behavior under selective conditions. We found that the balance of hormones contained in the selective medium modulated strongly the percentage of surviving explants in both transformed and control tissues. A detailed analysis of this influence does not exist, but it is probable that the different conditions favor the proliferation of different cell types, which in turn may be differentially susceptible towards the toxin. This hypothesis is further supported by the observation that the number of transformed cell colonies varies with the hormone level.

The nitrogen source may be equally important as the hormones. The presence or absence of nitrate and ammonium ions have a strong regulatory effect on key metabolic pathways, such as the nitrogen assimilation pathway (Rajasekhar and Oelmüller (1987)). The presence of a reduced organic N-source, such as amino acids or casein hydrolysate, prevents selection with phosphinotricin in rice tissues, by bypassing the inhibited enzyme, glutamine synthetase (Dekeyser et al. (1989)). This effect was not observed in our experiments with sunflower hypocotyl explants. In contrast, we found nitrate to induce resistance towards this herbicide, while the presence of casein hydrolysate was not disturbing. De la Haba et al. (1988) found that nitrate induces several enzymes in the nitrate assimilation pathway of sunflower, glutamine synthetase being among them. Glutamine synthetase as well as other nitrate assimilating enzymes are present at lower concentrations in cultured sunflower protoplasts than in tobacco protoplasts (Lenée and Chupeau (1989)), so that the increase provoked by increasing nitrate concentration may lead to a significant increase in glutamine synthetase activity, effectively outcompeting the toxic inhibitor. Our observation that paromomycin selection is not influenced by the N-source of the selection medium supports this view. Paromomycin selection is, however, modulated by a medium effect, and the analysis of the utilized media shows that this must be due to hormonal influences.

In conclusion, our experiments show that it is necessary and possible to optimize the conditions for selection of transformed tissues, and that it may be difficult to transpose the experience gained with one system for the use with another species or even another genotype. The adaptation of culture condition during selection may significantly enlarge the window between killing almost all of the non-transformed tissue and a maximal survival of the transformed cells.

References

1. De Block, M., Botterman, J., Vanderwiele, M., Dockx, J., Thoen, C., Goessele, V., Movva, N.R., Thompson, C., Van Montagu, M., and Leemans, J. (1987) 'Engineering herbicide resistance in plants by expression of a detoxifying enzyme', EMBO J. 6, 2513-2518.
2. Dekeyser, R., Claes, B., Marichal, M., Van Montagu, M., and Caplan, A. (1989) 'Evaluation of selectable markers for rice transformation', Plant Physiol. 90, 217-223.
3. De La Haba P., Agüera, E., and Maldonado, J.M. (1988) 'Development of nitrogen-assimilating enzymes in sunflower cotyledons during germination as affected by the exogeneous nitrogen source', Planta 173, 52-57.
4. Everett N.P., Robinson, K.E.P., and Mascarenhas, D. (1987) 'Genetic engineering of sunflower (*Helianthus annuus* L.)', Bio/Technol. 5, 1201-1204.
5. Hain R., Stabel, P., Czernilofsky, A.P., Steinbiss, H.H., Herrera-Estrella, L., and Schell, J. (1985) 'Uptake, integration, expression, and genetic transmission of a selectable chimaeric gene by plant protoplasts', Mol. Gen. Genet. 199, 161-168.
6. Hartmann C.L., Donald, P.A., Secor, G.A., and Miller, J.F. (1988) 'Sunflower tissue culture and its use in selection for resistance to *Phoma macdonaldii* and white mold (*Sclerotinia sclerotiorum*)', Proc. 12th Int. Sunflower Conference, Vol. II, Int. Sunflower Assoc. (ed.), Novi Sad, Yugoslavia, pp. 347-351.
7. Hauptmann, R.M., Vasil, V., Ozias-Akins, P., Tabaeizadeh, Z., Rogers, S.G., Fraley, R.T., Horsch, R.B., and Vasil, I.K. (1988) 'Evaluation of selectable markers for obtaining stable transformants in the *Gramineae*', Plant Physiol. 86, 602-606.
8. Horsch, R.B., Fry, J.E., Hoffmann, N.L., Eichholz, D., Rogers, S.G., and Fraley, R.T. (1985) 'A simple and general method for transferring genes into plants', Science 227, 1229-1231.
9. Lenée, P., and Chupeau, Y. (1989) 'Development of nitorgen assimilating enzymes during growth of cells derived from protoplasts of sunflower and tobacco', Plant Sci. 59, 109-117.
10. McDonnell, R.E., Clark, R.D., Smith, W.A., and Hinchee, M.A. (1987) 'A simplified method for the detection of neomycin phosphotransferase II activity in transformed plant tissue', Plant Mol. Biol. Rep. 5, 380-386.
11. Murashige, T., and Skoog, F. (1962) 'A revised medium for rapid growth and bioassays with tobacco tissue cultures', Physiol. Plant. 15, 473-497.
12. Negrutiu, I., Jacobs, M., and Cattoir-Reynaerts, A. (1984) 'Progress in cellular engineering of plants: biochemical and genetic assessment of selectable markers from cultured cells', Plant Mol. Biol. 3, 289-302.
13. Paterson, K.E., and Everett, N.P. (1985) 'Regeneration of *Helianthus annuus* inbred plants from callus', Plant Sci. 42, 125-132.
14. Rajasekhar, V.K., and Oelmüller, R. (1987) 'Regulation of induction of nitrate reductase and nitrite reductase in higher plants', Physiol. Plant. 71, 517-521.
15. Thompson C.J., Movva, N.R., Tizard, R., Crameri, R., Davies, J.E., Lauwereys, M., and Botterman, J. (1987) 'Characterization of the herbicide resisrance gene *bar* from *Streptomyces hygroscopicus*', EMBO J. 6, 2519-2523.
16. Towill L.E., and Mazur, P. (1975) 'Studies on the reduction of 2,3,5- triphenyl tetrazolium chloride as a viability assay for plant tissue cultures', Can. J. Bot. 53, 1097-1102.
17. Wilcox McCann A., Cooley, G., and Van Dreser, J. (1988) 'A system for routine plantlet regeneration of sunflower (*Helianthus annuus* L.) from immature embryo-derived callus', Plant Cell Tissue Org. Cult. 14, 103-110.

DIRECT GENE TRANSFER AND GENE RESCUE IN SUGARBEET PROTOPLASTS

K. LINDSEY and *P. GALLOIS
Leicester Biocentre, University of Leicester,
Leicester LE1 7RH, U.K. and
*AFRC Inst. of Arable Crops Research,
Dept. of Biochemistry, Rothamsted Exptl. Station,
Harpenden, Herts. AL5 2JQ, U.K.

ABSTRACT. A system has been developed for the stable transformation of protoplasts of sugarbeet (*Beta vulgaris*) using electroporation-mediated direct gene transfer. Conditions for the reversible permeabilization of protoplasts, and the subsequent uptake of foreign DNA, were optimized by following the uptake of a) a hydrophilic dye, phenosafranine and b) the *cat* and *gus* reporter genes in transient expression studies. The frequencies of protoplast division and colony development were optimized, and stable integration of the aminoglycoside phosphotransferase gene, after electroporation, was demonstrated by the resistance of protoplast-derived colonies to kanamycin and by Southern blot analysis. Preliminary experiments were carried out to develop a shotgun cloning strategy for the isolation of selectable genes from plant tissue, using a direct gene transfer approach

1. INTRODUCTION

1.1 *Transformation systems in plants*

Methods for introducing foreign DNA into plant cells can be classified broadly into those which employ the use of vectors and those which do not. The most important of the former type is *Agrobacterium*-mediated gene transfer (Klee et al., 1987a). This system exploits the fact that the soil-borne bacteria *A. tumefaciens* and *A. rhizogenes*, which are the causative agents of the crown gall and hairy root diseases respectively, exert their pathogenic effects by transferring specific plasmid sequences into plant cells to which they are attached, at wound sites. Genetic modification of the tumour-inducing (Ti) and root-inducing (Ri) plasmids to produce the binary (e.g. Bevan, 1984) and co-integrate (e.g. Rogers et al., 1988) vector systems has allowed the successful transformation and regeneration of over twenty commercially important plant species, without the oncogenic symptoms characteristic of transformation by wild-type strains of *Agrobacterium*.

Despite this achievement, the application of this approach to all plants is far from routine: the cereals, which in economic terms are the most important category of arable crop, are not natural hosts of

R.S. Sangwan and B.S. Sangwan-Norreel (eds.), The Impact of Biotechnology in Agriculture, 355–380.

the pathogen and currently are not susceptible to *Agrobacterium*-mediated transformation techniques (although it may eventually be possible to overcome this problem as more becomes known about the mechanism of T-DNA transfer and integration). Furthermore, the success of transformation of a given dicotyledonous species may not be readily reproducible for closely related species or even of different cultivars or varieties of the same species. The reasons for this are not entirely clear, but may relate to differences in regeneration capacity, in susceptibility to selective agents and/or in the plant-bacteria interactions (e.g. the ability of the plant cells to produce the phenolic signal molecules necessary to induce the T-DNA transfer process, Stachel et al., 1985).

A number of alternative strategies are evolving which avoid the use of *Agrobacterium* and are therefore of potential value for gene transfer to the cereals and other similarly recalcitrant species. These methods include:

- *injection of DNA*, either into single cells or organized structures (e.g. Reich et al., 1986; de la Pena, 1987);
- *DNA-coated microsphere transformation*, in which tungsten or gold particles carry foreign DNA into intact cells, driven by either explosive or electrostatic charge (e.g. Klein et al., 1988; McCabe et al., 1988);
- *chemically-mediated DNA transfer* to protoplasts, by either polyethylene glycol treatment (e.g. Krens et al., 1982; Lorz et al., 1985), calcium phosphate-DNA co-precipitation (Hain et al., 1985) or liposomes (e.g. Deshayes et al., 1985);
- *electroporation.*

It is this latter technique which we will discuss in detail in this article, with particular reference to DNA transfer to protoplasts of sugarbeet, *Beta vulgaris*. The intention is to describe a strategy for the development of a method for the stable transformation of plants based on electropermeabilization of protoplasts, and for the identification of factors which influence the efficiency of this process. Finally we will introduce an application of direct gene transfer other than simply for generating transgenic plants: namely in gene rescue and mutant complementation studies.

1.2 *Basic features of electroporation*

When microbial or animal cells or plant protoplasts are subjected to short (µs or ms duration) pulses, they may, under appropriate conditions, experience an increase in the permeability of the plasmamembrane to molecules which,they are hydrophilic, are normally unable to enter the cell. Nucleic acids are such molecules. Above a threshold field strength, electric pulses induce irreversible damage to membrane structure resulting in cell death. Below this threshold value, however, threre may occur a non-lethal permeabilization of the plasmamembrane termed 'reversible dielectric breakdown' (Zimmermann, 1986), which it has been suggested is due to the transient formation or enlargement of existing pores in the lipid bilayer (e.g. Sugar and

Neumann, 1984). In animal systems it has been demonstrated that the effects of electrical fields on membrane permeability are relatively long-lived (minutes) compared with the duration of the applied pulse (see Zimmermann, 1986) and similar conclusions can be made for plant protoplasts (discussed below for sugarbeet).

1.3 *The sugarbeet experimental system*

The experimental system to be described makes use of two types of protoplast of sugarbeet: those derived from liquid-cultured cells, originally produced from hypocotyl callus ('suspension protoplasts') and those derived from the leaf tissue of sterile shoot cultures ('leaf protoplasts').

Electrical pulses were delivered by either of two systems: the first generates rectangular pulses, delivered at a constant predetermined voltage for a predetermined duration (Fig. 1a), both parameters being precisely controlled by sophisticated electronics. In the second electroporation system, pulses were delivered from a bank of capacitors charged to a preset voltage. Here, pulses decay exponentially and the duration at a given field strength value is transient (Fig. 1b). The area under the graph of voltage vs. time depends predominantly upon the capacitance of the system and also the resistance of the solution between the electrodes in which the protoplasts are suspended (as well as, to a lesser extent, the impedance of the circuitry). Pulse duration is conventionally described as a time constant represented by the time taken for the voltage to decay to approx. 37% (in fact 1/e) of its initial value, and can be manipulated by altering the capacitance and/or resistance of the medium.

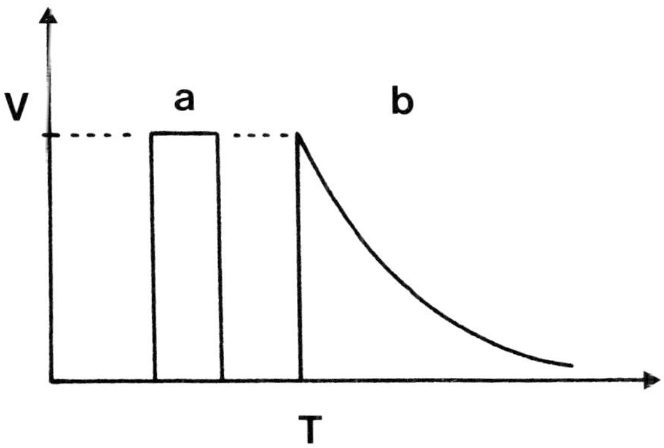

Fig. 1: Form of (a) rectangular and (b) exponentially decaying pulses. Dotted line represents initial voltage.

2. CHANGES IN PLASMAMEMBRANE PERMEABILITY IN ELECTROPORATED PROTOPLASTS AND CELLS OF SUGARBEET

2.1 *Phenosafranine accumulation*

While a large amount of work has been carried out into the effects of electric fields on animal cells (see Zimmermann, 1986) relatively little is known about the electropermeabilization of plant cells and protoplasts. The greatest interest in electroporation techniques for plants has been not in biophysical studies but in relation to genetic transformation. As a consequence, 'permeability' changes under different electrical conditions have been determined most commonly by functional assays, namely by measuring the level of expression of introduced reporter genes, such as in transient expression studies of the chloramphenicol acetyltransferase (*cat*) or β-glucuronidase (*gus*) genes. Such experiments are certainly valuable, and will be discussed in some detail below. However, the use of such techniques in isolation is not completely satisfactory, since it is a rather indirect determination of permeability changes: a lack of expression could be due to any one of a number of factors:

- there may be no uptake of DNA under the experimental conditions used;
- the gene construct may be non-functional, e.g. due to a mutation;
- the plasmid may be unstable within the protoplast;
- cellular factors may limit the level of expression of the gene construct.

To avoid these potential problems, and to determine more directly changes in plasmamembrane permeability *per se*, the uptake of dye molecules which are normally excluded by intact protoplasts and cells was measured under a range of electrical conditions. The objective here was to optimize the effects of eletroporation on membrane permeability as a prelude to DNA uptake studies, and also to obtain some information on the nature of the permeabilization process itself.

The dye molecule chosen for this work was phenosafranine (3,7-diamino- 5-phenylphenazinium chloride). This red compound is largely excluded from cells which possess a structurally intact outer membrane, and has been used as a viability stain (Widholm, 1972). The rationale of the approach was that, if protoplasts or cells were subjected to electrical pulses while suspended in a medium containing phenosafranine, successful permeabilization would be revealed as uptake of the dye (Lindsey and Jones, 1987a). Membrane resealing after electroporation, to maintain cell viability, could be determined by both the retention (net accumulation) of the dye after washing and also by a second staining with the viability indicator fluorescein diacetate (Widholm, 1972).

This technique was used to obtain information in three areas: first, in optimizing electropermeabilization conditions, by investigating the effects of a range of electrical parameters on phenosafranine accumulation; secondly, in determining the proportion of a population of protoplasts or cells which is permeabilized under a given set of electrical conditions; and third, in studying the kinetics of the permeabilization process.

2.2 *Influence of electrical parameters on permeabilization*

In order to quantify the permeabilization effects of electrical pulses, a spectrophotometric assay was developed to determine the amount of phenosafranine accumulated by a population of cells or protoplasts. The details of the method are to be found in Lindsey and Jones (1987a). Essentially, electroporation was carried out in the presence of a known amount of the dye, followed by a 45 min. incubation period to allow membrane resealing to take place (see below) and then three washes in culture medium to remove both extracellular dye and intracellular dye released by damaged cells. The remaining trapped phenosafranine was extracted in ethanol and its concentration determined spectrophoto metrically. The mean amount of dye accumulated on a per cell basis was thereby calculated, and results for each experimental treatment were expressed as the difference in phenosafranine content between electroporated populations and non-electroporated controls (which represent the background level of staining).

The effect of field strength on phenosafranine accumulation by suspension culture cells and suspension protoplasts of sugar beet is illustrated in Fig. 2. On applying 5 x 50 µs pulses over a series of field strengths, dye accumulation and retention increased in proportion to increased field strength values. The fact that there was observed uptake at all field strengths tested is consistent with the view that different populations of cells/protoplasts are permeabilized at different field strengths; it is known that cell size is a factor determining the voltage at which membrane breakdown occurs (Neumann et al., 1982; Zimmermann, 1986), and individuals in a culture of sugarbeet suspension cells or protoplasts vary enormously in size (between approx. 30-150 µm diameter for cells, 10-50 µm for protoplasts). Presumably a reflection of the difference in volume is the higher levels of phenosafranine accumulated by electroporated cells compared with electroporated protoplasts (Fig. 2).

Similar studies can be carried out into the effects of applied pulse number and duration (Lindsey and Jones, 1987a). For example, if 20 pulses (50 µs) were given at increasing field strength (Fig. 2b), it was found that the highest level of accumulation of dye, representing an optimum trade-off between reversible and irreversible permeabilization, occurred at a lower field strength (500-1000 V/cm) than if fewer (5) pulses were applied. Similarly, if multiple (5) pulses of duration greater than 50 µs were applied, phenosafranine accumulation was reduced, again presumably due to irreversible membrane damage.

3. VIABILITY OF ELECTROPORATED PROTOPLASTS AND CELLS

The reduced level of phenosafranine accumulated at high pulse number and duration is presumably a consequence of damage to the plasma membrane, such that post-pulse resealing is incomplete. This possibility was examined further by fluorescein diacetate staining for cell and protoplast viability (Widholm, 1972; Larkin, 1976). In this method, all cells and protoplasts take up the membrane-permeable fluorescein diacetate molecule, but intracellular esterase activity cleaves the diacetate moiety from the fluorescein which, being membrane impermeable, is retained and detectable under UV illumination only in protoplasts/cells which possess an intact (i.e. resealed) plasma-membrane.

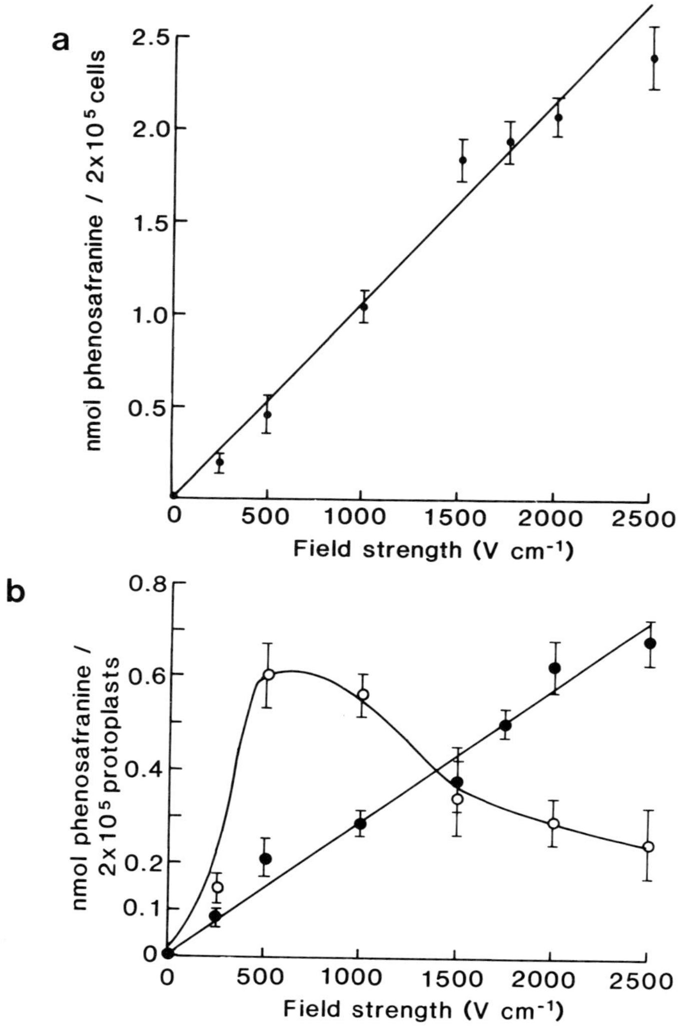

Fig. 2: Relationship between phenosafranine accumulation and field strength for cells (a) and protoplasts (b). For protoplasts, either 5 (closed circles) or 20 (open circles) pulses were supplied; cells were given 5 pulses.

Both the short term and longer term effects of electroporation on viability were investigated. If up to 10 x 50 μs pulses, of field strengths of up to 2500 V/cm were applied to intact cells of sugarbeet, there was observed no detrimental effect on viability, when measured at 2 hours after electroporation. If however 25 or more pulses were supplied at field strengths of more than 500 V/cm, viability at 2 hours was significantly reduced (Lindsey and Jones, 1987a).

Protoplasts were more susceptible to damage than cells. At field strengths of up to 500 V/cm there was no significant difference in the ability of the plasmamembrane to reseal, whether 5, 10 or 25 pulses (50 μs) were applied, but there was a loss in viability in about 10-15% of the population. If only 5 pulses were applied, increasing the field strength to 2500 V/cm had no further detrimental effect on viability as measured at 2 hours. Higher pulse numbers and field strengths did however cause more immediate damage to the resealing process such that only 12% of protoplasts given 25 pulses of 2500 V/cm exhibited fluorescence, and division could not be induced subsequently in non-fluorescing protoplasts. Some bursting of protoplasts under these more extreme conditions was observed.

More long term effects of electroporation on protoplast viability were determined as the proportion of the population staining with fluorescein diacetate 45 hours after electroporation (Lindsey and Jones, 1987b). The viability of the control (unpulsed) population at 45 hours was 85-90%, but the survival of protoplasts was reduced with increasing pulse number (1-10) and pulse duration (50 or 99.9 μs). Loss in viability due to the use of longer pulses could be reduced by increasing the interval between pulses from 0.5s to 9.9s. Effects of electrical conditions on plating efficiency is discussed later, in relation to the influence upon transformation efficiency.

The reasons why cells are more resistant to damage from multiple pulses, particularly at higher field strengths, are not clear: this phenomenon is in spite of the fact that the mean size of cells is greater than that of protoplasts, and on the basis of this alone might be expected to exhibit a lower, rather than a higher, field strength requirement for irreversible membrane breakdown. It may be that the aggregated nature of cell suspensions is protective, acting to diffuse the charge. Certainly it was found that there is an inverse relationship between cell density and the proportion of cells permeabilized, at densities greater than about 2 x 10^5/ml. A second possibility is that the cell wall-degrading enzymes used in protoplast isolation in some way affect membrane stability, but no direct evidence for this is available; however, Browse et al. (1988) have found that the lipid composition of *Arabidopsis* cells changes during protoplast formation, and this may influence the ability of the plasmamembrane to reseal after electroporation.

4. KINETICS OF PERMEABILITY CHANGES AFTER ELECTROPORATION

The kinetics of the pore resealing process after electroporation was investigated by studying the ability of permeabilized protoplasts and cells to retain phenosafranine during the post-pulse incubation period.

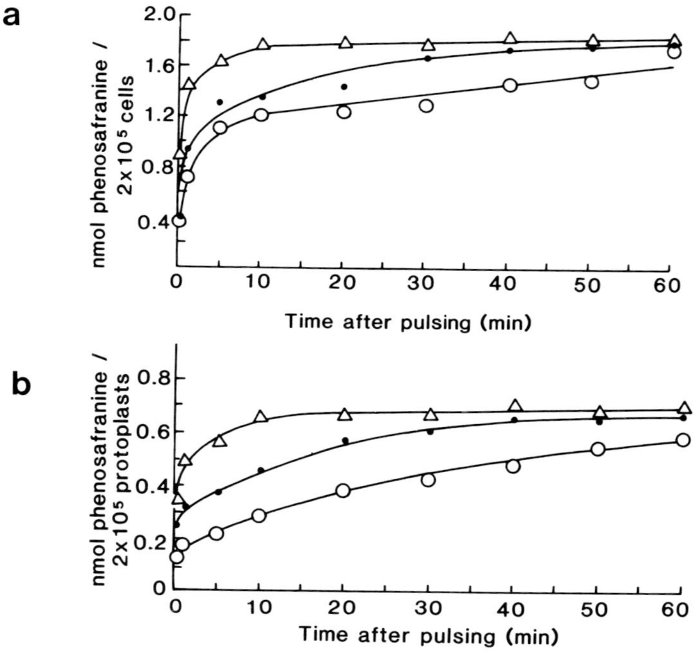

Fig. 3: Accumulation of phenosafranine by cells (a) and protoplasts (b), determined at intervals after electroporation, at 4°C (open circles), 25°C (closed circles) or 35°C (triangles).

The effect of temperature on membrane resealing was also examined. The experimental approach was to electroporate protoplasts and cells in the presence of phenosafranine and to wash populations at increasing time intervals after pulsing. The phenosafranine content at each time point was determined: relatvely low values would indicate incomplete resealing. Electrical parameters were chosen which induced relatively high levels of dye uptake but a minimal loss of viability in the population.

2 x 10^5 cells or protoplasts were therefore given a series of five pulses (50 µs, 2000 V/cm) at 25°C, and immediately incubated at either 4°C, 25°C or 35°C. The relationship between the incubation time after pulsing and the amount of phenosafranine accumulated after washing is illustrated in Fig. 3. The results show that 1) the pattern of phenosafranine accumulation and retention is similar for both cells and protoplasts at each of the three incubation temperatures; 2) by 60 minutes, similar levels of phenosafranine were accumulated and retained by cells and protoplasts respectively at each of the incubation temperatures; however 3) the retained phenosafranine content reached a maximum level, under the conditions tested, more rapidly at higher than at lower temperatures.That level of phenosafranine which was extractable from cells or protoplasts after 60 minutes of incubation at

4°C was retained after washing after approx. only 30 minutes (cells) and 25 minutes (protoplasts) at 25°C and after only 5 minutes (cells and protoplasts) at 35°C.

These data demonstrate that the time taken for the integrity of the plasmamembrane to be restored was relatively long (approx. 45 minutes at 25°C) compared with the duration of the pore-inducing electrical pulses (50 µs); and the sealing process was temperature-dependent, possibly due to differential membrane fluidity at the different temperatures. The results are consistent with evidence for the changes in the permeability of electroporated mouse thymocytes to the dye eosin (Zimmermann et al., 1980) and of chromaffin granules of the bovine adrenal medulla (Lindner et al., 1977), although the recovery period for sugarbeet cells and protoplasts (minutes) would appear to be longer than that in the chromaffin granules (seconds). The permeabilization ('pore formation') itself occurs relatively rapidly at 25°C, as indicated by the almost instantaneous electrically-induced fusion of adjacent sugarbeet protoplasts (Eady et al., 1988).

5. THE UPTAKE AND EXPRESSION OF FOREIGN DNA

The simple dye uptake experiments described here provide some useful information on electrically-induced changes in plasmamembrane permeability. In order to determine whether the conditions identified as permitting phenosafranine accumulation are also suitable for DNA uptake (as distinct from DNA integration into the genome), two types of study were undertaken: 1) the uptake of radiolabelled DNA and 2) the uptake of transiently expressed reporter genes.

5.1 *Uptake of radiolabelled DNA*

In the first study, sugarbeet protoplasts and cells were electroporated over a range of field strengths (5 pulses of 50 µs each) in the presence of [^{3}H]pABD1, a plasmid of 5.3 kb (Paszkowski et al., 1984), plus carrier (unlabelled) pABD1 and herring sperm DNA. After washing, the radioactivity associated with the protoplasts and cells was determined by scintillation counting, and the results are shown in Fig. 4. It was found that, for protoplasts, the accumulation of radioactivity increased with increasing field strength, in a qualitatively similar way to which phenosafranine was taken up. Under the conditions used, the highest level of radioactivity accumulated (by 2 x 10^5 protoplasts) represented approx. 1-2% of the total amount supplied. This result indicates that phenosafranine uptake is an acceptable model for optimizing conditions for DNA uptake by sugarbeet suspension protoplasts.

However, the pattern of accumulation of radioactivity by electroporated cells was quite different from that by protoplasts (Fig. 4). With increasing field strength the level of radioactivity associated with the cells declined rather than increased, from a relatively high level in unpulsed cells, which presumably is due to non-specific binding to the cell wall. The DNA was dissociated from the wall by the application of electric pulses, perhaps due to an as yet uncharacterized charge effect. This result demonstrates that the cell

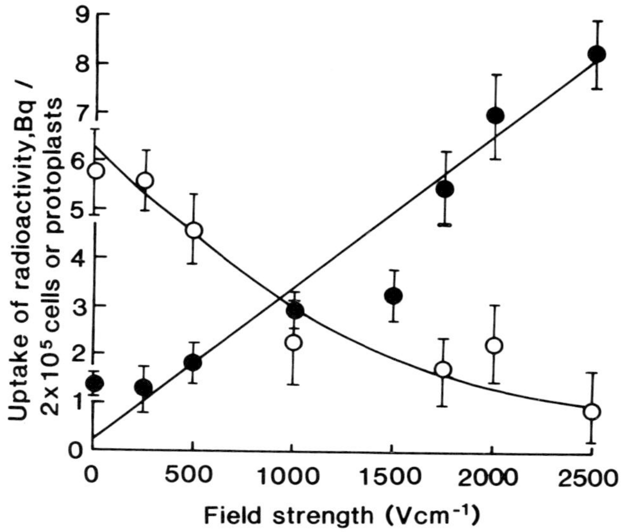

Fig. 4: Accumulation of [^{3}H]pABD1 by electroporated protoplasts (closed circles) and cells (open circles).

wall is a barrier to free DNA uptake, but does not preclude the possibility that some DNA can get into the cell: this experimental technique cannot distinguish between extracellular and intracellular DNA. The second type of study involving the transient expression of reporter genes, sheds more light on DNA uptake by both cells and protoplasts, and is now discussed.

5.2 *Transient expression of reporter genes*

The transient expression of reporter genes has been a valuable tool to study the uptake of foreign DNA and RNA by both animal cells (e.g. Gorman et al., 1982) and plant protoplasts (e.g. Fromm et al., 1985). The objective is to introduce genes into the cell, and specifically into the nucleus, and uptake is confirmed by assaying the gene product, which is usually an enzyme. The expression of the introduced reporter gene does not depend upon its integration into the genome and so there is no requirement for a selectable gene marker system. The fact that only a very small proportion of the introduced DNA will be integrated, the vast majority being free and subject to degradation, accounts for the transience of expression (see below). The three most commonly used reporter genes in plant systems include *cat*, encoding chloramphenicol acetyltransferase (Gorman et al., 1982), *gus* (*uidA*), encoding β-glucuronidase (Jefferson et al., 1987), and the genes encoding either bacterial (*lux*) or firefly (*luc*) luciferase (Schauer, 1988). A fourth reporter which may become increasingly used in transient assays is the *bar* gene, encoding phosphinothricin acetyltransferase (De Block et al., 1987) though this and other genes encoding readily assayable enzymes such as neomycin phosphotransferase, various opine synthases or β-

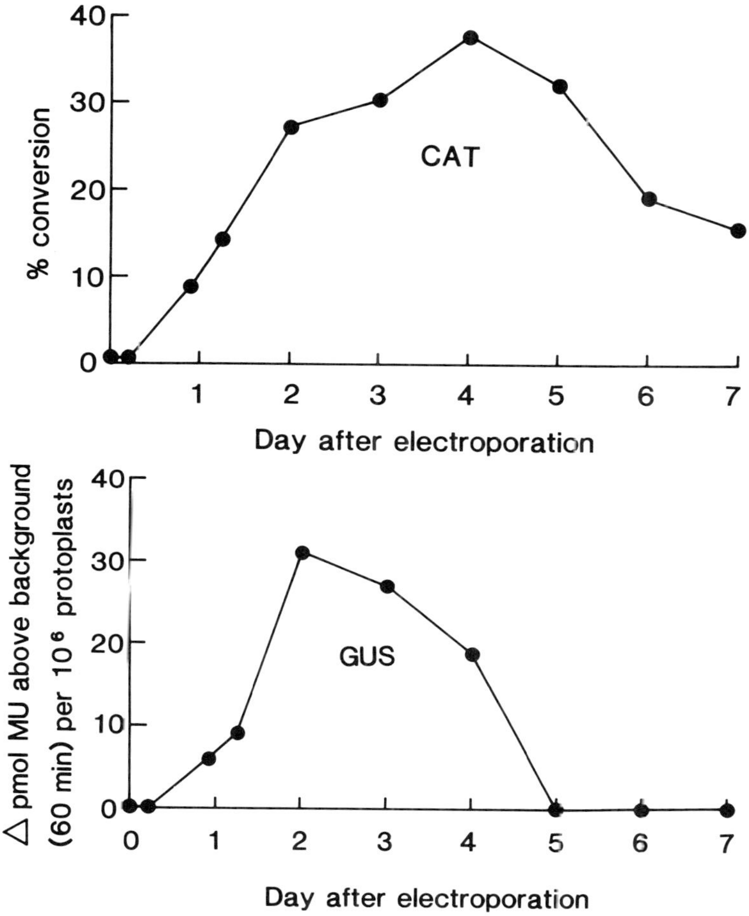

Fig. 5: Transient expression of CAT and GUS reporters.

galactosidase are more common as screenable markers for stably transformed tissues.

The applications of transient expression experiments can be listed as follows:

1) to determine whether, under a given set of direct gene transfer conditions, DNA is taken up into the cell and reaches the nucleus;
2) to determine whether a given gene construct is functional;
3) to determine the relative activity of a range of gene constructs, such as reporter genes fused to different promoters or 3' flanking regions, with or without introns etc.;
4) to determine the interaction between a given promoter region and protoplasts of a particular cell type (i.e. to study tissue-specific gene expression).

Unlike studies in which foreign genes are stably integrated into the chromosomes, transient expression avoids the problems of position effects which are known to influence the level and/or pattern of expression of an introduced gene and thereby complicate the interpretation of quantitative studies. As far as this article is concerned, however, the main point to consider is that, in combination with the work on phenosafranine uptake, transient expression studies can be used to obtain a clear picture of the effects of electroporation on changes in the permeability of the plasmamembrane to DNA molecules. This is considered to be a strong basis upon which to optimize a procedure for the stable transformation of protoplasts.

For transient expression studies with sugarbeet protoplasts two reporter genes were used, encoding CAT (p35SCN) and GUS (pCGUS) respectively, each on a pUC19-based plasmid. Both *cat* and *gus* coding regions were under the transcriptional control of the cauliflower mosaic virus 35S RNA gene (CaMV 35S) promoter and the nopaline synthase (*nos*) termination sequence. Expression of the genes was quantified as activity of the respective enzymes. The transient nature of CAT and GUS expression in sugarbeet suspension protoplasts is illustrated in Fig. 5; protoplasts were electroporated under conditions which induced a minimal loss of viability as measured 2 days after pulsing. CAT activity was found to reach a peak at d.4 after electroporation, while GUS activity peaked earlier, at d.2. The decline in activity is presumably due to the degradation of the plasmid, mRNA or protein. Since GUS enzyme activity shows a more rapid decline over the time course, there may be more rapid turnover of either the *gus* plasmid, transcript or mature enzyme than is the case for CAT. Since both p35SCN and pCGUS are similar in structure, size and configuration (being introduced predominantly in the supercoiled form), it is perhaps more likely that differences in enzyme activity relate to transcript or protein turnover, rather than plasmid instability.

5.3 *Factors influencing the level of transient gene expression*

It might be assumed, simplistically, that the level of transient expression of a reporter gene at a given time point after electroporation is a reflection of the amount of plasmid DNA taken up by a population of protoplasts. To investigate the extent to which this is a valid concept, a number of experiments were carried out to identify factors which influence expression, with a view to optimizing the efficiency of stable transformation.

5.3.1 *Effect of exogenous plasmid concentration*

To determine the effect of the amount of exogenously applied plasmid DNA (p35SCN) on the level of reporter (*cat*) gene expression (assayed at d.2), 2 x 10^5 suspension protoplasts were electroporated in the presence of different amounts plasmid under conditions inducing a minimum loss of viability over this period. It was found (Lindsey and Jones, 1987b) that detectable CAT activity was obtained when 0.01, but not when 0.001 μg of plasmid was used, and extractable enzyme activity increased with an increasing supply of DNA up to approx. 10 μg. Higher amounts of plasmid (10-100 μg) resulted in no further increase in CAT activity.

5.3.2 *Effect of protoplast density*

If less than 5 x 10^4 protoplasts were electroporated in the presence of 10 μg of p35SCN, no CAT activity was detectable (at d.2, Lindsey and Jones, 1987b). There was, however, a linear relationship between the number of protoplasts electroporated and extractable CAT activity up to a density of 10^6 protoplasts per sample (0.4 ml). If more than this number were electroporated as a single sample (i.e. 3 x 10^6 or 5 x 10^6 per 0.4 ml), then this linear relationship was found to break down, and CAT activity on a per protoplast basis was reduced compared with that of lower densities. Electroporation in the presence of 50 μg rather than 10 μg of plasmid did not change this pattern, indicating that, at these high protoplast densities at least, plasmid availability was not limiting. If, however, lower densities (e.g. 10^6 per sample) were electroporated separately and the samples pooled to make a final extract from 3 x 10^6 protoplasts, the linear relationship between CAT activity and protoplast number was largely restored. This indicates that at densities higher than approx. 10^6 per 0.4 ml sample, there was a lower proportion of permeabilized protoplasts in the pulsed population than if a lower density was electroporated, a result which was confirmed by dye uptake studies.

It can be concluded that DNA availability can be influenced in at least two ways: by the exogenous supply, and by the density of the protoplast population which in turn affects the proportion of that population which is permeabilized under a given set of electrical conditions. Based on the above observations, protoplasts were routinely electroporated at relatively low densities (3 x 10^5/sample) in the presence usually of 10 μg plasmid DNA (plus 50 μg carrier DNA).

5.3.3 *Effect of electrical parameters*

Electrical parameters can be expected to influence both the level of transient gene expression and also the efficiency of stable transformation of protoplasts in at least three ways:

1) by the influence on DNA uptake *per se*, i.e. by the effects on the duration or extent of non-lethal membrane permeabilization and on the electrophoretic transport of the charged DNA molecules through membrane 'pores';
2) by the influence on protoplast survival, up to and beyond 2 days after electroporation; and
3) by the interaction with protoplast size: different fractions of a population, heterogeneous in size, would be expected to respond differently to a given set of electrical conditions.

The relationship between rectangular pulse field strength, duration and length of interval between pulses with CAT expression in suspension protoplasts of sugarbeet was examined. Protoplasts were given 3 pulses of field strengths ranging from 250-2500 V/cm, of 50 μs or 99.9 μs duration with intervals between pulses of either 0.5 s or 9.9 s, and the results are summarized in Fig. 6. Highest CAT activity was obtained by applying relatively low field strength pulses (250-750 V/cm) of relatively long duration (99.9 μs), when 3 pulses were given with the longer interval between the pulses. Field strengths greater than 750

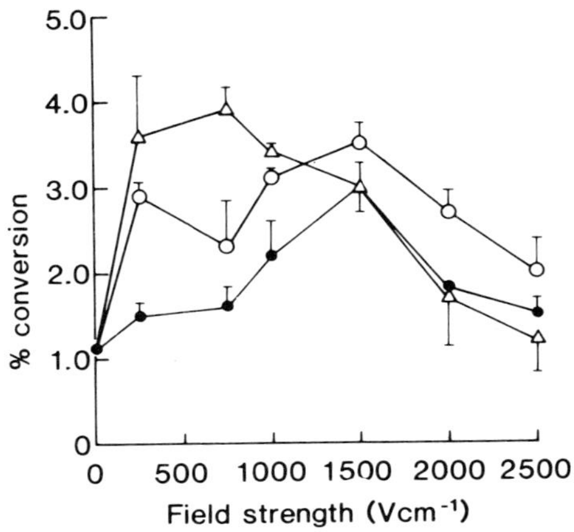

Fig. 6: Effect of pulse field strength, duration and interval between pulses on CAT activity in suspension protoplasts: 99.9 µs, 9.9 s interval (triangles), 50 µs, 9.9 s interval (open circles), 50 µs, 0.5 s interval (closed circles). 2 x 10^5 protoplasts assayed/sample.

V/cm with the longer pulses resulted in reduced transient expression, and under these circumstances was correlated with a reduced viability of the protoplast population.

Much longer pulses could be generated by the 'capacitor discharge' system described earlier. Using a capacitance of 200 µF, and with a medium resistance of 1300 Ω, the time constant of the applied pulse was approx. 260 ms. It was found that a single such pulse of between 275-325 V/cm gave equivalent CAT expression to multiple, shorter (99.9 µs) rectangular pulses (Table 1), even though protoplast survival over 2 days was slightly reduced in comparison. Multiple (2-5) pulses of 260 ms resulted in no detectable CAT activity, and was associated with a significant reduction in the viability of the protoplasts. A reduction of the time constant to 130 ms resulted in a slightly reduced level of transient CAT expression, but the survival rate was improved (Table 1).

When suspension protoplasts were electroporated over a range of field strengths, there was found to be no threshold voltage below which either phenosafranine uptake or DNA uptake (transient gene expression) was detectable. This is consistent with the view that at each of the conditions tested, provided viability of the protoplasts was maintained, at least a proportion of the population was permeabilized; and this is presumed to be related to the size distribution of suspension protoplasts.

Under optimum electrical conditions for transient CAT or GUS expression, phenosafranine uptake studies have demonstrated that no more than 50% of the surviving population are permeabilized (Table 1), thereby putting an upper limit on the proportion of the protoplasts which can be transformed. If the size distribution was more restricted, however, it might be expected that a higher proportion would be permeabilized under optimum electrical conditions, resulting in more DNA uptake, higher levels of transient expression and higher frequencies of stable transformation.

This possibility was examined by electroporating sugarbeet leaf protoplasts, which are morphologically much more homogeneous, though on average smaller (mean diameter 26 +/- 3 μm), than suspension protoplasts (mean diameter 36 +/-15 μm). The CAT activity profile of

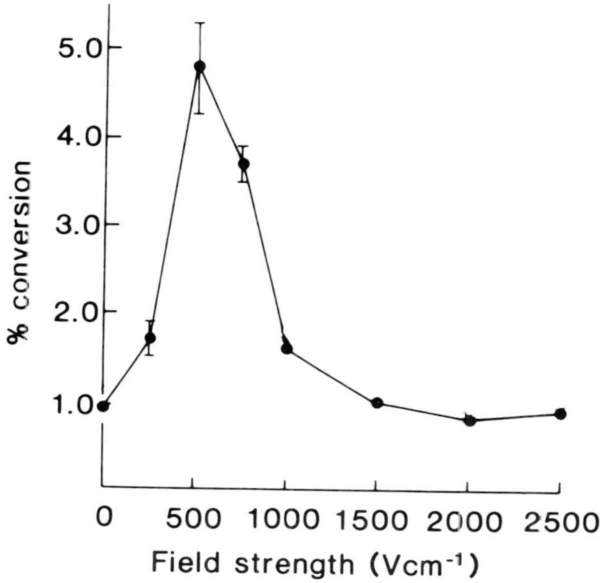

Fig. 7: Effect of pulse field strength and duration on CAT activity in leaf protoplasts. 8 x 10^5 protoplasts assayed/sample.

Field Strength	CAT Activity		% Viable		% Perm.	
V/cm	130ms	260ms	130ms	260ms	130ms	260ms
0	0.9	0.9	97.1	97.1	1.3	1.3
250	23.8	27.2	71.2	65.2	47.2	45.4
275	24.2	29.4	70.1	63.5	47.4	49.2
300	25.9	31.2	68.4	59.1	50.1	52.4
325	25.1	28.4	65.7	59.4	47.2	51.1
350	23.2	25.1	60.5	58.2	48.0	46.8
3 x 300	0.9	0.8	20.3	12.5	52.1	52.2

Table 1.
Effect of field strength and (exponentially decaying) pulse duration on CAT activity (10^6 protoplasts assayed at 48 h), % survival at 48 h and % of viable cells reversibly permeabilized after electroporation.

these protoplasts is illustrated in Fig. 7, and shows that the range of field strengths at which transient expression occurred was more restricted than for suspension protoplasts, as predicted. Furthermore, under optimum electrical conditions for transient expression, phenosafranine uptake studies revealed that 72% of the protoplasts were reversibly permeabilized, i.e. approx. 20% more than the more heterogeneous suspension protoplasts. It is, however, also possible that differences in the membrane characteristics of leaf and suspension protoplasts may influence their behaviour in electric fields.

Despite these results, which indicate that leaf protoplasts would be expected to take up more DNA on a per protoplast basis, the level of transient gene expression was typically 4-10 fold lower than for suspension protoplasts. This was a corollary of differential protoplast survival. It is suggested that this phenomenon is related to the physiological states of leaf and suspension protoplasts: the former failed to divide under the culture conditions used, while suspension protoplasts had a high plating efficiency (see below) and presumably were actively synthesizing RNA and protein during the 2 day culture period before the transient expression assay was carried out. Any gene with a constitutive promoter (such as the CaMV 35S) which is introduced into such active protoplasts might be expected to make more enzyme than the recalcitrant leaf protoplasts, and this possibility may contribute to the higher levels of transient expression observed. This is discussed in more detail in Lindsey and Jones (1987b).

6. DNA UPTAKE BY WHOLE CELLS

It would be advantageous in practical terms if DNA could be introduced into intact cells (as opposed to naked protoplasts). This would avoid some of the problems associated with the culture and regeneration of plants from protoplasts, and tissues which are particularly morphogenetic (such as embryos or meristems) could be transformed. The results of the dye uptake experiments demonstrate clearly that it is possible to reversibly permeabilize intact cells of sugarbeet, but the uptake of [^{3}H] pABD1 could not be detected, due to the insensitivity of the method. There is no doubt that the cell wall represents a major barrier to the free movement of DNA, but it could not be concluded from this type of experiment that no DNA uptake occurred.

It was considered that a more sensitive method to determine if DNA uptake can occur would be to electroporate whole cells in the presence of p35SCN, and assay for transient CAT expression. Suspension cells from three different lines of sugarbeet were therefore electroporated in the presence of the plasmid at a range of field strengths and cultured for 48 hours before analysis of CAT enzyme activity. It was found that in two cell lines no CAT activity above that of control treatments (electroporated with carrier DNA only) was detectable, even if five samples of 10^{6} cells per sample were pooled and analyzed. Significant CAT activity was, however, detectable in a third cell line, if certain criteria were met: maximum enzyme activity was obtained after electroporation at field strengths of between 250 and 750 V/cm

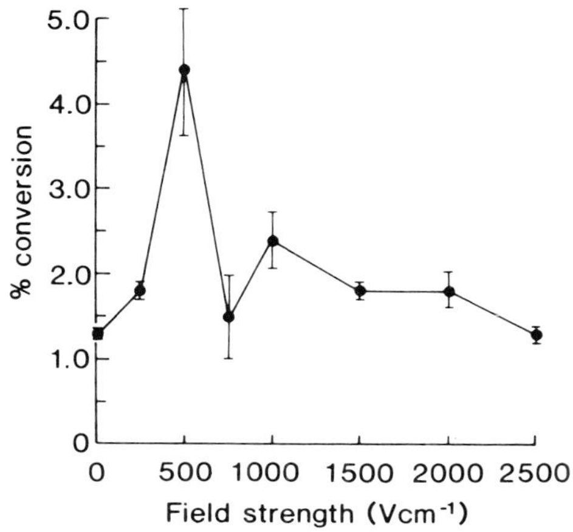

Fig. 8: Effect of field strength on CAT activity in intact suspension culture cells. 3 X 10^6 cells assayed/sample.

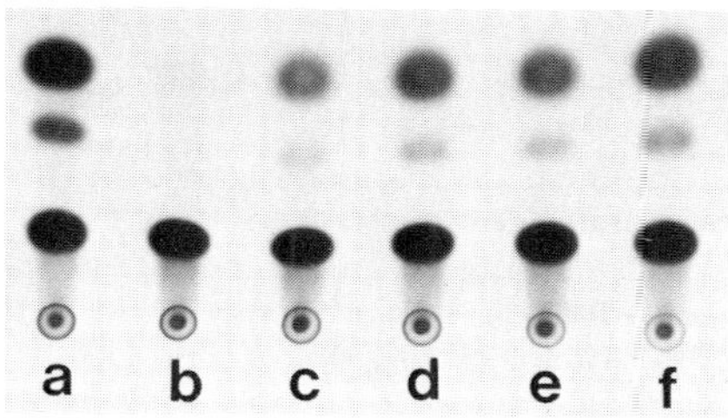

Fig 9: Effect of pectolytic enzyme treatment of cells on CAT activity after electroporation. a = protoplasts, b = untreated cells, c = cells, 0.3% macerozyme R-10, d = 0.75% macerozyme R-10, e = cells, 0.08% pectolyase Y-23, f = cells, 0.5% pectolyase Y-23.

(using 3 rectangular pulses of 99.9 µs each, Fig. 8) and the extracts of at least 2-3 x 10^6 cells was required to detect activity. This latter result indicates that the amount of DNA taken up by this cell line was at most 20-50 fold less than that taken up by protoplasts, in which introduced CAT activity is detectable in extracts of between 5 x 10^4 and 10^5 electroporated protoplasts, as discussed earlier. The basis of this calculation is that under the experimental conditions used the amount of DNA taken up is directly proportional to the level of extractable CAT activity.

The cell line in which DNA uptake and expression was detectable was composed of smaller cell aggregates than were the other two lines, with 70% of the aggregates comprising 2-10 cells compared with 75% comprising 30-50 or more cells in the 'non-expressing' lines. The possible significance of this observation is discussed shortly.

To determine whether the porosity of sugarbeet cells to DNA molecules could be increased by chemical modification of the wall, populations were treated with pectin-digesting enzymes and washed, before electroporation in the presence of p35SCN. It was found (Fig. 9 and Lindsey and Jones, 1987b) that all enzyme treatments of cells (i.e. of cells otherwise unable to take up detectable levels of DNA) resulted in significant levels of CAT activity, presumably therefore due to improved accessibility of DNA to the plasmamembrane. Treatment of cells with 0.5% w/v pectolyase Y-23 allowed approx. 40-50 % of the level of DNA uptake as occurred in protoplast 'controls'. It was also observed that pectinase treatments caused a decrease in the average size of cell aggregates.

These data suggest that apart from electrical conditions, two factors at least determine whether DNA can be taken up by whole cells. The first is that pectin itself may inhibit DNA transfer. In support of this, Baron-Epel et al. (1988) have implicated pectins as the major determinants of wall porosity in soybean cells, and they found that the size of trans-wall channels could be enlarged by treatment with pectinase but not with cellulase (cellulysin) or protease. Secondly, since the pectinase treatments of sugarbeet cells reduced the mean size of the cell aggregates, it is likely that a higher proportion of the population would be susceptible to permeabilization and therefore DNA uptake.

In relation to this, it can be speculated that the uptake of DNA by whole, undigested cells of sugarbeet (Fig. 8.) was possible because of the particular pectic content or composition of that cell line. A low pectic content might account for both the relatively small size of the cell aggregates which was observed and a relatively high wall porosity. Pectins and other carbohydrates are released from the walls of suspension culture cells (e.g. Scragg and Fowler, 1985), presumably to different extents in different cell lines, and it is argued that this heterogeneity of wall structure would be expected to determine to a large extent the differential capacity for DNA transport observed for the three sugarbeet cell lines. We have not, however, analyzed the pectin content of the sugarbeet cell lines to verify this possibility.

The transient expression studies described point to the importance of two general factors in determining DNA uptake by electroporation: first, the relatively easily controlled physical parameters, including DNA concentration, protoplast density and electrical conditions; and second, the currently less well defined physiological characteristics of the protoplasts themselves, such as size distribution, membrane characteristics, the presence of cell wall material etc.. In combination with dye uptake studies, it is possible to obtain a large amount of background information before undertaking stable transformation studies.

7. STABLE TRANSFORMATION OF PROTOPLASTS

Apart from a maximum level of DNA uptake by the protoplast population, a high frequency of stable transformation has two further major requirements:

1) a good protoplast plating efficiency, i.e. a high proportion of the

initial population should divide and form colonies from which plants can be regenerated, and
2) an efficient system for selecting transformed protoplasts against a background of non-transformants.

The successful culture of sugarbeet protoplasts has already been described for some genotypes (Bhat et al., 1985; Szabados and Gaggero, 1985), but the regeneration of sugarbeet plants from protoplast-derived callus has not yet been published.

7.1 *Protoplast culture*

We (Lindsey and Jones, 1989) optimized culture conditions for suspension protoplasts of sugarbeet by determining the division frequency on a range of media based on PGo salts (de Greef and Jacobs, 1979). The medium which allowed the highest division frequency, measured at d.10, comprised PGo salts supplemented with 0.1 mg/l 2,4-D, 0.01 mg/l BAP, 30 g/l sucrose and 9% w/v mannitol. It was also observed that if this medium was solidified with 0.6% w/v agarose, the division frequency was increased more than three-fold, to 37.2% of the population undergoing a first division at d.10 of culture. This immobilization procedure was used routinely. After d.15 of culture the agarose was floated on liquid medium lacking mannitol, and old medium was replaced by fresh at 7 day intervals. Initial protoplast density also influenced the subsequent division frequency, with best results being obtained at 10^5 protoplasts/ml (Lindsey and Jones, 1989).

7.2 *Establishment of selection conditions*

The most commonly used strategy for isolating stably transformed plant tissues is the introduction of modified bacterial genes conferring resistance to toxic antibiotics. The approach adopted was therefore to electroporate sugarbeet protoplasts in the presence of pHP23, a pUC18-based vector (4.6 kb) carrying the aminoglycoside phosphotransferase gene under the transcriptional control of both the CaMV 35S and 19S RNA gene promoters (J. Paszkowski, Zurich). This gene construct was expected to confer resistance of transformed protoplasts to the aminoglycoside group of antibiotics, including kanamycin. Preliminary experiments demonstrated that the proliferation of untransformed protoplasts under optimum agarose culture conditions was inhibited if 50-100 mg/l kanamycin was included in the medium from d.15 onwards (Lindsey and Jones, 1989).

7.3 *Transformation conditions*

To generate transformed colonies, pHP23 was introduced into protoplasts using either the rectangular pulse system (3 pulses of 99.9 μs, 500 V/cm) or the capacitor discharge system (1 pulse of 130 ms or of 260 ms, 290 V/cm). All sets of conditions had previously been found to give relatively high levels of transient reporter gene expression. For each set of electrical conditions, 10 x 10^6 protoplasts were electroporated in batches of 3 x 10^6, in the presence of 15 μg of plasmid DNA plus 50 μg sheared herring sperm carrier DNA. Protoplasts were then cultured in agarose at an initial density of 10^5/ml, for 15 days in the absence of antibiotics and subsequently in the presence of

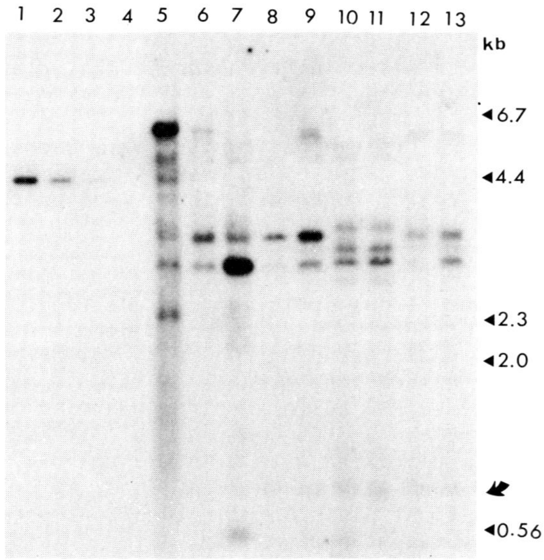

Fig. 10: Hybridization of ^{32}P-pHP23 with 10, 5 and 1-2 copy number reconstructions (lanes 1-3) and genomic DNA of untransformed (lane 4) and transformed (lanes 5-13) protoplast-derived callus colonies.

50 mg/l kanamycin. When protoplast-derived colonies were 2-3 mm in diameter, they were transferred to medium solidified by agar in the presence of 100 mg/l kanamycin, and on reaching a colony size of approx. 1 cm were considered as putative transformed lines of single protoplast origin. The frequency of kanamycin colony formation at 6 months for each of the three electrical conditions is described in Table 2: the highest frequency of colony formation was obtained if 130 ms were used to introduce DNA rather than 260 ms or 3 x 99.9 µs, and this was associated with a higher protoplast plating efficiency. Absolute transformation frequencies under these conditions were between 0.5 and 1 x 10^{-4} resistant colonies per initial number of protoplasts electroporated.

The presence of transforming DNA in colonies electroporated by both rectangular and exponentially decaying pulses was confirmed by Southern blot analysis (Fig. 10), using radiolabelled pHP23 as a probe. By comparison with the gene copy number reconstructions (lanes 1-3), it can be estimated that there are at least 10 copies of the plasmid per haploid genome in each transformed line. No hybridization was seen with DNA from untransformed sugarbeet cultured cells. The restriction patterns are interesting, in that a number of fragments of similar size are observed to be present in different cell lines, though there are stoichiometric differences. This pattern is difficult to explain, but suggests that integration in individual transformed lines is not random, and recombinational 'hotspots' may exist within the sugarbeet

	Number of Resistant Colonies			Absolute Transformation Frequency	Relative Frequency
	pHP23	PE	Control		
100 µF	587	51.7	7	6×10^{-5}	1×10^{-4}
200 µF	192	38.4	1	2×10^{-5}	5×10^{-5}
3x 99.9µs	254	42.4	11	2×10^{-5}	6×10^{-5}

Table 2.
Number of colonies produced at 6 months from an initial 10×10^{6} protoplasts per electroporation treatment (absolute transformation frequency), cultured in the presence of 100 mg/l kanamycin. PE = plating efficiency of transformed cultures at day 21. The relative transformation frequency (number of resistant colonies per number of surviving protoplasts) is given. This represents the data from a single comparative experiment, and similar transformation frequencies (from 3×10^{-5} to 10^{-4} (absolute) for 100µF pulses) and plating efficiencies are typical.

genome. This possibility requires further investigation. A second point of interest relates to lanes 10 and 11 of Fig. 10. Loaded onto these lanes is DNA from a single transformed line which, after an initial culture period of 3 months was divided into two and each daughter subclone was cultured separately for a further 8 months before DNA analysis. The identical restriction patterns reveal that no major detectable rearrangements of the transforming DNA had occurred during the 8 month culture periods.

8. APPLICATIONS OF DIRECT GENE TRANSFER

As was discussed at the beginning of this article, the major potential application of direct gene transfer techniques is in the production of transgenic plants. This approach has been carried through successfully using electroporation for a handful of species, including tobacco (Potrykus et al., 1985), *Brassica napus* (Guerche et al., 1987), maize (Rhodes et al., 1988) and rice (Toriyama et al., 1988). The regeneration of whole plants back from protoplasts of sugarbeet is yet to be published, although this situation may change in the near future with success being reported at this Amiens meeting (R.D. Hall et al., S.V.P., Wageningen).

What can we conclude, from the evidence summarized in this article, of the possibility of transforming whole or partially digested cells, for example from highly morphogenetic tissues? For whole cells, the frequency of stable transformation would appear to be limited by the low amount of DNA which is able to cross the cell wall (20-50 fold less or lower for cells than protoplasts of sugarbeet) and also by the proportion of the population which, having taken up DNA, integrates at least one copy of the intact gene into a region of the genome which permits production of a selectable enzyme. A highly efficient culture and selection system is therefore necessary for the successful recovery of stable transformants by this approach, and we have not achieved this for sugarbeet. The use of partially digested cells, which take up more exogenous DNA than do whole cells, would seem to be a more viable alternative. In our experience, however, partially digested suspension culture cell of sugarbeet divide at only a low frequency (less than 0.01% plating efficiency compared with 30-50% for suspension protoplasts), and the reasons for this are unclear. If, however, such culture problems can be overcome, the electroporation of partially digested tissues, of small cell number and high morphogenetic capacity such as proembryonic tissues, may become a feasible alternative to the electroporation of protoplasts.

An alternative application of direct gene transfer techniques, which has not yet been exploited or even discussed to any extent in the literature, is in gene rescue or gene complementation by shotgun cloning. The general approach is to introduce DNA from one plant carrying a selectable or screenable gene into protoplasts or cells of a plant or cell line lacking that gene; and then selecting or screening for colonies/plants which become transformed with that gene, as detected by the respective phenotypic change. If the introduced DNA is randomly fragmented, and each fragment is tagged with a known sequence before transformation, it should be possible subsequently to rescue the fragment carrying the gene of interest, using the tag DNA to probe a genomic library made from a selected colony. The success of this strategy depends upon a number of factors:

1) the size of the genome from which transforming fragments are obtained;
2) the copy number of the transforming gene in the original genome;
3) the selectable/screenable trait should be monogenic, or if polygenic, the genes should be closely linked;
4) the transformation efficiency of the 'host' cells;
5) the size of the fragments introduced.

The only report of a shotgun transformation for gene rescue using a plant transformation system is by Klee et al. (1987b). Here, a cosmid library was constructed from *Arabidopsis thaliana* resistant to the antibiotic kanamycin, due to the presence of a gene encoding the enzyme neomycin phosphotransferase, NPT-II. The cosmid library was introduced into *A. tumefaciens* which was inoculated onto kanamycin-sensitive *Petunia* explants, and a small number of resistant host plants were regenerated. This demonstrates that, in principle, shotgun transformation can work in plant systems, but the use of *Agrobacterium*-mediated vectors is in this case limited to the transfer of DNA from plants with small genomes, such as *Arabidopsis*.

By introducing genomic DNA fragments directly into protoplasts, some of these limitations may be overcome. As a model experimental system to test out this approach we have attempted to transfer selectable genes from the tobacco genome into sugarbeet protoplasts. Plants of *Nicotiana tabacum* SR1 were transformed with the neomycin phosphotransferase (*npt-II*) gene by *A. tumefaciens* using standard techniques, and genomic DNA from one kanamycin-resistant individual was purified and shown to contain approx. 20 copies of the *npt-II* gene.

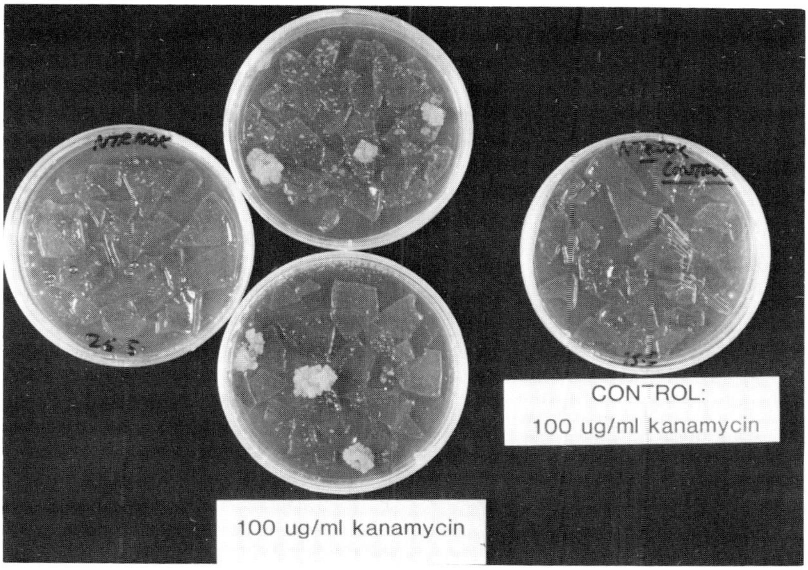

Fig. 11: Shotgun-transformed kanamycin-resistant colonies.

This DNA was then digested with EcoR1 and electroporated into sugarbeet suspension protoplasts (50 μg/10^6 protoplasts) using 1 x 130 ms decaying pulse (290 V/cm) per sample. In one experiment, a total of 25 x 10^6 protoplasts were treated in this way, and cultured in agarose in the presence of 100 mg/l kanamycin as described earlier. From 25 x 10^6 protoplasts, 17 kanamycin-resistant colonies were isolated (Fig. 11) and analyzed for the presence of the *npt-II* gene by Southern hybridization. The results (Fig. 12) show that, indeed, an internal fragment of the gene is present.

These data are preliminary, but do demonstrate that it is possible to use electroporation to transfer, and therefore ultimately to isolate, selectable genes from genomes significantly larger than that of *A. thaliana*. A more detailed account of this and related work will be reported elsewhere.

9. CONCLUSIONS

Our aim in this article has been to illustrate some of the features of electroporation by reference to the development of a system for the stable transformation of sugarbeet protoplasts. Protoplast culture and selection procedures for this species are now well developed, and the efficiency of transformation by electroporation is high enough that, before long, transgenic plants will be produced by this route. We further believe that a shotgun strategy for plant gene isolation and complementation studies in homologous systems is now a feasible proposition.

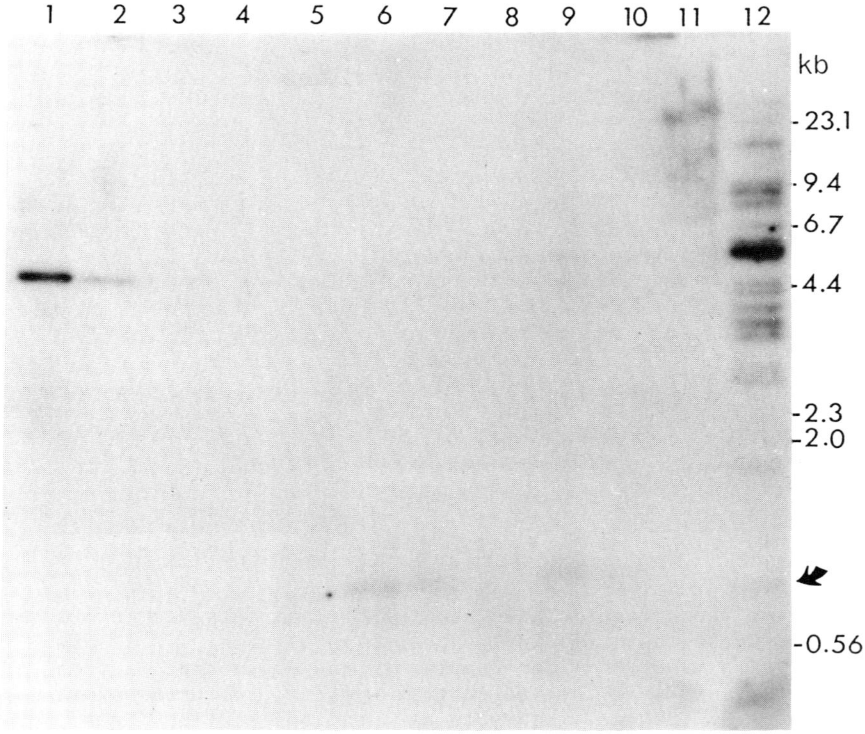

Fig. 12: Hybridization of a ^{32}P-labelled internal fragment of the *npt-II* gene with 10, 5 and 1-2 copy number reconstructions (lanes 1-3) and genomic DNA of untransformed (lane 4) and shotgun-transformed, kanamycin-resistant callus colonies (lanes 5-10). Lane 11 = λ size markers, lane 12 = transforming tobacco DNA.

REFERENCES

Baron-Epel, O., Gharyal, P.K. and Schindler, M. (1988). 'Pectins as mediators of wall porosity in soybean cells'. Planta 175, 389-395.

Bevan, M.W. (1984). 'Binary *Agrobacterium* vectors for plant transformation'. Nucleic Acids Res. 22, 8711-8721.

Bhat, S.R., Ford-Lloyd, B.V. and Callow, J.A. (1985). 'Isolation of protoplasts and regeneration of callus from suspension cultures of cultivated beets'. Plant Cell Rep. 4, 348-350.

Browse, J., Somerville, C.R. and Slack, C.R. (1988). 'Changes in lipid composition during protoplast isolation'. Plant Sci. 56, 15-20.

De Block, M., Botterman, J., Vandwiele, M., Dockx, J., Thoen, C., Gossele, V., Rao Movva, N., Thompson, C., Van Montagu, M. and Leemans, J. (1987). 'Engineering herbicide resistance in plants by expression of a detoxifying enzyme'. EMBO J. 6, 2513-2518.

De La Pena, A., Lorz, H. and Schell, J. (1987). 'Transgenic rye plants obtained by injecting DNA into young floral tillers'. Nature 325, 274-276.

Deshayes, A., Herrera-Estrella, L. and Caboche, M. (1985). 'Liposome-mediated transformation of tobacco mesophyll protoplasts by an *Escherichia coli* plasmid'. EMBO J. 4, 2731-2737.

Eady, C., Warren, G., Lindsey, K. and Jones, M.G.K. (1988). 'Electrofusion and electroporation of sugar beet (*Beta vulgaris* L.) protoplasts', in K.J. Puite, J.J.M. Dons, H.J. Huizing, A.J. Kool, M. Koorneef and F.A. Krens (eds.), Progress in Plant Protoplast Research, Kluwer Academic Publishers, Dordrecht, pp. 261-262.

Fromm, M.E., Taylor, L.P. and Walbot, V. (1985). 'Expression of genes transferred into monocot and dicot plant cells by electropporation'. Proc. Natl. Acad. Sci. USA 82, 5824-5828.

Gorman, C.M., Moffat, L.F. and Howard, B.H. (1982). 'Recombinant genomes which express chloramphenicol acetyltransferase in mammalian cells'. Mol. Cell. Biol. 2, 1044-1051.

Guerche, P., Charbonnier, M., Jouanin, L., Tourneur, C., Paszkowski, J. and Pelletier, G. (1987). 'Direct gene transfer by electroporation in *Brassica napus*'. Plant Sci. 52, 111-116.

Hain, R., Stabel, P., Czernilofsky, A.P., Steinbiβ, H.H., Herrera-Estrella, L. and Schell, J. (1985). 'Uptake, integration, expression and genetic transmission of a selectable chimaeric gene by plant protoplasts'. Mol. Gen. Genet. 199, 161-168.

Jefferson, R.A., Kavanagh, T.A. and Bevan, M.W. (1987). 'GUS fusions: β-glucuronidase as a sensitive and versatile gene fusion marker in higher plants'. EMBO J. 6, 3901-3907.

Klee, H., Horsch, R. and Rogers, S. (1987a). '*Agrobacterium*-mediated plant transformation and its further applications to plant biology'. Ann. Rev. Plant Physiol. 38, 467-486.

Klee, H.J., Hayford, M.B. and Rogers, S.G. (1987b). 'Gene rescue in plants: a model system for 'shotgun' cloning by retransformation'. Mol. Gen. Genet. 210, 282-287.

Klein, T.M., Fromm, M., Weissinger, A., Tomes, D., Schaaf, S., Sletten, M. and Sanford, J.C. (1988). 'Transfer of foreign genes into intact maize cells with high-velocity microprojectiles'. Proc. Natl. Acad. Sci. USA 85, 4305-4309.

Krens, F.A., Molendijk, L., Wullems, G.J and Schilperoort, R.A. (1982). 'In vitro transformation of plant protoplasts with Ti-plasmid DNA'. Nature 296, 72-74.

Larkin, P.J. (1976). 'Purification and viability determination of plant protoplasts'. Planta 128, 213-216.

Lindner, P., Neumann, E. and Rosenheck, K. (1977). 'Kinetics of permeability changes induced by electric impulses in chromaffin granules'. J. Membrane Biol. 32, 231-254.

Lindsey, K. and Jones, M.G.K. (1987a). 'The permeability of electroporated cells and protoplasts of sugarbeet'. Planta 172, 346-355.

Lindsey, K. and Jones, M.G.K. (1987b). 'Transient gene expression in electroporated protoplasts and intact cells of sugarbeet'. Plant Mol. Biol. 10, 43-52.

Lindsey, K. and Jones, M.G.K. (1989). 'Stable transformation of sugar

beet protoplasts by electroporation'. Plant Cell Rep. 8, 71-74.
Lorz, H., Baker, B. and Schell, J. (1985). 'Gene transfer to cereal cells mediated by protoplast transformation'. Mol. Gen. Genet. 199, 178-182.
McCabe, D.E., Swain, W.F., Martinell, B.J. and Christou, P. (1988). 'Stable transformation of soybean (*Glycine max*) by particle acceleration'. Bio/Technol. 6, 923-926.
Neumann, E., Schaeffer-Ridder, M., Wang, Y. and Hofschneider, P.H. (1982). 'Gene transfer into mouse lyoma cells by electroporation in high electric fields'. EMBO J. 1, 841-845.
Paszkowski, J., Shillito, R.D., Saul, M., Mandak, V., Hohn, T., Hohn, B. and Potrykus, I. (1984). 'Direct gene transfer to plants'. EMBO J. 3, 2717-2722.
Potrykus, I., Paszkowski, J., Saul, M.W., Petruska, J. and Shillito, R.D. (1985). 'Molecular and general genetics of a hybrid foreign gene introduced into tobacco by direct gene transfer'. Mol. Gen. Genet. 199, 169-177.
Reich, T.J., Iyer, V.J. and Miki, B.L. (1986). 'Efficient transformation of alfalfa protoplasts by the intranuclear microinjection of Ti plasmids'. Bio/Technol. 4, 1001-1003.
Rhodes, C.A., Pierce, D.A., Mettler, I.J., Mascarenhas, D. and Detmer, J.J. (1988). 'Genetically transformed maize plants from protoplasts'. Science 240, 204-207.
Rogers, S.G., Klee, H., Horsch, R.B. and Fraley, R.T. (1988). 'Use of cointegrating Ti plasmid vectors'. Plant Mol. Biol. Man. A2, 1-12.
Schauer, A.T. (1988). 'Visualizing gene expression with luciferase genes'. Trends Biotechnol. 6, 23-27.
Scragg, A.H. and Fowler, M.W. (1985). 'The mass culture of plant cells', in I.K. Vasil (ed.), Cell Culture and Somatic Cell Genetics of Plants, Vol. 2, Academic Press, Orlando, Fl., pp. 103-128.
Stachel, S.E, Messens, E., Van Montagu, M. and Zambryski, P. (1985). 'Identification of the signal molecules produced by wounded plant cells that activate T-DNA transfer in *Agrobacterium tumefaciens*'. Nature 318, 624-629.
Sugar, I.P. and Neumann, E. (1984). 'Stochastic model for electric field-induced membrane pores: electroporation'. Biophys. Chem. 19, 211-225.
Szabados, L. and Gaggero, L. (1985). 'Callus formation from protoplasts of a sugarbeet cell suspension culture'. Plant Cell Rep. 4, 195-198.
Toriyama, K., Arimoto, Y., Uchimiya, H. and Hinata, K. (1988). 'Transgenic rice plants after direct gene transfer into protoplasts'. Bio/Technol. 6, 1072-1074.
Widholm, J.M. (1972). 'The use of fluorescein diacetate and phenosafranine for determining viability of cultured plant cells'. Stain Technol. 47, 189-194.
Zimmermann, U. (1986). 'Electrical breakdown, electropermeabilization and electrofusion'. Rev. Physiol. Biochem. Pharmacol. 105, 175-256.
Zimmermann, U., Vienken, J. and Pilwat, G. (1980). 'Development of drug carrier systems: electric field induced effects in cell membranes'. Bioelectrochem. Bioenerg. 7, 553-574.

Cauliflower Mosaic Virus: Biology and Applications.

Karl Gordon
Friedrich Miescher-Institut, PO. Box 2543, CH-4002 Basel, Switzerland

The caulimovirus family

Of the hundreds of plant viruses described, most have RNA genomes. The caulimoviruses are among the very few which, in contrast, have DNA genomes. In addition to their interest as model systems for studying plant/virus interactions, the caulimoviruses have therefore also been the subject of much investigation as possible vectors for introducing foreign genes into plants. This review will explore some of the exciting complexities of caulimovirus biology and discuss real and possible uses as vectors.

The caulimovirus family of aphid-borne viruses contains at least nine members, infecting a variety of hosts from at least 10 families (Table 1). The type member, cauliflower mosaic virus, mainly infects members of the Cruciferae, although some strains can also infect members of the Solanaceae, e.g. *Datura* sp. (Shepherd and Lawson, 1981; Daubert, 1989). CaMV infects plants systemically, causing chlorotic mottling, vein clearing, leaf curling and stunting of new leaves. Analysis of infected leaves by electron microscopy revealed characteristic dense structures of irregular size and shape in the cytoplasm of infected cells. These inclusion bodies or viroplasms have been shown to contain virus particles embedded in a matrix composed of a protein which is virus encoded (Odell and Howell, 1980; Covey and Hull, 1981).

Caulimovirus particles contain a circular double stranded DNA genome of about 8000 bp in size. This DNA has an open circular conformation as a result of several gaps, which more detailed analysis has shown to actually be short overlaps (Franck et al., 1980). Cloned DNA lacking the overlaps is infectious if excised from the vector in such a way that the viral genome can be directly repaired upon entry into the plant cell (Brisson and Hohn, 1986). Furthermore, if the plasmid carries the complete viral genome and a terminal repeat of sufficient length, the viral genome can be excised from the plasmid in the plant cell by recombination, and then establish an infection (Lebeurier et al., 1982). The minimum repeat length required for initiating viral infection is less than 0.025 genome (200bp: K.G., unpublished).

Commencing with the Strasbourg isolate Cabb-S (Franck et al., 1980), several isolates of CaMV have now been completely sequenced (Table 1). Moreover the naturally occurring mutant CM4-184, which lacks ORF II, has been partly sequenced. It is in part related to the Strasbourg strain and in part to strain CM-1841 (Dixon et al., 1986). The sequences of three other members of the caulimovirus family, figwort mosaic virus

R.S. Sangwan and B.S. Sangwan-Norreel (eds.), The Impact of Biotechnology in Agriculture, 381–390.

(FMV), carnation etched ring virus (CERV) and soybean chlorotic mottle virus (SoyCMV) are also now known - see Table 1. The other members of the caulimoviruses are dahlia mosaic virus (DaMV), mirabilis mosaic virus (MiMV), strawberry vein banding

Table 1

Members of the caulimovirus family

Virus	hosts*	Reference
Viruses whose genomes have been sequenced		
CaMV	Cruciferae (some Solanaceae)	
isolate Cabb-S		Franck et al., 1980
CM-1841		Gardner et al., 1981
D/H		Balazs et al., 1982
Xinjing		Rongxiang et al., 1985
JI		J. Stanley (pers. comm.)
FMV	various spp. in Scrophulariaceae and Chenopodiaceae	Richins et al., 1987
CERV	*Dianthus caryophyllus* and other spp. in Caryophyllaceae	Hull et al., 1986
SoyCMV	Soybean and some other spp. in Leguminosae	Hasegawa et al., 1989
Other viruses		
DaMV	*Dahlia variabilis* and other spp. in Compositae; some spp. in Solanaceae, Chenopodiaceae and Amaranthaceae	
MiMV	*Mirabilis* sp. in Nyctaginaceae	
StVBV	*Fragaria* sp. in Rosaceae	
HLaV	Cruciferae	Richins and Shepherd, 1986
PCLSV		Hibi and Kameya-Iwaki, 1988

* Each of these viruses appears to have a host range limited to a few species in the family listed. See Shepherd and Lawson (1981) for details and further references.

virus (StVBV), horseradish latent virus (HLaV) and peanut chlorotic leaf streak virus (PCLSV).

Genetic organization of CaMV

In all isolates and viruses, the basic genetic organization of the virus is conserved, with six major open reading frames present in the same order on the same strand of DNA,and all packed closely in a tight head-to-tail arrangement. The amino acid sequences of the gene products from the essential ORFs I, III, IV, V and VI show significant but variable degrees of similarity among the four viruses (Richins et al., 1987; Hasegawa et al., 1989). In general, ORF V encodes the most highly conserved product, and ORF VI the most variable, apart from a short domain in the central region.

Information is slowly accumulating about the functions of these genes. There is some evidence that ORF I is involved in virus spread within the plant (Stratford and Covey, 1989); ORF II encodes a factor absolutely required for aphid transmissibility but otherwise dispensible (Howarth et al., 1981; Woolston et al., 1983); ORF III encodes a protein, found in the capsid, with DNA binding properties (Giband et al., 1986). The ORF IV product is the coat protein and the ORF V product includes a domain near the amino-terminus encoding a protease and two other domains encoding the reverse transcriptase/RNase H activities (Toh et al., 1985). ORF VI encodes a protein produced in large amounts which also appears to have several functions. It serves as the matrix protein of the inclusion bodies or viroplasms found in CaMV- infected cells and also as a factor which trans- activates the translation of downstream genes on polycistronic mRNAs (see below). Furthermore, it has been shown to play a role in determining virus host range and the symptoms developed upon CaMV infection (Daubert et al., 1984; Schoelz et al., 1986; Schoelz and Shepherd, 1989). Further evidence that the ORF VI product is involved in determining the host response comes from the work of Goldberg et al. (1987), Baughman et al. (1988) and Takahashi et al. (1989), who constructed transgenic tobacco plants expressing the ORF VI gene. These plants developed symptoms similiar to those normally induced by CaMV in host plants.

Essential domains on the virus genome have been identified by various mutagenesis and deletion studies. ORFs VII (Dixon et al., 1983) and II (Howarth et al., 1981) can be completely deleted without affecting the viability of the virus, although it is, of course, no longer aphid- transmissible in the case of ORF II. Furthermore, a segment of 147 bp in the leader can also be deleted, although the resulting virus has a longer latency period (Penswick et al., 1988).

The major transcripts hitherto identified are the 35S pregomic RNA and the 19S RNA. The former carries a 180 nt terminal redundancy and is the probable mRNA for most of the viral proteins. It also appears to be the template for reverse transcription during the viral replication cycle (see below). The 19 S RNA is a specific mRNA for the ORF VI encoded protein.

CaMV replication

The caulimoviruses replicate by reverse transcription of the genome- length 35S RNA (Guilley et al., 1983: Hull and Covey, 1983; Pfeiffer and Hohn, 1983). In outline, the replication model proposes that the gapped virion DNA enters the nucleus upon being unpackaged. It is then repaired and binds histones to form a minichromosome, which is then transcribed to yield the 35S RNA and other significant transcripts. These are then exported to the cytoplasm, where viral proteins are produced and the processes of reverse transcription, viroplasm formation and encapsidation occur. For details of the model, the experimental evidence and work on the reverse transcriptase, the reader is referred to reviews by Gronenborn (1987), Pfeiffer et al. (1987), Bonneville et al. (1988), Pfeiffer and Hohn (1989) and Hohn et al. (1990).

Transcription signals on the genome

The major viral promoter is that for the 35S transcript. Since it was originally regarded as being constitutively expressed in all cell and tissues in many different species, it has been widely used in plant genetic engineering. More recently, however, evidence accumulated which suggested that this promoter was indeed showing expression specificity in response to either the cell cycle or to specific tissues (Nagata et al., 1987; Jefferson et al., 1987; Benfey and Chua, 1989). Dissection analysis of the promoter in transgenic plants has now shown it to be a mosaic of elements, capable of conferring specifity of expression in different tissues (Benfey et al., 1989). The domain consisting of the 90 bp upstream from the start of transcription confers activity in the radicle of the embryo and in root tissue of mature plants and seedlings. The other domain extending from 343 to 90 bp upstream of the transcription start point leads to activity in the stems and leaves of plants and in the cotyledons of the embryo and seedlings.

The viral transcription control signals i.e. the 35S promoter and the poly(A) signal have been widely used both for transient expression of foreign genes in protoplasts and for expression of trans-genes in plants. Examples of transient expression plasmids are pCaMVCAT (Fromm et al., 1985) and the very convenient vector pDH51 (Pietrzak et al., 1986), where a multiple-cloning site has been introduced only 3 bp downstream from the original start of 35S RNA transcription. The 19 S promoter has also been used in gene expression studies, but is much less active than the 35S promoter (Lawton et al., 1987).

Gene expression and post-transcriptional regulation

The polycistronic nature of the 35S RNA, the presumed mRNA for ORFs VII to V, may present problems for expression of these genes, since eukaryotic mRNAs are usually monocistronic and the cellular translation machinery expresses downstream genes on di- or polycistronic mRNAs only weakly (Peabody and Berg, 1986; Bonneville et al., 1989). Analysis of frame-shift mutations within the non- essential ORFs VII and II showed that mutants in which these ORFs extended beyond the start of the adjacent essential downstream genes were non-infectious, whereas those in which the mutated ORF still terminated before the downstream gene remained infectious (Sieg and Gronenborn, 1982; Dixon and Hohn, 1984). These observations of the polar effects of such mutations gave rise to the relay-race model for translation of successive ORFs by termination and re-initiation.

There have been no reports of successful translation *in vitro* of CaMV 35S RNA obtained from infected plants. Individual CaMV proteins from all eight ORFs could be translated *in vitro* from *in vitro* transcripts covering individual ORFs (Gordon et al., 1988), but downstream CaMV genes on dicistronic transcripts were translated poorly.

More recently, it has been shown that the protein encoded by ORF VI, the only gene with an abundant own mRNA, is capable of trans-activating the expression of downstream genes on a polycistronic mRNA in transient expression studies in protoplasts (Bonneville et al., 1989; Gowda et al., 1989). The trans-activation effect, which actually requires the production of the ORF VI protein, is most dramatic when translation initiating at the start of ORF I is measured by using a reporter gene (CAT or GUS) fused to this initiation point and with ORF VII present as the upstream ORF. The presence or absence of the 600 nt leader upstream of ORF VII makes no difference to this trans- activation phenomenon.

This long (600 nt) leader upstream of ORF VII, the first gene which is known to be translated *in vivo* (Dixon and Hohn, 1984), has been shown to reduce translation of reporter genes placed downstream both *in vivo* and *in vitro* (Fütterer et al., 1989; Gordon et al., 1988). However certain sequences in the leader are also capable of stimulating translation in cells of host species, so that the overall reduction in gene expression attributable to the leader is not as great in host plant cells as in those from non-host species (Fütterer et al., 1989; 1990). Moreover, the ORF VI product is also able to trans-activate expression of genes downstream from the 600 nt long leader of the 35S RNA (J. Fütterer, unpublished).

A strategy for induced resistance

The phenomenon of transactivation could be employed to introduce an inducible virus resistance into susceptible plants (Bonneville et al., 1989). Transgenic plants would constitutively produce a polycistronic mRNA carrying a downstream gene for a resistance factor or toxin. Upon infection of the cell, by a caulimovirus, the virus encoded transactivator would cause sufficient expression of the downstream gene to block virus replication or spread, or to cause the infected cell to suicide.

Processing of viral gene products

A feature of ORF V disclosed by searches for sequence
similarities between this ORF and those of other retroviral pol genes is that the N-terminal domain encodes an aspartic protease. Mutagenesis experiments, in conjunction with assays for a specific protease activity, showed that this domain does indeed encode an aspartyl protease (Torruella et al., 1989). The enzyme appears to be involved in processing of the primary ORF translation product to yield what is presumably active protease with a monomeric size of about 20 kD, and then of that from ORF IV to yield among others, the major form of the mature capsid protein (44kD). This suggests that it might be possible to express a foreign gene as part of a larger polyprotein which is then processed to yield the mature form desired, if appropriate cleavage sites were included around the foreign gene inserted.

Using CaMV as a vector

Earlier experiments showed that insertions which result in a viral genome larger than about 8300 bp result in unstable constructions with rapid loss of the foreign sequences (Gronenborn et al., 1981; Daubert et al., 1983). This apparent limit on the maximum size of the DNA molecule may result from a packaging requirement.

The analysis of non-essential regions referred to above showed that three major regions could be deleted to make room for genetic payload; their sum is about 1000 bp. There have now been a number of attempts to use CaMV directly as a vector for foreign marker genes under this limit of 1000 bp in size. Successful expression has been reported of the following genes inserted in place of ORF II: a short (204 bp) bacterial gene for dihydrofolate reductase (DHFR) which confers methotrexate resistance to turnip plants Brisson et al., (1984); a 200 bp metallothionein gene (Lefebvre et al., 1987) and the 501 bp human-interferon ∝D gene (de Zoeten et al., 1989).

Other genes have been inserted in place of ORF VII, but their insertion proved unstable (de Zoeten et al., 1989). This cannot be due to a requirement for ORF VII, which is, as already noted, dispensible, but suggests that the virus is very sensitive to polar effects of insertions in the domain between the primer binding site and the start of the first essential ORF (I) . Intriguingly, Dixon and Hohn (1984) found that mutants which removed the complete ORF VII (including the AUG start codon) and replaced it with a non-coding region of 30 nt before the start of ORF I were also unstable, with selection for reversion to a shorter interval of no more than 14 nt between the primer binding site and the start of ORF I.

At least one gene, the npt II gene conferring kanamycin resistance, proved unstable wherever inserted in the virus genome - either at ORF VII or II (T. Hohn, unpublished). Although the reasons for this instability are not known, it may be that the base composition of this gene presents difficulties for the virus or that there are other polar effects on downstream gene expression, e.g. through secondary structure of the transcript.

Analysis of the stability of different constructs carrying foreign genes has disclosed the constraints applying which relate to the translation requirements of the viral genome (Brisson and Hohn, 1986; de Zoeten et al., 1989). The foreign gene needs to be fitted precisely into the position of a CaMV gene (eg ORF II), with the initiation codon being immediately downstream from the upstream ORF's termination codon, and its termination codon immediately before the start of the following gene. Intergenic regions at these points which were longer than the single nucleotide found naturally proved to be unstable, resulting in frequent loss of the inserted gene.

Agroinfection

Major restrictions on the use of caulimoviruses as vectors are imposed by the limited host range available for any one virus and their inability to become stably incorporated into the host genome. These limitations can be circumvented by using a virus in combination with *Agrobacterium*-mediated plant transformation, or "agroinfection" (Grimsley and Bisaro, 1987). The viral genome, in a greater than unit length form so that e.g. transcription of a complete 35S RNA would be possible, is placed between the T-DNA borders of the Ti plasmid and the chimeric DNA introduced into the plant genome via *Agrobacterium* transformation. Virus replication would then be possible within each cell, whether or not systemic spread was possible. Furthermore, this system is suitable for the study of some mutants which result in non-viable virus.

A variation on this theme involved placing the CaMV genome between the T-DNA borders such that the insertion site on the viral genome lay within an appropriately manipulated ORF II (Bakkeren et al., 1989). *Agrobacterium* containing such Ti plasmids was inoculated directly onto host (turnip) plants. Isolation of progeny virus DNA and cloning to analyse the resulting variability of T-DNA border sequences within ORF II has shed light on possible recombination mechanism(s) by which the T-DNA border sequences can function in transferring the T-DNA from the Ti plasmid.

Conclusions

CaMV has shown only limited potential as a vector for expressing foreign genes in plants due to the high probability of recombination and the constraints - some understood, some not - on the stability of genes inserted. Only ORF II is so far available as a - not always reliable - site for insertion.

Nonetheless, it has proved possible to express certain foreign genes satisfactorily when inserted at this site, and the system remains one to be considered for certain applications. Transcription signals from CaMV have, in contrast, found wide application in gene transfer systems, and it is likely that other features of this complex genome will also be of use in plant genetic engineering.

Acknowledgements

I am grateful to Drs. T. Hohn and J. Fütterer for discussions and reading the manuscript.

References.

Bakkeren, G., Koukolikova-Nicola, Z., Grimsley, N. and Hohn, B. (1989) "Recovery of *Agrobacterium tumefaciens* T- DNA molecules from whole plants early after transfer", Cell 57, 847-857.

Balazs, E., Guilley, H., Jonard, G., Richards, K. and Hirth, L. (1982) "Nucleotide sequence of DNA from an altered-virulence isolate D/H of cauliflower mosaic virus", Gene 19, 239-249.

Baughman G.A., Jacobs, J.D. and Howell, S.H. (1988) "Cauliflower mosaic virus gene VI produces a symptomatic phenotype in transgenic tobacco plants", Proc. Natl. Acad. Sci. USA 85, 733-737.

Benfey, P.N. and Chua, N.-H. (1989) "Regulated genes in plants", Science 244, 174-181.

Benfey, P.N., Ren, L. and Chua, N.-H. (1989) "The CaMV 35S enhancer contains at least two domains which can confer different developmental and tissue-specific expression patterns", EMBO. J. 8, 2195-2202.

Bonneville, J.M., Sanfaçon, H., Fütterer, J. and Hohn, T. (1989) "Posttranscriptional trans-activation in cauliflower mosaic virus", Cell 59, 1135-1143.

Bonneville, J.M., Hohn, T. and Pfeiffer, P. (1988) "Reverse transcription in the plant virus, cauliflower mosaic virus" in Domingo, E., Holland, J.J. and Ahlquist, P. (eds), "RNA genetics", CRC press, Boca Raton, vol 2, pp. 23-42.

Brisson, N., Paszkowski, J., Penswick, J.R., Gronenborn, B., Potrykus, I. and Hohn, T. (1984) "Expression of a bacterial gene in plants by using a viral vector", Nature 310, 511-514.

Brisson, N. and Hohn, T. (1986) "Plant virus vectors: Cauliflower mosaic virus", Meth. Enzymol. 118, 659-668.

Covey, S.N., and Hull., R. (1981) " Transcription of cauliflower mosaic virus DNA. Detection of transcripts, properties and location of the gene encoding the virus inclusion body protein", Virology 111, 463-474.

Daubert, S., Shepherd, R. and Gardner, R.C. (1983) "Insertional mutagenesis of the cauliflower mosaic virus genome", Gene 25, 201-208.

Daubert, S., Schoelz, J., Dibao, L. and Shepherd, R. (1984) "Expression of disease symptoms in cauliflower mosaic virus genomic hybrids", J. Mol. Appl. Genet. 2, 537-546.
Daubert S., (1989) "Sequence determinants of symptoms in the genomes of plant viruses, viroids, and satellites", Mol. Plant-Microbe Interactions 1, 317-325.
De Zoeten, G.A., Penswick, J.R., Horisberger, M.A., Ahl., P., Schultze, M. and Hohn, T. (1989) 'The expression, localization, and effect of human interferon in plants", Virology 172, 213-222.
Dixon, L. and Hohn, T. (1984) "Initiation of translation of the cauliflower mosaic virus genome from a polycistronic mRNA: evidence from deletion mutagenesis", EMBO J. 3, 2731-2736.
Dixon, L.K., Koenig, I. and Hohn, T. (1983) "Mutagenesis of cauliflower mosaic virus", Gene 25, 189-199.
Dixon, L., Nyffenegger, T., Delley, G., Martinez-Izquierdo, J. and Hohn, T. (1986) "Evidence for replicative recombination in cauliflower mosaic virus", Virology 150, 463-468.
Franck, A., Guilley, H., Jonard, G., Richards, K. and Hirth, L.,(1980) "Nucleotide sequence of cauliflower mosaic virus DNA", Cell 21, 285-294.
Fromm, M., Taylor, L.P. and Walbot, V. (1985) "Expression of genes transferred into monocot and dicot plant cells by electroporation", Proc. Natl. Acad. Sci. USA 82, 5824- 5828.
Fütterer, J, Gordon, K., Pfeiffer, P., Sanfaçon, H., Pisan, B., Bonneville, J.M. and Hohn, T. (1989) "Differential inhibition of downstream gene expression by the cauliflower mosaic virus 35S RNA leader", Virus Genes 3, 45-55.
Fütterer, J, Gordon, K., Sanfaçon, H., Bonneville, J.M. and Hohn, T. (1990) "A ribosome-shunt model for translation of the cauliflower mosaic virus pregenomic 35S RNA", submitted.
Gardner, R.C., Howarth, A., Hahn, P., Brown-Leudi, M., Shepherd, R.J. and Messing, J.(1981) "The complete nucleotide sequence of an infectious clone of cauliflower mosaic virus by M13mp7 shotgun sequencing", Nucl. Acids Res. 9, 2871-2888.
Giband, M., Mesnard, J.M. and Lebeurier, G. (1986) "The gene III product (p15) of CaMV is a DNA binding protein while an immunologically related p11 polypeptide is associated with virus", EMBO J. 5, 2433-2438,
Goldberg, K., Young, M., Schoelz, J., Kiernan, J. and Shepherd, R. (1987) "Single gene of CaMV induces disease", (Abstr.) Phytopathology 77, 1704.
Gordon, K., Pfeiffer, P., Fütterer, J. and Hohn, T. (1988) " *In vitro* expression of cauliflower mosaic virus genes, EMBO J. 7, 309-317,
Gowda, S., Wu, F.C., Scholthof, H.B. and Shepherd, R.J. (1989) "Gene VI of figwort mosaic virus (caulimovirus group) functions in posttranscriptional expression of genes on the full-length RNA transcript", Proc. Natl. Acad. Sci. USA 86, 9203-9207.
Grimsley, N. and Bisaro, D. (1987) "Agroinfection", in: "Plant DNA infectious agents", Hohn, T. and Schell, J. (eds), Springer, Wien, New York, pp. 87-107.
Gronenborn, B., Gardner, R.C., Schaefer, S. and Shepherd, R.J. (1981) "Propagation of foreign DNA in plants using cauliflower mosaic virus as vector", Nature 294, 773-776.
Gronenborn, B. (1987) "The molecular biology of cauliflower mosaic virus and its application as plant gene vector", in: Hohn, T. and Schell, J. (eds), "Plant DNA infectious agents". Springer, Wien, New York, pp. 1-29.
Guilley, H., Richards, K.E. and Jonard, G. (1983) "Observations concerning the discontinuous DNAs of cauliflower mosaic virus", EMBO J. 2, 277-282.
Hasegawa, A., Verver, J., Shimada, A., Saito, M., Goldbach, R., van Kammen, A., Miki,

K., Kameya-Iwaki, M. and Hibi, (1989) "The complete sequence of soybean chlorotic mottle virus DNA and the identification of a novel promoter", Nucl. Acids Res. 17, 9993-10013.

Hibi, T. and Kameya-Iwaki, M. (1988) CMI/ AAB Descriptions of plant viruses No.331.

Hohn, T., Bonneville, J.M., Fütterer, J., Gordon, K., Jiricny, J., Karlsson, S., Sanfaçon, H., Schultze, M. and de Tapia, M. (1990) "The use of 35S RNA as either messenger or replicative intermediate might control the cauliflower mosaic virus replication cycle", in "Viral genes and plant pathogenesis", Pirone, T. and Shaw, J. (eds), Springer, Wien, New York, in press.

Howarth, A.J., Gardner, R.C., Messing, J. and Shepherd, R.J. (1981) "Nucleotide sequence of naturally occurring deletion mutants of cauliflower mosaic virus", Virology 112, 678-685.

Hull, R., Sadler, J. and Longstaff, M. (1986) "The sequence of carnation etched ring virus DNA: comparison with cauliflower mosaic virus and retroviruses", EMBO J. 5, 3083-3090.

Hull, R. and Covey, S.N. (1983) "Does cauliflower mosaic virus replicate by reverse transcription?", TIBS 8, 119- 121.

Jefferson, R.A., Kavanagh, T.A.and Bevan, M.W. (1987) "GUS- fusion: β-glucuronidase as a sensitive and versatile gene fusion marker in higher plants", EMBO J. 6, 3901-3907.

Lawton, M.A., Tierney, M.A., Nakamura, I., Anderson, E., Komeda, Y., Dubé, P., Hoffman, N., Fraley, R.T. and Beachy, R.N. (1987) "Expression of a soybean β-conglycinin gene under the control of the cauliflower mosaic virus 35S and 19S promoters in transformed petunia tissues", Plant Mol. Biol. 9, 315-324.

Lefebvre, D.D., Miki, B.L. and Laliberté, J.F. (1987) "Mammalian metallothionein functions in plants", Biotechnology 5, 1053-1056.

Lebeurier, G., Hirth, L., Hohn, B. and Hohn, T. (1982) "In vivo recombination of cauliflower mosaic virus DNA", Proc. Natl. Acad. Sci. USA 79,2932-2936.

Nagata, T., Okada, K., Kawazu, T. and Takabe, I.(1988) "Cauliflower mosaic virus 35S promoter directs S phase specific expression in plant cells", Mol. Gen. Genet. 207, 242-244.

Odell, J.T. and Howell, S.H. (1980) "The identification, mapping and characterization of messenger RNA for p66, a cauliflower mosaic virus coded protein", Virology 102, 349-359.

Peabody, D. S. and Berg, P. (1986) "Termination- reinitiation occurs in the translation of mammalian cell mRNAs", Mol. Cell. Biol. 6, 2695-2703.

Penswick, J.R., Hübler, R. and Hohn, T. (1988) "A viable mutation in cauliflower mosaic virus, a retrovirus-like plant virus, separates its capsid protein and polymerase genes", J. Virol. 62, 1460-1463.

Pfeiffer, P., Gordon, K., Fütterer, J. and Hohn, T. (1987) "The life cycle of cauliflower mosaic virus", in: von Wettstein, D. and Chua, N.-H. (eds), "Plant Molecular Biology", Plenum, pp. 443-458.

Pfeiffer, P. and Hohn, T. (1983) "Involvement of reverse transcription and the replication of cauliflower mosaic virus: a detailed model and test of some aspects", Cell 33, 781-789.

Pfeiffer, P. and Hohn, T. (1989) "Cauliflower mosaic virus as a probe for studying gene expression in plants", Physiol. Plant. 77, 625-632.

Pietrzak, M., Shillito, R.D., Hohn, T. and Potrykus, I. (1986) "Expression in plants of two bacterial antibiotic resistance genes after protoplast transformation with a new plant expression vector", Nucl. Acids Res. 14, 5857- 5868.

Richins, R.D. and Shepherd, R.J. (1986) Phytopathol. 76, 749-753.

Richins, R.D., Scholthof, H.B. and Shepherd, R.J. (1987) "Sequence of figwort mosaic virus DNA (caulimovirus group)", Nucl. Acids Res. 15, 8451-8466.

Rongxiang, F., Xiaojun, W., Ming, B., Yingchuan, T., Faxing, C. and Kekiang, M. (1985) "Complete sequence of cauliflower mosaic virus (Xinjing isolate) genomic DNA", Chin. J. Virol. 1, 247-256.

Schoelz, J., Shepherd, R.J. and Daubert, S. (1986) "Region VI of cauliflower mosaic virus encodes a host range determinant", Mol. Cell. Biol. 6, 2632-2637.

Schoelz, J. and Shepherd, R.J. (1988) "Host range control of cauliflower mosaic virus", Virology 162, 30-37.

Shepherd, R.J. and Lawson, R.H. (1981) "The caulimoviruses", in "Handbook of plant virus infections and comparative diagnosis", Kurstak, E. (ed), Elsevier/Nth Holland Biomedical Press, pp. 847-878. Sieg, K. and Gronenborn, B. (1982) "Introduction and propagation of foreign DNA in plants using cauliflower mosaic virus as vector", NATO/FEBS Advanced course: Structure and function of plant genomes, p. 154.

Stratford, R. and Covey, S.N. (1989) "Segregation of cauliflower mosaic virus symptom genetic determinants", Virology 172, 451-459.

Takahashi, H., Shimamoto, K. and Ehara, Y. (1989) "Cauliflower mosaic virus gene VI causes growth suppression, development of necrotic spots and expression of defense-related genes in transgenis tobacco plants" Mol. Gen. Genet. 216, 188-194.

Toh, H., Kikuno, R., Hayashida, H., Miyota, T., Kugimuja, W., Inouye, S., Yuki, S. and Saigo, K. (1985) "Close structural resemblance between putative polymerase of a Drosophila transposable element 17.6 and pol gene product of MoMuLV", EMBO J. 4, 1267-1272.

Torruella, M., Gordon, K. and Hohn, T. (1989) "Cauliflower mosaic virus produces an aspartic proteinase to cleave its polyproteins", EMBO J. 8, 2819-2825.

Woolston, C.J., Covey, S.N., Penswick, J.R. and Davies, J.W. (1983) "Aphid transmission and a polypeptide are specified by a defined region of the cauliflower mosaic virus genome", Gene 23, 15-23.

SALINITY AND DROUGHT STRESS IN RICE

A. CAPLAN, B. CLAES, R. DEKEYSER., and M. VAN MONTAGU
Laboratorium voor Genetica
Rijksuniversiteit Gent
K.L. Ledeganckstraat 35
B-9000 Gent
Belgium

ABSTRACT. Understanding the physiological responses of plants to saline soils and drought is more important today than ever before since genetic engineering is providing techniques by which these responses can be altered. This paper reviews some of the current ideas about how osmotic and ionic stresses affect plants, and about physiological adaptations that sustain growth. Particular attention is focused on studies of the graminae. In closing, we summarize work begun recently on rice. We have isolated cDNA clones for a gene expressed primarily in root and sheath tissues of salt- or drought-stressed rice plants, and for a drought-induced peroxidase. Studies of these genes show that the molecular responses to adverse conditions can be tissue-specific, since neither gene was induced throughout the plant by either stress treatment. It is unclear whether this organ-restricted response is an evolutionary limitation or a useful adaptation in the physiology of rice plants. Answers to questions such as this may help molecular biologists develop more rational approaches to extending the adaptative capabilities of important crop plants.

1. INTRODUCTION

Inorganic ions such as NH_4^+, Ca^{+2}, Mg^{+2}, SO_4^{-2}, and PO_4^{-3} must always be imported into plants in order to provide raw materials and co-factors needed to synthesize and activate the organic components of the cell. Other ions such as K^+, Na^+, and Cl^- are imported to produce the ionic balance and osmotic potential that is optimal for different physiological processes. When the external concentrations are low, ions can be brought into cells at controlled rates through pumps and channels. When any one of these ions is in such excess that it alters the osmotic potential of the soil, then cells are confronted with a dilemma. If the salt is excluded, ion toxicity is minimized, but water must be imported against a free energy gradient. If, on the other hand, salts are freely absorbed so that the internal and external ion pools reach equilibrium, then many biochemical pathways may be inhibited. Some components of salt stress resemble drought. In particular, each condi-

R.S. Sangwan and B.S. Sangwan-Norreel (eds.), The Impact of Biotechnology in Agriculture, 391–402.

tion reduces the availability of free water in the cell, and simultaneously increases the concentration of endogenous salt. The two stresses differ in that water can still be absorbed from saline soils, but only at the risk of disrupting the ionic balance through uptake of salts, while during drought, water is simply unavailable outside the cell. Although there is no wholly adequate solution, plants have several ways to make temporary adjustments that protect themselves from both ionically-induced and drought-induced damage. Some of these adaptations can be applied equally to the two stresses, but additional changes may yet prove specifically designed to retard evaporation from drought-affected tissues, or counter toxic effects of different ions.

Understanding the nature of all these adaptative processes, particularly with regard to economically important crops, is more important today then ever before since these processes can now be altered by genetic engineers. The tools provided by molecular biology make it possible to re-examine plant stress to extend the work of the breeders who have identified the genes involved in plant protection and of the physiologists who have tried to dissect the biochemical pathways employed for adaptation. Once the limitations in the defenses are known, it should be possible to construct and introduce new genes designed to augment the weakest points in the adaptive processes.

This review will cover several aspects of osmotic stress, not with the intent of reviewing all the knowledge in this field (more details are presented in reviews by Greenway and Munns, 1980; Hanson and Hitz, 1982; Yeo, 1983), but in order to identify the nature of some of the problems the cells experience and some of the corresponding biochemical responses that molecular biologists may wish to manipulate. Equal weight will be given to studies of salt stress and of osmotic stress. Salt stress is most often imposed by additions of NaCl to the culture medium, but other salts can be equally effective. For technical reasons, drought or osmotic stress is often imposed in the laboratory by treating plants with polyethylene glycol or mannitol which reduce free water concentrations extracellularly. The closing section deals especially with salt stress in rice and how it is being investigated.

2. Nature of ionically-induced damage

Macromolecular assembly and enzyme activity associated with shaping and maintaining each cell can only proceed within the properly constituted ionic environment. The inorganic ions selectively neutralize charges on macromolecular surfaces and simultaneously permit formation of intramolecular bridges that determine the final conformation of many proteins. These same ions also determine the availability of free water around enzymes and their substrates and thus the rate of catalysis. Finally, ionic gradients, set up at considerable cost to the plant cell, constitute free energy gradients that can be tapped to direct the flow of organic molecules within and between cells.

An extracellular ion excess invariably disrupts the ionic balance intracellularly. With the influx of salt, proteins may denature or aggregate leading to a loss of function, gradient-driven pumps may reverse and thus block the normal redistribution of symported molecules, membrane fluidity and consequently, the activity of some membrane components, may change, and even the entry of water may be restricted. Some ions may have additional secondary effects. For example, increasing amounts of intracellular Na^+ can lead to decreases in the concentration of K^+ (Ben-Hayyim *et al.*, 1987; Binzel *et al.*, 1987). This, in turn, reduces the rate of photosynthesis (Pier and Berkowitz, 1987), and, based on studies with bacteria, can accelerate polysome decay and degradation of the freed ribosomal proteins (St. John and Goldberg, 1980). Salt-imposed stress has been shown to have an impact even before ions enter the cell. Extracellular Na^+ (or mannitol), for example, can leach Ca^{+2} from root cell plasmalemma, and as a result of membrane destabilization, increase K^+ efflux (Cramer *et al.*, 1985).

These are only the immediate problems facing the cell. If the stress is prolonged, normal maintenance processes are impaired because general protein synthesis (Hurkman and Tanaka, 1987) and metabolism (Criddle *et al.*, 1989) both decline. Denatured proteins may form inactive complexes with otherwise functional proteins. Enzymes may be poisoned when inorganic co-factors are displaced by incoming salts. If the plant is to adapt, each of these classes of damage must be repaired, or if possible, prevented. This is not likely to be possible through any single response to saline or dry conditions, but rather through a series of independent biochemical adaptations induced sequentially according to the type and severity of damage sustained.

3. Adaptive changes in cellular chemistry

Plant cells respond to salt and osmotic stresses by selectively increasing the concentration of a number of common metabolic compounds. It has been assumed that these molecules would not be synthesized in such great amounts if they did not in some way serve as osmoprotectants. In the simplest case, such chemicals would be used both to adjust the osmotic potential of the cell with biologically inert molecules and to preserve the structural integrity of the cell. Polyamines may be one member of this class. In oats, putrescine and spermidine are synthesized within two hours from the beginning of the osmotic stress, before there is any detectable loss in cell viability (Flores and Galston, 1984). These highly charged molecules tend to associate with many negatively charged structures, from tRNAs to membranes. Exogenously provided polyamines are known to inhibit protoplast lysis (Altman *et al.*, 1977), and could be acting similarly within osmotically stressed cells.

The most commonly considered example of a general osmoprotectant is free proline. There is an eleven-fold increase in the concentration of proline 20 hours after excised barley leaves are transferred to salt-containing medium (Chandler and Thorpe, 1987). However, even in the most dramatic cases such as with adapted cell lines of tobacco, this increase only accounts for 1-4% of the intracellular osmotic potential (Handa *et al.*, 1983; Binzel *et al.*, 1987; Moftah and Michel, 1987). In some cell types, this contribution might be even smaller. For example, high concentrations of proline are found in leaf epidermis and the vascular bundles (Zuñiga *et al.*, 1989), but not in mesophyll cells where it might protect the photosynthetic machinery. Often, there is also no linear correlation between internal proline levels and growth rates of isogenic cell lines on different salt concentrations (Hassan and Wilkins, 1988). More revealingly, although it can sometimes begin within hours of stress treatment (Steward and Voetberg, 1985), proline synthesis is not always induced rapidly, and in soybean seedlings only begins after injury is evident (Moftah and Michel, 1987). Furthermore, the proline concentration can remain high, even when stressed callus is returned to normal osmotic conditions and cultured for one month (Chandler and Thorpe, 1987). The implications from these studies is that proline is being used to minimize the effects of a particular form of cell damage, and not simply to adjust the intracellular osmotic potential.

Other organic compounds such as sucrose (Binzel *et al.*, 1987), and under some conditions, reducing sugars and malate, contribute as much as proline to the osmotic potential of stressed cells (Binzel *et al.*, 1987). Studies of different varieties of sorghum, however, have shown that the production of sucrose (Newton *et al.*, 1986), like that of proline (Bhaskaran *et al.*, 1985) is not consistently correlated with stress tolerance.

Osmotic adjustment does not depend solely on the production of organic molecules. In barley, the resulting changes in the intracellular ion pools activate a Na^+/H^+ exchange process in the tonoplast membrane. This activation is quite rapid, occurring within 15 minutes, and does not appear to require new protein synthesis (Garbarino and DuPont, 1989). In this way, the vacuoles can accumulate vast amounts of Na^+ and Cl^- (Binzel *et al.*, 1988) and by compartmentalizing the ions, protect many of the salt-sensitive cytoplasmic processes.

Although this cannot continue indefinitely, adaptation to salt stress appears to depend more on the rapid compartmentalization of ions than on any other single physiological change, including any significant reduction in the rate of entry of ions into the cell (Watad *et al.*, 1986; Binzel *et al.*, 1988). Adaptations involving membrane transport processes are ion specific and thus may explain how cell lines adapted for rapid growth on NaCl remain sensitive to other salts such as KCl and Na_2SO_4 (Hassan and Wilkins, 1988). If protection against one salt primarily involves the synthesis of general osmoprotectants such as proline and sucrose, then there should have been cross-protection against similar salts. The same argument can be applied to adaptation to external osmotic potential: it has been seen that not all cell lines adapted to growth on high salt were able to withstand increasing osmotic stress (Ben-Hayyim, 1987), so osmotic adjustment also does not necessarily depend on any single osmoprotectant.

4. Osmotically induced changes in protein populations

Many metabolic adjustments to reduced free water or increased ion concentrations may simply entail changes in reaction rates mediated by feed-back inhibition or post-translation modifications of existing molecular species. These are difficult to detect without prior knowledge of regulation of specific biochemical pathways. Lacking this insight, most studies of the molecular biology of stress-response employ two-dimensional gel electrophoresis to identify changes in the protein population brought about by treatments with salt or desiccation. The number of changes identified in this way varies considerably, in part depending on the choice of detection systems (radio-isotope incorporation vs. visualization of the proteins with dyes), and in part depending on whether the comparison is between temporarily stressed or fully adapted cell lines or plants. The protein pattern can be qualitatively more complex at the onset of stress than in adapted cells (Singh *et al.*, 1985; Winicov *et al.*, 1989). If we disregard species differences for the moment, in order to generalize, then between 8 and 25 proteins show notable increases in abundance in salt-stressed suspensions or roots. Between 4 and 75 proteins decrease in amount disproportionately (Singh *et al.*, 1985; Hurkman and Tanaka, 1987; Ramagopal, 1987; Winicov *et al.*, 1989). A numerically similar, but molecularly distinct set of proteins can be induced by drought or PEG treatments (Singh *et al.*, 1985; Vartanian *et al.*, 1987). There are also distinguishable differences in the proteins induced in different organs.

In some species which evolved to live in saline environments, the numerous changes in the protein composition of the cell can be part of switches in entire metabolic pathways to ones more capable of conserving water. The halophyte, *Mesembryanthemum crystallinum*, for example, switches from C3 to C4 photosynthesis when grown in adverse conditions. In the course of doing so, the ribulose-1,5-bisphosphate carboxylase small subunit and chlorophyll *a/b*-binding protein gene families are transiently down-regulated, while phosphoenolpyruvate carboxylase and pyruvate orthophosphate dikinase messenger populations increase 50-fold (Michalowski *et al.*, 1989).

The significance of particular alterations of the protein pattern in salt-sensitive species is much less clear. It is possible that some of the qualitative differences are exchanges of specialized members of a protein family for those isoforms no longer active in the altered osmotic environment, but, in at least one investigation, malate dehydrogenase and NADH nitrate reductase were equally sensitive to NaCl or KCl regardless of whether rice plants were grown with salt or without it (Yeo and Flowers, 1983). None of the proteins detected in the more careful surveys has the properties expected for components of an ion pump (Hurkman *et al.*, 1988).

One protein produced in considerable amounts in many different dicots has been termed osmotin (Singh *et al.*, 1987a, 1987b). There are at least two forms of this protein which differ in solubility. At least one of these exists in granular bodies in vacuoles (Singh *et al.*, 1987a). These proteins are related to several other differentially regulated molecules including a pathogen-induced protein, a fruit protein termed thaumatin, and a maize trypsin inhibitor. Such homologies are probably not fortuitous, but there is no evidence for functional similarities between any of the proteins. Perhaps a protease-inhibitor could be useful in regulating the rate of turnover of osmotically-denatured proteins.

Roots of salt-tolerant and salt-sensitive tomato plants seem to contain similar amounts of osmotin, or proteins like it, during normal growth (King *et al.*, 1986). The protein then accumulates further when plants are treated with sufficient salt to inhibit growth by 40%, but this increase is only seen after 10 days. This is quite late compared with the synthesis of osmoprotectants such as proline, but does begin at approximately the same time as salt-induced growth inhibition is overcome (Singh *et al.*, 1985). The protein might be thought of as an integral component of some actively metabolizing cell types, rather than as a generalized stress protein.

Desiccation should not be viewed as an unusual event in a plant's life cycle. Seeds go through a period of dehydration during the normal maturation process. At this time, a set of "late embryogenesis-abundant" or *lea* genes are induced and are thought to serve as macromolecular osmoprotectants (Baker *et al.*, 1988). Although the genes were originally isolated to study seed development, most, if not all, can be induced in seedlings and in mature plants when the appropriate osmotic or ionic stress is applied.

The Lea proteins fall into at least three subsets distinguishable by the type of amino acid repeats they contain (Dure III *et al.*, 1989). In general, all three subsets are very hydrophilic, not associated with any specific cellular structure or organelle, and rich in glycine, alanine, serine, and threonine. One particular subset found in cotton (Baker *et al.*, 1988), rice (Mundy and Chua, 1988), barley, maize (Close *et al.*, 1989), and presumably in all plants, has been termed "dehydrin". Despite differences in specific sequences, dehydrins have a characteristic organization: a long central run of serine residues followed by a lysine-rich domain and ending with a different lysine-rich region. The rice dehydrin family is rapidly induced by both NaCl and desiccation (Mundy and Chua, 1988), and therefore may be part of the primary adaptation process to osmotic stress.

Less can be said about the remaining stress-induced genes analyzed or isolated to date. One clone obtained from maize (Gómez *et al.*, 1988) encodes a glycine-rich protein containing an RNA-binding domain (Mortenson and Dreyfuss, 1989). This could play a regulatory role governing the translation of specific messenger RNAs, or could protect all mRNAs from damage or aggregation when water is limiting.

Protein denaturation is presumably a major problem in dehydrated cells. For this reason, it is interesting that water stress seems to induce *hsp70* (Heikkila *et al.*, 1984), a member of a family of proteins that facilitate protein transport into cell organelles by dissolving molecular aggregates (Chirico *et al.*, 1988). This activity may be important to ensure that there is no interruption of the flow of material into mitochondria and chloroplasts, or that irrevocably denatured proteins are rapidly shunted to vacuoles where they can be degraded (Canut *et al.*, 1986).

5. Abscisic acid mediates osmoregulation and gene expression

The plant epidermis is enclosed in a waxy layer that reduces evaporation. The only gaps in this barrier are stomata which can be opened or closed by expansion of flanking pairs of guard cells. In drought conditions newly synthesized abscisic acid (ABA) stimulates these guard cells to close in order to retard water loss. Correlations between ABA levels and drought tolerance have been demonstrated in several ways. For example, one tomato mutant, *flacca*, cannot close its stomata efficiently, is very sensitive to water deficit, and produces relatively little ABA during drought conditions (Bradford, 1983; Bray, 1988). Conversely, maize lines selected for drought tolerance contain up to 6 times more ABA than drought-sensitive lines during normal growth and 2.5 times more when drying (Larque-Saavadra and Wain, 1974). However, the effects of ABA are not confined to the regulation of stomata closure. ABA is produced throughout the leaf, not merely in the vicinity of the stomata (Harris *et al.*, 1988). ABA can accelerate osmotic adjustment in NaCl-treated tobacco cells, primarily by increasing the accumulation of sucrose and reducing sugars (LaRosa *et al.*, 1987). In soybean hypocotyls, the hormone inhibits localized cell expansion (Bensen *et al.*, 1988), perhaps so that water can be conserved for more essential physiological processes. Finally, ABA induces virtually all drought- and salt-induced genes discussed previously, sometimes even faster than stress induces them (Singh *et al.*, 1987b). Some genes, however, can be induced by drought even when ABA levels are not greatly changed (Bray, 1988). In other cases, drought-induced proline concentrations seem correlated with high levels of endogenous ABA, yet salt-induced proline concentrations do not (Steward and Voetberg, 1985). Thus, although ABA is undoubtedly the single most important regulatory molecule known to control the molecular response to water stress, it is not the only signal.

6. The stress response in rice

There are no outstanding differences between rice and any of the other salt-sensitive plants studied to date. Nevertheless, each species must be assumed to have its own design faults which must be identified before molecular engineers can alter salinity tolerance. For this reason, we have begun studying the molecular response of rice to salt-stress and drought with the intention of characterizing the activities of differentially regulated genes.

Mature rice plants begin showing signs of stress within three days of exposure to increased salinity (for example, 1% Murashige and Skoog salts or NaCl). Leaves begin to wilt and yellow at the tips and produce large amounts of at least eight new proteins in the roots. We have recently isolated and partially sequenced one 14.5-kDa protein that routinely appears in salt-stressed roots (Claes *et al.*, in preparation). Using an oligonucleotide probe based upon this sequence, we have isolated the corresponding cDNA and genomic copy for the salt-induced protein. The polypeptide is hydrophilic, rich in glycine, and negatively charged in amino acid residues, but quite distinct from any of the osmotically regulated gene families yet isolated from other species. Like the majority of these genes, the protein we have isolated lacks any apparent targeting signals that would direct it to organelles or the cell wall.

The gene has been given the mnemonic, *salT*. The *salT* transcript is present at a low level in the sheath and roots of hydroponically grown rice, but absent from leaf lamina. Northern analysis shows that the mRNA begins accumulating within 2-6 hours of salt treatment. Unexpectedly, the accumulation is greatest in the sheath, and after three days, still absent from the lamina. This induction pattern is also seen in drought conditions even though plants are dried uniformly.

The response pattern of the *salT* transcript closely resembles the pattern of Na^+ accumulation in salt-stressed plants. Yeo and Flowers (1982) showed that rice takes up Na^+ very quickly, but then establishes a stable gradient that concentrates the ion in the sheathes of the oldest leaves, and partially excludes it from the lamina of the youngest ones. Since it has been shown that water deficit can inhibit photosynthetic rate in spinach (Berkowitz *et al.*, 1983; Willeford *et al.*, 1989), the significance of this gradient may be to protect the metabolically essential parts of the plant from the toxic effect of excess ions. At the same time, this distribution could restrict the accumulation of salts in and around the meristematic tissues that give rise to the next generation, and thus permit some seeds to form in adverse conditions. The SalT protein might facilitate the establishment of this ion gradient, or participate in the process that protects cells from the damage that the gradient causes.

Another osmotically regulated protein in rice is the dehydrin, encoded by *rab*21 (Mundy and Chua, 1988). Like SalT, it begins accumulating rapidly after treatments with salt, drought, or ABA. This gene is induced in both roots and shoots of hydroponically grown plants when the stress is applied, but is also produced in rice embryos during normal seed maturation.

In rice, not all genes induced by drought seem to be induced by salt. For example, an anionic peroxidase has been cloned which is not strongly transcribed in any part either of healthy plants, or those stressed with 1% Murashige and Skoog salts (unpublished results). The gene is induced, however, in the sheath of drying plants, or after plants have been grown for three days in 5% PEG. From sequence comparisons with other anionic peroxidases (Lagrimini *et al.*, 1987), this rice protein might be exported to the cell wall. If so, the induction seen during drought conditions could indicate that this particular peroxidase is used to catalyze cell wall molecule cross-links in order to build a tighter barrier to retard water loss. Such an adaptation would not benefit salt-stressed plants.

Most of the results to date on the molecular biology of rice are fairly superficial in that they note changes without explaining them. The most difficult question to address at this time is whether any of the induced proteins contribute significantly to the adaptive process. The answers to this question will ultimately depend on increasing our knowledge of what distinguishes salt or drought tolerant varieties from sensitive ones. Although many tolerant varieties have been identified by breeders, the multigenic complexity of the traits has prevented the construction of isogenic lines needed for careful physiological evaluation. In one study (Aslam and Qureshi, 1989), tolerant and sensitive lines of rice showed several differences in the rates of ion transport, according to the ion, and the age of the plants investigated. Since the lines are not isogenic, it is not clear if such differences are responsible for the relevant phenotypes. A broader survey of different rice genotypes (Yeo and Flowers, 1983), has shown two components to tolerance. One permits Na^+ to accumulate without substantial cellular damage, whereas the other seems to permit growth in spite of the damage sustained from exposure to salt. The next step forward towards breeding or engineering tolerant lines of rice will come when we can distinguish the various osmotically regulated proteins by their contribution to one or the other of these two phenotypic classes.

7. References

Altman, A., Kaur-Sawney, R. and Galston, A.W. (1977) Stabilisation of oat leaf protoplasts through polyamine-mediated inhibition of senescence, Plant Physiol. 60, 570-574.

Aslam, M. and Qureshi, R.H. (1989) Ion transport in two rice varieties grown under saline conditions. Int. Rice Res. Newsletter 14, p. 3.

Baker, J., Steele, C. and Dure III, L. (1988) Sequence and characterization of 6 *Lea* proteins and their genes from cotton, Plant Mol. Biol. 11, 277-291.

Ben-Hayyim, G. (1987) Relationship between salt tolerance and resistance to polyethylene glycol-induced water stress in cultured citrus cells, Plant Physiol. 85, 430-433.

Ben-Hayyim, G., Kafkafi, Y. and Ganmore-Neumann, R. (1987) Role of internal potassium in maintaining growth of cultured *Citrus* cells on increasing NaCl and $CaCl_2$ concentrations, Plant Physiol. 85, 434-439.

Bensen, R.J., Boyer, J.S. and Mullet, J.E. (1988) Water deficit-induced changes in abscisic acid, growth, polysomes, and translatable RNA in soybean hypocotyls, Plant Physiol. 88, 289-294.

Bhaskaran, S., Smith, R.H. and Newton, R.J. (1985) Physiological changes in cultured sorghum cells in response to induced water stress, Plant Physiol. 79, 266-269.

Berkowitz, G.A., Chen, C. and Gibbs, M. (1983) Stromal acidification mediates *in vivo* water stress inhibition of nonstomatal-controlled photosynthesis, Plant Physiol. 72, 1123-1126.

Binzel, M.L., Hasegawa, P.M., Rhodes, D., Handa, S., Handa, A.K. and Bressan, R.A. (1987) Solute accumulation in tobacco cells adapted to NaCl, Plant Physiol. 84, 1408-1415.

Binzel, M.L., Hess, F.D., Bressan, R.A. and Hasegawa, P.M. (1988) Intracellular compartmentation of ions in salt adapted tobacco cells, Plant Physiol. 86, 607-614.

Bradford, K.J. (1983) Water relations and growth of the *flacca* tomato mutant in relation to abscisic acid, Plant Physiol. 72, 251-255.

Bray, E.A. (1988) Drought- and ABA-induced changes in polypeptide and mRNA accumulation in tomato leaves, Plant Physiol. 88, 1210-1214.

Canut, H., Alibert, G., Carrasco, A. and Boudet, A.M. (1986) Rapid degradation of abnormal proteins in vacuoles from *Acer pseudoplatanus* L. cells, Plant Physiol. 81, 460-463.

Chandler, S.F. and Thorpe, T.A. (1987) Proline accumulation and sodium sulfate tolerance in callus cultures of *Brassica napus* L. cv. Westar, Plant Cell Reports 6, 176-179.

Chirico, W.J., Waters, M.G. and Blobel, G. (1988) 70K heat shock related proteins stimulate protein translocation into microsomes, Nature (London) 332, 805-810.

Close, T.J., Kortt, A.A. and Chandler, P.M. (1989) A cDNA-based comparison of dehydration-induced proteins (dehydrins) in barley and corn, Plant Mol. Biol. 13, 95-108.

Cramer, G.R., Läuchli, A. and Polito, V.S. (1985) Displacement of Ca^{2+} by Na^{+} from the plasmalemma of root cells, Plant Physiol. 79, 207-211.

Criddle, R.S., Hansen, L.D., Breidenbach, R.W., Ward, M.R. and Huffaker, R.C. (1989) Effects of NaCl on metabolic heat evolution rates by barley roots, Plant Physiol. 90, 53-58.

Dure III, L., Crouch, M., Harada, J., Ho, T.-H.D., Mundy, J., Quatrano, R., Thomas, T. and Sung, Z.R. (1989) Common amino acid sequence domains among the LEA proteins of higher plants, Plant Mol. Biol. 12, 475-486.

Flores, H.E. and Galston, A.W. (1984) Osmotic stress-induced polyamine accumulation in cereal leaves, Plant Physiol. 75, 102-109.

Garbarino, J. and DuPont, F.M. (1989) Rapid induction of Na^{+}/H^{+} exchange activity in barley root tonoplast, Plant Physiol. 89, 1-4.

Gómez, J., Sánchez-Martínez, D., Stiefel, V., Rigau, J., Puigdomènech, P. and Pagès, M. (1988) A gene induced by the plant hormone abscisic acid in response to water stress encodes a glycine-rich protein, Nature (London) 334, 262-264.

Greenway, H. and Munns, R. (1980) Mechanisms of salt tolerance in non-halophytes, Ann. Rev. Plant Physiol. 31, 149-190.

Handa, S., Bressan, R.A., Handa, A.K., Carpita, N.C. and Hasegawa, P.M. (1983) Solutes contributing to osmotic adjustment in cultured plant cells adapted to water stress, Plant Physiol. 73, 834-843.

Hanson, A.D. and Hitz, W.D. (1982) Metabolic responses of mesophytes to plant water deficits, Ann. Rev. Plant Physiol. 33, 163-203.

Harris, M.J., Outlaw, W.H. Jr, Mertens, R. and Weiler, E.W. (1988) Water-stress-induced changes in the abscisic acid content of guard cells and other cells of *Vicia faba* L. leaves as determined by enzyme-amplified immunoassay, Proc. Natl. Acad. Sci. USA 85, 2584-2588.

Hassan, N.S. and Wilkins, D.A. (1988) In vitro selection for salt tolerant lines in *Lycopersicon peruvianum*, Plant Cell Reports 7, 463-466.

Heikkila, J.J., Papp, J.E.T., Schultz, G.A. and Bewley, J.D. (1984) Induction of heat shock protein messenger RNA in maize mesocotyls by water stress, abscisic acid, and wounding, Plant Physiol. 76, 270-274.

Hurkman, W.J. and Tanaka, C.K. (1987) The effects of salt on the pattern of protein synthesis in barley roots, Plant Physiol. 83, 517-524.

Hurkman, W.J., Tanaka, C.K. and DuPont, F.M. (1988) The effects of salt stress on polypeptides in membrane fractions from barley roots, Plant Physiol. 88, 1263-1273.

King, G.J., Hussey, C.E. Jr and Turner, V.A. (1986) A protein induced by NaCl in suspension cultures of *Nicotiana tabacum* accumulates in whole plant roots, Plant Mol. Biol. 7, 441-449.

Lagrimini, L.M., Burkhart, W., Moyer, M. and Rothstein, S. (1987) Molecular cloning of complementary DNA encoding the lignin-forming peroxidase from tobacco: molecular analysis and tissue-specific expression, Proc. Natl. Acad. Sci. USA 84, 7542-7546.

LaRosa, P.C., Hasegawa, P.M., Rhodes, D., Clithero, J.M., Watad, A.-E. A. and Bressan, R.A. (1987) Abscisic acid stimulated osmotic adjustment and its involvement in adaptation of tobacco cells to NaCl, Plant Physiol. 85, 174-181.

Larque-Saavedra, A. and Wain, R.L. (1974) Abscisic acid levels in relation to drought tolerance in varieties of *Zea mays* L, Nature (London) 251, 716-717.

Michalowski, C.B., Olson, S.W., Piepenbrock, M., Schmitt, J.M. and Bohnert, H.J. (1989) Time course of mRNA induction elicited by salt stress in the common ice plant (*Mesembryanthemum crystallinum*) Plant Physiol. 89, 811-816.

Moftah, A.E and Michel, B.E. (1987) The effect of sodium chloride on solute potential and proline accumulation in soybean leaves, Plant Physiol. 83, 238-240.

Mortonson, E. and Dreyfuss, G. (1989) RNP in maize protein, Nature (London) 337, 312.

Mundy, J. and Chua, N.-H. (1988) Abscisic acid and water-stress induce the expression of a novel rice gene, EMBO J. 7, 2279-2286.

Newton, R.J., Bhaskaran, S., Puryear, J.D. and Smith, R.H. (1986) Physiological changes in cultured sorghum cells in response to induced water stress. II. Soluble carbohydrates and organic acids, Plant Physiol. 81, 626-629.

Pier, P.A. and Berkowitz, G.A. (1987) Modulation of water stress effects on photosynthesis by altered leaf K^{+1}, Plant Physiol. 85, 655-661.

Ramagopal, S. (1987) Salinity stress induced tissue-specific proteins in barley seedlings, Plant Physiol. 84, 324-331.

Singh, N.K., Handa, A.K., Hasegawa, P.M. and Bressan, R.A. (1985) Proteins associated with adaptation of cultured tobacco cells to NaCl, Plant Physiol. 79, 126-137.

Singh, N.K., Bracker, C.A., Hasegawa, P.M., Handa, A.K., Buckel, S., Hermodson, M.A., Pfankovh, E., Regnier, F.E. and Bressan, R.A. (1987a) Characterization of osmotin. A thaumatin-like protein associated with osmotic adaptation in plant cells, Plant Physiol. 85, 529-536.

Singh, N.K., LaRosa, C., Handa, A.K., Hasegawa, P.M. and Bressan, R.A. (1987b) Hormonal regulation of protein synthesis associated with salt tolerance in plant cells, Proc. Natl. Acad. Sci. USA 84, 739-743.

St. John, A.C. and Goldberg, A.L. (1980) Effect of starvation for potassium and other inorganic ions on protein degradation and ribonucleic acid synthesis in *Escherichia coli*, J. Bacteriol. 143, 1223-1233.

Steward, C.R. and Voetberg, G. (1985) Relationship between stress-induced ABA and proline accumulations and ABA-induced proline accumulation in excised barley leaves, Plant Physiol. 79, 24-27.

Vartanian, N., Damerval, C. and de Vienne, D. (1987) Drought-induced changes in protein patterns of *Brassica napus* var. *oleifera* roots, Plant Physiol. 84, 989-992.

Watad, A.-E.A., Pesci, P.-A., Reinhold, L. and Lerner, H.R. (1986) Proton fluxes as a response to external salinity in wild type and NaCl-adapted *Nicotiana* cell lines, Plant Physiol., 81, 454-459.

Willeford, K.O., Ahluwalia, K.J.K. and Gibbs, M. (1989) Inhibition of chloroplastic respiration by osmotic dehydration, Plant Physiol. 89, 1158-1160.

Winicov, I., Waterborg, J.H., Harrington, R.E. and McCoy, T.J. (1989) Messenger RNA induction in cellular salt tolerance of alfalfa (*Medicago sativa*), Plant Cell Reports 8, 6-11.

Yeo, A.R. (1983) Salinity resistance: physiologies and prices, Physiol. Plant. 58, 214-222.

Yeo, A.R. and Flowers, T.J. (1982) Accumulation and localisation of sodium ions within the shoots of rice (*Oryza sativa*) varieties differing in salinity resistance, Physiol. Plant. 56, 343-348.

Yeo, A.R.,and Flowers, T.J. (1983) Varietal differences in the toxicity of sodium ions in rice leaves, Physiol. Plant. 59, 189-195.

Zuñiga, G.E., Argandoña, V.H. and Corcuera, L.J. (1989) Distribution of glycine-betaine and proline in water stressed and unstressed barley leaves, Phytochemistry 28, 419-420.

Acknowledgements

The authors like to thank M. De Cock for preparation of the manuscript. This work was supported by grants from the Rockefeller Foundation (#RF 86058) and the Services of the Prime Minister (U.I.A.P. #120C0187). R.D. is a Research Assistant of the National Fund for Scientific Research (Belgium).

PROSPECTS FOR THE GENETIC MANIPULATION OF METABOLIC PATHWAYS LEADING TO SECONDARY PRODUCTS

M. HOLDEN
Department of Botany, University of Edinburgh
King's Buildings, Mayfield Road, Edinburgh EH9 3JH

Introduction

Plants are cultivated not only as foodstuffs but also as sources of valuable chemicals which are widely employed as pharmaceuticals and also in the food flavouring and colouring industries. Many of these compounds are by their nature very expensive to produce and extract and many are difficult to synthsise chemically. Secondary products are often found in only low amounts in the whole plant and furthermore may be produced in only one tissue type or at only one stage of development. Many of the plants which yield desirable secondary metabolites can only be grown in tropical climates which results in the further cost of harvesting and importation. There is therefore considerable interest in increasing the productivity of plants which synthesise and accumulate useful chemicals. Conventional plant breeding has met with considerable success in increasing the percentage of the biomass composed of the desired product, and the length of the productive phase of the plant. In this respect many of the goals of the plant breeder trying to increase secondary product synthesis are the same as for breeders working with food crops, such as increased resistance to pathogens, defence against insects and tolerance of different climatic conditions. However it is notable that the primary objective of many food-plant breeders, the production of greater quantities of the desired tissue, (e.g. cereal grains or potato tubers) often results in a diminution in the quantity of accumulated secondary products. It is not merely anecdotal that present varieties of fruit and vegetables have less flavour compounds than older varieties, in many instances this is due to breeding for biomass production and not for secondary product accumulation. Therefore while the majority of aims for breeding plants for foodstuffs and for chemical accumulation are shared they differ in some respects.

As it is essential to be able to define the traits which a breeder is seeking to improve, a thorough knowledge of the biochemistry of the particular plant is necessary. In this respect, tissue culture is an invaluable tool, which enables controlled experiments to be carried out on defined tissue. It has also been suggested that tissue culture grown on a large scale can act as a source of the

R.S. Sangwan and B.S. Sangwan-Norreel (eds.), The Impact of Biotechnology in Agriculture, 403–417.

desired product (Fowler, 1977; see Yeoman *et al.*, 1989 for review).

There is some evidence that production of certain secondary products is limited because of weak expression or non-expression of regulatory genes. In some examples increased transcription leads to increased translation resulting in increased enzyme activity and more product synthesis. There therefore exists the prospect for increasing transcription of regulatory genes by e.g. increasing copy number or linking the gene to a strongly expressed promoter and both these techniques have been shown to result in increased enzyme activity in several systems. Similarly, if translation of key mRNAs is inhibited by e.g. antisense mRNA or artificial ribozymes specific enzyme activities can be reduced. Therefore there is considerable interest in the use of genetic manipulation to increase product yield. It is anticipated that whether the modified gene is finally expressed in whole plants, plant tissue culture or microorganisms will depend on the nature of the product of interest.

Prospects for the manipulation of gene expression in tissue cultures

Tissue cultures of plants have a number of advantages as the final site of expression of genes regulating secondary product accumulation. Foremost among these is the fact that where cells are cultured in bioreactors using similar technology to microbial fermenters, no regeneration of whole plants is necessary. It is a common observation that it is often difficult to obtain genetically transformed plants from transformed cells and this is especially true in many cereal plants which represent the majority of important crop plants. Some plants which are grown principally for their secondary product content are also difficult to regenerate from suspension cultures. In many instances the product of interest is tissue specific and/or developmentally regulated in the plant but by use of tissue cultures continuous or semi-continuous synthesis can be achieved. For example the production of capsaicin, the pungent principle of the chilli pepper, is limited to late stages of development of the fruit (Hall *et al.*, 1987) however cultures derived from stem tissue when grown in a bioreactor can be induced to synthesise this compound over prolonged periods (Mavituna *et al.*, 1987).

Cultures of many species release their products to the culture medium and this is particularly evident in immobilised cultures (for review see Hall *et al.*, 1988), this leads to greater efficiency in extracting the phytochemical, as plant tissue does not need to be processed. The prospect of continuous product removal may also increase synthesis by reducing any feedback inhibition of regulatory enzymes. Another advantage of cultured cells is, that limiting precursors to biochemical pathways or single step biotransformations can be added to the culture medium at any stage of the growth cycle which is not possible in whole plants. A final reason why there has been considerable industrial interest in the production of secondary products from cell culture is that production can be strictly controlled and is thus free from the political and economic vagaries of the international commodities markets.

Different types of cultures have been proposed as sources of fine chemicals, of which suspension cultures are the simplest. Suspension culture systems form the basis of one commercial process, the production of shikonin from cultures of *Lithospermum erythorhizon* a dye with mild antibiotic and anti-inflammatory properties (Tabata, 1985). Suspensions can be grown on a large scale in simple bioreactors similar to those used for microbial fermentation and cultures of *Catharanthus* have been grown up to 5,000 l (Schiel and Berlin, 1987) and *Nicotiana* cultures have been scaled up to 20,000 l (Hashimoto *et al.*, 1982) using such fermenters. Callus cultures, although they frequently synthesise more product than suspension cultures are not considered a viable basis for large scale processes due to difficulties in managing the tissue. Immobilised cells represent somewhat of an intermediate system between callus and suspensions, in that the cells retain close cell-cell contact while they can be successfully grown on a large scale in bioreactors (Mavituna, 1987).

Several configurations of bioreactor have been designed for growth of immobilised cells and the increased differentiation afforded by such systems represents a promising basis for genetically manipulated cells to be grown in bulk. Cultures exhibiting a greater degree of differentiation have also been considered as sources of fine chemicals such as root cultures, embryoid cultures, and meristemoid cultures. Another type of culture which may in future give rise to commercial processes is the growth of hairy roots. Cells can be transformed with T-DNA from the bacteria *Agrobacterium rhizogenese*. Transformed cultures grow very rapidly as a mass of root tissue and have also been shown to accumulate a range of products (e.g. Hamill *et al.*, 1986). This approach is particularly suitable for products which are synthesised in roots such as nicotine in *Nicotiana tabacum* and saponins in *Panax ginseng*. However not all species can be transformed and synthesis of compounds not produced in roots may prove difficult.

There are however two major problems which have slowed down the widespread adoption of cell cultures in the production of fine chemicals. The first is that the rate of synthesis in cultures is frequently not high enough to justify the high capital and labour costs involved. The second is that cells in culture are often genetically unstable and biosynthetic capacity can decline over time (Deus-Neumann and Zenk, 1984).

The problem of low productivity has until recently been tackled by largely empirical means. The simplest has employed the innate variability of cell cultures and has involved screening many cell lines derived from either single cells or small groups of cells. Thus a range of capsaicin synthesis in over 20 clones of *Capsicum frutescens* ranges from nil to greater than the whole plant (Lindsey and Yeoman, 1985). The composition of the medium also plays a part in determining yield and by modifications of the levels of nutrients and growth regulators further increases can be achieved (Lindsey,

1985). In the *Capsicum* model system synthesis has been shown to be further enhanced by cell immobilisation (Lindsey and Yeoman, 1984), precursor feeding (Yeoman *et al.*, 1980) and fungal elicitation (Holden *et al.*, 1988; Holden *et al.*, 1988b). Modification of secondary product genes in an undirected fashion by mutagenesis has also met with some success and flower colour has been modified in certain species by this technique (Broertjes and van Harten, 1988).

Increase in expression of designated genes

The rapidly advancing field of molecular genetics allows a more directed approach to be adopted for altering gene expression. The use of cell cultures in the production of desirable compounds is particularly suited to these methods as the frequently encountered difficulty in regenerating transformed plants does not represent a problem. Cells with enhanced expression of secondary product genes are also less likely to have a diliterious effect on viability and growth than transformants affecting primary metabolism. If a simple hypothetical pathway is considered (Fig. 1), it is clear that there are obvious ways to increase synthesis of the desired product "C". Increasing the activity of regulatory synthetic enzymes or decreasing the activity of enzymes leading to side branches or to product degradation might be expected to increase accumulation.

If manipulation of the genes controlling the synthesis of a specific enzyme is to be successful in increasing yield the enzyme should fulfill the following criteria.

- the enzyme must be rate limiting
- the rate of transcription must determine the rate of product formed
- product synthesis should be tissue specific and/or developmentally regulated
- product synthesis should be inducible
- the product should be reduced to the medium

The first two of these criteria are the most important. Although the concept of rate limiting enzymes is fraught with difficulties (Kacser and Burns, 1973), there are numerous examples where increased enzyme activity results in increased product synthesis. The second requirement is perhaps the most difficult to satisfy as even if increased transcription is achieved, factors such as post transcriptional modification of mRNA, mRNA transport, transport of the protein, compartmentation of enzyme and substrate, enzyme activation/deactivation and feedback inhibition can all influence final product synthesis. The other criteria should ideally be met due to the protocol outlined below.

The gene encoding the mRNA for capsaicin synthase, the enzyme that catalyses the final step in the capsaicin biosynthetic pathway (Iwai *et al.*, 1978) fulfills many of the above criteria. It is the end product of a pathway which is tissue specific and developmentally regulated (Hall *et al.*, 1987). Capsaicin synthase activity is inducible by treatment with a ß glucan preparation from the fungus *Gliocladium deliquescens* (Holden *et al.*, 1988) and the product is released into the medium. However little is known of the effect of rate of transcription on product synthesis and this is currently being investigated in our laboratory. The strategy which is being

Figure 1. Hypothetical pathway leading to the synthesis of desired compound C from precursor A via intermediate B. The pathway also has a side branch from B to X and a degradation product D.

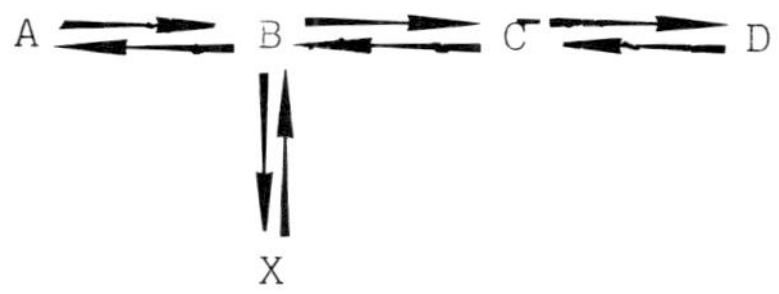

developed to study and alter the expression of genes of secondary metabolism is given below. This protocol may be applicable to many regulatory genes of secondary metabolism in other species

- extraction and purification of enzyme
- antibodies to pure enzyme(s) raised
- sequencing of enzyme to reveal unique sequences
- synthesis of labled oligonucleotide probes to unique sequences
- *in vitro* translation of total poly A+ RNA from cultures grown under inductive (elicitor treated) and non-inductive conditions (controls)
- electrophoretic separation of translation products and probing with antibodies
- creation of cDNA libraries from mRNA's from cultures under inductive and non-inductive conditions.
- screening of libraries with oligonucleotide probes
- creation of genomic DNA libraries with cDNAs to obtain parts of the whole gene.

This protocol shall allow isolation of the native gene encoding mRNA for a particular enzyme. If, mRNA transcript levels are high, cDNAs can be used to isolate mRNAs which can be translated *in vitro* and the products probed with the antibody (hybrid-select translation).

Once the gene has been isolated it is possible to anticipate altering its expression in cells. As capsaicin synthase activity is tissue specific and developmentally regulated, it is possible that its synthesis is regulated in a similar manner to other proteins that are found predominantly in one tissue. The synthesis of storage proteins which are present in large quantities have been widely studied and evidence exists that DNA sequences upstream from the region that codes for the protein partly control expression of mRNAs for barley storage proteins (Kreis *et al.*, 1986) and patatin, a class of potato storage proteins (Twell and Ooms, 1987). Patatin has also been shown to possess enzymic activity (lipid acyl hydrolase) (Racusen, 1984) and other proteins with enzymic function have also been shown to be regulated by upstream regions of DNA such as the polygalacturonase that is expressed during tomato ripening (Grierson *et al.*, 1987).

Similarly upstream DNA regions have been isolated that are involved in the induction of certain genes by environmental stimuli such as those that control proteinase inhibitor synthesis in response to wounding (Sanchez-Serrano *et al.*, 1987) and alcohol dehydrogenase induced in response to anoxia (Howard *et al.*, 1987). The possibility therefore exists of replacing such a (cis acting) developmentally regulated sequence with a constitutively expressed promoter such as that of the cauliflower mosaic virus 35S RNA gene (CaMV 35S) which has been shown to cause promotion of transcription of certain genes to which it has been ligated (Nagy *et al.*, 1985; Shah *et al.*, 1986). The differences in transcriptional activity of different promoters causes dramatic alterations in level of mRNAs under their control. For example, the CaMV 35S promoter has a 30 times greater strength than the nopaline synthase (nos) promoter when used to express the NPTII gene for kanamycin resistance in *Petunia* plants (Sanders *et al.*, 1987). As a result the enzyme levels of NPTII were 110 times higher in plants driven by the CaMV 35S promoter as opposed to the nos promoter. Duplication of enhancers of promoters has also been shown to greatly increase the transcriptional activity of the promoter (Kay *et al.*, 1987). Therefore "swapping" of promoter regions or enhancer regions offers a precise method to increase transcription of mRNAs for rate limiting enzymes.

A second approach to increase the specific activity of a regulatory enzyme is to increase the copy number of the genes responsible for its synthesis. Copy number of DNA sequences is often observed to influence the production of enzymes in many eucaryotic systems and also in some transformed plants (Potrykus *et al.*, 1985; Scott and Draper, 1987). Increasing copy number of selected genes linked to a selectable marker by use of *Agrobacterium tumefaciens* or another suitable vector could increase enzyme activity. However, although such transformants have shown stable inheritance of the trait, the random location within the genome of insertions causes difficulties, as transcription efficiency is largely determined by the site of the gene within the genome. However if a suitable selection technique is available, clones with high levels of expression can be obtained (Jones *et al.*, 1985).

Reduction in expression of designated genes

In a situation where the pathway leading to the desired product has side branches that compete for limiting precursors or where the product is enzymically degraded it is desirable to block the action of these enzymes. This is the case in capsaicin synthesis as over 50% of phenylalamine supplied to cultures becomes polymerised in the cell wall (Hall *et al.*, 1987) although capsaicin is not further metabolised. Inhibitors of lignin synthesis in other systems can increase availability of common precursors for other pathways but molecular techniques offer a more precise method to stop or reduce enzyme synthesis.

Antisense RNA is RNA transcribed in reverse orientation from "sense" or mRNA which binds to messenger sequences and prevents translation. The exact mode of action of antisense mRNA on protein synthesis is not known but it is thought that binding of sense and antisense RNA disturbs effective translation. This effect has been shown to be a regulatory mechanism in prokaryotes (Simons and Kleckner, 1988) and artificial antisense RNA can be exploited in plants (Green *et al.*, 1986). This requires the isolation of the gene encoding the enzyme of interest, the inversion of the coding region in respect to upstream sequences, ligation of reporter genes, insertion into a suitable vector and transformation of plants (Fig. 2).

The isolation of the gene of interest can be carried out in the same way detailed above for genes for which enhanced expression is the aim and transformation can also be performed in the same manner. One obvious group of target enzymes which it would be desirable to block translation of in such a manner in the capsaicin system are those involved in the conversion of free phenolic acids into lignin. Enzymes that catalyse the activation of free phenolic acids, the CoA-ligases have been well characterised in certain systems (Gross and Zenk, 1974). It would also be necessary to link the antisense coding region and reporter gene to an inducible promoter. Such promoters can induce activity in response to such environmental stimuli as heat shock (Spena *et al.*, 1985) UV light or wounding. Thus, in a bioreactor containing transformed cells of the chilli pepper it maybe possible to arrest or attenuate lignin synthesis thus favouring capsaicin production by changing cultural conditions at any desired stage of the growth cycle. In other systems where the desired product is synthesised but enzymically degraded or further metabolised it may be possible to block the synthesis of degradative enzymes by use of antisense RNA.

Another and even more recent technique to block specific gene expression which can be contemplated in the future is the employment of engineered ribozymes (Haseloff and Gerlach, 1988). These are RNA segments with specific endoribonuclease activities and, in theory, if the transcribed sequences of the target mRNA are known it should be possible to target engineered ribozymes against specified cleavage sites. The cleaved mRNA would not then be translated into active enzyme. The use of these techniques is at the moment merely hypothetical for increased secondary product synthesis in cultures but may in the future be important. Thus molecular genetics presents opportunities for increased or reduced expression of designated genes of secondary metabolism in cultures.

Figure 2. Translation in the presence and absence of antisense RNA.

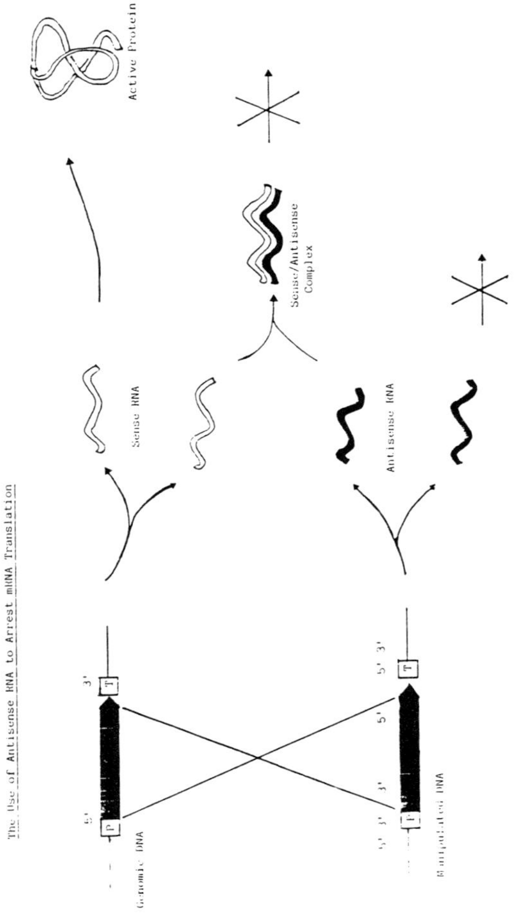

Prospects for the Manipulation of Secondary Product Gene Expression in Plants

The employment of transgenic plants in the production of valuable secondary metabolites has the innate advantage of lower labour and capital costs than cell cultures. If whole plants can be regenerated from transformed cells with altered secondary metabolism they can be used as "green bioreactors" to produce useful compounds grown on land surplus to traditional agricultural requirements.

There is also immense interest in modifying the synthesis of phytoalexins in response to pathogen attack. Use of transgenic plants offers the possibility of new disease resistant cultivars of many species. The methods detailed above can be employed to obtain transgenic plants with increased activities of synthetic enzymes and/or decreased activities of degradative or competing pathways. The same criteria for suitable target genes also apply. Some of the aims for increased gene expression are listed below:

to increase rate of synthesis
to extend the synthetic period
to obtain synthesis in other organs
to obtain synthesis in plants of other species

Where the product is developmentally regulated and/or tissue specific it may prove possible to enhance transcription by changing the promoter to a strongly expressed constitutively expressed one. Similarly, if the promoter is only weakly expressed it may be possible to enhance mRNA synthesis by use of a "stronger" promoter.

For example, berberine, which is an alkaloid with antimicrobial, uterine, anti-inflammatory, antileukemic and antineoplastic properties is extracted from rhizomes of *Coptis* species which take 5-6 years of growth before the product can be harvested (Kondo, 1976). This makes the cost of the rhizome very high and therefore a transformation that resulted in accelerated synthesis would be desirable. Similarly, it would be advantageous if higher levels of expression of genes for limiting enzymes could be obtained throughout the plant by changing promoter regions. At present, however, the enzymology of berberine synthesis has yet to be fully determined and further biochemical studies are therefore required.

As many plants which we use for extraction of fine chemicals are grown in tropical countries it would be of obvious advantage to transfer genes that encode rate limiting enzymes for particular products to related or even unrelated temperate plants. This approach would be possible when much of the biochemical pathway is present in a temperate plant but one or a few enzyme activities are absent.

In situations where the product of interest is a protein (although these cannot be classed as secondary products) and not a product of enzyme mediated reaction, another avenue exists to alter product synthesis in intact plants. For instance the potato plant *Solanum tuberosum* accumulates a family of immunologically identical glycoproteins known as "patatin" which represents up to 40% of total soluble protein (Racusen and Foote, 1980) in the tuber. It is possible to alter this synthesis in order to encode the production of selected mRNAs for other proteins. Therefore in theory it would be possible to transform cultured potato cells which would be regenerated into plants which yielded high levels of any desired protein.

It is even hypothetically possible that the protein prefferentially translated would have enzymic properties and the whole plant could act as "factory" for secondary product production. Certain enzymes of industrial importance are currently extracted from plant tissue, e.g. papain extracted from papaya latex and bromelain extracted from pineapple stems which are used extensively in the food industry (Godfrey and Reschelt, 1983) and a phosphodiesterase synthesised by cultured tobacco cells is currently available (Bethesda, Maryland). It might therefore be ultimately possible to transfer the regulatory genes to a temperate crop so that increased production would be possible.

Reduction in the expression of particular genes by the use of antisense RNA or artificial ribozymes as described above and elsewhere in this volume will also be important in altering secondary product formation in plants. The use of antisense RNA has already been employed to alter flower colour by blocking specific branches of metabolism (Van der Krol *et al.*, 1988). Commercial varieties of *Petunia* with novel colouration have now been obtained by blocking the translation of mRNAs encoding enzymes of flavonoid biosynthesis (Mol *et al.*, 1988). The attenuation chalcone synthase activity, a key enzyme of flavonoid synthesis, results in dramatic alteration to flower pigmentation in *Petunia* species. The employment of such "anti-gene" techniques to block pathways may also be employed to divert limiting substrates to a desired product (Lindsey, 1988) and also to reduce enzymic degradation of the product.

Certain secondary products, for instance the alkaloids in potatoes are toxic and therefore undesirable in plants; use of these methods could be used to block their synthesis. In this example it would be essential to link the antisense gene not to a constitutively expressed promoter but to a tissue (tuber specific) promoter as it is desirable to retain the alkaloids in the stem and leaves to deter insect and animal attack.

Prospects for Expression of Plant Genes in Microorganisms

Microbial systems offer the advantages over plant suspension cultures that they grow faster, are more shear resistant and the chemical engineering aspects are well understood. Heinstein (1985) has speculated that certain secondary product genes, suitably modified, may eventually be transcribed in bacteria or yeast. This is of course a proposition for the long term but one worth consideration none the less. Many enzyme systems are associated with the chloroplast membrane and therefore are not suitable targets for manipulation and it is probable that complex, multi-step pathways from plants cannot be engineered into microorganisms. It is however possible to anticipate the transformation of microorganisms with single genes that encode mRNAs for proteins that perform single step biotransformations of microbial metabolites.

Concluding Remarks

Although all of the above is highly speculative and perhaps premature, it is essential that these possibilities are examined and the fruitful ones adopted and the unsuccessful ones rejected. It is clear that the rapidly advancing area of molecular genetics has gifted us a very powerful tool by which to enhance or reduce expression of specific genes. These methods are particularly suitable to produce enhanced synthesis of desirable secondary products as many of them as produced in low quantities and are developmentally regulated and tissue specific. Whether altered genes are ultimately expressed in whole plants, to confer disease resistance, or to produce known or novel products, or incorporated into plant or microbial cells to be grown in bioreactors will probably depend on the nature of the specific compound. Given the limited success achieved in boosting secondary product biosynthesis by conventional techniques the adoption of genetic manipulation is clearly one way forward. It is surely only a matter of time before the problems are overcome and valuable new crops and cultures are produced.

Acknowledgments

The author would like to thank Glaxo Group Research Ltd. for funding current investigations and Mrs. E. Raeburn and Mrs. J. Summers for preparing the manuscript.

References

Balandrin, M.F. and Kloche, J.A. (1988) 'Medicinal, aromatic and industrial materials from plants', in Y.P.S. Bajaj (ed.) Biotechnology in Agriculture and Forestry 4: Medicinal and Aromatic Plants I, Springer-Verlag, Berlin.

Balandrin, M.F., Kloche, J.A., Wurtele, E.S. and Bollinger, W.H. (1985) 'Natural plant chemicals: sources of industrial and medicinal materials', Science **228**, 1154-1160.

Berlin, J., Beier, H., Fecker, L., Forche, E., Noe, W., Sasse, F., Schiel, O. and Wray, V. (1985) 'Conventional and new approaches to increase the alkaloid production of plant cell cultures, in K.-H. Neumann, W. Barz and E. Reinhard (eds.), Primary and Secondary Metabolism of Plant Cell Cultures, Springer-Verlag, Berlin, pp. 272-280.

Betz, B., Schäfer, E. and Hahlbrock, K. (1978) 'Light-induced phenylalanine ammonia-lyase in cell suspension cultures of *Petroselinum hortense*. Quantitative comparison of rates of synthesis and degradation', Arch. Biochem. Biophys. **140**, 126-135.

Broertjes, C. and Harten, A.M. (1988), Developments in Crop Science **12**, Elsevier, Amsterdam.

Cramer, C.L., Ryder, T.B., Bell, T.N. and Lamb, C.J. (1985) 'Rapid switching of gene expression by fungal elicitor', Science **227**, 1240-1242.

Deus-Neumann, B. and Zenk, M.H. (1984) 'Instability of indole alkaloid production in *Catharanthus roseus* cell suspension cultures', Planta Med. **50**, 427-435.

Fowler, M.W. (1977) 'Growth of cell cultures under chemostat conditions', in W. Barz, E. Reinhard and M.H. Zenk (eds.), Plant Tissue Culture and its Biotechnological Application, Springer-Verlag, Berlin.

Godfrey, T. and Reschelt, J. (1983) Industrial Enzymology, Macmillan, London.

Green, P.J., Pines, O. and Inou, Y.E. (1986) 'The role of antisense RNA in gene regulation', Ann. Rev. Biochem. **55**, 569-597.

Gross, G.G. and Zenk, H. (1974) 'Isolation and properties of hydroxycinnamate: CoA ligase from lignifying tissues of *Forsythia*', Eur. J. Biochem. **43**, 453-459.

Hahlbrock, K. and Grisebach, H. (1979) 'Enzymic controls in the biosynthesis of lignin and flavenoids', Annu. Rev. Plant Physiol. **30**, 105-130.

Hall, R.D., Holden, M.A. and Yeoman, M.M. (1987) 'The accumulation of phenylpropanoid and capsaicinoid compounds in cell cultures and whole fruit of the chilli pepper, *Capsicum frutescens* Mill', Plant Cell, Tissue and Organ Culture **8**, 163-176.

Hall, R.D., Holden, M.A. and Yeoman, M.M. (1988) 'The immobilization of higher plant cells', in Y.P.S. Bajaj (ed.), Biotechnology in Agriculture and Forestry **7**: Medicinal and Aromatic Plants I, Springer-Verlag, Berlin, pp. 136-156.

Hamill, J.D., Parr, A.J., Robins, R.J. and Rhodes, M.J.C. (1986) 'Secondary product formation by cultures of *Beta vulgaris* and *Nicotiana rustica* transformed with *Agrobacterium rhizogenese*', Plant Cell Rep. **5**, 111-114.

Haseloff, J. and Gerlach, W.L. (1988) 'Simple RNA enzymes with new and highly specific endoribonuclease activities', Nature **334**, 585-591.

Hashimoto, T., Azechi, J., Sugita, S. and Suzuki, K. (1982) 'Large-scale production of tobacco cells by continuous cultivation', in Fujiwara (ed.) Plant Tissue Culture, Maruzen, Tokyo, pp. 403-404.

Heinstein, (1985) 'Future approaches to the formation of secondary natural products in plant cell cultures', J. Nat. Prod. **48**, 1-9.

Holden, M.A., Lindsey, K., Hall, R. and Holden, P.R. (1987) 'The effect of immobilisation on the metabolic performance of plant cells', in C. Webb and F. Mavituna (eds.), Process Possibilities for Plant and Animal Cell Cultures, Ellis Horwood, Chichester, pp. 45-56.

Holden, M.A., Holden, P.R. and Yeoman, M.M. (1988)

'Elicitation of cell cultures', in R. Robins and M.J.C. Rhodes (eds.), Manipulating Secondary Metabolism in Culture, Cambridge University Press, Cambridge, pp. 57-66.

Holden, M.A. and Yeoman, M.M. (1988) 'Optimisation of product yield in immobilised plant cell cultures', in G.W. Moody and P.B. Baker (eds.), Bioreactors and Biotransformations, Elsevier, London and New York, pp. 1-11.

Howard, E.A., Walker, J.C., Dennis, E.S. and Peacock, W.J. (1987) 'Regulated expression of an alcohol dehydrogenase I chimaeric gene introduced into maize protoplasts', Planta **170**, 535-540.

Ikuta, A. and Itokawa, H. (1989) '*Coptis*: *in vitro* regeneration of plants and the production of berberine', in Y.P.S. Bajaj (ed.), Biotechnology in Agriculture and Forestry 7, Medicinal and Aromatic Plants II, Springer-Verlag, Berlin.

Iwai, K., Kap-Rang, L., Kobashi, M., Suzuki, T. and Oka, S. (1978) 'Intracellular localization of the capsaicinoid synthesizing enzyme in sweet pepper fruits', Agric. Biol. Chem. **42**, 201-202.

Jones, J.D.G., Dunsmuir, P. and Bedbrook, J. (1985) 'High level expression of introduced chimaeric genes in regenerated transformed plants', EMBO J. **4**, 2411-2418.

Kacser, H. and Burns, J.A. (1973) 'The control of flux', Symp. Soc. Exp. Biol. **27**, 65-104.

Kay, R., Chan, A., Daly, M. and McPherson, J. (1987) 'Duplication of CaMV 35S promoter sequences creates a strong enhancer for plant genes', Science **236**, 1299-1302.

Kondo, Y. (1976) 'Organic and biological aspects of berberine alkaloids', Heterocycles **4**, 197-219.

Kreis, M., Williamson, M.S., Forde, J., Schmutz, D., Clark, J., Buxton, B., Pywell, J., Marris, C., Henderson, J., Harris, N., Shewry, P.R., Forde, B.G. and Miflin, B.J. (1986) 'Differential gene expression in the developing barley endosperm', Phil. Trans. R. Soc. Lond. **B314**, 355-365.

Kuhn, D.N., Chappell, J., Boudet, A. and Hahlbrock, K. (1984) 'Induction of phenylalanine ammonia-lyase and 4-coumarate: CoA ligase mRNAs in cultured plant cells by UV light or fungal elicitor', P.N.A.S. **81**, 1102-1106.

Leete, E. and Louden, L.C.L. (1968) 'Biosynthesis of capsaicin and dihydrocapsaicin in *Capsicum frutescens*', J. Am. Chem. Soc. **90**, 6837-6841.

Lindsey, K. and Yeoman, M.M. (1984) 'The synthetic potential of immobilised cells of *Capsicum frutescens* Mill. cv. annuum', Planta **162**, 495-501.

Lindsey, K. (1985) 'Manipulation, by nutrient limitation, of the biosynthetic activity of immobilised cells of *Capsicum frutescens* Mill. cv. annuum', Planta **165**, 126-133.

Lindsey, K. and Yeoman, M.M. (1985) 'Immobilised plant cell culture systems', in K.-H. Neumann, W. Barz and E. Reinhard (eds.), Primary and Secondary Metabolism of Plant Cell

Cultures, Springer-Verlag, Berlin, pp. 304-315.

Lindsey, K. (1988) 'Prospects for the manipulation of complex metabolic pathways', in R.J. Robins and M.J.C. Rhodes (eds.), Manipulating Secondary Metabolism in Culture, Cambridge University Press, Cambridge, pp. 123-136.

McGarry, T.J. and Lindquist, I. (1986) 'Inhibition of heat shock protein synthesis by heat-inducible antisense RNA', P.N.A.S. **83**, 339-403.

Mavituna, F., Park, J.M., Wilkinson, A.K. and Williams, P.D. (1987) 'Characteristics of immobilised plant-cell reactors', in C. Webb and F. Mavituna (eds.), Plant and Animal Cells, Process Possibilities, Ellis Horwood, Chichester, pp. 92-114.

Mavituna, F., Wilkinson, A.K. and Williams, P.D. (1988) 'Production of secondary metabolites by immobilised plant cells in novel bioreactors', in G.W. Moody and P.B. Baker (eds.), Bioreactors and Biotransformations, Elsevier, London and New York, pp. 26-37.

Mol, J., Stuitje, A., Gerati, A., Van Der Krol, A. and Jorgensen, R. (1989) 'Saying it with genes: molecular flower breeding', Tibtech **7**, 148-153.

Nagy, F., Morelli, G., Fraley, R.T., Rogers, S.G. and Chua, N.-H. (1985) 'Photoregulated expression of a pea rbcS gene in leaves of transgenic plants', EMBO J. **4**, 3063-3068.

Potrykus, I., Paszykowski, J., Saul, M.W., Petruska, J. and Shillito, R.D. (1985) 'Molecular and general genetics of a hybrid foreign gene introduced into tobacco by direct gene transfer', Mol. Gen. Genet. **199**, 169-177.

Racusen, D. (1984) 'Lipid acyl hydrolase of patatin', Can. J. Bot. **62**, 1640-1644.

Racusen, D. and Foote, M. (1980) 'A major soluble glycoprotein of potato', J. Food Biochem. **4**, 43-52.

Ryder, T.B., Cramer, C.L., Bell, J.N., Robbins, M.P., Dixon, R.A. and Lamb, C.J. (1984) 'Elicitor rapidly induces chalcone synthase mRNA in *Phaseolus vulgaris* cells at the onset of the phytoalexin defense response', P.N.A.S. **81**, 5724-5728.

Sanchez-Serrano, J.J., Keil, M., O'Connor, A., Schell, J. and Willmitzer, (1987) 'Wound-induced expression of a potato proteinase inhibitor II gene in transgenic tobacco plants', EMBO J. **6**, 303-306.

Sanders, P.R., Winter, J.A., Barnason, A.C., Rogers, S.G. and Fraley, R.T. (1987) 'Comparison of cauliflower mosaic virus ^{35}S and nopaline synthase promoters in transgenic plants', Nuc. Acid. Res. **15**, 1543-1557.

Schiel, O. and Berlin, J. (1987) 'Large-scale fermentation and alkaloid production of cell suspension cultures of *Catharanthus roseus*', Plant, Cell, Tiss. Org. Cult. **8**, 153-161.

Schröder, J. (1977) 'Light-induced increase of messenger RNA for phenylalanine ammonia-lyase in cell suspension cultures of *Petroselinum hortense*', Arch. Biochem. Biophys. **182**, 488-496.

Schröder, J., Kreuzaler, F., Schäfer, E. and Hahlbrock, K. (1979) 'Concomitant induction of phenylalanine ammonia-lyase and flavone synthase mRNA in irradiated plant cells', J. Biol. Chem. **254**, 57-65.

Scott, R.J. and Draper, J. (1987) 'Transformation of carrot tissues derived from proembryonic suspension cells: a useful model system for gene expression studies in plants', Plant Mol. Biol. **8**, 265-274.

Shah, D.M., Horsch, R.B., Klee, H.J., Kishore, G.M., Winter, J.A., Tumer, N.E., Hironaka, C.M., Sanders, P.R., Gasser, C.S., Aykent, S., Siegel, N.R., Rogers, S.G. and Fraley, R.T. (1986) 'Engineering herbicide tolerance in transgenic plants', Science **233**, 478-481.

Simons, R.W. and Kleckner, N. (1988) 'Biological regulation by antisense RNA in prokaryotes', Annu. Rev. Genet. **22**, 567-600.

Spena, A., Hain, R., Ziervogel, U., Saedler, H. and Schell, J. (1985) 'Construction of a heat-inducible gene for plants. Demonstration of heat-inducible activity of the *Drosophila* Hsp70 promoter in plants', EMBO J. **4**, 2739-2743.

Tabata, M. (1985) 'Production of shikinon by plant cell cultures', in M. Zaitlin, P. Day and A. Hollaender (eds.) Biotechnology in Plant Science. Relevance to Agriculture in the Eighties, Academic Press, pp. 207-218.

Twell, D. and Ooms, G. (1987) 'The 5' flanking DNA of a patatin gene directs tuber specific expression of a chimaeric gene in potato', Plant Mol. Biol. **9**, 345-375.

Van Der Krol, A.R., Lewting, P.E. and Veenstra, J. (1988) 'An anti-sense chalcone synthase gene in transgenic plants inhibits flower pigmentation', Nature **333**, 866-869.

Yeoman, M.M. (1987) 'Techniques, characteristics, properties and commercial potential of immobilised plant cells', in I.K. Vasil (ed.) Cell Culture and Somatic Cell Genetics of Plants **4**, Cell Culture in Phytochemistry, pp. 197-215.

Yeoman, M.M., Holden, M.A., Corchete, P., Holden, P.R., Goy, S. and Hobbs, M.C. (1989) 'Exploitation of disorganised plant cultures for the production of secondary metabolites', in B.V. Charlwood (ed.) Secondary Products from Plant Tissue Cultures, Oxford University Press, Oxford (in press).

Yeoman, M.M., Miedzybrodzka, M.B., Lindsey, K. and McLauchlan, W.R. (1980) 'The synthetic potential of cultured plant cells', in F. Sala, B. Parisi, R. Cella and O. Ciferri (eds.) Plant Cell Cultures Results and Perspectives, Elsevier-North Holland, London and New York, pp. 327-343.

Section 3. Special topics

THE CYTOSKELETON: IMPORTANCE FOR PLANT CELL AND PROTOPLAST RESEARCH

CHRIS H. BORNMAN and IRINA STAXÉN
Cell Biology, Research Divison, Hilleshög AB
Box 302
S-261 23 Landskrona
Sweden

ABSTRACT. Cell and protoplast cultures are convenient systems for investigating the organization and role of microtubules in various cytological processes such as division, cell wall formation, and response to promoters and inhibitors of cell growth and development. Indirect immunofluorescence microscopy using anti-tubulin antibodies and other labelling techniques, now make it possible to study the rate and nature of the three-dimensional reorganization of the microtubular network in regenerable as well as recalcitrant protoplasts. This provides information as to the possible effects of genotype, isolation procedure and culture environment at the level of the intact protoplast.

Introduction

Development by animal cell biologists of techniques for antibody isolation, purification and labelling have also provided plant cell biologists with the incentive for studying the cytoskeleton in plant cells and protoplasts. Among the first to report the presence of tubulin-associated cellular components were Franke et al. (1977) who studied the mitotic apparatus of endosperm cells of <u>Leucojum aestivum</u> using anti-tubulin antibodies, and Lloyd et al. (1979) who used <u>Daucus carota</u> protoplasts to relate the organization of the cortical microtubule network to cell shape.

Today, the existence of the cytoskeleton in plant cells is no longer questioned, as overwhelming evidence for two of its components has accumulated rapidly. At least four components of the cytoskeleton are known from animal cell systems (for detailed treatment see Schliwa 1986 and

R.S. Sangwan and B.S. Sangwan-Norreel (eds.), The Impact of Biotechnology in Agriculture, 421–435.

Bershadsky and Vasiliev 1988): (1) microtubules (MT), (2) microfilaments (MF), (3) intermediate filaments (IF) and (4) microtrabeculae. In plant cell systems evidence for intermediate filaments is sparse and the question as to whether or not the microtrabecular lattice is a physical feature of the plant cell or an artefact of preparation is still open.

MICROTUBULES

The network of microtubules is the most extensively studied component of the cytoskeleton in plant cells. Microtubules are associated with four discrete features: cortex, preprophase banding, mitotic spindle and phragmoplast. From the nature of the appearance of the MT-cytoskeleton it is possible to discern the particular stage of a cell's cycle of division and DNA-synthesis. The different stages related to the cell cycle are summarized in Fig. 1.

In interphase the microtubules of the cortex are orientated transversely to the longitudinal axis of the cell. With the transition between interphase and mitosis, cortical microtubules begin to disassemble and two new structures make their appearance. The nucleus becomes surrounded by a dense net of perinuclear microtubules and in the cell cortex there is a condensation of microtubules in the equatorial plane of the nucleus, the socalled preprophase band (PPB). In some cells, such as dividing endosperm (Wick and Duniec (1984), cells in suspension culture (Simmonds 1986), and during division of the generative cell in pollen tubes (Palevitz and Cresti 1989), the PPB has not been observed. As the PPB appears to determine the future plane of cell plate formation, it is possible that its presence is not crucial to cells, such as those in suspension culture, that display unorganized growth.

During prophase the PPB becomes increasingly more condensed and, as the nuclear envelope is degraded, the mitotic spindle is formed. During telophase perinuclear microtubules are again observed, but this time it is the microtubules of the spindle that condense around the newly-formed telophase nuclei. As is well-known from differential interference contrast, scanning and transmission electron microscopical studies, as well as from fluorochrome- and gold-labelled antibodies, microtubuli also are involved in formation of the phragmoplast.

Existing information (Wick and Duniec 1983, Wick 1985) discount participation of microtubules of the interphase array in the synthesis of mitotic structures such as the PPB, prophase band, spindle or phragmoplast.

Cortical microtubules in the plant cell display a parallel orientation to the cellulosic microfibrils of the cell wall.

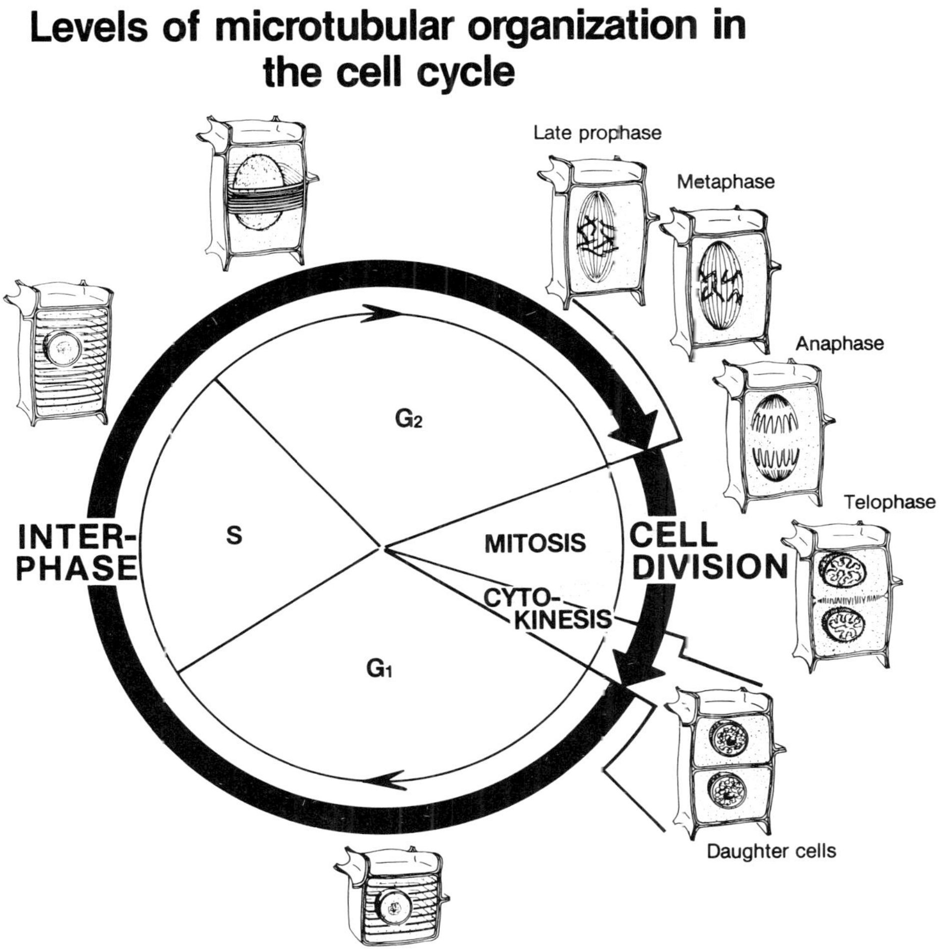

Figure 1. Levels of microtubular organization in the cell cycle. During most of interphase, microtubules form a cortical array but before the onset of mitosis become arranged in a narrow band that encircles the nucleus. The plane of this preprophase band corresponds to the forthcoming equatorial plate. During mitosis, microtubules occur in the form of a spindle but disappear with the separation and movement of the chromatids to opposite poles of the cell. New microtubules give rise to the phragmoplast, at the equator of which the cell plate forms. The cell plate differentiates toward and eventually joins the wall of the dividing cell. Cell division is completed when each daughter cell forms its own primary wall.

Microtubules have been suggested to orientate the deposition of microfibrils in the cell wall by directing the movement of the cellulose synthase complex (Varner and Lin 1989). Other reports show that deposition of the cell wall is not MT-dependent, as cell walls can be synthesized when microtubules are depolymerized (Hepler and Fosket 1971, Hardham and McCully 1982). The most illustrative example of the involvement of microtubules in cell wall deposition is that of the differentiation of cells of _Zinnia elegans_ into xylem tracheary elements (Burgess and Lindstead 1984, Falconer and Seagull 1985). Isolated cells of _Zinnia elegans_ were kept in suspension culture and allowed to differentiate into tracheary elements. From this study it was evident that microtubule reorganization precedes deposition of secondary cell wall material. As expected ontogenetically, tracheary elements first appeared with an annular to spiral secondary wall configuration, but could then evolve a reticulate to scalariform organization. The new pattern of microtubules was visible before the cells stained positively with the cell wall stain, calcofluor. Also, from this pattern it was possible to predict the future appearance of the cell wall. The transition in xylem development was accompanied by a reorganization of the cortical microtubules. The mechanism controlling the reorganization of the microtubules is not known. Falconer and Seagull (1985) suggest that it may take place via an aggregation of the cortical microtubules.

Cells differentiating into tracheary elements in suspension cultures are not constricted in their form by neighbouring cells. It is therefore not surprising that their form varies from elongated to spherical. Again, it is possible to see a relationship between microtubules, secondary cell wall deposition and cell shape. Elongated cells have parallel cortical microtubules and the cell wall is deposited in a band pattern, while spherical cells display random microtubules and the cell wall is deposited in a web pattern.

MICROFILAMENTS

The first demonstration of actin filaments in plant cells was that on the algae _Chara_ (Williamson 1974) and _Nitella_ (Palevitz and Hepler 1975). An extensive study of the organization of actin filaments during the cell cycle of _Daucus_ was done by Traas et al. (1987) who demonstrated the existence of a complex network of actin filaments throughout the cell cycle. They identified the following structures: a network surrounding the nucleus at interphase, thick cables in both transvacuolar strands and in the cytoplasm, a cortical network of thin cables, and a fine, transverse array in the cell periphery. Traas et al. (1987) compared the appearance of microfilaments in fixed versus unfixed cells. A more extensive network as well as more structures (e.g. microfilaments of the phragmoplast, spindle-associated actin filaments & PPB-actin filaments) were seen in unfixed cells, including an ample net of microfilaments in transverse orientation. The cortical filaments appear to be permanent during cell division and do not disassemble as do the microtubules. Other studies suggest that the orientation of cortical microfilaments is axial, not transverse as is that of microtubules (Parthasarathy 1985).

Microtubules and microfilaments compete as to their involvement in organelle movement in cyclosis. Treatment of cells with microtubule depolymerizers such as cryzaline and colchicine does not inhibit the movement of organelles, and so it has been assumed that microtubules are not involved in the process. At the same time there is evidence for the organelles being coated with cytoplasmic myosin. The actin filaments are then thought to be responsible for generating the force required for organelle movement in a manner similar to that generated during muscle contraction. Also, organelle movement is ATP-dependent and Ca^{2+}-sensitive, a strong indication for the involvement of the actin-myosin system in cyclosis. Another line of evidence for the dependence of organelle movement on microfilaments is that myosin-coated beads have been shown to move along actin fibres in a way that resembles organelle movement (Williamson 1986).

An instructive study of organelle movement has been done on the redistribution of chloroplasts in algae. In cells that were irradiated with UV-light the chloroplasts clustered in the wounded area. In cells treated with cytochalasin D, an actin depolymerizer, no movement of chloroplasts could be detected (Menzel and Elsner-Menzel 1989). Serlin and Ferrell (1989) treated the alga Mougheuotia with the microtubule stabilizer, taxol, and microtubule depolymerizers, colchicine and nocodazole. Their results clearly showed that depolymerization of microtubules speeds up chloroplast rotation while microtubule polymerization (taxol treatment) slows down or completely inhibits chloroplast rotation in response to light irradiation. The role of the microtubules would appear to be that of modulating the movement of organelles. (See Staiger and Schliwa 1987, for a review of microfilaments).

The evidence of the existence of myosin in plant cells is sparse. Myosin has been isolated from plant tissue but it seems unlikely that it assembles into filaments in vivo, as is the case in muscle cells. In other eukaryotic cells myosin monomers or oligomers coat the organelles, including the nucleus, which then move along the actin filaments. This is true even for plant cells as shown by staining of Nicotiana alata pollen tubes with anti-myosin antibodies.

INTERMEDIATE FILAMENTS

All evidence of the existence of intermediate filaments (IF) in plant cells is indirect and comes from biochemical work. Dawson et al. (1985) used an anti - animal intermediate filament antibodies to stain _Daucus_ _carota_ suspension cells. They found that the antigen co-distributes with microtubules in these cells. This does not necessarily mean that the antigen is present in a fibrous form, but could be in a depolymerized form associated with microtubules.

Anti-intermediate filament antibodies also recognize paracrystalline bundles in carrot cells that do not look like "normal" cytoskeleton components.

More evidence on the identity of the fibrillar bundles (FB) with intermediate filaments in plant cells is acculumating (Hargreaves et al. 1989a). Fibrillar bundles from carrot suspension cells polymerize in vitro under conditions that are used to repolymerize animal IFs. The carrot IFs repolymerized in vitro are associated laterally and are several micrometers in length as seen in the electron microscope. They agree in structure with the fibrillar bundles stained in carrot protoplasts with immunofluorescence. The role of IFs in animal cells is mainly to give stability to the cell and to integrate mechanically multicellular systems (via desmosomes). Their role in higher plant cells is not known, but it is improbable that it is the same as in animal cells as these functions are fulfilled by the cell wall.

MICROTRABECULAR LATTICE

Microtrabeculae were first reported in animal cells by Buckley and Porter (1975). They employed high-voltage electron microscopy for the study of whole cells. The image that emerged was different from that expected, namely, a complex matrix of fibres between 2-15 nm width (varying with the preparative technique used) to which polysomes were attached. Microtubules, organelles and membrane systems were suspended in the microtrabecular lattice, as were cytoplasmic enzymes. The microtrabeculae create a spatial separation of the cytoplasm.

The biochemical properties of the microtrabeculae have not been established. Wolosewick and Porter (1979) proposed that the system is built up of actin and coated with proteins. Microtrabeculae can be extracted with nonionic detergents together with the bulk of cytoplasmic proteins, but are not as stable as the other components of the cytoskeleton.

The microtrabecular network is not seen in resin embedded tissue; it is assumed that the resin and the microtrabeculae have the same electron scattering properties. However, the network is seen if the material is embedded in polyethylene glycol. Hawes et al. (1983), using the techniques of Porter's group, reported that microtrabeculae exist in both the cytoplasm and nucleoplasm of resin-free sections of plant cells.

Table 1 shows the occurrence of proteins in the different cytoskeletal components in plant cells.

TABLE 1. Occurence of proteins in different cytoskeletal structures in plant cells. Fibrillar proteins in plant cells are composed of vismetin, desmin and glial fibrillar acidic protein.

Fibrillar component	Cortex	Peri-nuclear micro-tubules	Pre-prophase band	Spindle	Phragmo-plast	Reference
Actin	+	+	+	+	+	Palevitz (1987) Partharsarathy (1985) Traas et al. (1987) Staiger & Schliwa (1987)
Myosin	+					Kato & Tonomura (1977)
Inter-mediate filaments and fibrillar bundles	±	+	+	+	+	Dawson et al. (1985) Hargreaves et al. (1989b) Parke et al. (1987)
Tubulin	+	+	+	+	+	Numerous

Dynamics of the cytoskeleton

PROTOPLASTS

As cortical microtubules have been implicated in cellular processes that affect shape and determination of growth, it is possible that lack or severe inhibition of cell regeneration in protoplast culture systems is associated with the effects of isolation or culture procedures on the cytoskeleton (Lee et al. 1989).

Reorganization of cortical microtubules is affected by plant growth regulators

Simmonds (1983) showed that the type of plant growth regulator present during the culture of protoplasts affects both the appearance and reassembly of microtubules, with an indication that microtubule reorganization is strongly auxin dependent. Roberts et al. (1985) found that ethylene induced oblique and longitudinal arrays and inhibited the induction of transverse arrays of microtubules in pea epicotyl and mungbeans hypocotyl cells, in contrast to the predominant pattern of a tightly compressed helix with microtubules transversely with respect to the cell's longitudinal axis.

Reinstatement of the microtubule lattice is a requirement for cell division

Hahne and Hoffmann (1984) compared Nicotiana glutinosa protoplasts isolated from a regenerative callus line with non-regenerative protoplasts isolated from a Hibiscus rosa-sinensis callus line. The Nicotiana protoplasts were rendered non-regenerative by treatment with colchicine while the Hibiscus protoplasts were stimulated to divide by treatment with dimethyl-sulphoxide (DMSO). The common feature for the protoplasts that were regenerative, that is Nicotiana and DMSO-treated Hibiscus protoplasts, was the presence of a well-developed cortical microtubule network. Thus far, it has not been possible to reproduce these results in other species, for example, Beta vulgaris, as addition of DMSO to the isolation or culture medium would be a relatively easy measure to take. To complicate the problem further, the mechanism of action of DMSO on both plant and animal cells is not known. Some of the reported effects of DMSO on cells are: permeabilization of membranes (Erdei et al. 1982), enhancement of cell division (Hahne and Hoffmann 1984), stabilization of microtubules, and lengthening of the period of the circadian clock in the pulvini of Phaseolus cocinneus (Mayer et al. 1985).

The major conclusion from this study was that the reinstatement by the plant protoplast of a cortical microtubule lattice is a prerequisite for subsequent cell wall resynthesis and cell division.

Presence and absence of the cytoskeleton in young and ageing tissues, respectively

Freshly-isolated protoplasts from the mesophyll of young or

expanding *Nicotiana* (Hahne and Hoffmann 1985) and *Medicago sativa* leaves (Dijak and Simmonds 1988) have limited numbers of cortical microtubule strands. However, after a period of only hours in culture there is an increase in the number of microtubules. For the onset of cell division, it appears to be necessary for the microbutule network of the protoplasts to be first re-established. The cytoskeleton in mesophyll protoplasts of mature rosette leaves of *Brassica* (Hahne and Hoffmann 1985) was found to be either absent or greatly reduced.

Not infrequently, it happens that photomicrographs of protoplast membrane ghosts are shown. Membrane ghosts are not necessarily representative for the organization of microtubules in the protoplast, as little is known about microtubule-membrane interaction. Were the microtubules anchored in the membrane at a few positions, then only part of the microtubule lattice will be visualised.

Cytological abnormalities during divisions of protoplasts

The inability of protoplasts of some species (among them economically important crops) to regenerate cells, callus and plants may be connected to the microtubule-cytoskeleton system. The consequence of abnormal spindle formation (Simmonds & Setterfield 1986, Meijer et al. 1988) is a high frequency of aneuploid or polyploid cells (protoplasts) that degenerate after the first divisions. Abnormalities that have been observed in microtubule organization are: one PPB per nucleus in homokaryons, one PPB per two nuclei in binucleated protoplasts, displaced PPB, lack of spindle formation, formation of polypolar spindles, and incomplete formation of the phragmoplast. Coupled to the aberrations in microtubule organization are karyogenic and cytogenic irregularities such as: no segregation of chromosomes, incomplete segregation of chromosomes, segregation of chromosomes into a number of smaller nuclei, formation at anaphase of chromatin bridges as a result of non-disjunction, and the inability of the chromosomes to organize at the equatorial plate. Irregularities during cell division are less frequent if the protoplast has first regenerated a cell wall before entering division.

Effect of electrostimulation

Exposure of protoplasts to an electric field can induce direct embryogenesis, an observation made on mesophyll protoplasts of *Medicago sativa* (Dijak et al. 1986, Dijak and Simmonds 1988). The protoplasts were exposed to a 0.02 V direct current for 17 h. After 17 h in culture the treated protoplasts displayed more numerous microtubule strands and more rapid cell wall deposition as compared to untreated protoplasts. After two days in culture the electric field-treated cells had a disorganized but very dense cortical microtubule network and there were few binucleated cells. The random organization of the microtubules in the embryogenic cells was maintained throughout the course of the experiment. Divisions were asymmetric and the resulting callus embryogenic. Untreated cells developed parallel cortical microtubule arrays by day 2 and underwent symmetric division, a feature of the cells of a non-embryogenic callus.

Asymmetric cell divisions that result in polarized growth apparently do not involve cortical microtubules.

Rearrangement of the cortical microtubule lattice (Hazewawa et al. 1989)

During isolation of protoplasts there is a disorganization of the cortical microtubules. After 20 h in culture, the microtubules start to reorganize. The reorganization, to parallel, starts at the centre of the protoplast (cell) and with the simultaneous deposition of cell wall material. In elongated cells the microtubules collapse after 3 weeks in culture. The disappearance of the cortical microtubules from these cells could be a consequence of senescence. Elongated cells that have been in culture for more than 1 week could not be induced to divide. The microtubules of these cells are in parallel arrays and orientated perpendicularly to the long axis of the cell.

Effect of protoplast fusion on the microtubule lattice

In a study by Hahne and Hoffmann (1986) protoplasts of Nicotiana glutinosa callus cells and protoplast of Daucus carota cell suspension were fused. Twenty hours after fusion with polyethylene glycol the cortical microtubule network of the two protoplasts had become integrated. The microtubules were found to reorganize from the centre towards the poles of the heterokaryon. When protoplasts were treated before fusion with compounds that depolymerize microtubules, such as oryzalin and colchicine, an increase in the rate of fusion was observed compared to cells treated with taxol, a microtubule polymerizer, but not as compared to the untreated controls. Even if there is a strong indication of the positive effect of microtubule depolymerization on fusion frequency, the effect and fate of the other components of the cytoskeleton, e.g. actin microfilaments, must be taken into account.

Effects of protoplast isolation and incubation procedures on the cytoskeleton

Lee et al. (1989) studied the effects of enzyme concentration and duration of incubation on viability, plating efficiency and organization of the cortical microtubule network in Vitis rotundifolia protoplasts. High enzyme concentration decreased viability and plating efficiency but left the cortical network in an apparently stable condition of random and parallel arrayed microtubules. Long incubation time, however, decreased plating efficiency and the frequency of parallel microtubules. Microtubule organization appeared to be more sensitive to duration of the incubation period and less to enzyme concentration.

MICROSPORES

The cytoskeleton in developing and redifferentiating microspores

Microspores of Brassica napus cultured at 25°C in vitro follow the normal course of development that leads to pollen grain formation. However, when first subjected for a 4-day period to an elevated temperature of 32.5°C, a fraction of the microspores can be made to redifferentiate and develop instead into haploid embryos.

Simmonds (1988) used immunofluorescence to gain more information about the nature of the interruption of the normal sequence of development. She found, in the first case, that microtubules in the late uninucleate stage of microspore development were associated with the perinuclear envelope and that they radiated into the cytoplasm. Migration of the nucleus to the microspore's periphery results in asymmetric division in the absence of preprophase band formation.

When first exposed to a period of high temperature, an unorganized microtubule array develops in the microspore. In late interphase a PPB of microtubules are formed and symmetric division follows, with nuclei of the daughter cells condensing to the same degree. The result is the development of embryogenic cells.

Concluding remarks

Investigations of the cytoskeleton provide additional means for studying phenomena of differentiation and development. In plant cell and protoplast culture, the culture environment has been known to induce genetic instability in the form not only of aneu- and polyploid cells but also of karyokinetic and cytokinetic abnormalities that presumably result from aberrant microtubule organization during the mitotic phase of the cell cycle. Immunofluorescent techniques make it possible to use components of the cytoskeleton such as the MTs not only as aids in interpreting aberrant cellular events, but also to follow switches in development, as exemplified by the redifferentiation of microspores to embryos.

Although the organization of MTs in different plant cells seems principally to be the same, information regarding the interaction of the cytoskeleton with the plasma membrane and the cell wall on the one hand and the nature of the microtubule organizing centres on the other, should lead to a better understanding of protoplast behaviour in culture.

REFERENCES

Bershadsky, A.D. and Vasiliev, J.M. (1988) 'Cytoskeleton', Plenum Press, New York. ISBN 0-306-42508-4.

Buckley, I.K. and Porter, K.R. (1975) 'Electron microscopy of critical point dried whole cultured cells', J. Microsc. 104, 107-120.

Burgess, J. and Lindstead, P. (1984) 'In-vitro tracheary element formation: structural studies and the effect of tri-iodobenzoic acid', Planta 160, 487-489.

Dawson, P.J., Hulme, J.S. and Lloyd, C.W. (1985) 'Monoclonal higher plant cells', J. Cell Biol. 100, 1793-1798.

Dijak, M., Smith, D.L., Wilson, T.J. and Brown, D.C.W. (1986) 'Stimulation of direct embryogenesis from mesophyl protoplasts of Medicago sativa', Plant Cell Rep. 5, 468-470.

Dijak, M. and Simmonds, D.H. (1988) 'Microtubule organization during early direct embryogenesis from mesophyll protoplasts of Medicago sativa L', Plant Science 58, 183-191.

Erdei, L., Vigh, L. and Dudits, D. (1982) 'Isolation of wheat cell line with altered membrane properties', Plant Physiol. 69, 572-574.

Falconer, M.M. and Seagull, R.W. (1985) 'Immunofluorescent and calcofluor white staining of developing tracheary elements in Zinnia elegans suspension cultures', Protoplasma 125, 190-198.

Franke, W.W., Sieb, E., Osborn, M., Weber, K., Merth, W. and Falk, M. (1977) 'Tubulin containing structures in the anastral mitotic apparatus of endosperm of the plant Leucojum aestivum as revealed by immunofluorescence microscopy', Cytobiologie 15, 24-48.

Hahne, G. and Hoffmann, F. (1984) 'Dimethyl sulphoxide can initiate cell divisions of arrested callus protoplasts by promoting cortical microtubule assembly', Proc. Natl. Acad. Sci. USA 81, 5449-5453.

Hahne, G. and Hoffmann, F. (1985) 'Cortical microtubular lattices absent from mature mesophyll and necessary for cell division?', Planta 166, 309-313.
Hahne, G. and Hoffmann, F. (1986) 'Cortical microtubules and protoplast fusion: effect and fate of microtubular lattices', Plant Science 47, 199-206.
Hardham, A.R. and McCully, M.E. (1982) 'Reprogramming of cells following wounding in pea _Pisum sativum_ roots. 1. Cell division and differentiation of new vascular elements', Protoplasma 112, 143-151.
Hargreaves, A.J., Goodbody, K.C. and Lloyd, C.W. (1989a) 'Reconstruction of intermediate filaments from a higher plant', Biochem. J. 261, 679-682.
Hargreaves, A.J., Dawson, P.J., Butcher, G.W., Larkins, A., Goodbody, K.J. and Lloyd, C.W. (1989b) 'A monoclonal antibody raised against cytoplasmatic fibrillar bundles from carrot cells, and its cross-reactivity with animal intermediate filaments', J. Cell Sci. 92, 371-378.
Hawes, C.R., Juniper, B.E. and Horne, J.C. (1983) 'Electron microscopy of resin-free sections of plant cells', Protoplasma 115, 88-93.
Hazewawa, S., Hogetzu, T. and Syono, K. (1988) 'Rearrangement of cortical microtubules in elongating cells derived from tobacco protoplasts - A time-course observation by immunofluorescence microscopy', J. Plant Phys. 133, 46-51.
Hepler, P.K. and Fosket, D.E. (1971) 'The role of microtubules in vessel member differentiation in _Coleus_-D', Protoplasma 22, 213-236.
Kato, T. and Tonomura, Y. (1977) 'Identification of myosin in _Nitella flexis_', J. Biochem. 82, 777-782.
Lee, N., Wetzstein, H.Y. and Bornman, C.H. (1989) 'Cortical microtubule organization in _Vitis_ protoplasts as affected by concentration of enzyme isolation medium and duration of incubation', Physiologia Plantarum 77, 27-32.
Lloyd, C.W., Slabas, A.R., Powell, J.A., MacDonald, G. and Badley, R.A. (1979) 'Cytoplasmatic microtubules of higher plant cells visualized with anti-tubulin antibodies', Nature 279, 239-241.
Mayer, W.E., Maier, M. and Flach, D. (1985) 'Osmotica, dimethyl sulphoxide, prahydroxymercuribenzoate, and cyanide change the period of the circadian clock in the pulvini of _Phaseolus coccineus_ L.'. Chronobiology International 2, 11-17.
Mejer, E.G.M., Keller, W.A. and Simmonds, D.H. (1988) 'Cytological abnormalities and aberrant microtubule organization during early divisions in mesophyll protoplasts cultures of _Medicago sativa_ and _Nicotiana tabacum_', Physiologia Plantarum 74, 233-239.

Menzel, D. and Elsner-Menzel, C. (1989) 'Actin based chloroplast rearrangements in the cortex of the giant coenocytic green alga Caulerpa', Protoplasma 150, 1-8.
Palevitz, B.A. and Hepler, P.K. (1975) 'Identification of actin in situ at the ectoplasm-endoplasm interface of Nitella. Microfilament-chloroplast association', J. Cell Biol. 65, 29-38.
Palevitz, B.A. (1987) 'Actin in PB of Allium cepa', J. Cell Biol. 104, 1515-1519.
Palevitz, B.A. and Cresti, M. (1989) 'Cytoskeletal changes during generative cell division and sperm formation in Tradescantia virginiana', Protoplasma 150, 54-71.
Parke, J.M., Miller, C.C.J., Cowell, I., Dobson, A., Dowding, A., Downes, M., Ducket, J.G. and Aderton, B.J. (1987) 'Monoclonal antibodies against plant proteins recognize animal intermediate filaments', Cell Motil Cytoskel 8, 312-323.
Partharsarathy, M.K. (1985) 'F-actin architecture in coleoptile epidermal cells', J. Cell Biol. 39, 1-12.
Roberts, N.I., Lloyd, C.W. and Roberts, K. (1985) 'Ethylene induced microtubule reorientations: mediation by helical arrays', Planta 164, 439-447.
Schliwa, M.L. 1986) 'The cytoskeleton: An introductory survey' Springer Verlag, Vienna. ISBN 3-211-81884-7.
Serlin, B.F. and Ferrell, S. (1989) 'The involvement of microtubules in chloroplast rotation in the alga Mougheotia', Plant Science 60, 1-8.
Simmonds, D.H., Setterfield, G. and Brown, D.L. (1983) 'Reorganization of microtubules in protoplasts of Vivia hajastana, Grossh. during the first 48 hours of culturing', Protoplasts 1983, Birkhäuser Verlag, Basel-Boston-Stuttgart.
Simmonds, D.H. (1986) 'Prophase bands of microtubules occur in protoplast cultures of Vicia hajastana Grossh.', Planta 167, 469-472.
Simmonds, D.H. and Setterfield, G. (1986) 'Aberrant microtubule organization can result in genetic abnormalities in protoplasts cultures of Vivia hajastana', Planta 167, 460-468.
Simmonds, D.H. (1988) 'The cytoskeleton in cultured cells and protoplasts' in Proceedings of the 2nd Canadian Plant Tissue Culture and Genetic Engineering Workshop.
Staiger, C.J. and Schliwa, M. (1987) 'Actin localisation and function in higher plants', Protoplasma 141, 1-12.
Tang, X., Hepler, P.K. and Scordlis, S.P. (1989) 'Immunochemical and immunocytochemical identification of a myosin heavy chain polypeptide in Nicotiana pollen tubes', J. Cell Sci. 92, 569-574.

Traas, J.A., Doonan, J.H., Rawlins, D.J., Shaw, P.J., Watts, J. and Lloyd, C.W. (1987) 'Actin filament is present in the cytoplasm throughout the cell cycle of carrot cells and associated with the dividing nucleus', J. Cell Biol. 105, 387-395.

Varner, J.E. and Lin, L.S. (1989) 'Plant cell wall architecture', Cell 56, 231-239.

Wick, S.M. and Duniec, J. (1983) 'Immunofluorescence microscopy of tubulin and microtubule arrays in plant cells. I. Preprophase band development and concomitant appearance of nuclear envelope-associated tubulin', J. Cell Biol. 97, 235-243.

Wick, S.M. and Duniec, J. (1984) 'Immunofluorescence microscopy of tubulin and microtubule arrays in plant cells II. Transition between the preprophase band and the mitotic spindle', Protoplasma 122, 45-55.

Wick, S.M. (1985) 'Immunofluroescence microscopy of tubulin and microtubule arrays in plant cells. III. Transition between mitotic / cytokinetic and interphase microtubule arrays', Cell Biol. Int. Rep. 9, 357-371.

Williamson, R.E. (1974) 'Actin in the alga, <u>Chara corallina</u>', Nature (London) 248, 801-802.

Williamson, R.E. (1986) 'Organelle movements along actin filaments and microtubules', Plant Physiol. 82, 631-634.

Wolosewick, S.J. and Porter, K.R. (1979) 'Microtrabecular lattice of the cytoplasmatic ground substance. Artifact or reality?', J. Cell Biol. 82, 114-139.

GLYPHOSATE TOLERANCE IN PLANT CELL CULTURES

M.L. RACCHI
Dipartimento di Genetica e di Biologia dei Microrganismi
Università degli Studi di Milano
Via Celoria 26, 20133 Milano
Italy

1. Introduction

In recent years the increasing interest in ecological problems has determined a demand for herbicides which do not harm the environment or human health. Glyphosate is particularly suitable in this respect; in fact the present knowledge of its behaviour and degradation gives no reason to suppose that this herbicide causes damage after application to the soil or elsewhere in the environment, but being a non-selective herbicide it is indiscriminate in its action against plants. To obtain glyphosate-tolerant crop species is therefore a target of great agronomic value. Various approaches have been adopted and to date successful results have been achieved by combinations of cell biotechnology and genetic engineering.

Glyphosate tolerance has been transferred to tobacco (Comai et al. (1985)), petunia (Shah et al. (1986)), tomato (Fillatti et al. (1987)), flax (Jordan et al. (1988)) and soybean (Hinchee et al. (1988)) using an Agrobacterium-mediated gene transfer system, so the production of glyphosate-tolerant crops by means of genetic engineering is now a reality. In this paper we shall consider mainly *in vitro* cell culture researches that play a basic role in the study of the mode of action of the glyphosate or have provided fundamental information for the genetic engineering of glyphosate tolerance in plants.

2. Herbicide Resistence in Cell Cultures

The achievement of herbicide-resistant crops by means of cell culture techniques is one of the most important targets of biotechnology in agriculture. Cell culture offers a number of advantages for selection and characterization of plant cell variants. High selection pressure can be applied to a large number of cells that can be screened for variants in a single experiment; moreover, the uniform methods of herbicide application greatly simplify selection procedures, as compared with those used when plants are grown in a open field.

Tissue culture also gives rise to an additional genetic variability, termed somaclonal variation (Larkin and Scowcroft (1981)), which favours the discovery of resistant mutants. Several reviews (Maliga (1984) Widholm (1984) Duncan and Widholm (1986)) have described plant cell selection and the use of culture-derived variant cells. In many cases the variants isolated have helped to elucidate biochemical, physiological or genetic mechanisms in plants. Attempts to achieve herbicide resistant crops by cell culture started many years ago and 1978 saw the first successes in achieving genetically stable full resistance to herbicides.

Chaleff and Parsons (1978) reported the selection of diploid tobacco (Nicotiana tobacum)

R.S. Sangwan and B.S. Sangwan-Norreel (eds.), The Impact of Biotechnology in Agriculture, 437–446.

suspension cell culture resistant to picloram and, in the same year, Radin and Carlson regenerated tobacco plants with bentazon and phenmedipham resistances. These successes greatly encouraged research on the potential of plant tissue culture for purposes of obtaining herbicide resistance.

Selection in cell culture can be effective if herbicide tolerance or resistance is based on a metabolic path expressed at cell level. On the other hand, herbicide-tolerance expressed in cultured cells will not necessarily be present at plant level. Cultured cells and whole plants represent different developmental states characterized by different patterns of gene expression. Many genes are obviously expressed both in cell cultures and in plants but many others are expressed only in one state or the other.

Herbicide-tolerance in plant cell culture can reveal different forms ranging from a temporary physiological modification to a permanent genetic change transmitted to the progeny of regenerated plants (Meredith and Carlson (1982)). Several herbicide-resistant plant cell lines have been isolated, but only in a few cases has a genetically stable full resistance been achieved and the plant regenerated, as in the case of the sulphonylurea tolerant mutant in tobacco (Chaleff and Ray (1984)) and the imidazolinones-tolerant mutant in maize (Anderson (1986)). In the absence of plant regeneration, a herbicide-tolerant cell culture is an important tool for studies of herbicide toxicity and mode of action, but in terms of crop improvement it is a very little importance.

Regeneration from plant cell cultures is still a major problem for the successful application of cell biotechnology to crop improvement but the recent results obtained in species considered recalcitrant, encourage optimism in this respect.

3. Glyphosate and EPSP Synthase

Since its introduction in 1971 (Baird et al. (1971)) glyphosate [N-(phosphonomethyl) glycine] applied usually as the isopropylamine salt, has aroused great interest, due to its unique herbicidal properties and widespread use. It is a non-selective, broad-spectrum post-emergence herbicide whose salts are highly water-soluble. It normally enters plants through the aerial, usually chlorophyl-containing part, and is extensively traslocated and accumulated in merismatic areas where it produces its phytotoxic effects (Sprankle et al. (1975a 1975c)). Glyphosate is rapidly abdsorbed by the soil (where it is pratically immobile) through the phosphoric acid moiety that competes for binding sites with inorganic phosphates.

Degradation of glyphosate may occur, to a minor extent, by chemical reaction, but it is mainly brought about by the microflora; degradation seems to be a co-metabolic process, thus microorganisms do not use the herbicide for growth (Torstensson and Aamisepp (1977) Torstensson (1982) Sprankle et al. (1975b)).

Glyphosate is a potent inhibitor of the shikimic acid pathway leading to aromatic amino acids and to many important secondary plant products.

The absence of this pathway in animals is an important factor in the low animal toxicity, of glyphosate, which is one of its most significant properties.

The action at the shikimic acid pathway level causes a deficit of aromatic aminoacid, the primary effect of which is the slowing down or cessation of protein synthesis and of the formation of various phenolic compounds which are responsible for bringing about cell death.

The reversal of glyphosate toxicity by aromatic aminoacids has been achieved in a variety of microorganisms and cultured plant cells. However, alleviation of the glyphosate-induced inibition of growth of an intact higher plant by means of aromatic amino acids has been demonstrated only in the case of Arabidopsis thaliana (Cole (1985)). The results of growth reversal experiments suggest that the herbicide acts somewhere along the shikimate pathway but do not explain its the biochemical mode of action.

Extensive studies on shikimate pathway enzymes and their interaction with glyphosate have led to the conclusion that the site of inhibition is the Enolpyruvil 3 phosphate synthase. EPSPS catalyzes the reversible transfer of the carboxyvinyl moiety of PEP to S3P yielding EPSP and Pi.

Specifically, the herbicide is a competitive inhibitor with respect to phosphoenolpyruvate (PEP) and an uncompetitive inhibitor with respect to shikimate-3-phosphate (S3P) in the EPSPS reaction (Boocock and Coggins (1983) Amrhein et al. (1983) Steinrucken and Amrhein (1980,1984)).

Because of glyphosate EPSPS is now one of the best characterised enzymes of the shikimic acid pathway.

Bacterial fungal and plant EPSPS have been purified; enzymes from bacteria and plants are monomeric with a molecular weight of 44.000-51.000 (Lewendon and Coggins (1983) Ream et al. (1988)). Biochemical characterization studies, carried out by the Monsanto group on enzymes from different sources, have revealed a high degree of similarity between EPSPS from a monocot, Sorghum bicolor, and the enzymes from dicots. Furthermore, isozymes for EPSPS have been reported in S. bicolor (Ream et al. (1988)) and in Pisum sativum, where a major plastidial and a minor enzyme of probable cytosolic origin have been described (Mousdale and Coggins (1985)). The existence of the isozymic form with different subcellular locations of some other shikimate pathway enzymes, "strongly" supports the hypothesis of Jensen that there are two spatially separate shikimate pathways, one localized in the plastid and the other in the cytosol; therefore, according Jensen, glyphosate has more than one target of action: cytosolic EPSPS, plastidial EPSPS and then cytosolic 3-deoxy-arabino-heptulosonate-7-phoshate (DAHP), first enzyme of the path (Jensen (1985)). It has recently been demonstrated that EPSPS is synthetized in the cytoplasm as a precursor form with an N-terminal transit peptide sequence that is required for its import into the chloroloplast (della Cioppa (1986)). Subsequently della Cioppa and Kishore (1988)) reported that this transfer step is inhibited by glyphosate; therefore, the mode of action of the herbicide includes an initial interaction with the precursor enzyme at cytosol level before its translocation into the chloroplast.

Chemical modification studies on EPSPS structure, have revealed that some aminoacidical residues are essential for the activity of the enzyme. In particular, researches on petunia EPSPS indicate that Lys and Arg residues, identified as Lys23 and Arg28, play an essential role in EPSPS activity and may represent a part of the active site of the enzyme. Sequencing data tend to confirm this possibility, revealing that amino acid sequences around these two residues are conserved in all EPSPS analyzed (Kishore and Shah (1988)).

4. Mechanisms of Glyphosate Tolerance

Herbicide-tolerance can be acquired by plants by a number of different mechanisms strictly based on the characteristics and mode of action of the herbicide On the basis of knowledge of the inhibiting effect of glyphosate on EPSPS, in order to obtain glyphosate tolerance in plants two approaches have been considered; overproduction of EPSPS and mutation of the EPSPS gene.

4.1 MUTATION OF THE EPSPS GENE.

Naturally occurring glyphosate insensitive forms of EPSPS have been isolated from microbes or, more recently, generated by site-directed mutagenesis of the EPSPS gene in petunia (Kishore and Shah (1988)).

Glyphosate-tolerant mutant EPSPS enzymes have been isolated in Salmonella typhimurium by Stalker et al. (1985), in Aerobacter aerogenes (Klebsiella pneumonie) by Amrheim and coworkers (Schulz et al. (1984)) and, finally, an altered EPSPS has been isolated by Kishore et al. (1986) in a E. coli mutant.

The EPSPS mutant gene of E. coli has been used in transformation experiments in tobacco but the transgenic calli do not express glyphosate tolerance (Kishore and Shah. (1988)) unless the bacterial enzyme is targeted to the chloroplast.

Glyphosate tolerance has been engineered in tobacco and tomato plants by Comai (Comai et al.(1985) Fillatti et al. (1987)) using the S. typhimurium aroA gene encoding glyphosate-insensitive EPSPS. Plants expressing the bacterial gene for cytosol-localized EPSPS, were tolerant to Roundup (the commercial form of glyphosate) sprayed at 1 and 0.84 Kg/ha for tobacco and tomato, respectively.

The results obtained using bacterial EPSPS genes are to a certain extent contradictory and pose interesting questions concerning the site of action of glyphosate and the isozymic form of EPSPS. It is beyond doubt however, that targeting the bacterial enzyme to the cloroplast should enhance the level of tolerance.

To overcome this problem, Monsanto researchers, to engineer glyphosate tolerance, used an EPSPS plant gene isolated from a glyphosate-tolerant petunia cell line (MP4-G) overproducing the enzyme (Shah et al. (1986)). In this case the presence of the "transit peptide" assures the transfer to the chloroplast of a precursor-EPSPS obtaining a high level of expression of petunia EPSPS in transgenic plants (Della Cioppa et al. (1986)). At present no data are available to show that the precursor EPSPS is the cytosolic isozymic form of EPSPS. To date, petunia (Shah et al. (1986)), soybean (Maud et al. (1988)) and flax (Jourdan and McHughen (1988)) have been successfully transformed for tolerance to glyphosate by means of the unmodified strongly expressed EPSPS gene of petunia.

4.2 OVERPRODUCTION OF EPSPS.

Several plant cells tolerating glyphosate have been described. The first report was that of Amrhein et al. (1983) on Corydalis sempervirens.

The cells adapted to growth in the presence of up to 10 mM glyphosate, exhibited a 10-30 fold increase in their EPSPS activity, while no differences were found between the properties of EPSP-synthases isolated from the control and glyphosate-tolerant cells.

Similar results were obtained by Nafzinger (1984) in carrots.

A stepwise selection resulted in the isolation of adapted cells on 25 mM herbicide with a 52 fold increase of tolerance to glyphosate. Adaptation was stable in the absence of glyphosate. Glyphosate sensitivities of EPSPS extracted from adapted and non-adapted carrot cells were virtually identical, 50% inhibition occurring in both cases with about 10 µM glyphosate. On the contrary EPSPS activity in adapted cells was 12-fold greater that in non-adapted cells.

Then, Smart et al. (1985), conclusively demonstrated that the glyphosate-tolerant cell line of C. sempervirens overproduced EPSPS by an increased rate of *de novo* protein synthesis, but whether the overproduction was the result of a gene amplification or of an alteration in transcriptional or translational regulation remained an open question.

Subsequently a glyphosate-tolerant petunia cell line was isolated by Steinrucken et al. (1986). Tolerance was the result of an increased EPSP synthase activity caused by an overproduction of a herbicide- sensitive form of the enzyme.

Biochemical characterization of the petunia enzyme established the molecular weight and its monomeric structure, similar to other bacterial and plant EPSPS previously characterised, and provided further evidence that EPSPS is the major target for glyphosate action in plants. The results obtained in C. sempervirens and Petunia hybrida, using immunoblot technique, demonstrated that enzyme protein overproduction is the cause for the high level of enzyme activity.

Enzyme overproduction has been shown to be a common and general mechanism by means of which cultured animal cells achieve resistance to metabolic inhibitors and in many cases gene amplification has been demonstrated to be the basis of enzyme overproduction (Stark and Wahl (1984)). In plant cells, resistance to a metabolic inhibitor has only in a few cases been associated with increased activity of their target enzymes, herbicide-resistance usually being mediated by structural alteration in the target protein. Glyphosate tolerance seems to represent an exception in this respect. In the tolerant petunia cell line, using EPSPS cDNA as a probe, genomic blot analysis showed that the gene encoding EPSPS is amplified in the tolerant cell line. Similar results were obtained in C. sempervirens but in this case only an increase of EPSPS at mRNA level was detected (Kishore and Shah (1988)).

Singer and McDaniel (1985) demostrated that glyphosate-tolerant plants can be obtained through a direct selection scheme in tobacco cell culture and/or somaclonal variation. Selection in the presence of 1 mM glyphosate allowed isolation of several tolerant cell lines many of them stable in the absence of the herbicide.

Cell lines selected for glyphosate-tolerance give rise to regenerated plants that could be tolerant or sensitive at whole plant level. Similarly cultured tissues from regenerated tolerant plants could exhibit herbicide tolerance or sensitivity. An epigenetic rather than a genetic change and/or a chimerical origin of the regenerating shoot have been proposed to explain these results. In addition, glyphosate tolerance has been expressed by cell lines selected for amitrole resistance. No genetic (plants were sterile) or biochemical analysis was carried out on the glyphosate-tolerant variants isolated, and so the mechanism of tolerance remains unknown.

Single step selection in the presence of 3 mM glyphosate has also been successful in cultured tomato cells (Lycopersicon esculentum x L. peruvianum hybrid) where a glyphosate-tolerant variant has been isolated (Smith (1986)).

The growth of a variant cell line was unaffected by 10 mM herbicide, revealing a tolerance 100 times greater than that of unselected cells.

Biochemical analysis at enzyme level has revealed an EPSPS specific activity 8-13 times greater than that of sensitive normal cells, while glyphosate is equally effective in inhibiting EPSPS activity.

The tolerance observed is therefore accounted by for an overaccumulation of an EPSPS with normal sensitivity to glyphosate.

Glyphosate-tolerance due to overproduction of EPSPS has recently been demonstrated also in Catharantus roseus (Cresswell et al. (1988)). A stepwise procedure, in this case, allowed the isolation of cells adapted to tolerate up to 10 mM glyphosate and able to retain tolerance also in the absence of the herbicide. An EPSPS activity 60 fold greater was present in the selected cells. The enzyme activity dropped to lower values when tested after 10 months of culture in the absence of glyphosate. These results can be explained by assuming a reversion of the tolerant cells and/or a reselection of non-tolerant cells.

In all the cases considered, irrespective of the selection procedure (single step or stepwise) applied, glyphosate tolerance was obtained by overproduction of the EPSPS target enzyme. Therefore, unlike in bacteria where glyphosate-insensitive forms of EPSPS have been isolated, to find cells which are tolerant as a consequence of a structural alteration in the target protein seemed difficult. This could be explaned by the presence of isozymic forms and the peculiarity of the active site of the enzyme.

5. Glyphosate Tolerance in Maize Tissue Culture

A naturally occurring glyphosate tolerance in maize tissue culture has recently been reported (Racchi et al. (1989)).

Cell cultures and calli of the cultivar Black Mexican Sweet (BMS) do not show a great reduction of growth in the presence of increasing doses of glyphosate up to 10 mM. The application of 0.1 mM of the herbicide on growing maize seedlings resulted, on the contrary, in a complete halt of root and shoot development (Fig. 1-2).

The unexpected tolerance observed is not related to the cultivar (BMS) utilized: similar

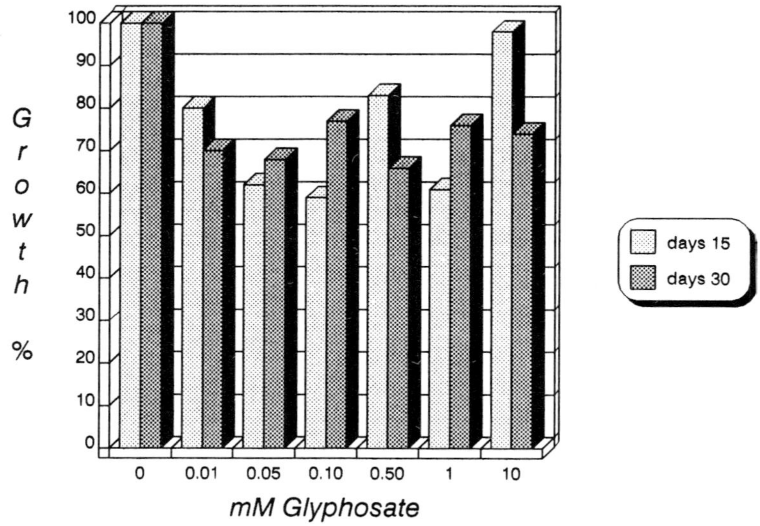

Figure 1. Glyphosate effect on BMS cells growth after 15 and 30 days of culture. The growth is expressed as % of the control.

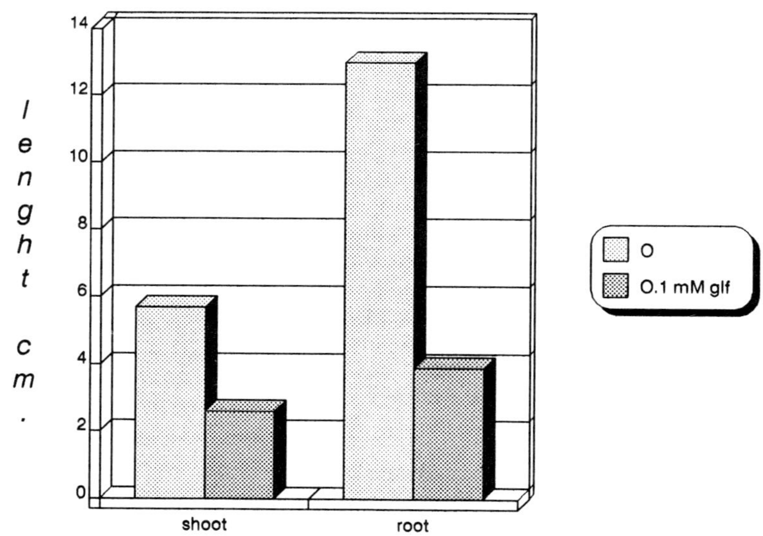

Figure 2. Glyphosate effect on shoot and root elongation of germinating BMS maize seeds.

results have in fact been obtained also with proliferating cultures of different hybrid strains and the tolerance is retained by the cultured cells even after several subcultures in the absence of the herbicide. The tolerance, however, is affected by the developmental state of the plant cell; thus embryogenic cultures are glyphosate-sensitive during the regeneration phase. The glyphosate tolerance observed in maize represents a quite intresting and rare event, in view of the sensitivity revealed by the cell cultures of other plants. In order to trow light on the biochemical basis of such results, the glyphosate inhibition kinetics of the target enzyme was investigated in plastid or cytoplasm-enriched extracts prepared from BMS cells. The results revealed that EPSP synthase activity of plastidial origin was severely affected by micromolar concentrations of glyphosate, whilst the enzyme coming from the cytoplasm-enriched fraction maintained a remarkable activity at 1 mM glyphosate (Fig. 3).

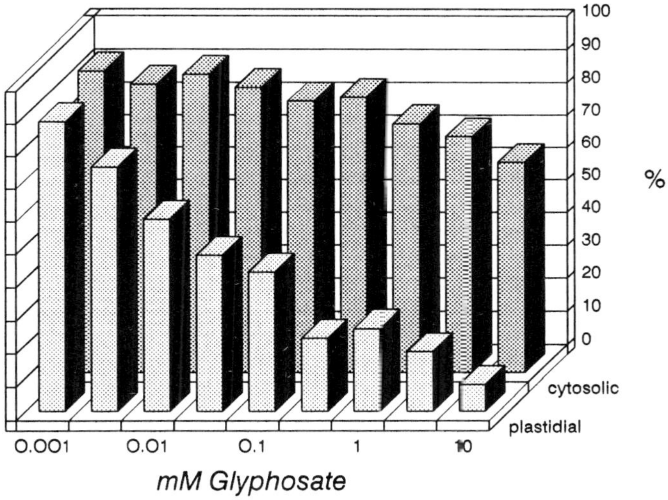

Figure 3. EPSP synthase activity in presence of different concentrations of glyphosate in plastid or cytoplasm enriched extracts prepared from BMS cells.

These results, confirmed in other maize cell lines, were not obtained in extracts prepared from cultured cells of carrot, where no tolerant EPSPS activity could be detected. The presence of two isoforms of EPSPS has been reported in monocots by Ream (Ream et al. (1988)) but no substantial differences in glyphosate sensitivity were described in these or in several other plant EPSP synthases.

The existence in maize of a cytosolic glyphosate-resistant EPSPS form may confer on cell cultures the ability to maintain levels of aromatic amino acids sufficient to sustain *in vitro* growth, thus explaining the tolerance to the herbicide.

Further evidence supporting the existence of two different EPSPS forms was recently obtained by means of the chromatographic separation of two EPSPS peaks from BMS cultured cell extracts. One of these was confirmed to be glyphosate-resistant and to have, as compared with the glyphosate-sensitive one, a decreased affinity for the substrate PEP (G. Forlani and E. Nielsen, unpublished results).

These data pose interesting questions concerning EPSP synthase; the presence of EPSPS isozymes has been revealed by different authors, but so far the existence of a glyphosate-insensitive EPSP synthase isoform able to confer tolerance of the herbicide on cultured cells has not been demonstrated. The presence of this glyphosate-insensitive form and its expression level need to be investigated at plant level. In any case it could have significant implications relating to the study of the mechanism of glyphosate tolerance at cell level and during the differentiation processes.

6. Acknowledgments

The Author's investigations were supported by the Italian Ministero della Agricoltura e delle Foreste project "Sviluppo di tecnologie avanzate applicate alle piante". The Author is grateful to E.Nielsen and G.Forlani for critical reading of the manuscript.

7. References

Amrhein N., Johanning D., Schab J. and Schulz A. (1983) Biochemical basis for glyphosate tolerance in a bacterium and a plant tissue culture. FEBS Letters 157, 191-196.

Anderson P. (1986) Cell culture selection of herbicide tolerant corn. Proc. Ann. Corn Sorghum Res. Conf. 41, 48-85.

Baird D.D., Upchurch R.P., Homesley W.B. and Franz J.E. (1971) Introduction of a new broad sprectrum post emergence herbicide class with utility for herbaceous perennal weeds control. Proc. Natl. Centr. Weed Central Conf. 26, 64-68.

Boocock M.R. and Coggins J.R. (1983) Kinetics of 5-enolpyruvylshikimate-3-phosphate synthase inhibition by glyphosate. FEBS Lett. 154, 127-133.

Chaleff R.S. and Parsons M.F. (1978) Direct selection *in vitro* for herbicide-resistant mutants of Nicotiana tabacum. Proc. Natl. Acad. Sci. U.S.A. 75, 5104-5107.

Chaleff R.S., Ray T.B. (1984) Herbicides resistant mutants from tobacco cell culture. Science 223, 1148-1151.

Cole D.J. (1985) Mode of action of glyphosate a literature analysis in E. Grossbard and D. Atkinson (Eds.) The herbicide glyphosate. Butterworths, pp. 48-73.

Comai L., Facciotti D., Hiatt W.R., Thompson G., Rose R.E. and Stalker D.M. (1985) Expression in plants of a mutant aroA gene from Salmonella typimurium confers tolerance to glyphosate. Nature Vol. 317, 741-744.

Cresswell R.C., Fowler M.W. and Scragg A.H. (1988) Glyphosate-tolerance in Catharanthus rosens. Plant Science 54, 55-63.

della Cioppa G., Bauer S.C., Klein B.K., Shah D.M., Fraley R.T. and Kismore G.M. (1986) Traslocation of the precursor of 5-Enolpyruvylshikimate-3-phosphate synthase into chloroplasts of higher plants *in .vitro* Proc. Natl. Acad. Sci. USA 83, 6873-6877.

della Cioppa G. and G.M. Kishore (1988) Import of a precursor protein into chloroplast is inhibited by the herbicide glyphosate. EMBO Jour.vol,7, 5, 1299-1305.

Duncan D.R. and Widholm J.M. (1986) Cell Selection for crop improvement. Plant Breeding Reviews, Vol. 4, 153-173.

Fillatti J., Kiser J., Rose J. and Comai L. (1987) Efficient transfer of a glyphosate tolerance gene into tomato using a binary Agrobacterium tumefaciens vector. Bio/Technology 5, 726-730.

Hinchee M.A., Connor-Ward D.V., Newell C.A., McDonnel R.E., Sato S.J., Gasser C.S., Fischoff D.A., Re D.B., Fraley R.T. and Horsch R.B. (1988) Production of transgenic soybean plants using Agrobacterium-mediated DNA transfer. Bio/Technology 6, 915-920.

Jensen, R.A. (1985) The shikimate/arogenate pathway: link between carbohydrate metabolism and secondary metabolism. Physiol. Plant. 66, 164-168.

Jordan M.C. and McHughen A. (1988) Glyphosate tolerant flax plants from Agrobacterium mediated gene transfer. Plant Cell Reports 7, 281-284.

Kishore G.m. and Shah D.M. (1988) Amino acid biosinthesis inhibitors as herbicides. Ann. Rev. Biochem. 57, 627-663.

Kishore G.M., Brundage L., Kolk K., Pagette S. et al. (1986) Isolation, purification and characterization of a glyphosate tolerant mutant *E.coli* EPSP synthase. Fed. Proc. Am. Soc. Exp. Biol. 45, 1506.

Larkin P.J. and Scowcroft (1981) Somaclonal variation: a novel source of variability from cell cultures for plant improvement. Theor. Appl. Genet. 60, 197-214.

Lewendon A. and Coggins J.R. (1983) Purification of 5 enol pyruvyl-shikimate-3 phosphate synthase from *Escherichia coli*. Biochem J. 213, 187-191.

Maliga P. (1984) Isolation and characterization of mutants in plant cell culture. Ann. Rev. Plant. Physiol. 35, 519-542.

Maud A.W. Hinchee, Connor-Ward D.V., Newell C.A., McDonnell R.E.,Sato S.J., Gasser C.S. ,Fischhoff, Re D.B.,Fraley R.T. Horsch R.B. (1988) Production of transgenic soybean plants using agrobacterium-mediated DNA transfer. Bio/Technology 6,915-922.

Meredith C.P. and Carlson P.S. (1982) Herbicides resistence in plant cell cultures. H.H. Le Baron and Gressel (Eds.) in Herbicides resistance in Plants, Wiley and Sons publisher, New York pp. 275-291.

Mousdale D.M. and Coggins I.R. (1984) Purification and properties of e enolpyruvylshikimate-3-phospate synthase from seedlings of Pisum sativum. L. Planta 169, 78-83.

Mousdale D.M. and Coggins I.R. (1985) Subcellular localization of the common shikimate-pathway enzymes in Pisum sativum L. Planta 163, 241-249.

Nafzinger E.D.,Widholm J.M., Steinrucken H.C.,Killmer J.L. (1984) Selection and characterization of a carrot cell line tolerant to glyphosate. Plant Physiol.76, 571-574

Racchi M.L., Forlani G., Pelanda R. and Nielsen E. (1989) Glyphosate effects on growth and EPSP synthase isozymes activity in cultured maize cells. Science for Plant Breeding, Vortrage fur Pflanzenzuchtg. 15,26-11.

Radin D.N. and P.S. Carlson (1978) Herbicide-tolerant tobacco mutants selected in situ recovered via regeneration from cell culture. Genet. Res. Camb. 32, 85.

Ream J.E., Steinrucken H.C., Porter C.A and Sikorski I.A. (1988) Purification and properties of 5 Enolpyruvlshikimate-3-phosphate synthase from dark grown seedling of Sorghum bicolor. Plant Physiol. 87, 232-238.

Schulz A., Sost D. and Amrheim N. (1984) Insensitivity of 5 enolpyruvylshikimic acid-3-phosphate synthase to glyphosate confers resistance to this herbicide in a strain of Aerobacter aerogenes. Arch. Microbiol. 137, 121-123.

Shah D.H., Horsh R.B., Klee H.I., Kishore G.M., Winter I.A., Tumer N.E., Hironaka C.M., Sanders P.R. Gasser C.S., Aykent S., Siegel N.R., Rogers S.G. and Fraley R.T. (1986) Engineering herbicide tolerance in transgenic plants. Science, Vol. 233, 476-481.

Singer S.R. and McDaniel C.N. (1985) Selection of Glyphosate-tolerant tobacco calli and the expression of this tolerance in regenerated plants. Plant. Physiol. 78, 411-416.

Smart C.C., Johanning D., Muller G., Amrhein N. (1985) selective overproduction of 5-enol-pyruvylshikimic acid 3-phosphate synthase in a plant cell culture which tolerates high doses of the herbicide glyphosate. J.Biol.Chem.260,16338-46.

Smith C.M., D. Pratt and Thompson G.A. (1986) Increased 5-enolpyruvylshikimic acid 3-phosphate synthase activity in a glyphosate-tolerant variant strain of tomato cells. Plant Cell Reports 5, 298-301.

Sprankle P., Meggit W.F. and Penner D. (1975a) Absorption, action and traslocation of glyphosate. Weed. Scie. 23, 235-240.

Sprankle P., Meggit W.F. and Penner D. (1975b) Adsorption, mobility and microbial degradation of glyphosate in the soil. Weed Science 23, 229-234.

Sprankle P., Meggit W.F. and Penner D. (1975c) Rapid inactivation of glyphosate in the soil. Weed Science 23, 224-228.

Stalker D.M., Hiatt W.R., Comai L. (1985) A single aminoacid substitution in the enzyme 5 enolpyruvylshikimate 3 phoshate synthase confers resistance to herbicide-glyphosate. J. Biol. Chem. 260, 4724-4728.

Stark G.R., Wahl F.M. (1984) Gene Amplification. Annu. Rev. biochem. 53,447-491.

Steinrucken H.C. and Amrhein N. (1980) The herbicide glyphosate is a potent inhibitor of 5-enolpyruvyl-shikimic acid-3-phosphate synthase. Biochem. Biophys. Res. Comm 94, 1207-1212.

Steinrucken H.C. and Amrhein N. (1984) EPSP synthase of Klebsiella pneumonie 2. Inhibition by gliphosate. EUR. J. Biochem. 143, 351-357.

Steinrucken H.C., Schulz A., Amrhein N., Porter C.A., Fraley R.T. (1986) Overproduction of 5-Enolpyruvylshikimate-3-phosphate Synthase in a glyphosate-tolerant *Petunia hibrida* cell line Arch. Biochem.Biophys.244,169-73.

Torstensson N.T.L. and Aamisepp A. (1977) Detoxication of glyphosate in soil. Weed Res. 17, 209-212.

Torstensson L. (1982) Decomposition of glyphosate in agricultural soils. Weeds and Weed control 23rd Swedish Weed Confer. pp. 385-392.

Widholm J.M. (1984) Selection and characterization of plant cell mutants for molecular biology studies. Newsl. Int. Assoc. Plant Tissue Culture 44, 2-6.

Section 4. Cryopreservation of plant cells

CRYOPRESERVATION OF PLANT CELL CULTURES.

THE IMPORTANCE OF PRETREATMENTS

SEITZ, U.[1], BANSPACH, D.[1], GÖLDNER, E.[2], REINHARD, E.[1]

1) Universität Tübingen, Pharmazeutisches Institut, Auf der Morgenstelle 8, D-7400 Tübingen

2) Boehringer Mannheim, Werk Penzberg, D-8122 Penzberg

ABSTRACT. Cryopreservation experiments were performed with cell suspension cultures of the following species: *Atriplex hortensis* L., *A. litoralis* L. (*Chenopodiaceae*), *Berberis wilsoniae* HEMSL. & WILS. (*Berberidaceae*), *Coleus blumei* BENTH. (*Lamiaceae*), *Daucus carota* L. (*Apiaceae*), *Digitalis lanata* EHRH. (*Scrophulariaceae*), and *Panax ginseng* C.A. MEY. (*Araliaceae*). The protocol comprises a preculture phase in nutrient media with enhanced osmolarity, treatment with cryoprotectants, slow freezing, storage at -196°C, rapid thawing, and post-thaw treatments. Biochemical capacities remain unchanged in frozen-thawed cultures. This was shown to be the case in *D. carota*, *D. lanata*, and *P. ginseng*. A survey on literature concerning this point is presented. Some alternatives to the standard cryopreservation procedure are given. Experiments are underway to investigate the physiological events during the preculture phase. As a consequence of osmotic stress cell viability decreased to a certain extent. A suspension of this process and even recovery have been observed in the cases of positive response. Further events such as the uptake of preculture additives and changes in the cytoplasm:vacuole volume ratio will be discussed.

1. INTRODUCTION

In the course of work with plant cell cultures the necessity has arisen for efficient storage methods. The

R.S. Sangwan and B.S. Sangwan-Norreel (eds.), The Impact of Biotechnology in Agriculture, 449–458.

routine techniques for culture maintenance used until now are expensive and time-consuming. Typically, a cell suspension culture would need to be transferred every 7-10 days, and a callus culture every 14-30 days. Furthermore, regular subculturing in the fast-growing state entails the risk of possible loss through contamination or equipment failure. The establishment of tissue culture collections also requires routine methods that would enable us to preserve the samples over long periods of time.

Cryopreservation , i.e. storage at the temperature of liquid nitrogen, is regarded to be the most suitable method for these purposes. The cryopreservation protocol comprises the following steps: preculture in media supplemented with osmotically active compounds, treatment with cryoprotective agents, slow freezing, storage at -196 °C, rapid thawing, and post-thaw treatments and recovery growth. The major problem in adopting this method more generally in tissue-culture laboratories is the fact that the protocol has to be optimized for each plant species in question. As a result, the number of plant species that have been cryopreserved successfully is still limited (Withers [18]).

As far as we know, the characteristics of plant cell cultures do not change during cryopreservation treatments. This point will be discussed on the basis of growth behaviour and biochemical capacities.

The cryopreservation protocol used in our investigations was based on the method published by Withers and King [19]. Examples are presented here to show the possibility of modifying this method and adapting it to various cell cultures. We have concentrated our attention on the preculture step, the treatment with cryoprotectants, and the thawing process.

Preculture in a medium supplemented with osmotically active compounds is found to enhance freeze tolerance markedly. The physiological events during this phase have not yet beén investigated thoroughly, but this treatment seems to induce a process which resembles stress hardening. In the course of these studies we investigated the influence of several preculture additives on cell viability and freeze tolerance. The objective of this work is to present a spectrum of possible alternatives for the optimization of cryopreservation protocols and to learn more about how to induce freeze tolerance in vitro.

2. MATERIAL AND METHODS

2.1. Plant species and culture conditions

Suspension cultures were used throughout these investigations. They were propagated in 300 ml Erlenmeyer flasks kept on a gyratory shaker at 25 °C in the dark.

TABLE 1. Culture conditions.

SPECIES	CULTURE MEDIUM SUBCULTURE REGIME	REF.
Atriplex hortensis	B5, 9-10 days	[8]
Atriplex litoralis	B5, 9-10 days	[8]
Berberis wilsoniae	MS, 10-11 days	[15]
Coleus blumei	B5, 7 days	[11]
Daucus carota	I2a, 7 days	[9]
Digitalis lanata	MS, 7 days	[14]
Panax ginseng	MS, 9-10 days	[13]

2.2. Cryopreservation protocol

The standard procedure was performed according to Withers and King [19]. Cells were taken from the beginning of the growth phase and suspended in nutrient medium supplemented with the respective additive (0,3 M). The incoculum was about twice the size of that used for routine subculture. Except for *P. ginseng* [13] the pre-culture period lasted 3 days.

The cryoprotectant mixture was prepared in culture medium. Final concentration: 0.5 M DMSO, 0.5 M glycerol, 1.0 M sucrose. The treatment was 1 h at 0-4 °C.

Freezing conditions. The cell suspensions were distributed into freezing ampoules (1 ml each) and kept on ice. The samples were frozen at a rate of 1 °C per min to -40 °C using the cooling apparatus Type BF-R 200/101, (Messer Griesheim, Düsseldorf, FRG). After 10-15 min at -40 °C the ampoules were immersed in liquid nitrogen.

Storage: in liquid nitrogen.

Thawing: Samples were thawed rapidly by agitating the ampoules in warm water (40 °C) In some experiments with *C. blumei* the samples were thawed at 60 °C or using a commercial microwave oven.

Post-thaw treatments: Immediately after thawing cell suspensions were spread on ca. 8 ml culture medium solidified with agar (0.7 %) in a 5 cm diameter Petri dish. Once growth had been reestablished, the cells were transferred into liquid medium and cultivated as usual.

2.3. Growth characteristics

Fresh and dry weight, packed cell volume, and cell number as well as medium components were determined as described previously [13].

2.4. Cell viability

In all experiments cell viability was routinely determined before and after freezing using fluorescein diacetate or phenosafranine staining [17]

3. RESULTS AND DISCUSSION

3.1. Conservation of cultures with special characteristics.

Daucus carota. Cells of an Afghan variety of *Daucus carota* accumulate large amounts of anthocyanin. The cell viability of frozen-thawed cells was high (60-70 % viable cells). The anthocyanin content of a culture which had been frozen was compared with that of a control which had not (Seitz [12]). The maximum values and the accumulation kinetics were both identical. This is in complete agreement with the results published by Doughall & Whitten [7], who investigated the anthocyanin content in frozen-thawed carrot cells for the first time.

Digitalis lanata. Cultivated cells of *D. lanata* do not produce cardenolides, but they are able to transform added cardenolides (biotransformation). A typical reaction is the 12ß-hydroxylation of ß-methyldigitoxin. The main part of the product (ß-methyldigoxin) is released into the culture medium. Cryopreserved and control cultures showed the same time course in ß-methyldigoxin production (Seitz et al. [14]). Moreover, the final yields were also identical. This was also valid after long-term storage (up to 4 years). Thus cell lines that have been selected for high productivity can be maintained in the frozen state.

TABLE 2. Freeze preservation of cell cultures with specific biochemical capacities

SPECIES	NATURAL COMPOUND	REF.
Catharanthus roseus	indole alkaloids	[5]
Chenopodium rubrum	betalaines	[20]
Coleus blumei	rosmarinic acid	[11]
Daucus carota	anthocyanin	[7,15]
Digitalis lanata	cardenolides (biotransformation)	[6,14]
Dioscorea deltoidea	steroids	[3]
Eschscholtzia californica	benzophen-anthridines	[20]
Lavandula vera	biotin	[16]
Panax ginseng	ginsenosides	[13]
Papaver bracteatum	chlorophylles	[20]
Thalictrum rugosum	isoquinolin-alkaloids	[20]

Panax ginseng. Cryopreservation experiments were performed using ginsenoside-producing cell cultures of *Panax ginseng*. We obtained viable cells after storage in liquid nitrogen. The protocol was optimized with regard to preculture conditions and cryoprotectants. The relatively low freeze tolerance of ginseng cells was enhanced from less than 10 % to about 40 % as expressed by cell viability immediately after thawing. Using HPLC analysis the ginsenosides Rg_1, Re, Rf, Rg_2, Rb_1, Rc, Rb_2, and Rd were determined. Neither ginsenoside yield nor ginsenoside pattern were influenced by cryostorage (Seitz and Reinhard [13]). Though less than 50 % of the cell population survived, no selection occurred. Additional information on this aspect was obtained by determining the growth behaviour of the cultures. The time courses of fresh and dry weight, cell number and packed cell volume proved unchanged by cryostorage (Seitz and Reinhard [13]).

A summary of plant cell cultures that have been investigated after cryostorage with regard to biochemical capacities is given in Table 2. All of the frozen-thawed cultures mentioned there have been proved to retain their biochemical characteristics.

3.2. Cryopreservation protocol

Many reports have shown that plant cells are only able to survive freeze-thaw processes if they have passed through a preculture treatment in a medium with enhanced osmolarity. Consequently, optimization of the protocol would always start at this point. Results are presented in paragraph 3.3.

Table 3. Comparison of two cryoprotectant mixtures in freezing experiments with *Atriplex litoralis* and *A. hortensis*. The percentage of viable cells was determined using fluorescein diacetate. A: at the end of preculture, B: after 1 h cryoprotection, C: immediately after thawing. DGS: DMSO, glycerol, sucrose; PGS: 1,2-propandiol, glycerol, sucrose.

Species	A	B		C	
		DGS	PGS	DGS	PGS
A. litoralis	92	80	78	54	52
A. hortensis	94	80	82	52	47

Cryoprotectant treatment. Without exception, plant cells require chemical cryoprotection. Mixtures are reported to be more suitable than single compounds. DGS, a mixture containing DMSO (0.5 M), glycerol (0.5 M) and sucrose (1.0 M) has been used successfully for a large number of species (Seitz [12], Withers [18]). On the other hand, we have found that with more than 10 different species including *D. lanata* and *P. ginseng* good results can be achieved with sucrose (1.0 M) as the sole cryoprotectant. The replacement of DMSO by 1,2-propandiol, successful with recalcitrant seeds [2], was investigated using two *Atriplex* species (PGS, Table 3). PGS and DGS proved to be of equal efficacy in terms of post-thaw viability. The general application of this compound, however, should be confirmed using other plant species.

Thawing and post-thaw treatments. In general, rapid thawing has been reported to be beneficial for the cells. This is confirmed by our results with *Coleus blumei* [11]. Cell viability could be enhanced by 10 % to 40 % if a 60 °C water bath was used instead of 40 °C (see Table 4). We also report here the successful thawing of plant cells using a microwave oven. Although thawing takes approx 3 min, cell viability was in the range of 40 %. This positive effect may be ascribed to a homogeneous rewarming of the samples under microwave irradiation.

TABLE 4. Effect of various thawing conditions on cell survival of *Coleus blumei*. (I): freezing rate 2 °C per min; (II): freezing rate 0.5 °C per min. The approximate duration of thawing is mentioned.

	post-thaw viability (%)		
thawing condition duration	water bath 45 °C 2-2.5 min	water bath 60 °C 1 min	microwave oven 3 min
I	30	40	40
II	20	30	30

A subsequent washing step to remove the cryoprotectant additives seems to be unnecessary and can even be deleterious. Therefore, the cell suspensions were normally spread on a semi-solid agar medium. As a rule, the cells need a recovery phase of two or more weeks before they can be transferred to normal culture conditions.

3.3. Cell viability during preculture and effectiveness of various additives.

Digitalis lanata cultures were used for an extensive study of this subject. *Digitalis* cells grown in normal culture medium showed only negligible changes in their viability rates over a 3-day period. As a consequence of preculture treatments the viability curves were characterized by a marked decrease during the first 24 hours. This was valid for trehalose, mannitol, sucrose and proline (Fig. 1) and also for sorbitol (data not shown). The reduction in viability was transient in all cases where the cells were able to tolerate the treatment (trehalose, mannitol, sucrose, melibiose). We consider this behaviour to be a consequence of stress, which was then compensated for by adaptation to the enhanced molarity of the medium. A more prolonged decrease in viability, the production of phenolic compounds and a browning of the culture were observed in the presence of proline or sorbitol.

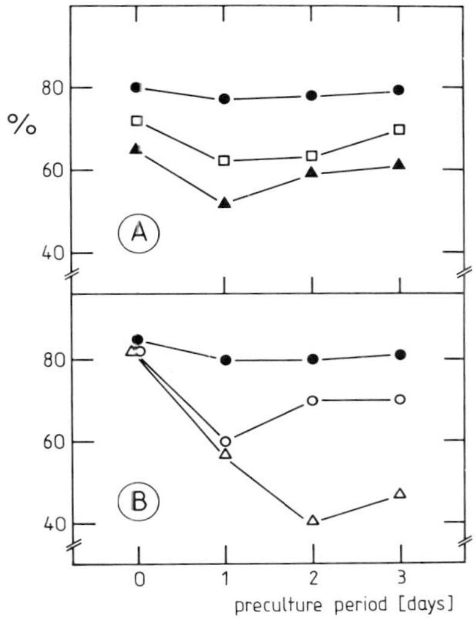

Figure 1. Viability of *Digitalis lanata* cells during a 3-day preculture period. Concentration of preculture additives: 0.3 M. Fluorescein diacetate was used for viability tests. A. control ●, sucrose □, mannitol ▲, B. control ●, trehalose o, proline △.

In a further series of experiments we investigated the influence of several preculture additives on the freeze tolerance of *Digitalis* cells. Besides the compounds which are commonly used as preculture additives, such as mannitol, we directed our attention to compounds which are accumulated in plants under stress conditions, such as proline. The best results were obtained when trehalose or mannitol were used (Fig. 2); the post-thaw viabilities were 66 % and 53 %, respectively. Furthermore, the cells were able to resume growth after a relatively short lag period. This is just one more example which demonstrates the broad applicability of mannitol. Trehalose has been used in cryopreservation experiments (Bhandal et al. [1]), but only very few examples are reported in literature. Melibiose and sucrose were equally effective with *Digitalis* cultures.

Neither proline nor sorbitol proved suitable for inducing freeze tolerance, even though post-thaw viability was relatively high. In both cases the cells were not able to resume growth.

With respect to the striking effect of preculture treatments on the freeze tolerance of cultured cells, it is surprising that very few reports have been published concerning the cellular processes that are initiated during this period (for a review see [18]). Pritchard and coworkers [10] considered the following changes to be the most important for survival: the cytoplasm : vacuole volume ratio, the water contênt, the accumulation of cytoplasmic solutes, and the osmotic behaviour of the cells.

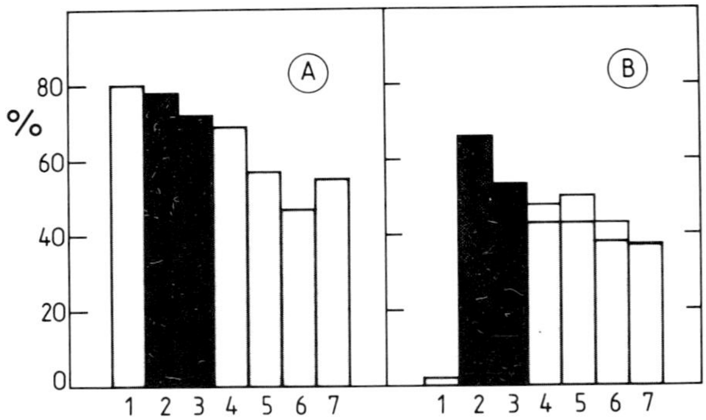

Figure 2. Viability (fluorescein diacetate) of *Digitalis lanata* cells at the end of the preculture period (A) and immediately after thawing. The following preculture additives (0.3 M) were used: 2) trehalose, 3) mannitol, 4) sucrose, 5) melibiose, 6) proline, 7) sorbitol. 1) represents the control.

Various types of stress, such as infection, UV light, and changes in the concentration of nutrients, were shown to enhance the level of phenylalanine ammonia lyase (PAL) activity (Camm and Towers [4]). In the search for an indicator of stress, PAL activity was tested in the course of preculture. In preliminary experiments PAL activity displayed an inverse pattern to the viability curves. In an additional set of experiments we intend to investigate the protein patterns of cells under preculture conditions.

With the type of experiments described here we attempt a closer look at cellular events during the preculture phase. In our opinion, such work will enable us to come closer to the point where it is possible to find the optimum conditions for a desired culture more easily and to develop generally applicable methods for cryostoring plant cell cultures.

4. ACKNOWLEDGEMENT

This work was supported by the Deutsche Forschungsgemeinschaft.

5. REFERENCES

[1] Bhandal, I.S., Hauptmann, R.M. and Widholm, J.M. (1985) Trehalose as cryoprotectant for the freeze preservation of carrot and tobacco cells, Plant Physiol. 78, 430-432.

[2] Boucaud, M-T. and Cambecedes, J. (1988) The use of 1,2-propandiol for cryopreservation of recalcitrant seeds: the model case of *Zea mays* imbibed seeds. Cryo-Lett. 9, 94-101.

[3] Butenko, R.G., Popov, A.S., Volkova L.A., Chernyak, N.D. and Nosov, A.M. (1984) Recovery of cell cultures and their biosynthetic capacity after storage of *Dioscorea deltoidea* and *Panax ginseng* cells in liquid nitrogen, Plant Sci. Lett. 33, 285-292.

[4] Camm, F.L. and Towers, G.H.N. (1977) Phenylalanin ammonia-lyase, Progr. Phytochem. 4, 169-188.

[5] Chen, T.H.H., Kartha, K.K., Leung, N.L. Kurz, W.G.W., Chatson, K.B. and Constabel, F. (1984) Cryopreservation of alkaloid-producing cell cultures of periwinkle (*Catharanthus roseus*), Plant Physiol. 75, 726-731.

[6] Diettrich, B., Popov, A.S., Pfeiffer, B., Neumann, D., Butenko, R. and Luckner, M. (1982) Cryopreservation of *Digitalis lanata* cell cultures, Planta Med. 46, 82-87.

[7] Dougall, D.K. and Whitten, G.H. (1980) The ability of wild carrot cell cultures to retain their capacity for anthocyanin synthesis after storage at -140°C. Planta Med. Suppl. 129-135.

[8] Gamborg, O.L., Miller, R.A. and Ojima, K. (1968) Nutrient requirements of suspension cultures of soybean root cells, Exp. Cell Res. 50, 151-158.

[9] Noé, W., Langebartels, C. and Seitz, H.U. (1980) Anthocyanin accumulation and PAL activity in a suspension culture of *Daucus carota* L., Planta 149, 283-287.

[10] Pritchard, H.W., Grout, B.W.W., Short, K.C. and Reid, D.S. (1982) The effects of growth under water stress on the structure, metabolism and cryopreservation of cultured sycamore cells, in Franks, F. and Mathias, S.F. (eds.) The Biophysics of Water, Cichester, pp. 315-318.

[11] Reuff, I. Seitz, U., Ulbrich, B. and Reinhard, E. (1988) Cryopreservation of *Coleus blumei* suspension and callus cultures, J. Plant Physiol. 133, 414-418.

[12] Seitz, U. (1987) Cryopreservation of plant cell cultures, Planta Med. 53, 311-314.

[13] Seitz, U. and Reinhard, E. (1987) Growth and ginsenoside patterns of cryopreserved *Panax ginseng* cell cultures, J. Plant Physiol. 131,215-223.

[14] Seitz, U. Alfermann, A.W. and Reinhard, E. (1983) Stability of biotransformation capacity in *Digitalis lanata* cell cultures after cryogenic storage, Plant Cell Rep. 2, 273-276.
[15] Seitz, U., Reuff, I. and Reinhard, E. (1985) Cryopreservation of plant cell cultures, in Neumann, K.-H., Barz, W. and Reinhard, E. (eds.), Primary and Secondary Metabolism of Plant Cell Cultures, Springer, Berlin Heidelberg, pp. 323-333.
[16] Watanabe, K., Mitsuda, H. and Yamada, Y. (1983) Retention of metabolic and differentiation potentials of green *Lavandula vera* callus after freeze-preservation, Plant & Cell Physiol. 24, 119-122.
[17] Widholm, J.M. (1972) The use of fluorescein diacetate and phenosafranine for determining viability of cultured plant cells, Stain Technol. 47, 189-194.
[18] Withers, L.A. (1985) Cryopreservation of cultured plant cells and protoplasts, in Kartha, K.K. (ed.), Cryopreservation of Plant Cells and Organs, CRC Press, Boca Raton, pp. 243-264.
[19] Withers, L.A. and King, P.J. (1980) A simple freezing-unit and routine cryopreservation method for plant cell cultures, Cryo-Lett. 1, 213-220.
[20] Ziebolz, B. and Forche, E. (1985) Cryopreservation of plant cells to retain special attributes, in Schaefer-Menuhr, A. (ed.), Advances in Agriculture and Biotechnology, Kluwer Acad. Publ., Dordrecht, pp. 181.

Section 5. Abstracts

REGENERATION OF FERTILE SUNFLOWER

M. FREYSSINET, B. PELISSIER and G. FREYSSINET
Rhône-Poulenc Agrochimie, Biologie Moléculaire et Cellulaire Végétale, BP 9163
69263 LYON Cédex 09, France.

A regeneration procedure is a prerequisite to apply genetic engineering techniques to Sunflower. As a first approach toward this problem we have used immature embryos to regenerate fertile sunflower. The embryos were collected 4 to 21 days after pollination (0.1 to 5 mm) and incubated on a Murashige and Skoog based medium containing BAP (0.5 to 1 mg/l). After incubation for two to three weeks on this medium somatic embryo-like structures appeared. They were isolated and transferred to a medium allowing formation of shoots and roots. The young plantlets were then transferred to soil, adaptated to greenhouse conditions and developed to maturity. These plants were self-polinated and set seeds about four months after the collection of the immature embryos. This technique has worked for the eight cultivars tested. Their progeny have been tested in the field for conformity. No variants have been observed. As a second approach we have used hypocotyl epidermis. This epidermis is composed of the epidermis (sensu stricto) plus 3 to 6 parenchyma layers. It shows a high level of totipotency. Depending on the sucrose concentration two kinds of observations were made :

1- Using 90g/l of sucrose, isolated embryos were obtained which do not germinate, but produce secondary embryos which develop to fertile plants.

2- Using 30g/l sucrose, early stage embryos germinated giving rise to fertile plants.

A cryopreservation study of Apple shoot apices (cv. Golden Delicious) and somatic embryos of Carrot (cv. Nanco and De Chantenay) cultured *in vitro* .

H. PAUL and B.S. SANGWAN.
Laboratoire Androgenèse et Biotechnologie, Université de Picardie, Faculté des Sciences , 33 rue Saint Leu, 80039 AMIENS, Cedex, France.

The apices obtained from in vitro cultured shoots of Apple on the multiplication medium, survive to freezing at -196°C because of the presence of cryoprotectors (glycerol, DMSO). The apices and stem cuttings were acclimated at -4°C in the darkness for the periods of 0 to18 hours and 0 to 6 weeks respectively. This acclimation coupled with cryoprotectors treatment allowed to obtain an acceptable survival rate. It seems that glycerol is more effective than DMSO. Presently, the survival is manifested only by formation of callus.

The petiole segments of Carrot cv. Nanco and De Chantenay bearing the somatic embryos were treated with the DMSO and were cryopreserved at low speed. Before freezing, the somatic embryos were dehydrated by exposing to sterile air flux. 5 or 10% of DMSO, the slow speed of freezing of 1, 2 or 5°C/min. and a dehydration of 30 minutes have allowed to obtain the survival and recovery of development of somatic embryos of cv. Nanco after rewarming. The recovery of development was made on the induction medium of somatic embryogenesis for a cryoprotection of 5 and 10% of DMSO, and on the development medium for 10% of DMSO for the cv. Nanco.

Whatever the tested treatment might be, the cv. De Chantenay did not show any sign of recovery of the development after rewarming.

Effect of sugars on vitrotuberization of *Solanum tuberosum* L. Cytological observation of microtubers.

L. LAVIEVILLE and B.S. SANGWAN

Laboratoire Androgenèse et Biotechnologie, Université de Picardie, Faculté des Sciences , 33 rue St Leu, 80039 Amiens Cedex FRANCE
G.I.E. des Syndicats de Producteurs de Plants de Pommes de Terre de la Région Nord, Zone Industrielle, Route de Tilloy, 62217 Beaurains FRANCE

Microtubers of potato were obtained *in vitro* by culturing nodal explants derived from axenic cuttings . Cultures were made during 5 weeks in darkness and in liquid medium of Murashige et Skoog without the addition of growth regulating substances. Glucidic requirement was tested using different sugars. Sucrose at the concentration of 80 g/l was found to be optimal for tubers production in 4 different varieties.

Using HPLC method, the contents of sucrose, glucose and fructose were determined in the cuttings, in the nodes cultured for tuberization and in the microtubers.

Morphological aspect of tubers obtained *in vitro* was controlled under Scanning Electron Microscopy (SEM). Semi thin sections realized on the very young tubers were also observed under light microscopy.

CYTOLOGY OF THE DEVELOPMENT OF BEET NECROTIC YELLOW VEIN VIRUS (B.N.Y.V.V.), AGENT OF RHIZOMANIA, IN YOUNG SUGARBEET ROOTS.

F.DUBOIS, R.S. SANGWAN and B.S. SANGWAN
Laboratoire Androgenèse et Biotechnologie ; Faculté des Sciences
33 rue Saint Leu , 80039 Amiens Cedex.

Sugarbeet sowings were made in soil infected with virulent strain of *Polymyxa betae* keskin. After different days of culture , the plantlets were fixed for light and electron microscopy , and ultrathin sections of roots, hypocotyles, leaves were made.

Electron microscopic observations showed that infection can be very precocious; four days after sowing certain seedlings, yet under soil, contains large quantities of viruses in their radicle.Viruses were mainly observed in the roots, rarely in the hypocotyls and never in the petiole and blade. Within the roots,viruses particles were frequently observed under different aspects, in the cortical parenchyma but also in the perivascular zone and in the vascular bundles.

Various aspects of virus infection through the fungal formations are described and the relationships between the precocity and the extension of infection are discussed.

DIRECT ORGANOGENESIS FROM THE CULTURED ZYGOTIC EMBRYOS OF MANIHOT (Manihot esculenta Crantz).

N. K. KONAN, R.S. SANGWAN, B.S. SANGWAN.

Laboratoire Androgenèse et Biotechnologie
Faculté des Sciences 33, rue Saint-Leu 80039 AMIENS Cedex.

Manihot or Cassava is one of the most important crop in the tropics. This is mainly because it provides a large diversity of food products such as, flour semolina etc... Recently it also became very competative for animals feeding, for the production of starche, glucose and dextrine in the european contries.

We have induced direct organogenesis from the mature zygotic embryos cultured *in vitro*. These explants were cultuvated in the dark (7 days) on MS medium containing 2,4 D (0,018 or 0,036 mmol) then transfered in light conditions (16 h of continious light) for 3 weeks on the same medium. The reduction in the 2,4 D concentrations simultaneously with BAP incorporation in the medium was a prerequisite for the induction of the adventitious buds. These buds were transfered on the elongation medium (MS + BAP + GA_3). Plantlets were observed 5 weeks later on this medium. These plants were then transferred to a greenhouse.

VITRO MANIPULATION OF CEREALS

Sabine Hartke and Horst Lörz
Max-Planck-Institut für Züchtungsforschung D-5000 Köln

The aim of our investigations is the establishment of a highly efficient and reproducible system for in vitro plant regeneration from cultured cells and protoplasts and the development of a reliable transformation protocol for barley, rice, maize , wheat and triticale. Protoplasts systems have been used for direct gene transfer experiments in barley, rice and triticale, testing different PEG-induced DNA uptake procedures, with plasmids containing the NPT II gene and/or the GUS gene as selectable and screenable marker, respectively. Integration of the foreign genes has been confirmed for barley and rice by Southern hybridisation. However, highly efficient plant regeneration from protoplasts and somaclonal variation among the regenerants is still the limiting factor for this transformation approach. Alternative transformation techniques which avoid the protoplast culture have been initiated, too. DNA delivery to cultured cells is attempted with electroporation, with high velocity microprojectiles, by pollen mediated transfer or by microinjection.

Remote somatic hybrids of higher plants.

Yu Gleba, Institute of Botany, Ukrainian Academy of Sciences, Kiev, USSR.

Somatic hybrids synthesized in Institute of Botany, Kiev will be described.

Remote cybrids. Using tobacco chlorophyll-less plastome mutant as a recipient, a number of remote cybrids have been constructed that combine *Nicotiana* genome with plastomes of *Atropa*, *Lycium*, *Duboisia*, *Physochlaine*, *Scopolia* and *Nolana*. All cybrids are functional (photosynthetically active and fertile). Some of them show different signs of nucleo-cytoplasmic incompatibilities including chlorophyll-deficiency, changes in thylacoid membrane protein spectra, all of them sort out cytoplasmic male sterility and flower malformations. Remote cybridization is proposed as an approach to construct new cytoplasms with cms.

Remote nuclear hybrids. Using different screening schemes, intergeneric intertribal hybrids have been produced in species combinations *Nicotiana* + *Atropa*, *Nicotiana* + *Solanum*, *Nicotiana* + *Physochlaine*, *Nicotiana* + *Datura*, *Nicotiana* + *Nolana*, *Medicago* + *Trifolium*. *Rauwolfia* + *Vinca*, *Rauwolfia* + *Rhazya*, *Rauwolfia* + *Catharanthus*. Fertile amphidiploids *Nicotiana* + *Atropa* as well as *Nicotiana* + *Physochlaine* will be described in more details. We regard those results as a first evidence that amphidiploid intertribal hybrids can be fully functional.

Remote asymmetric nuclear hybrids (gamma-hybrids).
Using nitrate-reductase deficient *Nicotiana plumbaginifolia* as a one partner and irradiation-inactivated (100-2000 Gy) wild-type cells of different species, fertile asymmetric hybrids were produced in combinations *N. plumbaginifolia* + *N. sylvestris* and *N. plumbaginifolia* + *Atropa belladonna*. These combine tetra- or hexaploid sets of *N. plumbaginifolia* chromosomes along with 5-50% of the donor's nuclear material.

Different aspects of remote somatic hybridization experiments will be discussed.

MUTANTS IN THE SYNTHESIS OF ASPARTATE-DERIVED AMINO ACIDS.

M. JACOBS, LABORATORIUM VOOR PLANTENGENETICA
VRIJE UNIVERSITEIT BRUSSEL (BELGIUM).

Over thirty mutants likely altered in enzymes of the synthesis of aspartate-derived (lysine, threonine, isoleucine and methionine) and related amino acids (leucine, valine) have been isolated and characterized (Bright *et al.*, 1985). This report deals with some of these biochemical mutants which have been obtained in our laboratory, specially with altered regulatory control.

The basic isolation procedure consists in using the inhibition exerted on growth due to the combinations of lysine + threonine and amino acid analogs such as s-aminoethyl cysteine. Relief of this inhibition by methionine in the first case or by the corresponding amino acid (lysine) in the second case indicate possible targets for the inhibitory effects, namely the key enzymes of the pathway aspartokinase (AK), dihydrodipicolinate synthase (DHDPS) and eventually homoserine dehydrogenase (HSDH). Resistance mutants have been obtained in various species by selecting at seed (barley, sorghum, *Arabidopsis*) or all level (*Nicotiana sylvestris,* carrot). They include regulatory mutants which have modified feedback control for AK and DHDPS and in the most of the cases show dramatic changes in soluble amino acid pools (increase in threonine or lysine).

We are presently cloning some of the concerned mutated genes; in particular a mutation (RAEC-1), due to a single dominant gene, obtained in *Nicotiana sylvestris,* is characterized by a DHDPS completely insensitive to lysine and as a consequence an overproduction of lysine (5-15X) is observed in the RAEC-1 mutant.

Such mutants can be very valuable for the knowledge of the genetic regulation of amino acid biosynthesis in plants. They are also of potential interest to improve the nutritional value of crops.

Bright, S.W., R.M. Wallsgrove and B.J. Miflin. in Plant Genetics, pp. 701-713 (ed. Alan R. Liss), New York (1985).

IN VITRO SELECTION FOR AUXIN AND CYTOKININ RESISTANCE WITH MUTAGENIZED SEEDS OF *NICOTIANA PLUMBAGINIFOLIA*

Didier LESUEUR, Rémy BITOUN, Michel CABOCHE, Michel LALOUE* and Marc JULLIEN
Biologie Cellulaire, INRA, 78026 Versailles, France.
*Physiologie Cellulaire Végétale, CNRS, 91190 Gif/Yvette, France.

10000 EMS mutagenized seeds of *Nicotiana plumbaginifolia* have been used to isolate naphtaleneacetic acid (NAA), indole-3-butyric acid (IBA) and cytokinin (zeatine) resistant mutants through "in vitro" selection on toxic concentration of these phytohormones. Six auxin resistant mutants have been isolated and classified in three complementation groups. One of the IBA resistant mutant (T1-217) is characterized by an improved and more synchronized germination compared to the wild type. Five zeatin resistant mutants isolated have been classified in five complementation groups. They show an hypertrophy of the hypocotyle and cotyledons when sawn on a high cytokinin medium whereas the wild type plantlets are severely inhibited. One of these mutants, Z5 expresses improved capacities for "in vitro" caulogenesis. In each of the mutants isolated, resistance was due to a recessive nuclear mutation at a single locus.

EFFECTS OF IMMOBILIZATION ON GROWTH AND TROPAN ALKALOID CONTENT OF *DATURA INNOXIA* CELLS

GONTIER, E., SANGWAN, B.S. and BARBOTIN, J.N.
Université de Picardie - Faculté des Sciences
Laboratoire de Génie Cellulaire
Androgenèse et Biotechnologie
33, rue Saint-Leu 80039 AMIENS Cédex

The physiological behaviour of free suspended and calcium alginate entrapped cells of *Datura innoxia* is studied. Sucrose consumption and tropan alkaloid content have been determined using H.P.L.C. Respiration and biomass were measured in connection with cytologic observations.

The fresh weight increasing for free and immobilized cells is correlated with an increase of respiration and a decrease of the sugar content (sucrose, glucose and fructose). In the case of immobilized cells, the sugar consumption does not seem to reduce drastically osmotical pressure in the medium during the first twenty days.

A modulation of the alkaloid content along the growth is observed and a maximum value is obtained for free cells at the end of the exponential phase.

SOMATIC EMBRYOGENESIS IN *PINUS CARIBAEA* AND CULTURE OF PROTOPLASTS ISOLATED FROM AN EMBRYOGENIC CELL SUSPENSION .

LAINE. E. & DAVID. A.
Laboratoire de Biotechnologies Végétales
UFR de Sciences, Université dePicardie
33, rue Saint-Leu. 80039 AMIENS Cedex. FRANCE

The goal of such studies is ultimately to produce transgenic coniferous trees. This will be highly advantageous due to the economic importance of conifers and the time required for the production of new genotypes by sexual crossing.

The first chalenge along the route is the establishment of a tissue culture system which may act as a target for transformation.

Immature zygotic embryos of *Pinus caribaea* were used as starting material. They were cultivated on various solid media containing auxin (2,4-D) and cytokinins (BA and kinetin). When the correct stage of development was associated with a convenient medium, only a few embryos produced typical translucent and mucilaginous embryogenic calli. These calli grew vigorously on media with reduced growth regulant levels. Culture in presence of abscisic acid stimulated the further evolution of proembryos: they protruded from the calli and developed further into somatic embryos bearing well differentiated cotyledons.

Cell suspension cultures initiated from embryogenic calli were used as protoplast source. These protoplasts divided actively and early stages of embryo formation were recovered.

Protoplasts have already been isolated from embryogenic cell suspension cultures. Among the species which were induced to form embryos, only three have been used previously : *Pinus taeda* (Gupta and Durzan 1987), *Pseudotsuga menziesii* (Gupta et al. 1988) and *Picea glauca* (Attree et al. 1987)

The persistence of embryogenic capacity of suspension cultures after protoplast isolation would enable the introduction of new genetic variation by direct gene transfer and regeneration of plantlets from transformed cells.

Up to now, transient expression of marker genes has been obtained in protoplasts isolated from embryogenic cell suspension culture of *Pseudotsuga menziesii* and *Pinus taeda* (Gupta et al. 1988) and*Picea glauca* (Bekkaoui et al. 1988; Wilson et al. 1989)

REFERENCES

P.K. Gupta and D.J. Durzan, Somatic embryos from protoplasts of loblolly pine proembryonal cells. Bio/Technology, 5(1987) 710-712.

S.M. Attree, F. Bekkaoui, D.I. Dunstan and L.C. Fowke, Regeneration of somatic embryos from protoplasts isolated from an embryogenic suspension culture of white spruce (*Picea glauca*). Plant Cell Reports (1987) 6:480-483.

P.K.Gupta, A.M.Dandekar and D.J. Durzan, Somatic proembryo formation and transient expression of a luciferase gene in douglas fir and loblollly pine protoplasts. Plant Sci., 58 (1988) 85-92.

F. Bekkaoui, M. Pilon, E. Lainé, D.S.S. Raju, W.L. Crosby and D.I. Dunstan, Transient gene expression in electroporated *Picea glauca* protoplasts. Plant Cell Reports (1988) 7:481-484.

S.M. Wilson, T.A. Thorpe and M.M. Moloney, PEG-mediated expression of GUS and CAT genes in protoplasts from embryogenic suspension cultures of *Picea glauca*. Plant Cell Reports (1989) 7:704-707.

Production of carrots (*Daucus carota* L.) with modified carotene levels by somatic embryogenesis.

N. PAWLICKI and B.S. SANGWAN.
UNIVERSITE DE PICARDIE
Faculté des Sciences Androgenèse et Biotechnologie
33, Rue Saint-Leu
80039 AMIENS Cédex FRANCE

Carotenes, from vegetables and fruits, are vitamin A precursors. Carrots (*Daucus carota* L.) are estimated to be one of the major source of carotenes. The existence of hyperpigmented carrots (higher carotene levels) would make extractions and industrial isolations easier, and could increase the dietary consumption of carotene and consequently vitamin A. By *in vitro* culture techniques it is possible to create a certain variability in plants. This spontaneous variability will be used to obtain taproots with higher carotene levels.

Petiole segments from aseptic seedlings of several varieties of carrots were cultured in Lin and Staba medium supplemented with 1mg/l 2,4D to induce embryogenic callus, then were subcultured on the same medium without 2,4D to develop embryos to plantlets. A few aseptic seedlings were irradiated (γ rays, 500, 1000 and 5000 rads) to increase spontaneous variability. After 3 months in a greenhouse, the taproots were harvested and the lower half parts were analysed for carotenes by spectrophotometry at a wavelength of 450 nm against petroleum ether. The results were compared with control taproots harvested from the greenhouse or field.

Following somatic embryogenesis, some taproots showed contents of carotene higher than control taproots of the same variety. To verify the transmission of the hyperpigmented characteristic to the descendance, the upper parts of taproots (with leaves) were vernalized for seed production.

The irradiations gave more voluminous callus with more embryos, but the evolution of embryos to plantlets was rare. Some plantlets could be regenerated, and their taproots were analysed. In some cases, higher carotene levels were found in comparaison with non irradiated plantlets. Also, after an irradiation of 5000 rads, the callus were more colored as compared to the control.

IN VITRO ADVENTITIOUS BUD FORMATION FROM IMMATURE APPLE COTYLEDONS cv. GOLDEN DELICIOUS

M. BELAIZI, R.S. SANGWAN and B.S. SANGWAN-NORREEL

Laboratoire Androgenèse et Biotechnologie, Université de Picardie, U.F.R. Sciences Fondamentales et Appliquées, 33, Rue Saint Leu, 80039 AMIENS Cédex FRANCE.

Immature cotyledons of cv. Golden delicious apple were collected 2 and 3 months after the anthesis, and cultured *in vitro* on Murashige and Skoog medium (1962) supplemented with naphthaleneacetic acid (NAA), 2,4-dichlorophenoxyacetic acid (2,4-D) and benzylaminopurine (BA).

Explants were placed in dark during 4 weeks, in a climate chamber at 27°C. Several media sequences were tested with different combinations and concentrations of growth regulators. The concentrations tested were 0-53.0 µM for NAA, 0-45.0 µM for 2,4-D and 0-8.8 µM for BA. The highest frequency of adventitious bud formation was observed on a medium with 31.8 µM NAA and 4.4 µM BA.

These buds were then subcultured on a multiplication medium with 4.4 µM BA at 24 °C under a 16 h light /8 h night dark-cycle ensured by a Sylvania coolwhite fluorescent lamps at an intensity of 20 µM m^{-2} s^{-1}. Rooting was realised after six subcultures with 4.9 µM IBA, the quality and rate of rooting were improved with a dark treatment of 7 days. On the other hand, root initiation occurred directly on the cotyledon with the highest concentration of NAA (31.8 µM), without bud formation.

An efficient system of plant regeneration *via* somatic embryogenesis or organogenesis in Pea (*Pisum sativum* L.)

FLANDRE Frédéric, TETU Thierry and SANGWAN Brigitte S.
Laboratoire Androgenèse et Biotechnologie,
Université de Picardie, Faculté des Sciences
33 rue St Leu, F-80039 AMIENS, FRANCE

Organogenesis:

In Pea (*Pisum sativum* L.), direct organogenesis occurred from zygotic immature embryos cultured on the Murashige and Skoog (MS) medium supplemented with 13,3 µM benzyladenine (BA) + 10 µM naphthalene acetic acid (NAA) + 0,2 µM triodobenzoïc acid (TIBA). Adventitious buds were obtained in high frequency at the basis of the cotyledon. A minimum of 4-5 weeks was necessary to initiate shoot/bud formation

Rooting:

In order to achieve rooting, the more developped stems were excised and cultured on MS medium supplemented with B5 vitamins (Gamborg, 1968) + 5,3 µM NAA + 0,6 mM myo-inositol . As soon as roots were initiated, (15 days), plantlets were transferred on the MS (hormone-free medium). Rooting was not influenced by the size of the explants at the time of culture. A second subculture on the same medium have permitted to obtain flowers, and finally regenerated plantlets were transferred in green house in order to set seeds. The position of the flowering nodes was evaluated and was compared with the breeders data.

Embryogenesis:

Direct somatic embryogenesis was initiated from immature zygotic embryos on MS supplemented with 43 µM NAA + 60 µM arginin + 15 µM thiamine chlorhydrate + 40 µM nicotinic acid. Embryogenic capacity was significantly influenced by genotypes and sizes of the immature embryos at the time of culture. After 4 or 5 weeks, somatic embryos were separated from the cotyledons and plated on the germination medium (MS + 15 µM indolbutiric acid + 2,2 µM benzyladenine + 0,1 M KNO_3). Somatic embryos were also obtained with Picloram (3-4-5 trichloropicolinic acid, a synthetic auxin compound).

Cytological and biochemical variations in relation to the position of fruits on trees

P. MARGUERY*° and SANGWAN, B.S.°

-*CETAFRUITS 19 Grand place, 62000 ARRAS
-°Université de Picardie, Faculté des Sciences
Androgenèse et biotechnologie
33 rue St Leu, 80039 AMIENS. FRANCE

In 1987 and 1988, we have observed on Apple trees (*Malus domestica* Borkh., of Golden delicious cultivar) the main differences between the fruits from the same inflorescence (central or peripheric) or from the wood of 1 or 2 years old.

In this study, it appears that the peripherical fruits on the two woods were smaller than the central ones due to the decrease of cortical cells diameters. However, if apples obtained from the youngest wood were lighter than the others, it's because the number and diameter of their cells is significatively lower.

The dry matter of the fruits from the same inflorescence is equal, but higher by 6% for the apples obtained from the youngest wood. The level of principals sugars of the apple fruit metabolism (sucrose, fructose, glucose and sorbitol) were higher in peripherical fruits, there was no differences between the two carrying woods. Generally, the error of each sample is lower than the classical sampling on fruit trees.

At present, the testing of new treatments on fruit trees is difficult because of the requirement of high numbers of apples (about 30 by test) in order to study the numerous parameters such as numbers and diameters of cells. For limiting those difficuties, our sampling method seems to be more practicable and can be used with 5 to 10 fruits.

Special thanks to Mr H. VANOYE for providing us the apple orchards.

In vitro Multiplication of Female Date Palm (*Phoenix dactylifera* L.) -
J.B. Chowdhury, D.R. Sharma, Neelam Yadav and V.K. Chowdhury, Department of Genetics, Haryana Agricultural University, Hisar - 125004 (INDIA).

The commercial propagation of date palm is through offshoots only, although these species can be multiplied by seeds also. Date palm being dioecious tree, the seed progeny, however, would have large number of mal plants and the female plants too may be genetically heterogenous. This necessitates the propagation of date palm through offshoots. Unfortunate number of offshoots in mother palm trees is also limited. Use of plant tissue culture techniques for fast multiplication of date palm, therefore is being considered by several tissue culture laboratories.

A method has been developed to regenerate large number of plants fr callus established from axillary buds and shoot tips of 3-4 year old offshoots in two varieties of date palm - *Medjool* and *Khadravi*. Best results were achieved when these ex-plants were cultured on a modified MS medium containing activated charcoal (0.3%), Na $H_2PO_4.2H_2O$ (170 mg/l), $KH_2PO_4.2H_2O$ (200 mg/l), 2,4-D (100 mg/l), BAP (5 mg/l) and thiamine Hcl (1 mg/l). Whole plants have been regenerated through somatic embryogenesis. Cell suspensions were also initiated from leaf ex-plants obtained from intact offshoots. These could be further used for regenerating plants. This technique has great potential for multiplying superior genotypes which have lost the ability of normal offshoot formation.

The regenerated plants were transferred to pre-sterilized soil containing farm yard manure and clay - sand mixture in small pots. High humidity, regular fungicide application and 2-3 hours of sun light every day were some of the major requirements for proper growth of the young plants. New leaves and roots appeared in 1-2 months in about 60% of the transplanted plants. These were then transferred to field conditions.

L. BOGORAD and S. R. RODERMEL, Department of Cellular and Developmental Biology, Harvard University, Cambridge, MA 02138 USA

ALTERATION OF THE EXPRESSION OF RUBISCO by ANTISENSE DNA

One copy of the gene rbcL, encoding the large subunit of ribulose bisphosphate carboxylase, is present in each chloroplast chromosome. Leaf cells may contain as many as 3.000-4,000 or so copies of this single gene. On the other hand, the small subunit of RUBISCO is encoded by a family of nuclear genes. The mRNAs for the large and small subunits of RUBISCO are among the most abundant messages in the cell.

We have found that young tobacco leaves of transgenic plants carrying copies of an antisense cDNA of the 5' end of tobacco RUBISCO small subunit gene may contain as little as about 10% of the normal level opf small subunit mRNA. Phenotypes of such transgenic plants will be discussed.

PLANT GENE EXPRESSION IN RESPONSE TO PATHOGEN INFECTION AND OXIDATIVE STRESS

D. Inzé and M. Van Montagu

Laboratorium voor Genetica, Rijksuniversiteit Gent, B-9000 Gent (Belgium)

One of the most important defense reactions of plants against pathogens is the so-called hypersensitive reaction (HR). This reaction is characterized by a rapid necrosis of the infected cells and is always accompanied by the induction of the so-called "pathogenesis-related" or PR proteins.

The mechanism of induction of the hypersensitive reaction remains speculative. ß(1,3)-Glucanases have been implicated in the release of elicitors, known to induce the HR response. Microsequence analysis of extracellular and vacuolar ß(1,3)-glucanase isoforms, induced in leaf tissue of *Nicotiana tabacum* infected with *Pseudomonas syringae* allowed us to design oligonucleotide probes specifically directed towards the different isoforms. Screening of a cDNA library with these oligonucleotides and in turn screening of a genomic library resulted in the isolation of two genes encoding a secreted ß(1,3)-glucanase and a vacuolar ß(1,3)-glucanase, respectively. The structure and expression analysis of both genes will be presented.

In order to evaluate the role of ß(1,3)-glucanase in the defense reaction of plants towards pathogens, transgenic plants have been constructed which overexpress an extracellular ß(1,3)-glucanase isoform. These plants are morphologically normal and fertile. Currently, we are investigating the behavior of the transgenic plants after infection with different pathogens.

There is some evidence to suggest that a membrane-bound oxidase is directly responsible for the necrosis of infected plant cells. The activity of this enzyme results in the production of superoxide radicals ($O_2^{\cdot-}$) which, directly or indirectly (via hydroxyl radical formation), cause severe oxidative damage which finally results in cell death. The importance of superoxide radicals in the hypersensitive cell death is further supported by our recent finding that the superoxide dismutases (MnSOD and FeSOD) are strongly induced by infection of tobacco plants with *P. syringae*. It is possible that these enzymes restrict the toxic effect of superoxide radicals surrounding the hypersensitive zone. The initial microsequencing of the MnSOD protein has recently allowed us to isolate the corresponding genes, by which the above hypothesis could be tested. Additionally, we were able to isolate a cDNA clone encoding a FeSOD, the first to be isolated from eukaryotes.

Interestingly, cell culture results also in the high expression of MnSOD, but not of FeSOD. This effect is dependent on the presence of a metabolizable sugar and is strictly correlated with the respiration rate. Since the MnSOD was shown to be located in the mitochondria, it was suggested that there is a specific need for protection against the superoxide radicals produced as a result of enhanced mitochondrial respiration.

Since numerous other stress conditions, such as sun scald, waterlogging, infection with *Cercospora*, etc. are thought to be mediated by superoxide radicals, the possible protective role of SOD during these conditions will be discussed.

Progress and Problems in the Quantitative and Qualitative Analysis of Chimeric Gene Activity in Transgenic Plants. Richard A. Jefferson. Institute of Plant Science Research, Cambridge Laboratory, Trumpington, Cambridge, UK, CB2 2JB.

Since the development and dissemination of the GUS gene fusion system several years ago, there has been a great deal of activity in construction and analysis of gene fusions in transgenic plants and in bacteria and fungi of agricultural and industrial importance. In this presentation I will illustrate some of our uses for the system in analyzing gene action in the potato, both in the laboratory and in the field, using quantitative measurements of GUS activity in extracts, and qualitative assessments of GUS activity in cells, tissues and explants using histochemical methods. While the system has been very successful, in our analysis of several thousand transgenic plants we have encountered several conceptual and technical limitations to design, execution and interpretation of these experiments. I will discuss strategies for and progress towards the development of new methods that will help overcome these limitations and enhance our understanding of gene action in plants, and in the complex communities of organisms that occur in the environment. In addition, some of the technical difficulties of the GUS system, such as occasional and unexpected background activities in plant tissues, have been traced to the presence of certain microorganisms in the assayed material - often unavoidable - whose contribution can now be minimized or eliminated.

Genetic Engineering of Legume Plants for Improving Symbiotic Nitrogen Fixation. D.P.S. Verma, Department of Molecular Genetics and The Biotechnology Center, Ohio State University, Columbus, OH, USA

Legume plants of agronomic importance (e.g., soybean, pea) are difficult to regenerate from tissue culture. Although Agrobacterium-mediated transformation has been successful in many legume plants, due to the difficulty in regeneration, progress in introducing foreign genes in this group of plants has been slow. Due to the advent of particle bombardment technology, it has now become possible to introduce foreign genes in tissues which maintain the regeneration potential. Using this approach, soybean has been successfully transformed. In addition, pasture legumes, like Lotus can be easily transformed with both A. tumefaciens and A. rhizogenes. The latter allows direct regeneration of shoots from transformed roots. We have used this approach to introduce soybean nodulin genes in *Lotus*. In addition, we have introduced maize alcohol dehydrogenase gene fused with the GUS-reporter gene and directly demonstrated relative level of oxygen in infected and uninfected cells of nodules. This difference in the level of oxygen may be responsible for cell-specific nodulin gene expression.

In tropical legume nodules, proline biosynthesis appear to play an important role in ureide production. We have isolated a gene controlling the last step in the proline biosynthetic pathway. In addition, proline may be important for maintaining high osmoticum in the infected cells. Introduction of this gene under the control of a strong tissue specific promoter should allow enhancement of proline biosynthesis and thus help increase efficiency of nitrogen fixation.

BIOCHEMICAL ASPECTS RELATED TO MORPHOGENESIS IN ZEA MAYS L.

M.A. Santos, I. Glaparols, X. Figueras, A.F. Tiburcio and J.M. Torné.
Centro de Investigation y Desarrollo. (C.S.I.C). Jordi Girona 18-26.
08034 Barcelone -SPAIN-

In order to study some molecular aspects relating to totipotency and regenerative capacity in maize calli, our group has established a callogenic system with different morphogenic characteristics, obtained from immature embryos of the same genotype. This system is formed by three types of callus MC (organogenic callus); EC (embryogenic callus) and RC (root-inducting callus). They were all cultivated in 2mg/l of 2,4-D (2,4-dichlorophenoxyacetique acid) and N6 (1985) or MS (1962) basal media.

In this communication we present a biochemical characterization of these three types of calli, using the following techniques: total proteins (Bradford, 1976, isoenzymatic analysis (PAGE); amino acid analysis (HPLC-PICO TAG system, polyamines (TLC) and pigments (Arnon, 1949).

The largest total protein content of these calli appears in EC, followed by MC and RC.

With regard to isoenzymes: Esterase bands present differences between EC, MC and RC. A glutamate dehydrogenase band which is not present in EC or RC appears in MC. EC and RC peroxidase bands are different from MC bands.Whereas calli do not differ significantly in structural amino acid composition some differences appear in the concentration of certain freee amino acid such as: asparagine and alanine for MC; proline and glutamic acid for EC; V-amino butiric acid for Mc and EC; alanine and glutamic acid for RC.

Polyamine concentration is greater in RC than in EC and MC with the axception of DAP, which is higher in EC than in MC, but is not present in RC.

Chlorophyll b and carotenoid relationship is : EC>MC>RC. In the case of chlorophyll b this relationship is: MC>EC>RC.

Taking into account all these biochemical characteristics, we may conclude that our *"IN VITRO "* system is suitable for studying the different physilogical and molecular characteristics of the distinct morphogenic patterns which are present in plant tissue and cell culture and, specially, in a recalcitrant plant such as maize.

CRYOPRESERVATION: THE BEHAVIOUR OF PLANT CELLS AND CRYOPROTECTIVE SOLUTIONS DURING THE FREEZE-THAW CYCLE

J. DEREUDDRE and C. GAZEAU, Laboratoire de Physiologie des Organes Végétaux après Récolte, CNRS, 4ter Route des Gardes, 92190 Meudon, France et Laboratoire de Cryobiologie Végétale, Université P. et M. Curie, 12, rue Cuvier, 75230 Paris Cedex 05, France.

The development of cryopreservation has been considerable during the past decade. However, while cryopreservation of most cell suspension cultures can be used routinely without major problems, difficulties still remain when freezing organized and/or macroscopic structures (meristems and somatic embryos). These difficulties will be solved by the development of fundamental research on the mechanisms involved during the different steps of the cryopreservation process.

In order to ensure survival of the cells after freezing in liquid nitrogen, it is necessary to prevent the formation of intracellular ice. Intracellular crystallisation can be avoided during the slow-cooling step by the dehydration of the cells. However, during the second step of cooling, from the prefreezing temperature (-40°C in most cases) down to -196°C, residual unfrozen water may crystallise or turn into amorphous solid water. During thawing, recrystallisation may occur, producing deleterious alterations of cell organisation. To avoid this damaging effect, rapid thawing is generally required to achieve high survival rates. However, in some cases, successful cryopreservation can be achieved independently of the cooling or the warming rate.

To increase our knowledge on the resistance of plant cells and organs to liquid nitrogen temperature, we focussed our research on the behaviour of cells and cryoprotective solutions during a freeze-thaw cycle.

During a freeze-thaw cycle of cryoprotected cells of *Catharanthus*, ice crystals do not seem to appear in the periprotoplasmic space between the cell wall and the plasmalemme: the cell wall represents an obstacle to the propagation of ice crystals. In the case of protoplasts, resistance to ice inoculation depends on the integrity of the plasma membrane.

Differential scanning calorimetry allows thermal analysis of cryoprotective solutions during cooling and thawing. On cooling, thermograms of cryoprotective solutions displayed progressive disappearance of the exothermic peak of crystallisation and the formation of a glass transition, as a function of their concentration. On warming, thermograms of these solutions at intermediate concentrations showed three successive events: the glass transition , an exothermic peak corresponding to the crystallisation (or devitrification peak) and a melting peak. The first two events disappeared at low concentrations; at high concentrations, on the contrary, the glass transition was the only change occurring in the solution.

This difference between cooling and warming thermograms seemed to be due to the crystallisation process which depends both on formation of ice nuclei and growth of crystal.

These results are discussed in terms of resistance of cells to the temperature of liquid nitrogen.

CRYOPRESERVATION OF GERMPLASM

Y.P.S. BAJAJ

3 Editor, Biotechnology in Agriculture & Forestry
A-137 New Friends Colony, New-Delhi 110065, INDIA

The success of any crop improvement program depends on the extent of genetic variability in the base population, however there is a depletion of the naturally occurring pool of germplasm. There are more than 20,000 plant species which are endangered, rare and threatened with extinction in their natural habitats. Owing to depletion of germplasm, there is an urgent need to generate and preserve genetic ressources by methods other than conventional. In this regard cryopreservation (-196C) of in vitro cultures offers an alternative. During the last decade significant work has been done in this area, and complete plants have been regenerated from freeze-preserved cultures of a number of agricultural crops and trees, noteable among these are wheat, rice, potato, cassava, chickpea, sugarcane, pea, strawberry, pines, spruce, cilpalm, etc. Cryopreservation may be of immediate use in the following areas :

1. Vegetatively Propagated Crops - To maintain the germplasm of vegetatively prapagated crops such as potato, the plants have to be grown in the nurseries annually, mreover they are exposed to pests and pathogens. Cryopreservation offers lon,g-term storage.

2. Recalcitrant Seeds - The large-sized seeds of many plantation crops and fruit trees are sensitive to temperature qand humidity, their embryos abort, and thus the germplasm cannot be stored trough seed.

3. Storage of Somaclones - The storage of cell cultures is opf immense importance for its use in large quantities for the production of medicinal compounds and other pharmaceutical products from in vitro generated high-yielding somaclones (see Bajaj 1988).

4. International Exchange of Germplasm - Some of the rare, elite and other desirable genetic stocks can be frozen and transported to various locations. This would be like the international exchange of semen.

The implications of biotechnology of freeze-preservation of cell and tissue cultures in agriculture and forestry will be discussed.

Current Plant Science and Biotechnology in Agriculture

1. H.J. Evans, P.J. Bottomley and W.E. Newton (eds.): *Nitrogen Fixation Research Progress.* Proceedings of the 6th International Symposium on Nitrogen Fixation (Corvallis, Oregon, 1985). 1985 ISBN 90-247-3255-7
2. R.H. Zimmerman, R.J. Griesbach, F.A. Hammerschlag and R.H. Lawson (eds.): *Tissue Culture as a Plant Production System for Horticultural Crops.* Proceedings of a Conference (Beltsville, Maryland, 1985). 1986 ISBN 90-247-3378-2
3. D.P.S. Verma and N. Brisson (eds.): *Molecular Genetics of Plant-microbe Interactions.* Proceedings of the 3rd International Symposium on this subject (Montréal, Québec, 1986). 1987 ISBN 90-247-3426-6
4. E.L. Civerolo, A. Collmer, R.E. Davis and A.G. Gillaspie (eds.): *Plant Pathogenic Bacteria.* Proceedings of the 6th International Conference on this subject (College Park, Maryland, 1985). 1987 ISBN 90-247-3476-2
5. R.J. Summerfield (ed.): *World Crops: Cool Season Food Legumes.* A Global Perspective of the Problems and Prospects for Crop Improvement in Pea, Lentil, Faba Bean and Chickpea. Proceedings of the International Food Legume Research Conference (Spokane, Washington, 1986). 1988 ISBN 90-247-3641-2
6. P. Gepts (ed.): *Genetic Resources of* Phaseolus *Beans.* Their Maintenance, Domestication, Evolution, and Utilization. 1988 ISBN 90-247-3685-4
7. K.J. Puite, J.J.M. Dons, H.J. Huizing, A.J. Kool, M. Koorneef and F.A. Krens (eds.): *Progress in Plant Protoplast Research.* Proceedings of the 7th International Protoplast Symposium (Wageningen, The Netherlands, 1987). 1988 ISBN 90-247-3688-9
8. R.S. Sangwan and B.S. Sangwan-Norreel (eds.): *The Impact of Biotechnology in Agriculture.* Proceedings of the International Conference 'The Meeting Point between Fundamental and Applied *in vitro* Culture Research' (Amiens, France, 1989). 1990.

KLUWER ACADEMIC PUBLISHERS – DORDRECHT / BOSTON / LONDON